Geography of Sub-Saharan Africa

Second Edition

Samuel Aryeetey-Attoh, Editor
University of Toledo

Pearson Education, Inc. Upper Saddle River, NJ 07458

Library of Congress Cataloging-in-Publication Data

Geography of Sub-Saharan Africa / Samuel Aryeetey-Attoh, editor--2nd ed.
    p. cm.
    Includes bibliographical references and index.
    ISBN 0-13-061025-9
    1. Africa, Sub-Saharan--Geography. I. Aryeetey-Attoh, Samuel.

DT351.9.G46 2003
916.7--dc21
                                              2002025125

Executive Editor: Dan Kaveney
Assistant Editor: Amanda Griffith
Editorial Assistant: Margaret Ziegler
Production Editor/Composition: Preparé, Inc.
Executive Managing Editor: Kathleen Schiaparelli
Assistant Managing Editor: Beth Sturla
Marketing Manager: Christine Henry
Art Director: Jayne Conte
Cover Designer: Bruce Kenselaar
Cover Image: © Tim Davis, Getty Images, Inc.; © Oldrich Karasek, Peter Arnold, Inc.
Manufacturing Manager: Trudy Pisciotti
Assistant Manufacturing Manager: Michael Bell
Manufacturing Buyer: Lynda Castillo
Vice President of Production and Manufacturing: David W. Riccardi

© 2003 by Pearson Education, Inc.
Pearson Education, Inc.
Upper Saddle River, NJ 07458

All rights reserved. No part of this book may be reproduced, in any form or by any means, without permission in writing from the publisher.

Printed in the United States of America
10 9 8 7 6 5 4 3 2 1

ISBN 0-13-061025-9

Pearson Education LTD., *London*
Pearson Education Australia PTY, Limited, *Sydney*
Pearson Education Singapore, Pte. Ltd.
Pearson Education North Asia Ltd., *Hong Kong*
Pearson Education Canada, Ltd., *Toronto*
Pearson Educación de Mexico, S.A. de C.V.
Pearson Education—Japan, *Tokyo*
Pearson Education Malaysia, Pte. Ltd.

CONTRIBUTORS

Samuel Aryeetey-Attoh, Ph.D
Professor and Chair
Department of Geography and Planning
University of Toledo
Toledo, Ohio 43606

Ibipo Johnston-Anumonwo, Ph.D
Associate Professor
Department of Geography
State University of New York at Cortland
Cortland, New York 13045

Barbara Elizabeth McDade, Ph.D
Associate Professor
Department of Geography
University of Florida
Gainesville, Florida 32611

Godson Chintuwa Obia, Ph.D
Professor and Associate Dean
Department of Geology and Geography
and College of Sciences
Eastern Illinois University
Charleston, Illinois 61920

Joseph Ransford Oppong, Ph.D
Associate Professor
Department of Geography
University of North Texas
Denton, Texas 76203

William Yaw Osei, Ph.D
Associate Professor
Department of Geography
Algoma University College
Sault Ste Marie, Ontario
Canada P6A 2G4

Ian E. A. Yeboah, Ph.D
Associate Professor
Department of Geography
Miami University of Ohio
Oxford, Ohio 45056

BRIEF CONTENTS

PREFACE		xix
Chapter 1	Introduction	1

PART I
PHYSICAL ENVIRONMENTAL CONTEXT

Chapter 2	The Physical Environment	12
Chapter 3	Human-Environmental Impacts: Forest Degradation and Desertification	48

PART II
SOCIOCULTURAL CONTEXT

Chapter 4	Historical Geography of Sub-Saharan Africa: Opportunities and Constraints	78
Chapter 5	Political Landscape of Sub-Saharan Africa: From Instability to Democratization?	105

| Chapter 6 | Culture, Conflict, and Change in Sub-Saharan Africa | 134 |
| Chapter 7 | Population Geography of Sub-Saharan Africa | 165 |

PART III
DEVELOPMENT CONTEXT

Chapter 8	Geography and Development in Sub-Saharan Africa	193
Chapter 9	Transport and Communication in Sub-Saharan Africa: Digital Bridges Over Spatial Divides	232
Chapter 10	Urban Geography of Sub-Saharan Africa	254
Chapter 11	Geography, Gender, and Development in Sub-Saharan Africa	298
Chapter 12	Medical Geography of Sub-Saharan Africa	324
Chapter 13	Agricultural Development in Sub-Saharan Africa	363
Chapter 14	Industry, Business Enterprises, and Entrepreneurship in the Development Process	406

PART IV
GLOBAL CONTEXT

| Chapter 15 | Looking Ahead: Prospects for Africa in a New Global Economy | 425 |

RECOMMENDED WEB SITES ON AFRICA	**443**
PHOTO CREDITS	**448**
INDEX	**449**

CONTENTS

PREFACE   xix

Chapter 1   **Introduction**   1
                  SAMUEL ARYEETEY-ATTOH

     The Size and Locational Dimensions of Sub-Saharan Africa   2
     The Regions of Sub-Saharan Africa   2
     The Physical-Environmental Context of Sub-Saharan Africa   5
     The Sociocultural Context   5
     The Development Context   8
     The Global Context   10
     Conclusion   11

PART I
PHYSICAL ENVIRONMENTAL CONTEXT

Chapter 2   **The Physical Environment**   12
                  WILLIAM YAW OSEI AND SAMUEL ARYEETEY-ATTOH

     Introduction   12
     Physical Landscapes   12
         The Plateau Continent   14
         The Rift Valley System   18
         Coastlines   19

Continental Shelf	21
Rivers	21
Drainage Basins	22
Climate	23
Pressure Systems and Wind Movement	24
Latitudinal Effect	27
Maritime Effect versus Continentality	28
Ocean Currents	28
Altitude	29
Lake Effect	30
Climatic Classifications	30
Vegetation	34
Vegetation Types	35
Rain forest	35
Woodlands	37
Savanna grassland	38
Sahel shrubland	40
Desert and semidesert	40
Mediterranean (Cape and Karoo shrubland)	41
Afromontane vegetation	41
Soils	41
Soil Distribution	42
Conclusion	45
Key Terms	46
Discussion Questions	46
References	46

Chapter 3 Human-Environmental Impacts: Forest Degradation and Desertification   48
WILLIAM YAW OSEI

Introduction	48
Deforestation in Sub-Saharan Africa	49
Definitions and Issues	49
Deforestation Trends	51
Causes of Deforestation	54
Agriculture	54
Logging	57
Fuel-wood consumption	57
Other Causes	59
Effects of Forest Degradation	61
Response Strategies to Deforestation	64
Desertification in Sub-Saharan Africa	66
Definition of Desertification	67
Spatial Coverage of Dry Lands in Sub-Saharan Africa	68
Causes of Desertification	68
Strategies to Combat Desertification	70

Conclusion	73
Key Terms	74
Discussion Questions	74
References	75

PART II
SOCIOCULTURAL CONTEXT

**Chapter 4 Historical Geography of Sub-Saharan Africa:
Opportunities and Constraints** — **78**
IAN E. A. YEBOAH

Introduction	78
Indigenous Heritage of Sub-Saharan Africa	78
Sub-Saharan Africa: Cradle of Humans	79
The Era of Ancient Civilization	79
Iron, Islam, and Medieval Civilizations	82
Bantu Migrations in Sub-Saharan Africa	84
The Era of Modern Kingdoms	84
Implications of the Indigenous Heritage of Sub-Saharan Africa	86
The Islamic Influence in Sub-Saharan Africa	86
The Spread of Islam in Sub-Saharan Africa	87
Islam and the Indigenous Heritage	88
Islam and the Western Heritage	89
The Western Influence in Sub-Saharan Africa	89
The Period of Initial Contact	89
The Enslavement of Sub-Saharan Africans to the Americas	91
The Age of Land Exploration in Sub-Saharan Africa	91
The Colonial Period, 1884 to the 1960s	96
The Economic and Cultural Impacts of Colonialism	97
The Political and Social Impacts of Colonialism	99
Independence from the West	101
Conclusion	102
Key Terms	103
Discussion Questions	103
References	103

**Chapter 5 Political Landscape of Sub-Saharan Africa:
From Instability to Democratization?** — **105**
IAN E. A. YEBOAH

Introduction	105
Sub-Saharan Africa Prior to the Early 1990s: A Landscape of Political Instability and Chaos	106
Causes of Political Instability in Sub-Saharan Africa	110
The Democratic Republic of Congo: a microcosm of political chaos and instability	117

Scenarios for Political Stability in Sub-Saharan Africa 120
Is There Hope for Sub-Saharan Africa? 122
    The Democratization Process in Sub-Saharan Africa 123
        Democratization in Malawi: societal pressures from the church and civil rights activists 123
        Ghana's democratization process: international influences 125
        The example and potential role of South Africa 127
        Political stability of Botswana: indigenous and modern values in government 128
Conclusion 131
Key Terms 132
Discussion Questions 132
References 132

Chapter 6   Culture, Conflict, and Change in Sub-Saharan Africa   134
JOSEPH RANSFORD OPPONG

Introduction 134
Cultural Geography 135
    Culture Change 136
    Obstacles to Cultural Diffusion 137
Elements of African Culture 138
    Religion in Sub-Saharan Africa 139
        Christianity 139
        Islam 141
        Traditional religions 142
    Family and Kinship Relations 143
    Language and Society 144
        Niger–Kordofanian family 145
        Nilo–Saharan 145
        Khoisan 145
        Afro–Asiatic (Semitic–Hamitic) 145
Land Tenure 148
    Family Land 148
    Communal Land 148
    Stool Land 149
    State Land 149
    Individual Private Ownership 149
Adornment, Dress Forms, and Symbolism in African Culture 150
Colonialism and Diffusion of Non-African Culture 158
Modernization and Cultural Conflict 160
    Modernization and the Aged in Sub-Saharan Africa 160
    Values and Expectations in African Society 161
Conclusion 162

Key Terms	162
Discussion Questions	163
References	163

Chapter 7   Population Geography of Sub-Saharan Africa — 165
SAMUEL ARYEETEY-ATTOH

Introduction	165
General Population Trends	166
Population Density and Distribution	166
The Dynamics of Population Change	171
Fertility Levels and Trends	171
Causes of high fertility	172
Cultural determinants of fertility	173
Mortality Levels	175
Africa and the Demographic Transition	176
Age Composition and Dependency Burdens	177
International Migration and Refugee Issues	180
International Migration Trends	180
Refugee Migration	182
Government Policy and Family Planning	185
Sovereign Rights of Nations and Human Rights	186
Integrative Family Planning Policies	186
Community and Market-Based Strategies and Social Marketing	187
Female Empowerment	187
Effective Targeting of At-Risk Populations	188
Partnerships with NGOs and the Private Sector	188
Programs for Refugees	188
Conclusion	189
Key Terms	189
Discussion Questions	189
References	190

Part III
Development Context

Chapter 8   Geography and Development in Sub-Saharan Africa — 193
SAMUEL ARYEETEY-ATTOH

Introduction	193
Defining and Measuring Development in Sub-Saharan Africa	194
Economic Dimensions of Development	194
Human and Social Dimensions	198
Technological Dimensions	200

Geographic Patterns of Dualism 201
    International Dualism 202
        The dual national economy (core–peripheral disparities in African countries) 203
    Urban Dualism 205
    Rural Dualism 205
Theoretical Explanations of Core–Peripheral Disparities 207
    Effect of Colonialism 207
    Economic Theories 207
    Structural-Institutional Theories 211
    Sociopsychological Theories 212
Development Strategies 214
    Growth with Equity 214
    Basic-Needs and Poverty-Reduction Strategies 215
    Intermediate or Appropriate Technology 216
    Self-Reliance Strategy 216
    Interdependent Development 218
Multilateralism 220
    Bilateralism and the Africa Growth and Opportunity Act 220
    Regional Integration and Pan-Africanism 220
    Multilateralism: Structural-Adjustment Programs and Nongovernmental Organizations 225
        Structural-adjustment programs 225
        The role of nongovernmental organizations 227
Conclusion 228
Key Terms 228
Discussion Questions 229
References 229

Chapter 9   Transport and Communication in Sub-Saharan Africa: Digital Bridges Over Spatial Divides   232
JOSEPH R. OPPONG

Transportation Systems in Africa 232
    Road Transportation 233
        Rural transportation 233
        Urban transportation 235
        The Pan-African highway 237
    Railways 239
        West Africa 240
        Congo basin and margins 240
        East Africa 241
        Southern Africa 241
    Air Transport 241

	Navigation	244
	Telephone Communications in Africa	245
The Digital Divide and Africa		247
	Distance Learning	251
Conclusion		252
References		252

Chapter 10 Urban Geography of Sub-Saharan Africa   254
SAMUEL ARYEETEY-ATTOH

Introduction	254
Historical Evolution of African Cities: Precolonial Cities	254
The Internal Structure of African Cities	258
The Internal Structure of Accra, Ghana	265
Current Urbanization Trends in Sub-Saharan Africa	269
Components and Determinants of Urban Growth	273
Gender migration	274
Urban–rural linkages	274
Consequences of Urban Growth	275
Urban primacy	275
The urban informal sector	276
Housing	279
Land management	282
Infrastructure provision, service delivery, and urban management	283
Water supply and sewage disposal	283
Solid-waste disposal	285
Public transportation	286
Solutions to Problems in African Cities	286
Microlevel Strategies: The Case for Effective Urban Management	287
Macrolevel Strategies: Regional Policies	291
New towns	291
Growth poles	292
Intermediate and small service centers	292
Conclusion	293
Key Terms	294
Discussion Questions	294
References	294

Chapter 11 Geography, Gender, and Development in Sub-Saharan Africa   298
IBIPO JOHNSTON-ANUMONWO

Understanding the Geography of Sub-Saharan Africa from a Gendered Perspective	298
Geographic Scholarship of Gender in Africa	300

Mapping the World of African Women: What Counts? 300
Women and Development in Sub-Saharan Africa 301
    Adverse Effects of Capitalist Development on African Women's Access to and Use of Productive Resources 301
Agriculture and Rural Environmental Degradation 302
    Human-Induced and Environmental Influences on Agricultural Production 304
    Unequal Allocation of Resources within the Household 305
Demographic Trends and Settlement Patterns: Geographical Contexts 306
    Population, Fertility, and Mortality Trends 306
    Displacement and Migration 308
    Differential Access in Urban Settings 309
Geography and Gender Inequality 311
    Change and Crisis 311
    Structural-Adjustment Programs and African Women 314
African Women as Agents of Change 315
    Geopolitics and Women's Initiatives 316
    Women's Role in Restoring the African Environment 316
Engendering the Geographic Study of Sub-Saharan Africa 318
    Gender and Development: Rethinking the Connections 318
    African Development: Recognizing and Empowering Women 319
    A Gendered Agenda for Future Geographic Research 320
Key Terms 321
Discussion Questions 321
References 322

Chapter 12   Medical Geography of Sub-Saharan Africa   **324**
JOSEPH R. OPPONG

Introduction 324
Disease Ecology of Sub-Saharan Africa 326
    Disease Terminology 326
Endemic Diseases of Sub-Saharan Africa 329
    AIDS—The African Catastrophe 329
        Structural adjustment and HIV–AIDS 332
        Can Nigeria avoid a major AIDS explosion? 333
    Other Diseases 334
        Malaria 334
        Trypanosomiasis 335
        Yellow fever 336
        Schistosomiasis 337
        River blindness 337
        Guinea worm (dracunculiasis) 338
        Buruli ulcer 338
        Ebola 340

Sanitation and Health in Sub-Saharan Africa	341
The Health-Care System	344
Spatial Imbalance in Health Services	348
The Rainy-Season Accessibility Problem	348
Health-Facility Locational Strategies and Location–Allocation Models	349
Bypassing and Utilization of Health Facilities	350
Technology and Health Care	351
Drugs, fake drugs, and health care in Africa: killers or curers?	351
African traditional medicine: dangerous, anachronistic, or a panacea?	352
Colonialism and Traditional African Medicine	353
Strengths and Weaknesses of Traditional Medicine and Biomedicine	354
Economic Crisis and Health in Sub-Saharan Africa	355
The Bamako Initiative and Health in Sub-Saharan Africa	357
Conclusion	358
Key Terms	359
Discussion Questions	359
References	360

Chapter 13   Agricultural Development in Sub-Saharan Africa   363
GODSON C. OBIA

Introduction	363
Farming Systems	365
Traditional Farming Systems	365
Shifting cultivation	369
Slash and burn agriculture	370
Intercropping	371
Rotational bush fallow	371
Permanent farming	371
Compound farming	371
Mixed farming	372
Pastoralism	372
Nomadism	372
Transhumance	372
Cash-Crop and Commercial Farming	375
Cash-crop production	375
Tree cropping	375
Commercial farming	379
Quasi-commercial production	379
Large-scale commercial agriculture	379
Production Trends in Sub-Saharan Africa	380

Problems of Agricultural Production	388
Rural Infrastructure	388
Institutional Factors	388
Land Degradation	389
Inappropriate Technologies	390
Problems with green revolution	390
Research and Development	391
Governments, International Agencies and Agricultural Markets	394
The World Trade Organization and Agriculture in Sub-Saharan Africa	395
Policy Options	397
Upgrading Rural Infrastructure	397
Education	398
Developing Appropriate Technologies	398
Developing Alternative Energy Sources	399
Agro-Based Industries	400
Community-Based Cooperatives	400
Integrated Rural Development	401
Conclusion	402
Key Terms	402
Discussion Questions	402
References	403

Chapter 14 Industry, Business Enterprises, and Entrepreneurship in the Development Process — 406
BARBARA ELIZABETH MCDADE

Introduction	406
Trends in Industry and Manufacturing	407
Problems with Industrial and Manufacturing Enterprises	409
Spatial Imbalance in Large-Scale Industries	409
Weak Inter- and Intrasectoral Linkages	409
Highly Protectionist Policies	410
Lack of Diversification and Expansion	411
Weak Institutional Capacity	411
Public Versus Private Enterprises in Sub-Saharan Africa	412
The Background to State-Owned Enterprises	412
Expanding the Private Sector in Sub-Saharan Africa	414
African Entrepreneurship and Small-Scale Industry	416
Characteristics of African Entrepreneurs	416
Small-Scale Enterprises	417
Conclusion	421
Key Terms	423
Discussion Questions	423
References	423

PART IV
GLOBAL CONTEXT

Chapter 15 Looking Ahead: Prospects for Africa in a New Global Economy   **425**
SAMUEL ARYEETEY-ATTOH

Globalization and Africa: Introduction	425
Meaning of Globalization	426
The Globalization Process: Opportunity or Threat?	427
Impacts of World Trade Negotiations on Africa	430
Negotiating Trade Reforms in Agriculture	432
Trade-Related Aspects of Intellectual Property Rights (TRIPS)	433
General Agreement on Trade and Services (GATS)	434
Trade Negotiations on Manufactured and Industrial Goods	435
Other Trade-Related Issues	436
Africa's Prospects for Greater Integration into the Global Economy	437
Conclusion	440
Key Terms	441
References	441

RECOMMENDED WEB SITES ON AFRICA   **443**

PHOTO CREDITS   **448**

INDEX   **449**

PREFACE

This text represents a systematic approach to the multifaceted aspects of the physical and human geography of Sub-Saharan Africa. It is designed to expose undergraduate students with little or no geographic training to a variety of contemporary ideas, theories, and concepts in African geography and to their applicability to "real-world" situations. This second edition begins with an overview of the size and the locational and regional dimensions of the African continent. The book is then divided into four parts, organized into 14 chapters that depict the physical-environmental, sociocultural, and development dimensions of Sub-Saharan Africa and the region's prospects for global integration.

NEW TO THE SECOND EDITION

The intent of this second edition is to provide as comprehensive a coverage as possible on relevant geographical issues pertaining to contemporary events in Sub-Saharan Africa. As a result, we have provided a broader and more up-to-date perspective of maps, illustrations, and data sets. New information and data have been added, but all pieces of information from the previous edition that continue to be relevant to contemporary Africa have been preserved. In addition, we have added two new chapters to the second edition: a chapter that evaluates the condition of transportation, communication, and digital networks, and a concluding chapter that assesses Sub-Saharan Africa's prospects for global integration and for expansion into the world trade market. We have also reorganized the chapters to conform to the four parts for consistency and clarity. The chapter on political landscapes has been thoroughly revised to reflect ongoing democratic reforms sweeping across the continent. The chapter on medical geography has an expanded section on HIV–AIDS to reflect the growing dilemmas and complexities associated with the pandemic. All other chapters have been thoroughly reviewed and revised where appropriate, to ensure that the research and information are kept current.

Wherever possible, we also dispel any misconceptions and misinterpretations about Sub-Saharan Africa. It is hoped that the stereotypical perceptions of the region as a "dark" continent are fading. We present a region with a rich and diverse physical and human resource base and a resilient cultural heritage. We acknowledge the significant contributions that previously ignored segments of African society have made (and that also have been slighted in other volumes), such as the informal sector, women, traditional medicine, self-help organizations, traditional belief systems, and indigenous political institutions. Despite recognizing these contributions, we nonetheless acknowledge the fact that parts of Sub-Saharan Africa continue to be mired in crises: poor health care, poverty, debt burdens, gender inequality, ethnic disputes, and political ineptness.

ACKNOWLEDGMENTS

Since the publication of the first edition, we have heard from numerous colleagues, who have provided support and encouragement along the way. In addition, several reviewers took time out of their busy schedules to provide constructive comments and suggestions. We thank Helen Aspass of Virginia Commonwealth University, Tarek Joseph of the University of Michigan at Dearbon, Jeffrey Popke of East Carolina University, and Henry Owusu of the University of Northern Iowa. Over the years, we have benefited from the insights of David Diggs of Central Missouri State University, Hari Gabarran of East Tennessee State University, Girma Kebbede of Mount Holyoke College, Ezekiel Kalipeni of the University of Illinois-Champaign, and Emmanuel Mbobi of Kent State University. We would also like to thank the professional staff of Prentice Hall for their untiring effort and support. Special mention goes to the executive editor, Dan Kaveney, for his enthusiastic and unyielding support, and to the editorial assistant, Margaret Ziegler, for guiding us through the procedural and organizational process.

In addition, we are grateful to copyeditor Edith Baker, photoresearcher Jane Sanders, and production manager Fran Daniele for keeping this project on track.

Lastly, I would like to thank my wife, Antoinette, and our children, Naa Dedei (Annette), Naa Okaikor (Annabelle), Nii Okoe (Stefan), and Naa Akwele (Sasha), for their encouragement and understanding throughout the several years during which I have been working on this project.

Samuel Aryeetey-Attoh
Geography and Planning
University of Toledo

1

Introduction
Samuel Aryeetey-Attoh

Sub-Saharan Africa has a rich and diverse physical and human resource base and a resilient cultural heritage. For most people who reside outside of Africa, the region evokes a variety of images; some of them, unfortunately, are influenced by stereotypes, and others are based on little information about the region. The information in this text provides as broad a perspective as possible, in an attempt to improve our understanding of the multidimensional and complex aspects of the physical, social, and development geography of Sub-Saharan Africa. Throughout our discussion of the region, however, we must always remember that much remains hidden, for Africa is a vast and varied region, home to over half a billion multi-ethnic people.

    Sub-Saharan Africa can best be described as a ***Land of Contrasts***. It is characterized by such contrasts as extreme wealth and abject poverty, pristine forests and voluminous rivers interspersed among degraded and desert landscapes, mineral-rich subterrains buried under destitute landscapes, majestic mountains hovering over deep river gorges and extensive plains, and European values and institutions superimposed on traditional cultural landscapes. The tremendous variety that is the African continent can be difficult to convey. Some African countries are among the poorest in the world, with limited agricultural or industrial resources. Others are making significant progress and have the resource potential for even greater growth. The four parts of this text set forth the geographical basis for the development of this diverse and increasingly important part of our world.

THE SIZE AND LOCATIONAL DIMENSIONS OF SUB-SAHARAN AFRICA

Africa is the second largest continent in the world. Its land area, at 11.8 million square miles (30.5 million square kilometers), is about three times larger than that of the United States of America, and it accounts for about 20% of the earth's land surface. The Sub-Saharan region alone is about 9.8 million square miles (24.3 million square kilometers). Africa's north–south extension stretches about 4800 mi (7725 km), from northern Tunisia to Cape Agulhas at the southernmost point of the Republic of South Africa. From east to west, it extends 4500 miles (7242 km), from Dakar in Senegal to the tip of Somalia in the Gulf of Aden, spanning four time zones. Sub-Saharan Africa comprises 42 countries located on its mainland and 7 island countries: Madagascar, the Comoros, Seychelles, Mauritius, and Réunion in the Indian Ocean, and Cape Verde, and São Tomé and Principe in the Atlantic Ocean (Figure 1.1). A land of contrasts, Africa has a mix of large and small countries. Sudan (966,000 square miles) is the largest country in Sub-Saharan Africa and is $3\frac{1}{2}$ times the size of Texas. Gambia is the smallest country on mainland Sub-Saharan Africa, with an area of 4200 square miles.

Some unique features of Africa's relative location are that it is located in all four hemispheres and that no other landmass on the earth sits so squarely astride the equator. Much of Sub-Saharan Africa's territory (about 70%) lies between the tropic of Cancer and the tropic of Capricorn. Climates are, therefore, mainly tropical; the major exception is the extreme southwestern Cape region of South Africa, which is Mediterranean. Sub-Saharan Africa is bordered on the east by the Indian Ocean, which historically has been a migratory route for Malay–Polynesian settlers in Madagascar, for Chinese and Indian settlers in the Seychelle Islands and in the Republic of South Africa, and for Middle-Eastern merchant settlers in Zanzibar, Kenya, and Uganda. The southern African region is centrally located in a pivotal area of the Southern Hemisphere relative to three major zones of peace: the Indian Ocean, which in 1971 was declared by United Nations as a "zone of peace"; the Antarctic region, which, under the provisions of the 1959 treaty, was restricted to research and scientific activities (South Africa has research stations in Antarctica); and Latin America, which was declared a nuclear-weapon-free zone under the 1967 Treaty of Tlatelolco. Southern Africa is also a major alternative route for international shipments of oil and minerals. The South Atlantic is also home to two strategically located islands—St. Helena and Ascension. The former houses a U.S. airforce base and a major meteorological station; the latter is a transoceanic fuelling station and has an international satellite telecommunications service. The North Atlantic coast of west Africa, near Cape Verde, is a source region for tropical cyclones (hurricanes) that move westward into the Caribbean Sea. It is also a major source of swordfish and tuna for several coastal west African nations.

THE REGIONS OF SUB-SAHARAN AFRICA

Sub-Saharan Africa can be divided into four regions: west, east, central, and southern (Figure 1.1). The *west African* countries are Benin, Burkina Faso, Cape Verde Islands, Côte d'Ivoire, Gambia, Ghana, Guinea, Guinea-Bissau, Liberia, Niger, Nigeria, Senegal,

The Regions of Sub-Saharan Africa

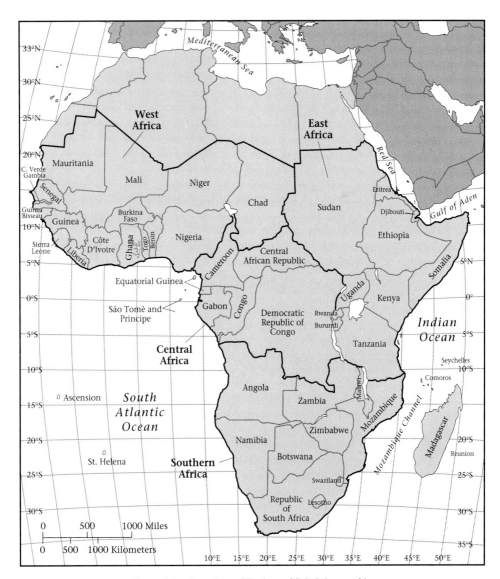

Figure 1.1 Countries and Regions of Sub-Saharan Africa.

Sierra Leone, Togo, and Chad. Chad's locational affiliation can be debated; it is caught in transition between west, east, and, arguably, central Africa. The region accounts for 36% of the Sub-Saharan Africa's land area and 38% of its current population. Like all the other regions, west Africa exhibits considerable geographical, cultural and economic diversity. The climate and vegetation transitions from a humid equatorial rainforest zone along the coast to a semiarid and steppe zone in the interior Sahel regions. Low-lying plains are interspersed with mineral-laden hills, such as the Jos Plateau (tin) in Nigeria and

Fouta Djallon (diamonds, bauxite) and Mt. Nimba (iron ore) in Guinea. Culturally, west Africa has more discrete language groups than any other African region. The languages include Wolof and Fula in Senegambia; Mande (Mali and Guinea), Mossi (Burkina Faso), and Hausa (northern Nigeria and Ghana); and the Guinean languages Kru (Liberia), Akan (Ghana), and Yoruba (western Nigeria). From a religion standpoint, coastal west Africa is predominantly Christian; the interior is predominantly Muslim. Economically, the drought-prone Sahel States (Mauritania, Mali, Chad, and Niger) have lower levels of development when compared to the coastal states—Côte D'Ivoire, Ghana, and Nigeria. An oil belt that extends from Abidjan, Côte d'Ivoire to Luanda, Angola contributes to the economic development of bordering states.

East Africa includes Burundi, Djibouti, Eritrea, Ethiopia, Kenya, Somalia, Sudan, Tanzania, and Uganda. It constitutes 26% land of Sub-Saharan Africa's land territory and 37% of its population. East Africa has majestic mountains (Kilimanjaro, Kenya, Ruwenzori, the Ethiopian highlands) juxtaposed against great rift valleys, extensive plains (Serengeti) and scenic lakes—including Lake Victoria, the world's second largest freshwater lake. Culturally, most of its languages originate from three major family groups—Bantu, Semitic–Hamitic (or Afro-Asiatic), and Nilotic. There is a mix of religions: Christianity; Islam (given the geographic proximity to Asia); Coptic Christians and Falasha Jews in Ethiopia; and a minority of Hindus in Kenya. The economies of east Africa are primarily agricultural; the principal cash crops include coffee, tea, and sisal. Horticultural and floricultural activities are beginning to emerge. Also ecotourism is a major industry, given the prevalence of national game parks and reserves in the savanna and forest zones.

Central Africa includes the Central African Republic, Congo, Equatorial Guinea, Gabon, and the Democratic Republic of Congo (formerly Zaire), plus Cameroon as a transition between west and central Africa. The region accounts for 17% of Sub-Saharan Africa's land territory and 12% of its people. The physical geography is dominated by the tropical rainforest and the humid equatorial climate. The majority of countries practice Catholicism, which is strongly influenced by traditional religions. The majority of languages can be traced to the Bantu subfamily. Central African countries, especially Gabon, the Congo Republic, and the Democratic Republic of Congo, are well endowed with minerals ranging from diamonds and iron ore to uranium and manganese. However, oil is the principal source of foreign exchange. Aside from timber reserves, the rainforest is a haven for such diverse crops as cocoa, oil palm, coffee, plantains, bananas, yams, and cassava.

Mainland *southern Africa* encompasses 10 states (Angola, Botswana, Lesotho, Malawi, Mozambique, Namibia, South Africa, Swaziland, Zambia, Zimbabwe) that account for 25% of Sub-Saharan Africa's land area and 13% of its population. The physical environment is just as diverse as in the other subregions. The Great Escarpment rims the higher parts, separating the interior plateau from the coastal lowlands. The Zambezi River cuts through deep gorges and thunders over the spectacular Victoria Falls. The Orange River, on the other hand, flows through the southern sections of the vast Kalahari and Namib Deserts, forming rich alluvial diamond beds at the mouth of the river. The Cape Fold Mountains form a range in the southwest sections; on the other end of the southeast coast lie the Drakensberg Mountains. Southern Africa is predominantly Christian, with a mix of Hinduism and Buddhism in South Africa. The majority of languages are derived from the Bantu subfamily; the Khoisan family is the oldest of all the language groups. The region is experiencing a transformation in its social, political and economic structures. Angola and Mozambique are recovering from more than 20 years of civil strife and seeking a peaceful transition to democratic and economic reform. Namibia just recently gained its independence from South Africa and is coping with the challenge of social and political reconciliation. Botswana has transformed its economy from one of the poorest in the world to one of the richest in Africa and is now being examined as a model of political stability. Zimbabwe gained its independence much later than most African countries and is still wrestling with the issues of land reform and industrial transformation. Finally, South Africa in the mid-1990s instituted a new constitution that guarantees civic and political liberties to all its citizens. South Africa potentially could be a major player, not only in the economic and political transformation of the southern Africa region, but also in that of the rest of Sub-Saharan Africa.

THE PHYSICAL-ENVIRONMENTAL CONTEXT OF SUB-SAHARAN AFRICA

The first part of the text begins with an overview of the physical-environmental geography of Africa, then continues with an analysis of the impacts of human activities on forest degradation and desertification. This is an appropriate starting point, because analyzing the physical landscapes, climates, vegetations, and soils of the African continent provides us with a basis for understanding where human and economic activities are located. Contrasting highland eastern and southern Africa with lowland western and central Africa enables us to learn more about the associated climatic, topographic and pedologic effects. The distribution of rainforests in central Africa and of savanna grasslands in west, east, and parts of southern Africa help us appreciate the cultural ecology of the societies that inhabit these regions. The varied and complex physical geography of Africa portrayed in this chapter gives us a perspective of the natural and physical assets and liabilities inherent in the region.

Chapter 3 examines the harmful effects of human activities on ecologically fragile areas of the African rainforest, savanna, and Sahel. The twin problems of forest degradation and desertification have caused significant loss of wildlife habitats, increase in soil degradation, and reduction in grain and crop production. Prolonged droughts continue to threaten the ecological stability of the African Sahel, a transition zone between the savanna grassland and the Sahara Desert. Consequently, the incidence of poverty and famine has increased in this region, prompting international agencies like the United States Agency for International Development (USAID) to develop early warning systems to monitor agricultural and economic conditions in the region. Chapter 3 suggests that countering desertification and food insecurity requires a multifaceted planning and management approach with an emphasis on reforestation and revegetation, soil-conservation and soil-fertility improvement, water conservation, and organizational initiatives.

THE SOCIOCULTURAL CONTEXT

In Part II of the text, the sociocultural geography of Sub-Saharan Africa is discussed within the context of the region's precolonial and colonial history, its varied political systems of governance, its diverse cultures, and its rapidly growing populations. The legacy of Africa's triple heritage (indigenous, Islamic, and European heritage) provides an important framework in our understanding of the diverse social, political, and economic systems that characterize contemporary Africa. We learn in Chapter 4 that the indigenous heritage is rooted in the ancient, medieval, and modern kingdoms of Africa. Precolonial Africa experienced a long and rich history; however, much of its history, traditions, customs, and artifacts were destroyed during the colonial era, and many misconceptions about African cultures and institutions arose and became entrenched. It was habitual for colonialists to deny any social and political achievements to Africans themselves. Some scholars at the time even questioned whether native technologies were really indigenous to Africa or external agents imported them. Other Westerners even went as far as denying "civilization" status to early African urban centers on the basis that their peoples lacked writing or organized social and political structures. Geographers and historians are now reconstructing Africa's past out of evidence from folklore, poetry, archeological sites, agricultural cropping systems, art, and architecture. We now recognize that the ancient civilizations were characterized by rules of social behavior, codes of law, and organized economies.

The chapter further discusses the influences of Islam, which was introduced to Africa as early as 700 AD through three diffusion waves. The first occurred through contact between Arabian traders and the people along the coast of east Africa and the surrounding islands of Zanzibar, Pemba, and the Comoros. The second occurred around 900 AD, through trans-Saharan routes between northern Africa and the western Sudanic civilizations; the third occurred via trade between Egypt and Arabia and within the horn of Africa.

Chapter 4 divides the Western influence on Sub-Saharan Africa into four main periods: the period of initial contact, the period of enslavement, the age of land exploration, and the colonial period. The conclusion of the Berlin Conference in 1884 ushered in the formal era of colonialism in Africa, an era characterized by the superimposition of foreign values on indigenous African institutions. Chapter 4 provides perspectives on the range of colonial policies instituted by the British, French, Belgians, and Portuguese. The chapter concludes with an evaluation of the cultural and economic implications of colonialism. The paternalistic and exploitative nature of colonialism, coupled with the efforts of nationalistic and humanitarian organizations, essentially brought an end to colonialism and ushered in the era of political independence for most African countries in the late 1950s and early 1960s.

Colonialism also had a detrimental impact on the political systems of African countries. Chapter 5 provides an overview of these political systems. The colonial authorities superimposed their respective political systems on traditional African political structures, thus creating much confusion. At independence, African countries had at least two types of political systems, one based on European law and one on African customary law. Mozambique's (a former Portuguese colony) legal system is based on Portuguese civil law and customary law. Some countries, such as Sudan, Mali, Senegal, Mauritania, Chad, and Niger, have a third legal system to contend with—Islamic law. Sudan, for example, has a combination of English common law, traditional customary law, and Islamic law. The latter applies to all residents of the Sudan's northern states regardless of their religion. As a result, Sudan continues to be plagued with religious and ethnic persecution and civil unrest. Chapter 5 is divided into two parts. The first part focuses on the causes of political instability from the beginning of independence to the early 1990s. The second part assesses recent political events (since the early 1990s) and focuses on the emerging optimism in relation to the new wave of democratization. The chapter argues that, in spite of the positive indications of change, the old order of political instability and chaos still exists in some ways in many countries. Thus, the democratic transition in Sub-Saharan Africa is both an incomplete and a reversible process.

The complexity and diversity of Sub-Saharan Africa's cultures are presented in Chapter 6 in a format that demonstrates the distribution of languages, religions, ethnicity, and family relations in the African landscape. One of the most intriguing aspects of Sub-Saharan Africa's cultural geography is the more than 1000 diverse languages that exist in the region. Most of these languages do not have a written tradition; there are one million or more speakers of each of approximately 40. These numerous languages are grouped into six major family groups: Niger–Congo, Semitic–Hamitic, Nilo–Saharan, Khoisan, Malay–Polynesian, and Afrikaans. Also intriguing is the mix of religions. Islam is concentrated in the northern regions, the horn, and the coastal corridors of Kenya, Tanzania, and Mozambique. There are also significant Muslim minorities in the central regions of several west African states. Christianity is prevalent in the central and southern sections of Africa: a strong Roman Catholic presence in Rwanda, Burundi, and the Democratic Republic of Congo; an Anglican presence in Ghana, Nigeria, and Kenya; and Presbyterians in Malawi. The Coptic Christian church in Ethiopia dates back to 332 AD and has resisted pressures from Islam since the seventh century. In Cameroon, Kenya, Tanzania, and the Democratic Republic of Congo, over 50,000 people practice the Bahai religion. Judaism has a strong following in the Gondar region of Ethiopia, where the Falashas (Black Jews) were converted to the faith by a large number of Jews who immigrated to Ethiopia between the first and seventh centuries AD. Small communities of Africans practice other religions, such as Buddhism and Hinduism. Traditional African religions also have a strong influence on Christianity and Islam to the extent that a syncretism or blending of religions occurs. Chapter 6 further provides examples of family and kinship relations with respect to the patrilineal and matrilineal societies in Africa. Examples of shared customs and traditions and land-tenure arrangements are provided, along with the expression of social and political value systems through cultural symbolism.

Chapter 7 focuses on the crisis associated with rapidly growing populations in Africa. Africa has the highest fertility rates (5.7 children) of any region in the world. The mid-2001 population estimate for Sub-Saharan Africa was 673 million people, or 11% of the world's population. The concern in this region is not only the high fertility and birth rates, but also the growing proportion of young dependents (44%), the relatively short time it takes the population to double (28 years), and the high infant mortality rates (94 infant deaths per 1000 in 2001). There are deeply rooted sociocultural factors that explain high fertility, such as early marriages, ethnic rivalries, and child fosterage. There are also variations in fertility level based on educational status, place of residence (urban–rural), and economic status. Chapter 7 suggests that a comprehensive set of strategies needs to be employed in order to curtail rapid population growth, including community-based and social marketing strategies, integrative family planning, female empowerment, and a broad set of socioeconomic development policies.

THE DEVELOPMENT CONTEXT

Part III portrays the multidimensional and interconnected nature of the development process in Sub-Saharan Africa by examining the broad perspectives of development (Chapter 8); the transportation, communication and digital infrastructure (Chapter 9); the urban sector (Chapter 10); the gender disparities (Chapter 11); the health status (Chapter 12); and the key production sectors, agriculture (Chapter 13) and industry (Chapter 14). For over 40 years, African governments have experimented with a variety of development models in an attempt to improve the standards of living of their constituents. The postindependence era in the early 1960s came with high expectations from Africans who looked toward greater opportunities for educational advancement and employment mobility. By the mid-1970s and -1980s, most African economies had become stagnant, as a result of misguided economic policies, political instability, corrupt and uncaring governments, benign neglect of the poor, and a host of structural and institutional factors.

Chapter 8 begins by examining the economic and human dimensions of development. The statistical evidence indicates that, with a few exceptions, African countries are characterized by relatively low incomes, slow economic growth, limited capital investment, high debt burdens, low levels of human development, and high rates of poverty. The development situation in Africa is exacerbated by the enormous disparities in living standards between core-urban areas and rural-peripheral regions. The underlying social, structural, and institutional factors explaining these low levels of development and widening disparities are further examined. Chapter 8 argues that an appropriate way to solve these development problems is through an interdependent process or a series of partnerships between local/rural, national, regional, and international organizations. In addition, the chapter evaluates the impact of nongovernmental organizations and World Bank/IMF-sponsored structural adjustment programs on the social, human, and economic development of Africa.

An important dimension that provides a foundation and support system to development efforts is the network of transportation, communication, and digital technologies. Chapter 9 evaluates the status of roads, railways, waterways, air transportation, and telecommunication networks in African countries. Given Africa's problems with various modes of transportation, the chapter assesses the extent to which developments in telecommunication and digital technologies can contribute to overcoming the physical divisions between communities. The chapter provides examples of the explosion of wireless technologies, telecenters, and internet cafes in Africa, but also acknowledges the digital divide between urban and rural areas and between Africa and the rest of the world.

Chapter 10 examines the evolution of urban societies, the spatial pattern or internal structure of cities, the cause-and-effect relationships associated with the rapid growth of cities, and the urban management and planning strategies adopted by African governments. Over all, about 220 million people, representing 34% of Sub-Saharan Africa's total population, reside in urban areas. While Sub-Saharan Africa is considered to be among the least urbanized regions, its urban population growth rate is among the highest in the world. Urban growth rates have remained at least twice as high as the population growth rates, which average 2.5% to 3.0%. A growing concern is the inability of African cities to provide the necessary urban services and industrial and administration resources to accommodate the needs

The Development Context

and expectations of their growing populations. Consequently, numerous cities lack the capacity and capability to provide affordable housing, effectively dispose of solid waste and waste water, adequately deliver water and sanitation services on a regular basis, and effectively cope with a growing urban informal sector. To devise appropriate solutions, African governments need to address the urban crisis via a two-pronged approach. First, appropriate strategies need to be designed at the microlevel that address the effective management and planning of African cities. Second, appropriate macrolevel and regional strategies need to be devised to maximize the benefits for people outside the dominant city and to retard the flow of migrants to urban areas.

Chapter 11 presents a number of scenarios that help improve our understanding of Africa's development problems from a gendered perspective. We learn that there are important variations among African countries in government policies, economic conditions, cultural values, and attitudes about women that make it almost impossible to generalize about gender issues. The chapter provides a historical account of gender-based inequalities associated with the development process in Sub-Saharan Africa and documents the experiences of rural women, especially their important roles in sustaining the household economy in an increasingly fragile environment. The chapter argues that African women have never remained passive in the face of either patriarchal domination or imperial domination, yet their dynamism, resistance, struggles, and achievements are often underrepresented in written and visual accounts of the continent, which depict African women stereotypically as helpless mothers of dying children. As a result, the last two sections of the chapter are devoted to recognizing the political, social, and economic accomplishments of African women and to identifying ways to conceptualize and study the human geography of Sub-Saharan Africa in such a way that gender differences and human-welfare disparities are reduced.

Chapter 12 examines the spatial variations of disease ecology and health care delivery in Africa. A social development issue that continues to draw the attention of the global community is the incidence of endemic diseases in Africa, most particularly the pandemic of HIV/AIDS. Chapter 12 examines how endemic diseases such as malaria, sleeping sickness, yellow fever, bilharzias, and HIV/AIDS are related to economic and environmental conditions and to cultural and behavioral practices in Africa. For example, the prevalence of HIV in rural areas is fueled by poverty, food insecurity, gender inequality, migration, war, and civil conflict and is accelerated by migration, trade, and refugee movements. Chapter 12 indicates that there are severe disparities in the delivery of health care between urban and rural areas. As a result, rural residents rely on traditional healers and are at times susceptible to itinerant drug vendors. Other issues discussed in Chapter 12 include the advantages and disadvantages of traditional medicine, the impact of modern technologies on health-care delivery, the impact of structural adjustment programs, the effects of privatization on the economics of health care, and the introduction of cost-recovery and user-fee programs in the African health sector. The lack of adequate health-care facilities and the continuing problems associated with endemic and pandemic diseases are having a devastating impact on the economic and social lives of Africans. From a development perspective, this translates into lower production levels in both the agricultural and the industrial sectors, as HIV/AIDS continues to affect working populations and children.

Sub-Saharan Africa's slow economic growth is attributable not only to a deteriorating physical infrastructure, significant gender disparities, and a lack of adequate health care, but also to a number of structural problems associated with such key productive sectors as agriculture and industry. Chapter 13 evaluates agricultural production trends in Sub-Saharan Africa and sheds light on regional deficiencies in the relative value of exports and imports. It provides examples of the major farming systems associated with the traditional subsistence sector and the commercial and cash-crop sectors. It also discusses the problems associated with declining production trends—a deteriorating rural infrastructure, land degradation, inadequate research and development, inappropriate technologies, misguided pricing and marketing policies, and discriminatory and restrictive institutional arrangements. In addition, the effects of structural-adjustment programs and of World Trade Organization (WTO) regulations on agriculture are examined. The chapter concludes by evaluating such policy options as upgrading rural infrastructure, developing appropriate technologies and agro-based industries, increasing community-based cooperatives and participation, and integrating rural development projects.

The problems confronting African business and industry are discussed in Chapter 14. The chapter begins with an examination of industrial and manufacturing trends in Sub-Saharan Africa. This is followed by an overview of the major problems confronting industries, including the uneven pattern of industrial development, the institution of highly protective import-substitution policies, and weak intersectoral linkages. Among the policy options considered to improve the business and industrial climate are the issues of privatization and the role of African entrepreneurship and small-scale industries.

THE GLOBAL CONTEXT

Part IV concludes with an evaluation of Africa's prospects in a new global economy. It begins with a review of the multidimensional components of globalization, then continues with an evaluation of the opportunities and threats that globalization presents for Sub-Saharan Africa. Because international trade is an essential aspect of the globalization process, Chapter 15 devotes an entire section to the impact of new World Trade Organization regulations and negotiations on Sub-Saharan African agriculture, services, manufactured and industrial goods, indigenous knowledge, and labor standards. Of particular concern is the increasing marginalization of African economies from the perceived benefits of globalization and the attendant problems of poverty, socioeconomic inequality, and sheer neglect. Although prospects for global integration perhaps appear bleak, Chapter 15 argues that a number of success stories from countries like Uganda and Mozambique demonstrate the region's ability to marshal its human and physical resources to transcend social and economic impediments. Chapter 15 advocates for a more humane globalization strategy, one by which Sub-Saharan African countries can simultaneously take advantage of the benefits of globalization, preserve their cultural heritage, and develop the capacity to promote democratic reforms to consolidate their political and economic sovereignty.

CONCLUSION

The intent of this book is to provide a panoramic view of the physical and human geography of Sub-Saharan Africa. We hope that our reorganizing the text into four parts will help students become aware of the physical, environmental, sociocultural, and development complexities that confront Sub-Saharan Africa. Students must also understand that Sub-Saharan Africa is inextricably linked to the global economy and is just as vulnerable to global regulatory mechanisms as other regions of the world. The chapters that follow set forth a basis for the development of this diverse and increasingly important part of our world. We attempt to provide insight into the challenges and progress of black Africa and to illustrate the varied approaches that are being pursued in the quest for development and global integration.

PART I

Physical Environmental Context

2

The Physical Environment
William Yaw Osei and Samuel Aryeetey-Attoh

INTRODUCTION

Physical geography emphasizes the natural environment in terms of associated landforms, climate, vegetation, soils, fauna, and resources identified with the natural landscape, such as minerals. Physical geographers examine the distribution of these characteristics over the earth's surface and the processes that modify them. For example, geomorphologists study the underlying processes and forces that mold and shape landforms; climatologists analyze the climatic processes and controls that explain the distribution of climatic regions; and biogeographers study the distribution of plants, soils, and animals, including the processes that produce vegetation and faunal patterns over the earth's surface. These physical environmental characteristics are addressed here within the context of Sub-Saharan Africa.

PHYSICAL LANDSCAPES

Africa is a very old continent; its physical geography has been greatly influenced by major geologic events in the earth's history. The theory of *plate tectonics*, for example, provides scientific explanations about Africa's geophysical characteristics and its relative location with respect to the continents of the world. As summarized by Chernicoff and Venkatakrishnan (1995), plate-tectonics theory is built on the framework that the earth's lithosphere consists of large, rigid plates that move in response to convective currents in the heat-softened layer of the asthenosphere, a zone within the earth's upper mantle.

Such movements have created the earth's ocean basins, continents, and mountains, all of which ride on the shifting plates. The movements have affected the latitudinal position of the African continent over geologic time and have led to the evolution of major landforms, such as the East African Rift Valley. The theory of plate tectonics incorporates and synthesizes earlier theories, such as *continental drift* and *sea-floor spreading*.

An understanding of the theory of plate movements and their relationship to earth-surface processes began with the work of the German meteorologist Alfred Wegener in 1912. Wegener hypothesized that the world's continents once consisted of a supercontinent called *Pangaea*. Pangaea was made up of two major segments: a northern portion known as Laurasia, which later formed North America and Eurasia; and a southern segment called Gondwanaland, from which Africa and the Southern continents, including India, were constituted. Africa lay at the heart of Gondwanaland (Figure 2.1).

The central thesis of the continental drift theory was that faulting and rifting eventually led to the fragmentation of Pangaea and that continents began to drift away from Gondwanaland, with Africa moving north and east to its current location. Several standard texts in geology and geography provide evidence that supports such a breakup and drifting of the continents. For example, Best and DeBlij (1977) document the close association between Africa's geomorphology and the continental drift sequence, examples being the similarity between rock types and geological features in Africa and in South America and that between Madagascar's paleontology and southeast Africa's. They also point to the fact that Africa was the only continent that possessed a vast internal drainage system with rivers flowing into large basins without any outer coastline. Furthermore, Africa is the only continent without a major linear mountain system. As the other continents drifted apart, their collisions with other plates caused the formation of elaborate mountain systems, such as the Rockies in the United States, the Andes in South America, the Himalayas in southern Asia, and the Alpine system in southern Europe. The theory of sea-floor spreading generally provided scientific clues to the mechanisms by which the rigid earth's plates moved and to the resultant tectonic activities associated with folding, faulting, mountain building, volcanism, and the formation of new crustal materials.

The breaking off of Africa from Gondwana led to a series of rifting and volcanic activities, resulting in the uplift of the continent to form broad plateau surfaces edged by a series of escarpments at the coastal margins, especially in the west coast of southern Africa (Summerfield, 1996). The migration of Africa from Gondwana positioned the bulk of the continent in its current tropical location. This location, by and large, includes the semipermanent high-pressure belts centered between Latitude 20° and 30° N and S, where dense descending air gives rise to the hyperarid environment of the Sahara desert. Toward the tip of Southern Africa is a geologic belt known as the Karoo Basin. Rocks exposed at the surface of this belt once were part of a continental swamp and lake basin that was covered by thick deposits of river-borne sediments converted to sedimentary rocks during the Gondwana era. These rocks are rich in fossils of mammallike land reptiles, with a second discovery of their kind only in Russia near the Ural mountains (Levin, 1996; Stanley, 1999). The rifting associated with the breakup of Gondwana caused repeated flows of fluid molten rock (lava)

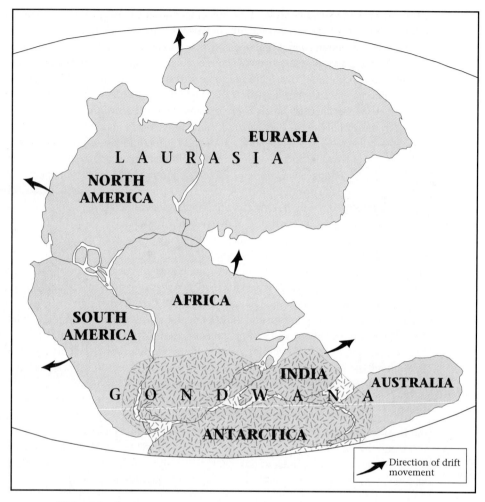

Figure 2.1 Gondwanaland. Adapted from Best, A. & DeBlij, H. (1977). *African Survey*, New York: Wiley.

from the earth's interior, burying the original Karoo landscapes. The depths of the lava floods at times exceeded 1000 m (Levin, 1996). Therefore, the Karoo's unique landscapes of today, with their unique flora, fauna, and geological formations, owe their uniqueness to the geological past. Their formations have attracted visitors, scientists, and agronomists alike.

The Plateau Continent

Much of Sub-Saharan Africa is a plateau region that tilts downward from the east to the west. This plateau is fragmented and scoured by major river systems, leaving large gorges and several undulating surfaces. Africa stands at an average elevation of about 3300 ft

Physical Landscapes

(1000 m) above sea level—higher than any other continent except Asia (Moores and Twiss, 1995). The southern and eastern sections of Africa are usually referred to as *Highland Africa*, which averages about 4000 to 5000 ft above sea level (1219 to 1524 m). The western section of Africa, *Lowland Africa*, averages 1000 to 1500 ft above sea level (305 m to 457 m) (Figure 2.2).

Within Highland Africa, there are several prominent mountain landscapes. Most of South Africa is rimmed by a part of the *Great Escarpment* that borders a narrow coastal belt. The plateau

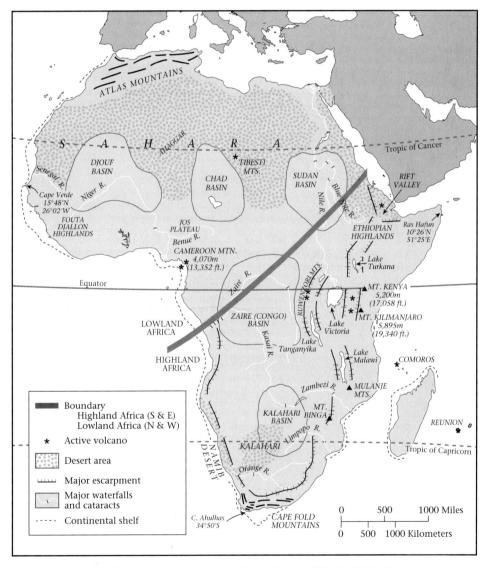

Figure 2.2 Major Landforms in Africa. Adapted from Griffith, I. (1994). *The Atlas of African Affairs*, 2d ed., Witwatersrand, South Africa: University of Witwatersrand Press.

Figure 2.3 Wildebeest graze below the majestic Mount Kilimanjaro.

here varies in altitude from about 2000 ft (610 m) to 8000 ft (2440 m). In the south-west corner of the Cape province lie the folded Cape Ranges, which reach 7632 ft, or 2325 m (Griffith, 1994). At the eastern rim of South Africa are the Drakensberg mountains, which rise over 11,000 ft (3354 m). The plateau extends further north into Zimbabwe, where the eastern highlands and the Inyangani mountains rise up to 8500 ft (2590 m). In Malawi, the Zomba Plateau rises to about 7000 ft (2134 m), and Mount Mulanje, its highest mountain, elevates to over 10,000 ft (3048 m). Zambia features the Nyika Plateau (7100 ft, or 2164 m), and Mozambique's highest point, Monte Binga (8000 ft, or 2438 m), is located along its mutual border with Zimbabwe.

Further north from Zimbabwe and Malawi is the extensive East African Plateau. This plateau features the two highest points in Africa, Mount Kilimanjaro (19,340 ft, or 5895 m) and Mount Kenya (17,058 ft, or 5200 m). Mount Kilimanjaro (Figure 2.3) is an extinct volcano whose summit is marked by the Kibo crater. Mount Kenya also has volcanic origins and commands a prominent place in the folklore and mystical tales of local people. The Ruwenzori mountains, sometimes dubbed the Mountains of the Moon, reach their highest elevation on the Margherita Peak, at 16,673 ft (5109 m) the third highest mountain in Africa.

Sharing the western branch of the Rift Valley is another prominent mountain chain, the Virunga Mountains, which are volcanic in origin. The highest point occurs on the Karisimbi peak, at about 14,787 ft (4507 m). Both the Ruwenzori and the Virunga are rich in unique flora and fauna and attract a stream of trekkers and ecotourists. Virunga has a thriving number of mountain gorillas, perhaps the largest concentration in the world. The mountain gorillas are under the threat of extinction from war and conflict in Rwanda and Democratic Republic of Congo. Marauding fighters as well as displaced populations are destroying their habitats. Desperate soldiers poach them for food, and some have even per-

Physical Landscapes

ished in cross fires or been taken away for sale. Finally, the Ethiopian Massif, which has as its highest point Ras Dashen (about 15,000 ft, or 4573 m) marks the northern limits of the eastern Africa's highland system.

The western sections of Sub-Saharan Africa are not entirely low-lying plateau regions. Mount Cameroon (13,352 ft, or 4070 m), the Jos Plateau in Nigeria (5840 ft, or 1780 m), the Fouta Djallon Highlands in Guinea, and the Ahaggar and Tibesti massifs are examples of major highlands that rise above the low plateaus.

The plateau continent is underlain primarily by Precambrian rock that dates back to more than 600 million years ago. For example, rocks as old as 3500 million years have been recorded in South Africa (Buckle, 1978). About one-third of Africa's surface consists of outcrops of Precambrian origin (Mountjoy and Hilling, 1988). These are crystalline rocks that have metamorphosed in many sections as a result of immense heat, pressure, and chemical changes and, therefore, contain a wealth of minerals. (Refer to Discussion Box 2.1.) There is also evidence of intense folding, as is the case with the cape ranges of south Africa. The Katangan rocks of the Zaire–Zambian copper belt, the diamond and gold mines of south Africa, and the bauxite of Guinea all lie within Precambrian rock foundations. Relatively younger sedimentary formations (less than 570 million years in age) that surround the cratons have also proven valuable. Jurassic–Cretaceous (200 to 100 million years ago) formations in the sandstone regions of the Sahara are known for their water-bearing qualities. The oil-bearing Cretaceous marine sediments (144–65 million years ago) in northern and western Africa and the Karoo series in southern Africa provide coal.

DISCUSSION BOX 2.1: ROCKS AND MINERAL RESOURCES IN SUB-SAHARAN AFRICA

Sub-Saharan Africa is well endowed with significant mineral reserves, most of which are embedded in various rocks, including sediments of broken-down (weathered) rocks. Some minerals, such as gold, exist in rocks as elements. Most of the other minerals, however, exist in chemical association with others (as chemical compounds) and are extracted by means of physical and chemical techniques.

The effects of past geological processes (such as folding, faulting, uplift, metamorphism, and erosion) cause different minerals to occur in different concentrations at various locations. For example, the old, geologically stable shield areas of Africa (cratons), where rocks can be dated to the Archean era (more than 3 billion years ago), are rich in chromium and asbestos. South Africa and Zimbabwe are world producers of chromium, which is an essential element of stainless steel. The Bushveld craton complex of South Africa is noted for platinum (which is resistant to corrosion) and vanadium (which is used in nuclear applications and to toughen steel). The old shield areas of west Africa and the Zaire–Angolan region also store minerals such as gold, diamonds, and manganese. Areas between the older cratons, which have been affected by mountain building over the past

1.2 billion years, are rich in copper, lead, zinc, and cobalt. In parts of Africa where tropical vegetation existed in large swamps during various geologic cycles, burial and lithification (compaction and cementation) transformed organic materials into sedimentary rocks containing coal. Nigeria, South Africa, Zimbabwe, Swaziland, and Zambia have large coal deposits. Oil and gas deposits are also associated with the younger sedimentary rocks along linear zones of the Atlantic front stretching from the Niger Delta to the Democratic Republic of Congo.

Africa supplies some of the most strategic minerals to world markets. South Africa, Botswana, Gabon, Nigeria, the Democratic Republic of Congo, and Zambia generate substantial revenues from the sale of minerals. In 1987, Africa exported about $4.6 billion worth of minerals. Thus, the sector has the potential to boost industrialization and stimulate technological innovation and income generation. However, if mining is to play a major role in sustaining development in Africa, governments and policymakers need to critically evaluate the sector's current structure and operations. Because African countries lack the technical and financial capacity to conduct geological surveys, they continue to rely on multinational corporations for assistance. In some instances, foreign companies have as much as 70% to 80% ownership. Another problem arises from the fact that higher grade ores that are pollution free are being depleted. For example, Nigeria has reserves of lignite, or brown coal, which is the lowest grade of coal ores.

The environmental cost of mining, in terms of atmospheric and water pollution and of the large scale of surface destruction, must be contained. Environmental rehabilitation should be considered part of a normal mine operation. Mining operations that have the technological capability must recover and recycle as much ore as possible.

The Rift Valley System

Another unique aspect of Africa's physiography is its Y-shaped Rift Valley (Fig. 2.4), which begins in the north with the Red Sea and extends through Ethiopia to the Lake Victoria region, where it divides into an east and a west segment and continues southward through Lake Malawi and Mozambique. Its total length, including the Red Sea extension, is estimated to be about 6000 mi, or 9600 km (Best and DeBlij, 1977). Physiologically, the Rift Valley system within east Africa can be divided into four main sections (Buckle, 1978): (1) the Ethiopian rift, stretching from the Afar triangle south to Lake Turkana (Rudolf); (2) the eastern rift in Kenya and Tanzania, including the branches in which lie Lake Eyasi and the Kavirondo Gulf, located east of Lake Victoria; (3) the western rift, from Lake Albert (Mobutu) to Lake Tanganyika; and (4) the Malawi rift, bounding Lake Malawi and the Shire valley. The Urema trough of Mozambique and the Luangwa valley of Zambia also form part of the Malawi system. The average width of the rift ranges between 20 mi (32 km) and 50 mi (80 km) wide, with walls as steep as 3000 ft (914m) (Griffith, 1994). In the Western rift, the

Physical Landscapes

Figure 2.4 The Y Rift Valley in East Africa is characterized by deep depressions and steep cliffs.

Ruwenzori, a block mountain within the rift, rises over 16,404 ft (5000 m); further south, the floor of Lake Tanganyika lies about 2133 ft (650 m) below sea level. Buckle (1978), on the other hand, noted that successive infills of basalts from past volcanic eruptions along the rims of the rift system have elevated the floors of some lakes in Kenya—an example being Lake Naivasha, which rises to about 5906 ft (1800 m) above sea level.

The Rift Valley's formation is explained by plate tectonics. It was created by faulting, as tensional forces began to pull the eastern sections of Africa away, causing the Rift Valley to subside. It is speculated that the eastern part of Africa will eventually break off and drift away from the African continent. An important feature associated with the Rift Valley is the Great Lakes system of east Africa (with the exception of Lake Victoria). Especially unique are the *elongated lakes* that occupy the deep trenches, such as Lake Malawi, Lake Tanganyika, Lake Turkana, and Lake Albert. Lake Victoria, the world's second largest lake in terms of area, is nestled between the two arms of the Rift Valley (Figure 2.2). It is not actually in the Rift Valley system itself.

The rift belt and the offshore islands of Réunion, the Comoros, and the Canaries, constitute the major volcanic belts of Africa. There are several explosive craters around the Uganda–Zaire border of transition (Figure 2.5).

Coastlines

Coastlines are dynamic zones where the ocean comes into contact with the land. Cooke and Doornkamp (1990) list both the positive and the challenging attributes of a coastline. They offer points of trade transfers and port-related activities fostering industrialization and urbanization. Coastlines also provide resource opportunities, such as sand and gravel for the construction industry, food (including fish and seaweed), and such amenities as beaches for

Figure 2.5 A Crater Lake in Uganda.

recreation and tourism. Yet coastlines are vulnerable to such hazards as floods, pollution, erosion, shoreline retreat, and human interference that can require costly responses.

Africa has a total coastline of about 18,641 miles (30,000 km). This represents the shortest coastline among all the continents in terms of the proportion of area to the length of coastline (Orme, 1996). For example, Asia has 43,496 miles (70,000 km) of coastline, but it is only about 1.5 times the size of Africa. Africa's coasts tend to be straight and smooth, with very few indentations, unlike the Scandinavian peninsula of northern Europe or the northeastern coast of North America, which have several deep fjords or river valleys. Another feature of Africa's coastlines is that they are exposed to erosion by longshore drifts (currents moving parallel to or near the shoreline). In west Africa, for example, several sandbars and lagoons front the coasts of Nigeria, Ghana, and Senegal. This was especially problematic during colonial times, when ships had to dock some distance from the coast, and surf boats would be employed to transport cargo to and from the coast.

Orme (1996) suggested two reasons for the short nature of the coastline: (1) in spite of the plateau nature of the continent, rugged mountainous coasts are rare; rather, the coasts are dominated by lowlands with sandy beaches, and (2) there are no large offshore islands (except Madagascar) to serve as natural barriers offshore and mitigate direct oceanic impacts on the mainland coast. Not surprisingly, Africa has very few natural harbors. Most of its harbors are human constructed (artificial), at considerable expense. Among the significant artificial harbors are the ports of Dakar (Senegal), Abidjan (Côte d'Ivoire), Tema (Ghana), Durban (South Africa), and Mogadishu (Somalia).

Physical Landscapes

A few good examples of natural harbors in west Africa are Freetown (Sierra Leone) and Banjul (Gambia). The southwestern shore along the Atlantic Ocean provides some major ports: Lobito and Luanda in Angola and Libreville in Gabon. Important railway terminal ports in Mombasa (Kenya), Maputo and Beira (Mozambique), Dar es Salaam (Tanzania), and Cape Town (South Africa) were all developed from sheltered natural harbors. Over all, Africa's coastlines provide both opportunities and challenges for sustainable development.

Continental Shelf

The continental shelf is that part of an ocean that extends from the shore to the transition zone (demarcating the open ocean) where depth increases rapidly. Generally, the continental shelf is shallow (below 425 ft, or 130 m) and deepens gradually, with an average gradient of 1 in 500, as noted by Pritchard (1979). Worldwide, the average width of the continental shelf has been estimated at more than 40 mi (65 km). The shallow continental shelf allows for much penetration of the sun's energy, thus providing a conducive habitat for fish. The continental shelves are among the richest fishing grounds in the world.

Africa is among the continents least endowed in terms of the breadth of its continental shelf. Except along the Namibian–Angolan stretch of coastline, the Ghanaian and Guinean coasts of west Africa, and the Mauritanian–Moroccan coasts, the African continental shelf extends for only a short distance from the coastline before the abrupt drop to the ocean depths. Countries with extensive continental shelves (United States, Peru, Chile, Argentina) have taken advantage of the opportunities they offer, such as offshore oil drilling, mineral exploration, and fishing. In Africa, illegal fishing by large-capacity foreign fleets from the European Union, Japan, and Russia has contributed to a dwindling fishing stock. A major disadvantage of very deep shorelines is the poor development of beaches and other shoreline resources that can benefit recreation and tourism. Furthermore, most of the continental shelves are being intensively exploited for oil. Active oil and gas drilling occurs in Gabon, Nigeria, Angola, and Ghana. The potential for major accidents associated with marine contamination demand environmental safeguards in the offshore drilling business.

Rivers

About one-third of Africa is desert; another third is drained by five major river systems: Nile, Zaire, Niger, Zambesi, and Orange. Other important rivers are the Limpopo, Senegal, and Volta. (See Figure 2.2.) An estimated 90% of the total annual volume of water flow from African rivers enters the Atlantic slope, and nearly 10% enters the Indian Ocean (Adams et al., 1996).

Most of Africa's rivers have a peculiar flow. Their upper courses are unrelated to the coast they exit from (DeBlij and Muller, 1994). For example, the Niger originates from the Fouta Djallon highlands by flowing in a northeast direction towards the Djouf Basin; then it bends in a southeast direction before turning south toward the west African coast. The Zambezi also begins its route towards the Indian Ocean by flowing in a southerly direction, then turning northeast before exiting in a southeasterly direction. This irregularity explains why only 48% of African rivers have direct drainage to the oceans (Mountjoy and Hilling, 1988).

Figure 2.6 The Congo River is Africa's largest potential source of hydroelectric power.

It is hypothesized that the original course of these rivers was toward the huge drainage basins of the Djouf and the Kalahari. Eventually, uplifts caused the basins to release a considerable amount of water, thereby channelling a new course for these rivers to exit into the sea.

African rivers tend to have limited navigability (along major stretches of their course) because of interruption by falls and rapids that usually mark the edge of a plateau surface. The Congo River (Figure 2.6) is navigable up to 85 miles (137 km) inland; then its course is broken by a series of rapids up to the Stanley pool (Pritchard, 1979).

The volume of flow of many of the rivers varies from season to season, with peak flows during rainy season and low flows during the season with low rainfall amounts. The Nile, which benefits from two major sources for its water, is able, to some extent, to compensate for this disadvantage from the confluence of the Blue Nile and the White Nile onwards. Water from the Lake Victoria source reaches the lower Nile system with different timing from that of the Blue Nile. By far, the Congo River experiences the least flow fluctuations, as a result of its location in a humid equatorial environment. Rivers in the savanna areas often dwindle almost to their beds during the dry season.

African rivers have much potential for hydroelectric power generation; it is due to their steep falls and swift flow. The Congo River carries the second largest volume of water in the world and has enormous potential for hydroelectric power. Programs aimed at harnessing the hydroelectric potential of African rivers must also account for potential environmental and human impacts. African rivers also contain significant species of marine life, with inland fishing being a major commercial activity.

Drainage Basins

Africa has five broad, shallow drainage basins (ranging from 1000 ft, or 305 m, to 3000 ft, or 915 m), separated by plateaus and mountain ranges. They include the Djouf, Chad, Sudan, Congo, and Kalahari Basins. (See Figure 2.2.) These basins were formed by tectonic forces that warped parts of the plateau downward. Those parts became repositories of marine sediments eroded from plateau surfaces and deposited by rivers that converged toward these basins. This provides further credibility to the continental drift theory, which characterizes Africa as arising from a vast, internal drainage area with no coasts. As Gondwanaland fractured and gave way, water was released from these huge internal lakes, finding its way toward the sea. These basins are surrounded by mountain ranges, such as the Fouta Djallon, which borders the Djouf, and the East African Highlands, which separate the Congo Basin from the coastal plains on Tanzania's east coast. The Sudan Basin is sandwiched between the Ethiopian massif in the east and the Dar Fur and Ennedi ranges in the west.

The Chad, Kalahari, and Djouf are internal drainage basins. They receive runoff (surface running water) from streams of all sizes, but there is no exit to the sea. The Chad Basin is centered on an impoundment of Lake Chad as the final recipient of surface water. The lake receives water from rivers that originate from the wetter southern regions of Nigeria, Cameroon, and the Central African Republic and from seasonal streams in the north. Lake Chad provides significant water resources to the surrounding countries—Chad, Niger, Nigeria, and Cameroon. However, droughts, high evapotranspiration rates, and increased human demands have diminished its water levels. With these threats, it is possible that the lake may eventually end up as a large wetland fringed by salt flats.

The Kalahari Basin lacks surface water in much of its southern sections. A third of its northern section, however, receives perennial stream flow from Angola, mainly from the Okavango River. Most of the surface running water terminates in the large Okavango Swamp in northern Botswana. The swamp overflows on occasion and channels water eastward to form small seasonal lakes. Some water is also diverted to the Makgadikadi (Makarikari) salt pans along the Zimbabwe border (Europa, 2000). The Okavango Swamp, the temporary lakes, and the salt flats are significant habitats for wildlife and could potentially provide a basis for ecotourism.

The Djouf Basin is the driest, from its proximity to desert climates; it does not receive any permanent streams. Runoff during the rainy season ends up in shallow ponds and marshes that evaporate in the summer, leaving salt marshes and plains. The structure of the Congo Basin, on the other hand, allows the exit of surface water through stream flow by way of a narrow westward outlet through the Congo River. The Congo (Zaire) River is, for that matter, a major water outlet for a large basin area–hence, its large volume. Similarly, the southern sections of the Sudan Basin divert water into the Nile, yet the northwestern section is internally drained.

CLIMATE

About 70% of Africa's territory lies between the tropic of Cancer and the tropic of Capricorn. African climates are, therefore, mainly tropical (with the exception of the extreme southwestern Cape region of South Africa and the northwestern African coast). While the climate of Sub-Saharan Africa is described as tropical, a degree of diversity occurs. This is mainly a result of certain climatic controls that differ from region to region, but work to modify climatic types from place to place. Six main factors affect the climatic patterns of Sub-Saharan Africa: (1) pressure systems and winds, (2) latitudinal location, (3) maritime and continental influences, (4) ocean currents, (5) altitude, and (6) the lake effect.

Pressure Systems and Wind Movement

Two dominant air-pressure systems, the *Intertropical Convergence* low-pressure system and the *Subtropical High-Pressure System* (STHP), dictate surface wind patterns and also influence rainfall and temperature regimes in Africa. The idealized model in Figure 2.7 shows the distribution of pressure belts around the world. Wind movement is usually determined by pressure gradients between areas. When the difference in pressure between adjacent areas is large enough, winds blow from high-pressure to low-pressure areas, pushed by a pressure gradient force. At the narrow equatorial belt, high average temperatures generate a thermally induced low-pressure system. In this zone, the air rises and is usually cooled adiabatically; if the cooling continues to the dew-point temperature of the air, condensation takes place, resulting in high rainfall and the unique ecosystems of equatorial regions around the world. As the winds rise from the equator and move toward the poles at high altitudes, they are deflected into the subtropics (at or around latitudes 25° to 40° N and S) by the

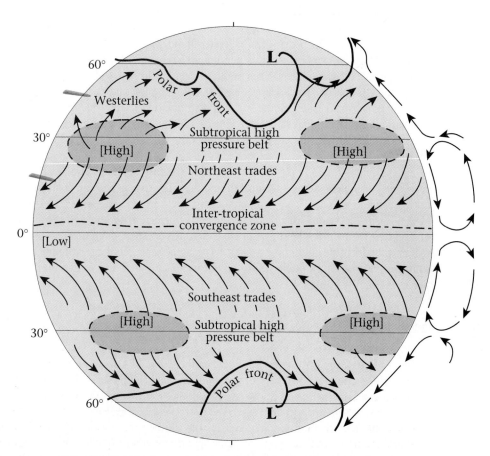

Figure 2.7 Global Pressure Belts and Surface Winds. Adapted from Strahler, A. & Strahler, A. *Modern Physical Geography*, 2d ed., New York: Wiley, 86.

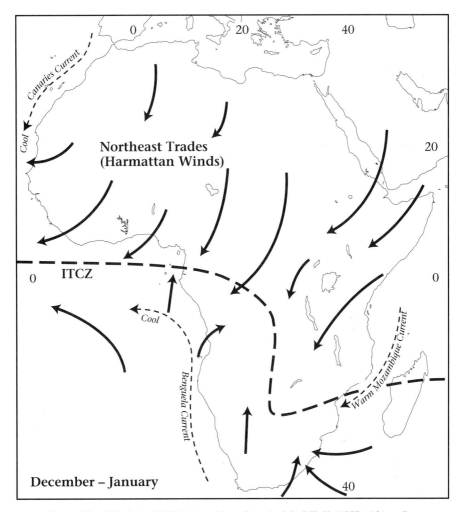

Figure 2.8 Shifts in the ITCZ. Adapted from Best, A. & DeBlij, H. (1977). *African Survey*, New York: Wiley.

earth's rotation; there they descend toward the surface, where they diverge toward the lows. As high-pressure systems sink toward the earth's surface, they generate calm and dry conditions. It is no coincidence that most of the world's semiarid and desert regions straddle areas of the subtropical highs (between latitudes 25° and 40°N and S). This is evidenced by the location of the Sahara, Arabian, Thar, and Gobi Deserts in the Northern Hemisphere, and the Atacama, Kalahari, and Great Sandy Deserts in the Southern Hemisphere.

The subtropical highs generate two major, consistent wind patterns that affect the climates in most of Sub-Saharan Africa. These are termed the Northeast Trade Winds, which originate from the north subtropical highs, and the Southeast Trades, which originate from the south subtropical highs (Figure 2.8). The two winds blow toward each other, so the zone of transition where they meet is defined as the *Intertropical Convergence Zone* (ITCZ).

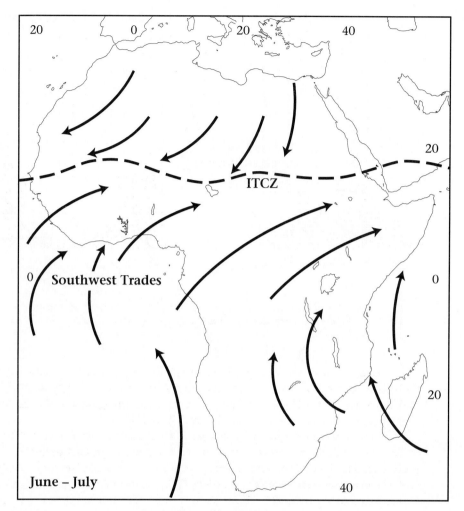

Figure 2.8 *(Continued)*

The location of the ITCZ shifts with the seasonal movement of the sun, as differential heating takes place on the African continent. During the summer season in the Northern Hemisphere (beginning June 21 or 22, summer solstice) when maximum heating occurs around latitude 23° N, the ITCZ shifts inland toward the tropic of Cancer and brings along with it rain-bearing southwesterly winds from the Atlantic Ocean (Figure 2.8a). The earth's rotation causes winds in the Southern Hemisphere to deflect to the left and winds in the Northern Hemisphere to deflect to the right. Therefore, southeasterly winds approaching the ITCZ from the Southern Hemisphere change direction to become southwesterlies when they cross the equator. During this same time of year, it is winter season, with relatively dry conditions, in the Southern Hemisphere. Figure 2.9 illustrates the difference in rainfall regimes between Lagos, Nigeria (located 6°27′ N 3°24′ E) and Johannesburg, South Africa (located 26°14′ S 28°09′ E).

Climate

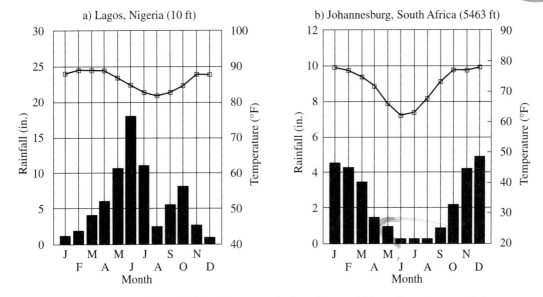

Figure 2.9 Rainfall and Temperature Regimes: Lagos and Johannesburg. Data compiled from Pearce, E. & Smith, G. (1990). *The Time Books World Weather Guide*, New York: Random House.

The diagram shows extremely high rainfall for Lagos, which is significantly affected by the southwesterly winds in June and July, when average rainfall measures 18.1 in. (452 mm) and 11 in. (275 mm), respectively. In Johannesburg, on the other hand, average rainfall amounts are as low as 0.3 in. (7.5 mm) in June and July.

From the autumnal equinox (September 22) until the vernal equinox (March 21), regions in the Southern Hemisphere receive more insolation from the sun. On December 21 or 22, the noon sun is directly overhead at the Tropic of Capricorn (23.5° S). At this time, sunlight is concentrated in the Southern Hemisphere, and the ITCZ moves toward the southern part of Africa. Figure 2.8a shows that the ITCZ skirts the west coast, then bends southward along the eastern margins of the Democratic Republic of Congo, and finally cuts through Zambia and Mozambique and then through Madagascar. As the ITZC moves southward, the influence of the dry northeasterly airmass prevails over west Africa. These trade winds are referred to as "Harmattan winds"; they blow across the Sahara Desert and bring dusty, hazy conditions to most of west and central Africa. In southern Africa, on the other hand, the prevalence of the ITCZ brings rainfall in the summer. Again, a comparison of Johannesburg and Lagos shows higher average monthly rainfalls in December and January for the former (4.5 in., or 113 mm) and lower measurements for the latter (approximately 1 in., or 25 mm).

Latitudinal Effect

The effect of latitude on climate is mainly defined by the location of a region with respect to the solar angle. The equator is straddled by a large portion of the African continent. Areas between latitudes 5° N and S of the equator receive high amounts of solar energy all year

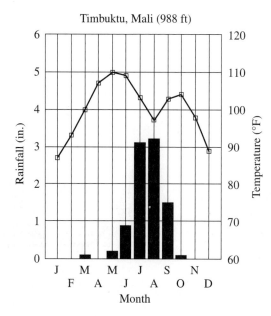

Figure 2.10 Rainfall and Temperature Regimes. Timbuktu.

round from the constantly high angle of the noonday sun. The average amount of solar energy received at the surface decreases from the equator to the poles. Temperatures in Lagos, Nigeria range from an average daily high of 89° F (31.6° C) in April to a low of 82° F (28° C) in August. The temperatures in Johannesburg, which is located in a subtropical environment, are relatively lower, ranging from an average daily high of 62° F (17° C) in June to 79° F (26° C) in December.

Maritime Effect versus Continentality

Proximity to such large bodies of water as the oceans has a modifying effect on climate. Land generally heats and cools faster than the oceans, which are in constant motion and are also deeper and hence capable of distributing heat within their volume. If we compare coastal Lagos (Figure 2.9a) with inland Timbuktu, Mali (Figure 2.10), we see that the latter has a higher temperature average and annual range than the former. Average daily temperature maximums in Timbuktu can reach as high as 110° F (43° C) in the summer. Daily temperature ranges are also high, averaging 34° F (17° C) in March. The lack of cloud cover partially explains the amount of heat loss during the day in Timbuktu. Furthermore, rain-bearing southwesterly winds lose their moisture to increasing arid conditions by the time they get to Timbuktu. This partly explains the low rainfall.

Ocean Currents

Ocean currents move warm water poleward and cold water toward the equator, thereby exerting a modifying influence on temperature. In the east coast of Africa, the westward-

Climate

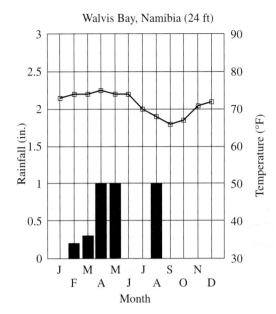

Figure 2.11 Rainfall and Temperature Regimes. Walvis Bay, Namibia. Data compiled from Pearce, E. & Smith, G. (1990). *The Time Books World Weather Guide*, New York: Random House.

flowing North Equatorial Current of the Indian Ocean divides to flow northward as the Warm Monsoon Drift and southward as the Warm Mozambique Current. On the west coast, the currents flow toward the equator as the Cool Canary Current from the north and the Cool Benguela Current from the south. The Cool Benguela Current, for example, partially accounts for the cool, dry influence along the Namibian and Angolan coasts. It cools air masses that pass over it; when they reach the heated land surface, they release little or no moisture. This explains the narrow strip of the Namib desert.

Figure 2.11 shows the cool and dry conditions for Walvis Bay, Namibia. Average monthly rainfall barely reaches 0.3 in. (8 mm) all year round. Temperatures are also relatively moderate, ranging from average daily maximums of 75°F (24°C) in April to average daily minimums of 66°F (19°C) in August.

Altitude

Air temperature decreases with altitude at an average rate of 3.5°F per 1000 ft (6.4°C per 1000 m). In Addis Ababa, Ethiopia, which sits at about 8000 ft (2438 m) above sea level, temperatures are relatively moderate. As Figure 2.12 shows, average daily temperature maximums range from 69°F (21°C) in July to 77°F (25°C) in May. Average daily temperature minimums can fall as low as 55°F (13°C) in January.

Both Kenya and Tanzania have a number of mountain ranges that rise to between 7000 ft (2167 m) and 10,000 ft (3096 m). In these regions, temperatures fall low enough for frost to develop; at the higher elevations, such as on Mount Kilimanjaro in Tanzania (19,340 ft, or 5985 m) and Mount Kenya (17,958 ft, or 5200 m), there are permanent snowcaps and small glaciers.

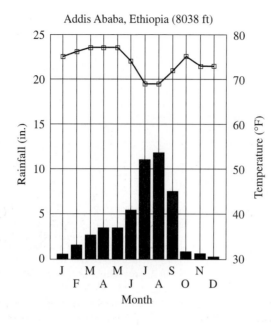

Figure 2.12 Rainfall and Temperature Regimes. Addis Ababa. Data compiled from Pearce, E. & Smith, G. (1990). *The Time Books World Weather Guide*, New York: Random House.

Lake Effect

Lakes have a modifying effect on temperature and rainfall as moisture from the lake is carried by prevailing winds to the land. Lake Victoria in east Africa has an effect on the microclimates of surrounding communities. The climograph for Entebbe (Figure 2.13a) shows the effect of Lake Victoria when compared with that for Kampala (Figure 2.13b), which is only a few miles inland from the lakeshore. Entebbe's average rainfall in April is 10.1 in. (26 cm), which is high compared with the 6.9 in. (18 cm) in Kampala. Entebbe records higher average rainfalls for all but two months of the year (August and September).

Climatic Classifications

Such climatic controls account for the varied temperature and rainfall regimes classified in Figure 2.14, which also define the climatic belts. The climatic regions shown are based on the Koppen system of classification, which employs a number of letter symbols describing precipitation and temperature. The climatic pattern shows a logical progression from humid conditions along the equator to subhumid, semiarid, and arid conditions away from the equator.

Humid Equatorial climates (Af) are confined to the eastern equatorial portions of Democratic Republic of Congo, the northern sections of Congo and Gabon, the narrow coastal strips of Cameroon and southeast Nigeria, and much of Liberia and Sierra Leone (Figure 2.14). Mean monthly temperatures here hardly fall below 64.4° F (18° C), and rains are consistent all year round, with every month recording more than 2.4 in. (6 cm). This belt conforms closely to the extent of the rainforest regions depicted in Figure 2.15.

Climate

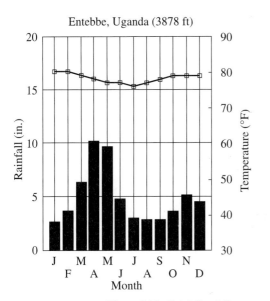 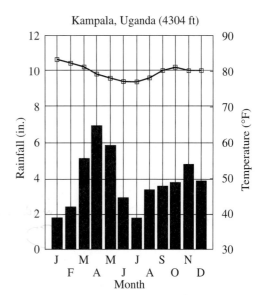

Figure 2.13 Rainfall and Temperature Regimes. Entebbe and Kampala. Data compiled from Pearce, E. & Smith, G. (1990). *The Time Books World Weather Guide*, New York: Random House.

Inland, away from the Af zone, rainfall decreases, and dry winter seasons occur (*Aw*). Several countries along west Africa's coast and others—such as the Central African Republic, southern Sudan, Uganda, Rwanda, Burundi, northern Angola, southern Congo and Gabon, and much of Tanzania and Democratic Republic of Congo—are subject to these subhumid conditions. During the year, there are two periods of concentrated precipitation interspersed with two dry seasons. This zone coincides with the wooded and grassland savannas of the region.

The rest of Sub-Saharan Africa is characterized predominantly by dry climates (B climates). *Semiarid climates (BSh)* are prevalent in the Sahel (located on the southern border of the Sahara Desert) and in Botswana and eastern Namibia in southern Africa. Semiarid conditions also prevail in the eastern sections of Ethiopia and Kenya and along a narrow western band in Madagascar. Annual rainfall is usually under 20 in. (51 cm), and mean annual temperatures are high. In N'djamena, Chad, which is located in the moister sections of the Sahel, average monthly rainfall exceeds 2 in. (5 cm) in only four months of the year: June (2.6 in., or 7 cm), July (6.7 in., or 17 cm), August (12.6 in., or 30 cm), and September (4.7 in., or 12 cm). The average maximum daily temperatures range from 87°F (31°C) in August to 107°F (42°C) in April. The fragile ecology of the African Sahel is constantly being threatened by desertification and soil degradation (as described in Chapter 3).

The Sahara region accounts for much of the *desert climate* of Africa *(BWh)*. Other areas include the Ogaden Desert of east Ethiopia, Somalia, a minute portion of southwest Madagascar, the Kalahari Desert, and the coastal strips of Namibia and Angola. Temperatures are very high, as is symbolized by the letter **h**, and the daily ranges can exceed 60°F (35°C). Parts of the central Sahara receive less than 4 in. (10 cm) of rainfall. Along the southwest coast of Namibia (in the Namib Desert region), temperatures are cooler (less than 64.4°F, or 18°C), from the effect of the cool Benguela ocean current.

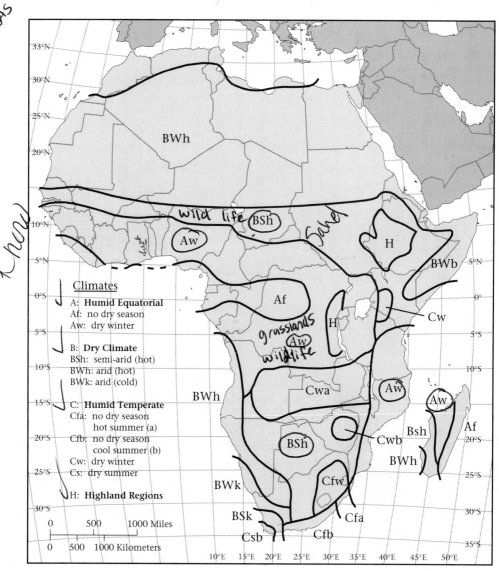

Figure 2.14 Climatic Regions in Sub-Saharan Africa. Adapted from Renwick, W. H. & Rubenstein, J. M. (1995), *People, Places, and Environment: An Introduction to Geography*, Englewood Cliffs, NJ: Prentice Hall.

Humid temperate climates (C) are present in isolated areas. Along the south–central coast of South Africa, there is adequate precipitation all year round with *cool summers (Cfb)*. The coastal strip east of this belt has similar rainfall characteristics, but with *hot summers (cfa)*. The southwest corner of South Africa's Cape province experiences a Mediterranean climate, which is characterized by *dry summers* and *wet winters (Csb)*. Inland, in the eastern sections of South Africa and in central Zimbabwe, dry winters and

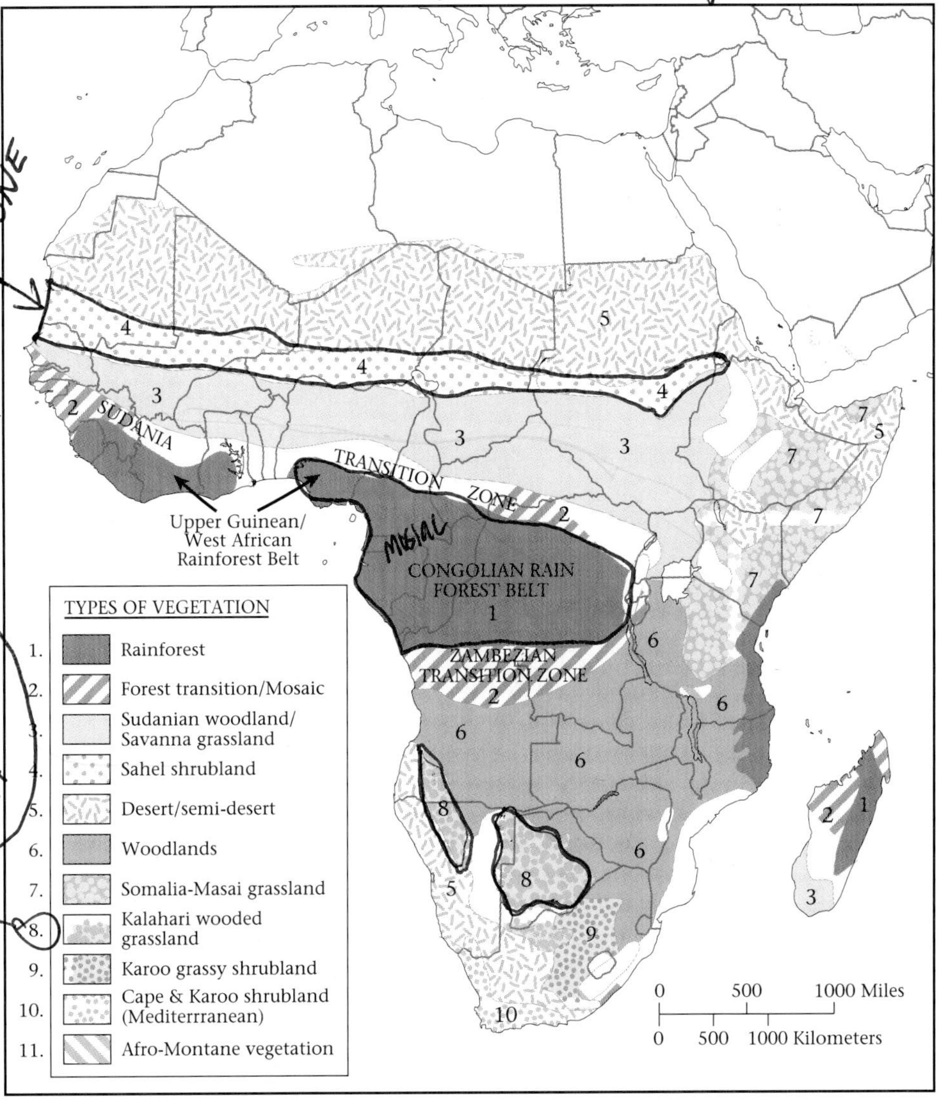

Figure 2.15 Vegetation of Sub-Saharan Africa. Modified from IUCN/UNEP (1986). *Review of the Protected Areas System in the Afrotropical Realm*, Geneva, Switzerland & Cambridge, UK: JUCN, Gland, 36.

cool summers prevail *(Cwb)*. Further inland, dry winters and hot summers *(Cwa)* affect eastern Angola, the Shaba province of the Democratic Republic of Congo, and much of Zambia.

Highland regions in Africa experience a range of climatic conditions. In the Kenyan Highlands, for example, the climate markedly changes from subtropical to temperate to polar as one moves from the lower elevations to the snowcapped surface of its peak. As

Figure 2.14 shows, the Ethiopian Highlands and the Mount Kenya and Mount Kilimanjaro regions are influenced by temperate climates and dry winters *(Cw)*.

VEGETATION

An underlying characteristic of the ecological setting of Sub-Saharan Africa is the rich diversity of plant cover that is organized into complex vegetation formations. The vegetation systems of the region, considered on a long-term basis, are a reflection of intricate combinations of past, present, and continuing processes that are ecological and human in nature. In the short term, however, contemporary factors tend to have the most effect and the most immediate influence on the nature, type, distribution, and stability of vegetation. A knowledge of the vegetation dynamics of Sub-Saharan Africa's diverse plant formations is necessary for the promotion of vegetation maintenance and enhancement, for the protection of associated wildlife and species diversity, and for the sustainable utilization of vegetation resources and vegetation space.

The large variety of plant species in Sub-Saharan Africa appears to defy simple and effective identification and categorization, which is necessary for local management and planning initiatives. However, through continuous field work in plant identification and categorization, a wealth of information has been acquired that makes African vegetation systems some of the most rigorously studied (as evidenced by White, 1983, and International Union of Conservation of Nature and Natural Resources (IUCN), 1986, 1990, 1992).

It is possible to organize Sub-Saharan Africa's plant systems into various vegetation zones, at varying spatial scales, to meet specific objectives. For example, in terms of the familial affinity of plants, the highest level of floristic (plant) organization to which most plants belong is the *Afrotropical Biogeographical Realm*. This large region of plants encompasses all of Sub-Saharan Africa. Plants of the region are unique from other such floristic kingdoms in terms of the dominance and relative absence of various diagnostic floristics. For example, the plant family Caesalpiniaceae is affiliated with most forest types in the region. Conifers and palms are poorly represented in most of the lowland forests, and Acacias dominate the open grassland woodland zones (savanna). This vegetation realm is associated with a rich and diverse fauna, having 38 families of fauna. Over 1500 species of birds and 2500 species of butterflies have been identified in its plant ecosystems (White, 1983; Collar and Stuart, 1988; IUCN, 1992).

The Afrotropical Biogeographical kingdom, however, appears too large and complex for any immediate local and regional applicability, especially because of intraregional variations in plant type and density at the species level. More useful categorizations have been developed at the biome level. A *biome* is an extensive unit of vegetation (plant) cover, together with its associated animal life, and it usually corresponds with particular climatic and soil types. Such a unit of classification has made it possible to categorize plant formations of Africa into various types of forests, grasslands, and deserts. White (1983), building on earlier work on vegetation mapping in Africa, has provided a comprehensive and useful classification system that overcomes some of the simplified and highly summarized divisions of vegetation in the past. He captures much of the complexity of vegetation in Africa in a reasonable number of classes. Figure 2.15 portrays the major vegetation zones of Sub-

Saharan Africa according to White's scheme, and it constitutes the framework for discussing the spatial distribution of vegetation in this section. Some of the existing nomenclature on vegetation descriptions in Africa (such as Udo, 1982) is utilized in conjunction with White's to meet the needs of the nonspecialist.

Vegetation Types

The major vegetation types of Sub-Saharan Africa include the rain forest, the woodlands, the Sudanian woodland or grassland, the Sahel bushland, the semidesert or desert, the Mediterranean (Cape and Karoo shrubland), and the Afromontane. These vegetation belts, depicted in Figure 2.15, need not necessarily correspond exactly to current conditions in the field, because several human factors, including agricultural and forest land use, tend to transform or undermine their core characteristics. At best, the belts depict areas with certain requisite natural conditions, such as climate and soil, that make it possible for attendant vegetation forms to develop to certain climax characteristics. They also depict areas where human management techniques have aided in maintaining the original characteristics of vegetation systems. The existing vegetation belts, however, are a true representation of the diversity of the African environment and also reflect the current dynamic processes of natural and human interactions.

A close association exists between climate and vegetation distribution in Sub-Saharan Africa (Figures 2.14 and 2.15). Areas along the wet-humid equatorial belt, where the annual soil-moisture balance is positive, tend to have high and closed forests. On the other hand, areas with relatively low annual precipitation levels and longer dry seasons tend to exhibit poorer forests, woodlands, and savanna, depending upon the degree of moisture stress. Vegetation zonation, depicting a deterioration from forests to grassland in a south (coastal) to north (interior) stacking, is well represented in west Africa.

Rain forest   The rain forest generally refers to all forests in the humid and semi-humid areas of Sub-Saharan Africa. It is, therefore, a complex vegetation region encompassing areas of perennially moist forests, ringed and interspersed by forests that are semideciduous in nature. Much of the true moist rain forest belt is confined to the lowland equatorial zone of Africa, where heavy rainfall and high temperatures help to promote rapid plant growth. The current rain forest belt covers an estimated 7% of the total land area of Sub-Saharan Africa and about 20% of rain forests worldwide. Although it accounts for a relatively small proportion of the region's vegetation, the rain forest is the richest ecologically, with more than 50% of the continent's plant and animal species. The west African coastal and central African basin zones alone are estimated to contain over 8000 plant species, about 80% of which are native (endemic) to the region. This zone is also estimated to contain about 84% of the African primate species, 68% of the passerine birds, and 66% of the butterfly species (IUCN, 1992).

The tropical rain forest is generally dominated by broad-leaved evergreen tree species, and it has a high diversity of species, which are structured into layers (Figure 2.16). The topmost layer of the vegetation structure consists of emergents that rise up to 165 ft (50 m) and, in some cases, as high as 300 ft (90 m). A middle canopy layer, made up of trees about 80 to 115 ft high (25 to 35 m), forms a continuous cover as their crowns interlock, and this

Figure 2.16 A Rainforest Region in West Africa. Notice the canopy in the upper layer.

tends to deprive the lower layers of direct sunlight. Trees of up to 50 ft (15 m) form the lowest layer. In their true formations, they exhibit very sparse undergrowth, except where trees fall or fires remove the excessive shade. Lianas, and other vascular epiphytes, link the forest floor with the canopy. In the relatively drier semideciduous belts, the structure of the vegetation is simple. Emergent trees tend to show a degree of deciduousness, with the lower canopies maintaining evergreen characteristics. Because of potential gaps in these forests, the undergrowth tends to be denser than in the moist tropical forests.

The tropical rain forests occur mostly in two major formations: the west African or Upper Guinean rain forests, and the central African or Congolian rain forests. A third, less prevalent formation exists along the lowlands of coastal east Africa and the east coast of Madagascar (Figure 2.15). Generally, the Upper Guinean and Congolian belts exhibit close similarities in most of their biota, indicating a past link that ensured significant exchange of species. The Upper Guinean rain forest is terminated by a dry zone of savanna woodland around the Dahomey Gap, located between the coast of eastern Ghana and the Benin–Nigerian border. The density and species diversity in this region are generally small compared with those in the Congolian belt.

The Congolian belt is the largest stretch of rain forest, with an area of about 918,680 sq mi (2,380,000 km^2). It is characterized by uniformly high temperatures and rainfall. The western equatorial belt (Cameroon, Gabon, and Equatorial Guinea) has the densest, most diverse, and most extensive evergreen vegetation in this zone. An estimated 600 species are endemic to Cameroon and Bioko (formerly Fernando Po), and over 1000 species are endemic to Gabon and Equatorial Guinea (IUCN, 1990, 1992; Collar and Stuart, 1988).

The area of tropical rain forests in eastern Africa is estimated at only 3831 sq mi (10,000 sq km) covering altitudinal ranges from near sea level to about 10,000 ft (3000 m) on Mount Kilimanjaro (Lovett and Wasser, 1993). True rain forests in east Africa are restricted to the coastal areas of Kenya and the Indian Ocean islands of Pemba, Zanzibar, and Mafia (Tanzania). The eastern Madagascar coast has an important stretch of tropical forests. Another fragmented stretch of forests occurs on the wetter ancient crystalline highlands of Usambara, Uluguru, and Uzungwa, along with smaller patches in Tanzania and the Taita Hills of Kenya, collectively referred to as the *Eastern Arc forests* (Lovett and Wasser, 1993).

Forests also occur on the slopes of the high volcanic mountains of Tanzania (particularly Kilimanjaro and Meru), Kenya, and southwestern Uganda. The forest ecosystems of southwestern Uganda are affiliated with those of the Congolian rim, which they border.

Generally, the forests of east Africa are relatively drier and less diverse than the Guineo–Congolian forests. They also have significant anthropogenic influences. Their isolation, fragmentation, and old age cause the eastern African forests to tend to exhibit a high degree of endemism. For example, along the eastern Madagascar coast, the flora is estimated to contain about 6100 species, 79% of which are endemic. Such unique attributes of the eastern African forest, together with their small extent, demand priority conservation management.

Other than their ecological value, the west African and central African rain forests account for most of the high-valued timber species of Sub-Saharan Africa. Other assets include several wildlife species of economic value, medicinal resources (a preserve for unique traditional societies), a pristine agricultural environment, and a setting for the now growing activity of ecotourism. These forests are being significantly threatened by land-use conversion and modification (Chapter 3). The Guinean forest, in particular, faces the most serious threat of extinction, because of its relative accessibility by large population concentrations and industrial activities in west Africa. High agricultural population densities constitute a significant threat to all the forests of east Africa, which are already patchy.

Woodlands The woodlands constitute the largest vegetative region in Sub-Saharan Africa. Toward the equator, they merge imperceptibly into the drier forests; in their southern boundaries, they merge with the savanna grasslands. The woodlands generally lie within the interior subhumid tropics. The equator end forms part of the relatively narrower *forest transition/mosaic zone*, where annual rainfall ranges between 40 and 64 in. (100 and 160 cm). In the larger northern (Sudanian) and southern (Zambezian) zones, rainfall generally ranges between 30 and 50 in. (80 and 125 cm) a year. A period of aridity of up to six months is followed by a summer concentration of rainfall.

The woodland proper is made up of open stands of trees, with at least a 40% coverage of tree crowns. The trees are at least 16 ft (5 m) tall, with increasing ground cover of herbaceous plants (UNESCO, 1973). Tree density, size, height, and cover generally deteriorate toward the savanna belt. At their equatorial ends, both the northern and southern section of the vegetation belt have remnants of forests that are mostly impoverished compared with the tropical forests. These forests are present as "cloud forests" (or Dembos) in the Angolan escarpment, and they extend into short, drier forests and thickets in the more continental interiors. Patches of forest mosaics also occur along the southern coast of Somalia and the coasts of Kenya, Tanzania, and Madagascar. In the Guinean zone, most of the dry forests exist in patches on hilly sites and riparian zones. The preexisting forests in this belt have been destroyed as a result of long human occupancy. Farming and bush fires are the major vegetation modifiers.

The core of the woodland proper is made up of the Guinea savanna woodland in west Africa and the Zambezian region in the southern belt. Most of the tree species are deciduous and semideciduous. Relatively shorter trees (generally below 25 ft, or 8 m) are interspersed with grass and other herbaceous plants. The dominant species in the Guinean–west African section include the African fan palm (*Borassus aethiopicum*), the shea butter tree

(*Butyrospemun parkii*), and the silk cotton tree (*Bombax petandrum*). The baobab (*Adansonia digitata*), African rubber (*Landolphia* spp.), and the gum arabic (*Acacia senegal*) are common to both the Sudanian and Zambezian sections of the vegetation belt (IUCN/UNEP, 1986).

In the Zambezian zone, the three main types of woodlands are the miombo, the mopane, and the Zambezian undifferentiated woodlands. The miombo woodlands are characteristic of the high plateaus and escarpments and are dominated by the *Brachystegia* species. The miombo can grow to be 33 to 65 ft in height (10 to 20 m) where the soils are deep and moist. The mopane are found in the drier parts of the woodlands (Figure 2.17). The dominant species is the *Colophospermum mopane*, which extends as far as Botswana, excluding the regions of the Kalahari sands. The Zambezian undifferentiated woodland occurs north of the Limpopo River, covering a smaller area with very few or no miombo species. It is floristically diverse compared with the miombo and mopane. Much of the woodlands have undergone human modification. In localized areas, soil constraints and swampy conditions, along with microclimatic conditions, are a factor.

Savanna grassland The savanna or grassland belt delimits a vegetation zone where herbaceous plants and grasses dominate the landscape. Savanna vegetation is extensive in the Sudanian, Zambezian, and *Somalia–Masai plains*, as well as in the Indian Ocean coastal belt. (See Figure 2.15.) In west Africa, it is commonly referred to as the Sudan savanna. In both the Sudanian and Zambezian zones, grasslands may also occur in flood plains, river valleys, and seasonally waterlogged depressions. The fadamas in northern Nigeria and southern Niger are floodplains covered with tall and dense grasslands and sparse tree

Figure 2.17 The Mopane Woodlands of Zimbabwe.

Vegetation

populations. In the Zambezian zone, the floodplain grasslands (Dembo grasslands) develop in seasonally waterlogged depressions, while shorter grasslands develop over seasonally waterlogged Kalahari sands. The Dembo grasslands thrive in relatively nutrient-rich, well-structured, and well-drained soils of the depressions and thus tend to be taller than the shorter grasslands where soils tend to be coarse-textured, sandy, and poorer in nutrients and often face extremes of excess and highly deficient soil-moisture conditions.

The savanna belt shows a degree of complexity in terms of the type, variety, height, and density of grasses and of tree combinations. Its vegetation ranges from perennial tall grasses with about 10% to 49% of tree cover to areas of true open grassland with less than 10% of trees, which are relatively small in stature and height. The savanna region includes the drier ends of the woodlands toward the semidesert environments, except where local soil conditions and climatic anomalies place them within climatically wet regions. Precipitation in the zone ranges from about 10 to 50 in. (25 to 125 cm) per year. A dry season of more than six months is common in most parts of the region. Most of the vegetation species exhibit strong adaptations to fire and moisture stress, such as baobabs, acacias, and mopane species.

The open vegetation and the grassy cover of the savanna region provide expansive habitats for both livestock and wildlife development. Traditional pastoralists, such as the cattle-herding Fulani of west Africa and the Masai of east Africa, have used the dry savannas (which have fewer tsetse flies than are common in the wetter woodlands) to further their age-old pastoralist cultures. The savanna belt supports most of the large mammals of contemporary Africa, especially where human encroachment is minimal. For example, the Serengeti Plain, in the Somalia–Masai region of east Africa, is credited with having some of the largest concentrations of wildlife in the world (White, 1983; also see Figure 2.18).

Figure 2.18 The Serengeti Plains of Tanzania.

Existing patches of forests in the savanna zone, and observed invasions of dry-forest species on long-abandoned fields, are indications that the savannas are vulnerable to human activity. The evidence, however, suggests that, while this could be true in most of the close-settled regions of the forest transition mosaics, much of the savannas are climatic and edaphic climax in origin. The diversity of savanna fauna and flora, especially in the Zambezian region, indicates that there have always been extensive areas of savanna. This implies that the continent was never completely forested in its past. Documentation on pollen cores from various sites in east Africa suggests a relatively open grassy vegetation about 35,000 to 11,000 years before present, which was followed by the extension of more closed forests (IUCN/UNEP, 1986).

Sahel shrubland The Sahel shrubland is a clearly defined region in the north where the Sahel constitutes a transition zone between the Sudanian grassland region and the Sahara Desert, stretching from Mauritania and Senegal in the west to southern Sudan and Ethiopia in the east. On a larger scale, it includes the southern edges of the Zambezian region and areas bordering on the *Kalahari wooded grasslands* and the *Karoo grassy shrublands*. Ecologically, it is a species-poor belt. Out of an estimated 1200 plant species in the region, only about 3% are considered endemic.

The Sahel (which in Arabic means border) is characterized by very low and unreliable rainfall patterns. It has an extended dry season (up to nine months) with annual rainfall between 10 and 20 in. (25 and 50 cm). Perennial soil-moisture constraints affect both plant growth and species diversity. In most areas, rain-fed agriculture is not feasible without an irrigation supplement. Depending upon the severity of moisture stress and local soil conditions, there is some presence of scrub forest, bushland, wooded grassland, and short and tufted grasses. Most of the grass withers during the drought season, exposing the surface to soil compaction and erosion. Acacias (for example, *Acacia senegal*) are more common here. Such African staples as the pearl millet (*Penniselum americanum*) and sorghum (*Sorghum bicolor*) are believed to have been domesticated here. It also offers a large ecological setting for pastoral farming. The ecological vulnerability of this region has resulted in a series of human disasters over the years, by way of droughts and severe food shortages (Chapter 3). Also, from its proximity to the Sahara Desert, it is actively affected by desert processes such as dune encroachment.

Desert and semidesert The desert vegetation is centered largely around the Sahara in the north and the Kalahari and Namib Deserts in the south. These regions suffer from extreme aridity, with rainfall generally below 10 in. (25 cm). Significant climatic stress affects vegetation development; semideserts grade into the grasslands and shrubs of the Sahel, the Somalia–Masai region, and the Karoo–Namib region, where seasonal increase in soil moisture can promote vegetation growth.

Within the core of the true deserts, the lack of soil moisture restricts vegetation to a few sites where water might occur near the surface, such as in sheltered and intermittent water courses and oases. In such zones, perennial plants might survive. Most of the plants and fauna are drought resistant, with both permanent and temporary adaptations. The date palm is a common cultivated crop, and there is limited cultivation of wheat and millet. Some ephemeral vegetation develops and completes its life cycle shortly after a brief, torrential downpour of rain.

The Namib Desert is floristically more diverse than the Sahara. However, pollen and anthropogenic evidence indicate that most parts of the Sahara were covered by woodland vegetation in wetter periods about 5000 to 10,000 years ago.

Mediterranean (Cape and Karoo shrubland)   A few anomalies exist in Sub-Saharan Africa's vegetation pattern. In the southwestern part of South Africa's Cape province, Mediterranean vegetation occurs as part of the Cape and Karoo shrubland vegetation system. This vegetation enclave is very rich in plant-species diversity but poor in associated fauna. The vegetation consists of low trees and shrubs that have developed adaptations to long periods of aridity, such as thick leathery leaves, needlelike leaves, and long roots. In the Cape Mediterranean, the dominant native vegetation is *fynbos*, which is a sclerophyllous shrubland. Important native food plants in this zone include wild yams and grain sorghum. Because of prolonged human occupancy, much of the native vegetation on the lowlands has been lost.

Afromontane vegetation   The montane or highland vegetation develops over isolated elevated regions of Sub-Saharan Africa and is affected by local temperatures, rainfall, and soil conditions. The result is a vertical zonation of vegetation belts corresponding to the vertical ecological characteristics of the slopes of such highlands. This type of vegetation is restricted to the Cameroon and Guinea highlands of west Africa; the Ethiopian, Kenyan, Tanzanian, and Albertine Rift highlands (Kivu–Ruwenzori) of east Africa; and the Drakensberg highlands (Thabana–Ntlenyana) in southeast Africa. The montane regions are estimated to carry about 4,000 plant species, 80% of which are endemic (IUCN, 1990).

Montane forests generally develop at lower levels, between 3900 and 8200 ft (1200 and 2500 m), where adequate precipitation [between 50 in. and 100 in. (1250–2500 mm) per year], temperature conditions, and relatively deep soils promote tree growth. Such conditions vary with altitude as the quality of the forests and diversity of species generally decreases with altitude. Most of the highland areas that offer fertile agricultural soils have been heavily farmed and are, therefore, susceptible to accelerated erosion.

SOILS

Agriculture is a dominant activity in Sub-Saharan Africa, so an understanding of the nature and distribution of soils provides a basis for assessing patterns of agricultural land use and opportunities for soil utilization and conservation. The variable environmental conditions in the region have altered the rate and nature of soil development, their physical and chemical properties, and their relative suitability for certain uses. Sub-Saharan African soils, for that matter, are not uniform. They have been shaped by natural soil-forming factors, which include climate, vegetation (and other living organisms), parent material, and surface topography. Human factors are equally important in shaping the soils of the region.

The soil types of the region are closely related to the climatic belts. An estimated 44% of Africa's soils have been classified as arid, 8% as semiarid, and 44% as humid, based on their moisture–temperature balance (World Resources Institute, 1992). Climate influences the type and rate at which rocks are physically and chemically broken down to provide the inorganic material (parent material) around which most of the soils develop—a process referred to as *weathering*. The deeply weathered parent material releases minerals and nutrients from native rocks into the soil solution for ready uptake by plants. However, heavy rains, accompanied by continued downward movement of water through the soil, drain the soil of essential minerals—a process known as *leaching* that leaves humid soils relatively impoverished.

Vegetation cover and other living organisms provide the organic matter of soils. The litter of organisms, when decomposed, becomes soil humus. This not only releases mineral nutrients, but also improves the physical property, texture, and structure of the soil, thereby improving its water absorption and retention capacity. In a general sense, therefore, under their pristine status, newly cut forests in the humid regions of Sub-Saharan Africa tend to be more fertile than soils in the arid environments. Soil nutrients are closely held in balance with the standing vegetation. Once this is removed, the nutrient cycling between the soils and the vegetation is disrupted, causing further soil impoverishment in humid forest regions.

Soils of the subhumid environments tend to retain a relatively higher proportion of their organic matter, in response to reduced rates of leaching, wherever human use has not affected the natural characteristics. Living organisms, such as termites, transfer chemical nutrients and organic matter throughout segments of the soil in which they operate. In addition, they physically influence the surface characteristics and bulk density of soils through the mounds they construct. Termite activities are common in the semihumid to arid environments, particularly in the Zambezian region (read Sattaur, 1991).

Most of the soils of Sub-Saharan Africa are formed over different rock-parent materials, with different mineralogical and physical attributes operating in concert with other soil-forming factors to generate different types of soils. The parent material imparts several attributes to soils, such as their texture and mineral status, especially when the soils are not very old. The volcanic slopes of the Cameroonian mountains and the highlands of east Africa, for example, tend to support lush vegetation and are good agricultural soils, because they developed over basaltic rocks containing high amounts of iron–magnesium minerals. However, most of Africa's soils are very old, and their relationships with the parent materials tend to be secondary, since other ongoing processes have undermined that linkage. Therefore, soil conservationists need to understand other processes, in addition to parent materials, when analyzing African soils.

Soil Distribution

Sub-Saharan Africa's soils have been surveyed and mapped under different classification systems, each meeting a set of specific objectives. Different soil nomenclatures, such as *swampy* soils, *forest* soils, and *lateritic* soils, are widely used (Udo, 1982 and Behrens, 1989). The Food and Agricultural Organization of the United Nations (FAO) and the United Nations Educational Scientific and Cultural Organization (UNESCO) published a classification scheme in 1989 that attempted to incorporate familiar and traditional soil names, such as chernozems, podsols, planosols, and lithosols (Gerrard, 2000). Notably, the FAO/UNESCO scheme did not include other familiar terms, such as brown forest soil, podzolic, prairie, lateritic, and alluvial, to avoid confusion due to different use in different countries.

Hundreds of soil groups exist in Africa, so it is better to utilize universally understood and scientifically verifiable classification systems that consolidate soils into logical, hier-

Soils

archical categories. The Soil Taxonomy System, a comprehensive classification system developed by the United States Department of Agriculture and used in conjunction with other systems relevant to Africa, is effective for analyzing patterns of soil distribution worldwide. Figure 2.19 summarizes the major soil groups of Sub-Saharan Africa. Their distribution may be generalized under two headings: the soils of the humid ecological zones, and the soils of the arid and semiarid ecological zones.

Among the soils of the humid region are the *oxisols*, which form in the intensely weathered moist equatorial belt and underlie areas between the western Congolian and the

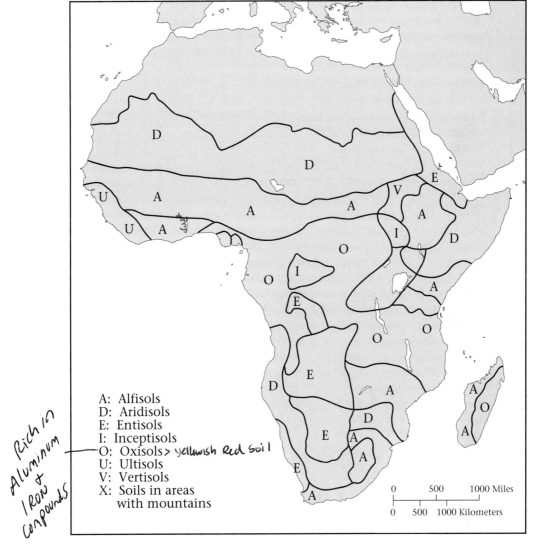

Figure 2.19 Soils of Sub-Saharan Africa. Adapted from Brady, N. (1990). *The Nature and Properties of Soils*, New York: Macmillan, 571.

east African coast. About 40% of the total land area in the humid tropics of Africa is underlain by oxisols (Brady, 1990). The surface horizons are characterized by a mixture of iron and aluminium oxides, which give them a reddish color. Much of the soil nutrients have been lost to lower horizons through continuous leaching, especially where the vegetation has been removed, leaving soils with low base nutrients for plant growth. In spite of the relatively poor soils, the areas covered by the oxisols carry large plantation crops of rubber and oil palm. In the southeastern Nigeria–Cameroon region, oxisols support large agricultural populations. Oxisols require delicate attention to meet the challenges of leaching associated with the humid environments and of overcultivation without adequate consideration given to soil management.

Ultisols are mostly developed in moist to subhumid climatic areas under forest conditions and woodlands; they are represented in Liberia, Guinea, and Sierra Leone and in sections of Kenya, Uganda, Zaire, Rwanda, and Burundi. They are relatively less weathered than the oxisols and are reddish and yellow in their surface horizons, from a greater representation of iron and aluminium. Ultisols are suitable, in west Africa, for rubber, coffee, and banana plantations, especially where fertilizer application and other modern agricultural management techniques are available. In the Congolian rim of Burundi, Uganda, and western Kenya, ultisols support some of the densest agricultural populations where tea, rubber, and banana crops are grown.

Alfisols are located in the subhumid to semiarid areas, ranging from the forest mosaics to segments of the savannas. Their humid zones tend to be more weathered. They are characterized by a light color in the upper horizon, low organic matter content, and a high clay content. They tend to maintain a high to medium nutrient status, compared with oxisols. Alfisols are represented by three suborders that differ mostly in terms of their soil-moisture regime and degree of weathering: udalfs form in humid sections, ustalfs underlie most of the woodlands and the savannas, and xeralfs underlie the arid areas of the southern tip of South Africa. Alfisols are widely cultivated, supporting a range of crops from humid to arid regions, including cocoa, rubber, bananas, maize, cassava, sorghum, and millet. In Côte d'Ivoire and Ghana, alfisols support such cash crops as cocoa, rubber, coffee, banana, and oil palm. Fruit crops, such as oranges and avocado pear, are well represented in the humid lowland forest sections. In the drier belts of the alfisols, cotton, groundnuts, tobacco, cassava, sorghum, and millet are farmed. Pastoralism is also a major activity in the drier sections.

Vertisols are confined to the subhumid to arid areas of southeast Sudan and western Ethiopia. They have a dark clay content that makes them difficult to work, because they shrink considerably under dry conditions and expand and become sticky under wet conditions. They offer potentially good soils with the right soil- and crop-management approach (Brady, 1990). When they dry, organic matter builds up in cracks that develop, thus enhancing the quality of the subsoil. They respond well when cultivation is supplemented with irrigation agriculture. In Sudan and Ethiopia, cotton, corn, sorghum, and millet are widely produced.

Soils over most of Sub-Saharan Africa's dry zones are classified as *aridisols*. Their relatively low biomass densities, coupled with high temperatures, result in low organic matter content and the concentration of hard layers of salt that build up near the surface. In other parts of the world, such soils have been widely utilized for agricultural purposes, with new technologies in irrigation and crop development. In Sub-Saharan Africa, crops such as date palms, sorghum, and millet are restricted to certain parts of the aridisol belt, and nomadic grazing is quite extensive. Irrigation opportunities and the development of drought-resistant crops can make the aridisols productive for agriculture.

Two other categories of soil orders depicted in Figure 2.19 are *entisols* and *inceptisols*, which have weak horizon and profile development. Entisols are still in an immature stage of the soil evolutionary process; in hilly areas, therefore, they tend to be shallow and stony (Cruickshank, 1972). Their deficiency in surface organic layers restricts the potential for agricultural production. Entisols are common in the Sahara Desert, the Kalahari wooded grassland border areas, the arid coastal strip of southwest Africa, and the highlands of east Africa.

Inceptisols are young soils, but have a relatively well-developed soil profile compared with entisols. They are usually developed in humid regions as well as regions with significant glacial, wind, and volcanic ash deposits (Muller and Oberlander, 1984). They are concentrated in the Niger Delta and adjoining areas and in the central Zaire Basin. In the Niger Delta, rice, cassava, maize, and oil palm are grown, especially where proper drainage facilities are installed to reduce the detrimental effects of waterlogging. In areas where poor drainage prevails, the surface of humid inceptisols becomes a breeding ground for such insect-borne diseases as malaria.

Sub-Saharan African soils are vulnerable to a range of surface conditions and human activities. As a result, moisture deficiency, waterlogging, accelerated erosion, and the poor retention of nutrients remain the major challenges for communities and governments to confront. This calls for a reexamination of priorities that is more sensitive to the ecological value of soils and for more efficient methods that promote effective soil management, soil improvement, and soil fertilization. A number of these issues are discussed in Chapter 3.

CONCLUSION

Sub-Saharan Africa's physical environment is defined by the dynamic interaction of several natural processes. The complexity and diversity of its physical landscapes and its climatic, vegetational, and biogeographical attributes offer a number of assets on the one hand and some inherent liabilities on the other. Included among the assets are the H.E.P. potential of its rivers, the scenic and economic value of its lakes, the biodiversity and commercial value of the rain forest region, the rich volcanic soils of east Africa, and the value of minerals embedded in pre-Cambrian rocks. On the other hand, the narrow and straight coastlines limit opportunities for natural harbors and the short continental shelves restrict potential for offshore oil exploration and fish breeding. Furthermore, the leached soils of the rainforest and the semidesert and desert environments inhibit agricultural development. A major threat to the physical environment, however, is the extent of human activity in ecologically sensitive areas. The magnitude of this threat will depend on the ability of Sub-Saharan African communities and governments to develop the appropriate response strategies to manage and conserve their fragile environments. These issues are addressed in Chapter 3.

KEY TERMS

Plate tectonics
Continental drift
Sea-floor spreading
Rift valley
Pangaea
Continental shelf
Gondwanaland
Escarpment
Subtropical high pressure
Intertropical convergence zone
Harmattan winds
Ocean currents

Lake effect
Tropical rain forest
Savanna grassland
Sahel
Montane vegetation
Maritime effect
Northeast trades
Southeast trades
Biome
Oxisols
Vertisols
Inceptisols

DISCUSSION QUESTIONS

1. Identify and discuss the major characteristic features of Sub-Saharan Africa's physical landscapes.
2. Describe how plate tectonics and continental drift have influenced the physical geography of Sub-Saharan Africa.
3. Identify the major climatic controls that influence the distribution of climatic belts in Africa.
4. What role does climate play in the spatial distribution of vegetation zones in Africa?
5. Outline the major characteristics of the rain-forest region in Africa.
6. To what extent does the physical-environmental geography of Sub-Saharan Africa relate to the distribution of human and economic activity?
7. Describe the distribution of soils in Sub-Saharan Africa, and relate it to the distribution of agricultural productivity.

REFERENCES

ADAMS, W.M., GOUDIE, A.S., & ORME, A.R. eds. (1996), *The Physical Geography of Africa*, New York: Oxford University Press.
ALLAN, T. (1993). *Deserts the Encroaching Wilderness*, New York: Oxford University Press.
BEST, A. & DEBLIJ, H. (1977). *African Survey*, New York: Wiley.
BRADY, N. (1990). *The Nature and Properties of Soils*, New York: Macmillan.
BUCKLE, C. (1978). *Landforms in Africa*, Hong Kong: Longman.
CHERNICOFF, S. & VENKATAKRISHNAN, R. (1995). *Geology*, New York: Worth.

References

Collar, N. & Stuart, S. (1988). *Key Forests for Threatened Birds in Africa*, ICBP Monograph, no.3. Cambridge, England: ICBP.

Cooke, R.U. & Doornkamp (1990). *Geomorphology in Environmental Management*, Oxford, UK: Oxford University Press.

Cruickshank, J. (1972). *Soil Geography*, Newton Abbot, UK: David and Charles.

De Blij, H. & Muller, P. (1994). *Geography: Realms, Regions, and Concepts*, New York: Wiley & Sons.

Europa Publications. (2000). *Africa South of the Sahara*, London: Europa Publications, Ltd.

Gerrard, J. (2000). *Fundamentals of Soils*, London: Routledge.

Griffiths, I. (1994). *The Atlas of African Affairs*, 2d ed., Witwatersrand, South Africa: University of Witwatersrand Press.

International Union for Conservation of Nature (IUCN) & Natural Resources/United Nations Environmental Program (1986). *Review of the Protected Areas System in the Afrotropical Realm*, Geneva, Switzerland & Cambridge, UK: IUCN, Gland.

IUCN (1990). *Biodiversity in Sub-Saharan Africa and Its Islands: Conservation, Management and Sustainable Use*, Geneva, Switzerland & Cambridge, UK: IUCN, Gland.

IUCN (1992). *The Conservation Atlas of Tropical Forests Africa*, Singapore: Simon & Schuster.

Levin, H. (1996). *The Earth through Time*, 5th Edition, New York: Saunders College Publishing.

Lovett, J.C. & Wasser, S.K. (1993). *Biogeography and Ecology of the Rain Forests of Eastern Africa*, Cambridge, UK: Cambridge University Press.

Moores, E.M. & Twiss, R.J. (1995). *Tectonics*, New York: W.H. Freeman.

Mountjoy, A. & Hilling, D. (1988). *Africa: Geography and Development*, Totowa, NJ: Barnes & Noble.

Muller, R. & Oberlander, T. (1984). *Physical Geography Today: A Portrait of a Planet*, New York: Random House.

Orme, A.R. (1996), "Coastal Environments," in *The Physical Geography of Africa*, Adams, W.M., Goudie, A.S., & Orme, A.R. eds., New York: Oxford University Press, pp. 238–267.

Pearce, E. & Smith, G. (1990). *The Time Books World Weather Guide*, New York: Random House.

Pritchard, J. (1979). *Landforms and Landscape in Africa*, London: Arnold.

Renwick, W.H. & Rubenstein, J.M. (1995). *People, Places, and Environment: An Introduction to Geography*, Englewood Cliffs, NJ: Prentice Hall.

Sattaur, O. (1991). "Termites Change the Face of Africa," *New Scientist* (January 26):27.

Stanley, S.M. (1999). *Earth System History*, New York: W.H. Freeman.

Strahler, A.N. & Strahler, A.H. (1983). *Modern Physical Geography*, New York: Wiley.

Summerfield, M.A. (1996). "Tectonics, Geology, and Long-Term Landscape Development," in *Physical Geography of Africa*, Adams, W.M., Goudie, A.S., & Orme, A.R., eds., New York: Oxford University Press, pp. 1–17.

Udo, R. (1982). *The Human Geography of Tropical Africa*, London: Heinemann Press.

UNESCO (1973). *International Classification and Mapping of Vegetation*, Paris: UNESCO.

White, F. (1983). *The Vegetation of Africa: A Descriptive Memoir to Accompany the UNESCO/AETFAT/UNSO Vegetation Map of Africa*, Paris: UNESCO.

World Resources Institute (1992). *World Resources, 1992–1993*, New York: United Nations Development Program.

3

Human-Environmental Impacts: Forest Degradation and Desertification

William Yaw Osei

INTRODUCTION

Sub-Saharan Africa possesses rich and diversified human and environmental resources that offer opportunities for sustained growth and development. Components of the natural environment—forests, soils, rocks, climate, vegetation, wildlife, and water bodies—provide opportunities for the cultural and material development of the region. For example, volcanic soils in east Africa provide ample opportunity for agricultural development; outcrops of Precambrian rocks have enhanced opportunities in the mining sector; and climatic and natural amenities offer opportunities for recreation and tourism. Generally, the cultural and moral institutions of societies, the level of technological development, and the demands placed on scarce resources are all part of a delicate balance that shapes and molds the human environment.

Views on human–environment relationships have been central in most writings. House and William (1981) observed that both Aquinas in the *City of God* and Plato in his *Republic* concerned themselves with the human–land ratio. Malthusian theory made explicit the dynamics involved in the human–environmental relationship and the possible implications for humanity. Were (1989) noted that land degradation has serious health implications for Africa, ranging from poor nutrition to poor sanitary environments. Homer-Dixon and Nazli (1993) further argued that critical shortages of resources and consequent environmental stresses are likely to generate violent socioeconomic and political conflicts, far beyond existing experiences, because of the rapidity with which they occur.

Ongoing and naturally occurring changes in our ecological systems, such as the establishment of new species, changes in microfauna and microclimates, and soil evolution, all add new dimensions to the human–environment matrix. These dynamic natural changes require some attention because they potentially affect the degree to which humans can effectively establish baselines for comparing environmental functionality. The World Conservation Strategy and the United Nations Conference on Environment and Development (1993) stress the need to redirect the human–environmental relationship toward a balanced framework to ensure a sustainable coexistence to benefit both human development and ecological needs.

Recent accounts from Sub-Saharan Africa have emphasized the negative human influence on the environment. Reports of drought, famine, and rapid vegetation modification, for instance, point to a possible breakdown in the natural environment, a process that will magnify and undermine the ecological cycle in the region and beyond. It is, however, difficult to design effective response mechanisms and support systems without an adequate understanding of the specific dynamics of culture–environment relationships in Sub-Saharan Africa. While it is mentioned that the problems of the region are well known and overstated, the sheer size and diversity of its people and environments imply a multifaceted approach toward devising appropriate solutions.

Among the several human–environmental issues that have emerged recently, the twin processes of forest degradation and desertification have gained the most attention internally and globally. These two processes are considered to be crucial for the long-term ecological viability of Sub-Saharan Africa and the world. They also directly affect the economic, social, and cultural existence of several million people.

This chapter analyzes the causes and effects of deforestation and desertification. Deforestation and desertification have negative impacts on *ecological* and *socioeconomic* systems. Ecological impacts affect natural variables such as species diversity, vegetation space, soil maintenance, and water retention. Socioeconomic impacts relate to food production, income-earning opportunities, social inequalities, settlement patterns, human mobility, and the general human security of a region. In this chapter, the analysis of deforestation precedes the analysis of desertification.

DEFORESTATION IN SUB-SAHARAN AFRICA

Definitions and Issues

Deforestation ranks among the most pressing global environmental concerns today. Attention has shifted more to the developing tropical countries, where forest loss and its implications continue unmitigated. Concerns for tropical forest deforestation are based on a number of reasons. The moist tropical forests possess a global heritage in biological diversity and contain some of the most concentrated assemblages of plants and animals. (See Chapter 2.) The biological and socioeconomic significance of these forests is yet to be fully understood. Tropical forests are also important for maintaining global atmospheric processes. They capture carbon dioxide in the atmosphere as part of the photosynthesis process and store it in plant tissues, helping to minimize the buildup of atmospheric carbon-dioxide and greenhouse-gas emissions. Estimates by the Intergovernmental Panel on Climate Change indicate that deforestation of tropical forests was responsible for between 20 and 30% of global human-caused greenhouse-gas emissions in the 1990s (Bonnie, *et al.*, 2000). Forests are a source of food, fiber, and medicine for local peoples and provide tradable goods and services for export.

There is no consensus on the definition of forests. According to the Food and Agricultural Organization (FAO, 2000) more than 650 definitions of forest types were assembled from 132 developing countries for a UN-sponsored Forest Resources Assessment 2000 project. In spite of such limitations, the FAO has provided a standard definition of forests that incorporates different perceptions and interpretations of forests around the world. It defines *forests* as lands of more than 0.5 hectares (1.235 acres) with a tree canopy cover of more than 10% that are not primarily under agricultural or urban land use (FAO, 2000). The FAO further distinguishes between a natural forest and a plantation forest. A natural forest is a subgroup of forests made up of tree species known to be native to the area. Plantation forests can be established through afforestation efforts on lands that previously did not support virgin forests or through the reforestation of land that previously supported forests that were lost to various land-use activities.

Deforestation, on the other hand, is defined as the conversion of forest to another land use or the long-term reduction of tree canopy cover below the 10% threshold (FAO, 2000). The FAO has attempted to provide standard criteria for identifying deforestation. Even where trees in a forest are massively cut for commercial and related purposes, as long as the minimum threshold of 10% crown coverage is retained, the term deforestation does not officially apply. At the other end of the scale, an area of productive agricultural land with a large proportion of tree coverage will be classified as deforested due to the presence of agricultural land use.

In addition, the FAO makes a distinction between *deforestation* and *forest degradation*. Any change within a forest class—for example, from a closed to an open forest—having negative impacts on forest sites and general productivity is considered degradation. For instance, if an area of forest is badly damaged by logging practices but has not been converted to another land use and still retains at least 10% crown coverage, it is designated as degraded rather than deforested. (Read FAO, 1993 for further details.) Such a distinction is significant for designing planning and policy strategies. Under many interpretations, any activity in a closed forest that could potentially affect species numbers, stand characteristics, and disturbance is considered to be forest loss, even where complete conversion or loss of contiguous forest space has not occurred.

With such a broad usage, the term "deforestation" can be applied equally to significant ecological hazards in woodlands and savannas. In addition, accurate information on forest-land use is difficult to generate at the field level, because of costs incurred for routinely collecting and processing such information. Therefore, most data on deforestation in Sub-Saharan Africa tend to be extrapolations based on estimates. This reality leads to speculations and anecdotes about the true state of the forests of Africa. Recent work by Fairhead and Leach (1996b; also, Nyerges and Green, 2000 and Sedjo, 1999) in six west African countries challenged conventional estimates for deforestation as being sometimes overblown. Fairhead and Leach observed that the reconstruction of past forest cover of an area tends to be fraught with errors and, ultimately, overestimation. They argued that most of the forest patches of the West African Guinea Savanna belt, for example, do not necessarily reflect earlier and original forests but rather human land-use practices, such as the planting of useful trees around villages in the past that have survived to the present. They do not necessarily reflect remnants of past contiguous forest space, and forests in the true sense might never have been in existence at all.

The distinction between deforestation and degradation, while innovative, appears to be complacent and unrealistic, especially with respect to the situation in many parts of Sub-Saharan Africa, where degradation often progresses to deforestation. Forests often lose their core characteristics even before the 10% crown-coverage threshold is breached. The FAO definition also seems to be more in favor of the forest products sector than of biological diversity considerations and other alternative uses, such as food and essential rural livelihood supports.

In this analysis, deforestation is used in its restrictive sense to refer to the long-term loss of forest vegetation as a result of land-use types and processes that reduce or replace forest space, retard forest-growth functions, or suppress on-site forest growth. It is also acknowledged that forests can be reestablished and expanded naturally through effective human-management response mechanisms. In those regions of Africa south of the Sahara where nonintensive, nonpermanent agricultural systems dominate the landscape, flexible solutions can be devised to assist with forest reclamation efforts. For such areas, it may seem premature to include all types of forest conversion within the negative context of deforestation. The forests could be cut without being completely lost. The Red Mountain of Ethiopia, a once highly deforested hill slope, has been symbolically renamed the "Green Mountain," because of a successful reforestation program coordinated through a partnership between the local community and the Canadian Physicians for Aid and Relief. Out of the 350,000 trees planted in a 5-year period, 80% survived, even on sites where trees would not seem likely to grow. (See Seymour, 1997 for details.)

Deforestation Trends

Irrespective of the definition employed, deforestation rates in Sub-Saharan Africa are relatively high and are cause for concern. Estimates from the FAO indicate that, even though deforestation rates in Africa are generally comparable to those in Latin America and Asia, the west African region recorded the highest rates of forest loss in the world—about 2.1% per year between 1981 and 1990, 1.0% between 1990 and 1995 (Table 3.1).

Coastal west Africa has experienced the most drastic loss of moist lowland tropical forests. Côte d'Ivoire, Nigeria, Liberia, Togo, and Sierra Leone all had relatively high rates of deforestation between 1990 and 2000 (Table 3.2). Rwanda had the highest rate of deforestation (3.9%) in mainland Sub-Saharan Africa during that same time period. Ghana and Nigeria once accounted for the largest extent of moist tropical forests in the region, but they now have less than 15% of the original forest vegetation left. Most of the existing forests

TABLE 3.1 Estimates of Tropical Forest Area and Rate of Deforestation

	Forest Area, 1980 1000 ha	Forest Area, 1990 1000 ha	Annual Percent Change 1981–90	Annual Percent Change 1990–95*
Africa	650,000	600,100	−0.8	−0.7
West Sahelian Africa	41,900	38,000	−0.9	−0.7
East Sahelian Africa	92,300	85,300	−0.8	−0.7
West Africa	55,200	43,400	−2.1	−1.0
Central Africa	230,100	215,400	−0.6	−0.6
Tropical Southern Africa	217,700	206,300	−0.5	−0.8
Insular Africa	13,200	11,700	−1.2	−0.8
Latin America	923,000	839,900	−0.9	−0.5*
Total Tropical South America*				−0.6
Asia	310,800	274,900	−1.2	−0.6
World*				−0.3

Sources: World Resources Institute. (1992). *World Resources: 1992–1993*, New York: UNDP and World Bank. World Resources Institute. (1988). *World Resources: 1998–1999*, New York: UNDP and World Bank.
*Food and Agriculture Organization of the United Nations (FAO). (1999). *State of World's Forests*, Rome: FAO.
Food and Agriculture Organization of the United Nations (FAO). (2000). *State of World's Forests*, Rome: FAO.

are in protected reservations, which tend to be isolated and patchy. It is important to note that, given its limited forest land base and ecoclimate, an annual rate of decline in forest area of 0.7% for the Sahelian forest region is even more significant. (See Table 3.1.)

Limited uses, such as logging, are still permitted in some of the ecologically most productive reservations. Liberia is the only country in west Africa with a little over 20% of its forests left. However, recent conflicts have caused significant damage. Several military offensives have occurred in forest zones, and resources such as timber have been exploited to finance war-related activities.

The Congolian belt of the moist tropical forests, which centers around Gabon, Congo, the Democratic Republic of Congo, and Cameroon, has the lowest rate of deforestation in Africa, with more than 40% of its original forests still intact. This region has some of the lowest densities of rural populations, and any form of forest exploitation is very localized. Most of the southern and eastern African regions have lost their moist forests. The only exception is Madagascar, which still has about 15% of its original forest intact.

It is somewhat erroneous to apply aggregate deforestation data as wholesale baselines for comparing magnitudes and impacts of forest loss between regions. Perrings and Lovett (1999) argue that, in the 1980s, individual countries in Asia and Latin America actually experienced higher deforestation rates when compared with Africa. Currently, the largest proportion of the remaining tropical moist forests in the world is in South America and Asia. Therefore, a deforestation rate of 0.4% per year for tropical South America (between 1990 and 2000) seems proportionately more significant than Africa's 0.8% in terms of actual forest-land cover. Furthermore, the Amazonian forests contain the richest and most diverse species of all moist forests per unit area. Losses here would tend to have more significant

Deforestation in Sub-Saharan Africa

TABLE 3.2 Forest Data: Selected African Countries and Major World Regions

Country	Land Area 000 ha	Total Forest Area 000 ha (2000)	Percent of Forest Area	Annual Percent Change 1990–2000
Benin	11,222	2650	23.6	−2.3
Cameroon	47,544	23,858	50.2	−0.9
Central African Republic	62,297	22,907	36.8	−0.1
Congo	34,200	22,060	64.5	−0.1
Côte d'Ivoire	32,246	7117	22.1	−3.1
Democratic Republic of Congo	226,705	135,207	59.6	−0.4
Equatorial Guinea	2805	1752	62.5	−0.6
Gabon	26,767	21,826	81.5	0.0
Ghana	23,854	6335	26.6	−1.7
Guinea	24,586	6929	28.2	−0.5
Guinea-Bissau	6312	2187	60.5	−0.9
Kenya	58,038	17,096	29.5	−0.5
Liberia	11,137	3481	31.3	−2.0
Madagascar	58,704	11,727	20.0	−0.9
Mauritius	203	16	7.9	−0.6
Mozambique	78,425	30,601	39.0	−0.2
Nigeria	91,283	13,517	14.8	−2.6
Réunion	251	71	28.3	−0.8
Rwanda	2633	307	11.7	−3.9
Seychelles	45	30	66.7	0.0
Sierra Leone	7174	1055	14.7	−2.9
South Africa	121,758	8917	7.3	−0.1
Togo	5679	510	9.0	−3.4
Uganda	20,421	4190	20.5	−2.0
United Republic of Tanzania	88,166	38,811	44.0	−0.2
Zambia	74,135	31,246	42.1	−2.4
Zimbabwe	38,775	19,040	49.1	−1.5
AFRICA	3,008,412	649,866	21.6	−0.8
ASIA	3,166,709	542,116	17.1	−0.1
EUROPE	2,276,233	1,039,513	45.7	0.1
SOUTH AMERICA	1,783,696	874,194	49.0	−0.4
WORLD	13,182,522	3,856,159	29.3	−0.2

Source: Adapted from Food and Agriculture Organization of the United Nations (FAO). (2000). *State of the World's Forests*, Rome: FAO.

implications for biodiversity and conservation than losses in the relatively impoverished African and Asian moist forests. Effective comparison of deforestation rates also depends on data accuracy. For several African countries, the moist forests are almost depleted in off-reserve locations, and deforestation rates basically apply to types of forests and woodlands other than true moist forests.

Irrespective of the interpretation of deforestation rates, the African tropical forest region is under tremendous stress from a variety of human and socioeconomic factors, which will now be examined.

Causes of Deforestation

The causes of deforestation are complex, and they vary by place and time. It is generally agreed, however, that it is mostly human induced—which is not to discount the importance of such physical causes as natural fires, floods, volcanic eruptions, pests, and disease vectors. These physical factors have been part of the evolution of most vegetation systems in Sub-Saharan Africa, and most organisms have developed adaptations of survival where they have not been severely disturbed. Moreover, their occurrences tend to be localized.

Three major human factors—agriculture, logging, and fuel-wood consumption—are consistently mentioned in the literature as the major causes of deforestation in Sub-Saharan Africa (Cline-Cole, 1997; El-Farouk, 1996; Osei, 1993; FAO, 1993; Ramakrishnan, 1992). These are considered major causes because they are continuous and are capable of affecting large areas. Other causes relate to infrastructural needs and development-oriented issues: the construction of roads, utilities, and rights-of-way and the clearing of land for storage dams and mining activity. Indirect causes include war and conflicts, population displacement, poverty and international debt servicing, and new international monetary policies linked to the World Bank's structural-adjustment instruments.

Agriculture As the major occupation in Sub-Saharan Africa, agriculture provides the most formidable challenge to forest retention. In areas such as the eastern Congolian rim (Rwanda and Burundi, western Uganda), the west African zone (Ghana, Côte d'Ivoire, and eastern Nigeria), and the east coast of Madagascar, the demand for fresh land for food cropping undermines forest preservation efforts.

In the food crop sector, the bush-fallow system of cultivation creates immense pressures on standing vegetation (as described in Chapter 13, on agriculture). The growing food demands from growing populations cause fallow periods to be reduced, thus limiting opportunities for soil replenishment and successful plant transition to a secondary forest stage. Where fallow periods are reduced, poorer woodlands and derived savanna (rather than forests) tend to colonize abandoned lands.

Arable land is scarce in particular areas. In countries such as Liberia, the Democratic Republic of Congo, Kenya, Somalia, Mali, and Chad, the proportion of arable land to total land area is estimated to be less than 4%. (See Table 3.3.) The majority of the population is rural and agrarian, so the potential for land scarcity and the probability of encroachment on "free" and "unused" forest lands are very high. Furthermore, many small and marginal farmers who employ the bush-fallow method do not practice any soil-improvement or stabilization techniques. Even in areas of low population densities, the practice degrades large areas of land (Figure 3.1).

In South Africa, arable land accounts for under 11% of the total land area, yet forests cover only about 4% (UNDP, 1994). This is mainly due to topographic and climatic limitations; however, agricultural activity has had its impact. Under the Homelands Act of South African apartheid, the majority of Africans were placed on marginal agricultural lands, creating artificial land shortages. A similar situation developed in Zimbabwe, a factor that eventually led to recent conflicts between the commercially viable large farm owners and the government of Zimbabwe. Elsewhere in Africa, the large concentration of subsistence farm-

TABLE 3.3 Extent of Arable and Forest Land for Selected Africa Countries

	Land Area (Million Hectares)	Arable Land (as percentage of Land Area)	Irrigated Land (percentage of Land Area)	Forest Area (as percentage of Land Area)
Benin	11.1	12.7	(.)	31.8
Burkina Faso	27.4	13.0	(.)	24.3
Burundi	2.6	43.7	5.0	2.6
Cameroon	46.5	12.8	(.)	53.0
Chad	125.9	2.5	(.)	10.2
Comoros	0.2	35.0	(.)	15.7
Côte D'Ivoire	38.1	7.6	2.0	24.0
Ethiopia	110.0	12.0	1.0	24.7
Gambia	1.0	17.8	7.0	16.2
Ghana	23.0	5.0	(.)	35.4
Kenya	57.0	3.4	2.0	4.1
Liberia	9.6	1.3	1.0	18.3
Madagascar	58.2	4.4	29.0	27.0
Mali	122.0	1.7	10.0	5.7
Mauritania	102.5	0.2	6.0	4.8
Niger	126.7	2.8	1.0	1.6
Nigeria	91.1	31.6	3.0	13.4
Rwanda	5.2	34.4	(.)	22.6
Senegal	19.3	27.1	3.0	30.9
Somalia	62.7	1.6	11.0	14.5
Sudan	237.6	5.2	15.0	19.0
Togo	5.4	25.3	(.)	29.6
Uganda	20.0	25.1	(.)	28.1
Zaire	226.8	3.2	(.)	77.0

Source: United Nations Development Program (UNDP). (1994). *Human Development Report*, New York: Oxford University Press.

ers, coupled with the practice of bush-fallow techniques and extensive grazing, exerted considerable pressure on the fragile agricultural land. Overcultivation and soil degradation have resulted in accelerated erosion. Under conditions of prevalent aridity in most parts of the country, existing woodlands and transition forests are unable to regenerate. Limited forest regions are confined to the densely populated narrow stretches of the extreme south and the east coast. (See Figure 2.12.) The degradation of soils, vegetation, and forests, therefore, continues to present significant challenges for ecologists in the new South Africa.

Other than the food-crop sector, plantation agriculture (such as cocoa, coffee, tea, tobacco, and rubber), which accounts for the bulk of Sub-Saharan Africa's export and commercial crops, has effectively monopolized the best agricultural lands. Plantation crops tend to be semipermanent and require certain optimal climatic and soil conditions to operate profitably, so large areas of prime forest land have been allocated solely for their use. Ironically, most plantation crops are not consumed locally, and most of the proceeds from sales in foreign markets are externalized. Consequently, the cash crop has failed to provide a sustained basis for local growth, thus stifling food production in rural economies and increasing food-import dependency rates. Ultimately, the rural poor must invade forest areas to ensure access to basic resources, including food, for their sustenance.

Figure 3.1 Denuded Hillsides in Uganda Affected by Agricultural Activity.

According to Avery (1995), African economies must aspire toward creating high-yielding farming systems along the same lines as those in developed economies. The haphazard expansion of underutilized, low-yielding cropland continues to undermine global biodiversity. A balance between agricultural productivity and environmental management is therefore required. It is important to note that clearing for agriculture does not always lead to the complete disappearance of forests. Most of the traditional farming systems incorporate fallow systems, and, as noted by the FAO (1993), the tree covers eventually grow back on previously farmed areas. However, this benefit is diminished when high population densities limit opportunities for land rotation and long fallow periods.

The challenge of agricultural improvement requires bold commitments by African policy makers. Traditional knowledge and practice in agriculture must be complemented by modern methods to consider new contexts of increasing population demands. Land consolidation, buyouts, and skills development for the nonfarm sector, together with substantial farm supports and subsidies, will improve farm-level productivity and move economically inefficient farmers into the nonfarm sector. This suggestion is quite radical, considering the prevailing view in the region of farming as a way of life rather than as an income-generating occupation. Whether the political climate, socioeconomic reality, or employment infrastructure will cater to the consolidation and depopulation of farms is a matter for debate. In Zimbabwe, the government's quest to fragment large blocks of commercial farms to benefit landless blacks has generated conflict at home and condemnation from the international community. Generally, the government's idea might be appealing to landless blacks. On closer examination, however, issues related to investment, productivity, and fair compensation at market value are discounted for political expediency. A balance between equity in land ownership on the one hand, and fairness and national productivity on the other, ought to be worked out to keep the farm sector viable.

The expansion of the nonfarm employment sector is necessary to alleviate pressure on forest zones. A recent policy experiment announced in Ghana to organize the country into agricultural production zones targeting basic staples foods, with associated agroprocessing units, will help in mitigating the problem. A second policy initiative, using Ghana's armed forces for afforestation programs, is also commendable. Sustained government commitment and funding without unnecessary negative political interference could allow planners and forest managers to devise appropriate economic and ecological frameworks to make these programs work. In this regard, and for continuity and efficiency purposes, the private sector needs to provide some leadership. A productive agricultural sector is a must for forest retention.

Logging Selective logging affects forest structure and species composition in many ways. It causes much collateral damage to the forest floor and adjacent canopies even if it involves the removal of few trees. Selective logging also creates gaps in previously well-structured rain forests and allows the invasion of foreign weeds. Furthermore, possible soil compaction and erosion can cause localized logged areas to generally support lower quality vegetation in otherwise rich, forested environments.

Logging activity has encouraged the construction of roads into previously inaccessible forest areas. In Côte d'Ivoire, for every 10 km^2 of forest area logged, 10 km of road is constructed (Chege, 1994). Such roads have been widely used by farmers to access new areas of farmland, thus expediting the rate of forest loss. In Gabon, where the rate of deforestation is low, and where logging is done in isolated spots, there is growing concern about the future effects of logging. (For details, see IUCN, 1992.) Generally, as a country strives to attain self-sufficiency in food production, direct clearance of land for logging routes could set the stage for an assault on the forests. In Cameroon, most logging occurs on state lands, but usually the postlogging forest is subject to usage by local people.

In spite of the negative impacts of logging roads on forest ecosystems, Pardo (1985) argues that they often provide crucial benefits in isolated rural areas. The roads provide access to distant markets and enhance opportunities for rural nonfarm employment. It is now standard practice in Cameroon for logging licenses to include a clause requiring logging companies to provide local services and facilities. The law permits local residents access to compatible forest resources in allotted concessions, on the condition that exploitation be undertaken in a low-impact, traditional manner. Integrating forest-harvesting plans with programs that benefit local people helps negate the detrimental impacts of irrational forest-land use.

Although selective logging might not lead directly to the loss of contiguous tracts of forest space, it does undermine forest-preservation efforts and also sets the tone for other direct onslaughts, such as road construction and agricultural land use.

Fuel-wood consumption The harvesting of wood for fuel has the potential for negative ecological impacts. In countries such as Burkina Faso, Chad, Ethiopia, Somalia, and Tanzania, wood accounts for over 90% of total national energy consumption (Eckholm, *et al.*, 1984). A study conducted by Cline-Cole (1997) in Northern Nigeria found wood to generate about 87.3% of the household sector's gross energy needs on a heat-equivalent basis. Governments have been slow to develop alternative and new forms of energy to meet current demand levels. Between 1979 and 1989, commercial energy production in Sub-Saharan Africa was only 6%, while energy requirements increased by as much as 52% (World Resources Institute, 1992). Traditional fuels, of which wood is the principal component, provided 43.7% of Mozambique's total energy needs in 1980 and, by 1997, traditional fuels supported 91.4% of total energy requirements (UNDP, 2001).

Because most African countries have high rates of population growth, and because wood provides the cheapest and most reliable source of fuel, demand pressures on wood resources are high and, in some cases, excessive. Figure 3.2 illustrates the dynamics involved in wood-fuel systems under unrestrained use. Sinn's (1988) study in Burkina Faso revealed that no firewood could be found within an 80 km radius of the capital city of Ougadougou and that it cost more to heat a cooking pot than to fill it. Harrison (1987) observed that, in cities such as Ougadougou, Bujumbura, and Nairobi, families spend between 20% and 30% of their monthly incomes on fuel wood and charcoal.

Fuel-wood consumption has been assumed, in some cases, to be the leading cause of deforestation. Some scholars are now critically examining the common overstatements and unsubstantiated reports of existing fuel-wood studies (Osei, 1987). Harrison (1987) concluded

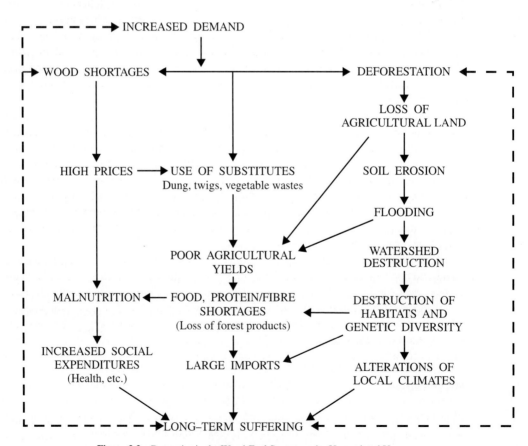

Figure 3.2 Dynamics in the Wood-Fuel System under Unrestricted Use.

that, in most parts of Africa, fuel-wood shortage is the result rather than the cause of deforestation. He agrees, however, that, once deforestation has passed a certain threshold, direct cutting for fuel needs can have an impact. In a practical sense, fuel needs can lead directly to deforestation if the standing vegetation, instead of the dead wood, is removed for fuel.

A study in Ghana indicated that, under the normal, traditional collection system, deadwood, instead of live trees, is harvested for fuel uses (Osei, 1993). This is especially true in areas where wood is still plentiful, such as in the sparsely settled areas of the Congolian and Guinean belts. In these regions, agricultural land use, fallow farms, and standing forests, at various successional stages, generate deadwood through attrition. In such cases where the collection of deadwood is prevalent, fuel wood might not have a significant, direct impact on forest degradation. Therefore, it is difficult to develop a causal relationship between fuel-wood consumption and the loss of forest trees. On the other hand, in cases where the vegetation is degraded to the extent that the standing vegetation is no longer able to produce deadwood through the natural attrition process, the threat of forest degradation becomes real.

Innovations such as the motorized chain saw, and opportunities in distant urban markets for highly prized, freshly cut and processed forest trees, have led to indiscriminate tree cutting. As the range of species sought increases, and as buyers and sellers increase, the standing forests will come under serious threat. The charcoal-manufacturing industry, for example, has had the most significant effects, especially in the transition and woodland zones of Sub-Saharan Africa. Here, large areas are directly cleared to provide raw wood inputs for charcoal supplied to the large urban areas. Charcoal is substituted for fuel wood in African cities because it is easier to use, contains more heat per unit weight than fuel wood, and is cheaper to transport over long distances to urban centers (Grainger, 1990).

Because fuel wood is a critical energy requirement in rural households, it will continue to affect the socioeconomic and cultural lives of end users negatively. For example, in many parts of Sub-Saharan Africa where firewood gathering is generally a female preserve, increasing distances to sources of firewood, given current shortages, affect a woman's overall socioeconomic and health status (Fig. 3.3). The disproportionate amount of time devoted to searching for wood reduces the time available for other economically productive and self-fulfilling activities. The health implications of labor-demanding, firewood-related activities cannot be underestimated. Gender-related issues must be an integral part of any comprehensive energy policy in the region.

The fuel-wood energy situation also has sociocultural impacts, as long-held local taboos and traditions are discounted. Trees previously regarded as taboo by particular clans are now used routinely. Almost every type of tree that can burn is used. Above all, local groves (Kayas in east Africa), once considered the abode of gods and ancestral spirits, have fallen prey to energy-dependent households. Such groves formed the basis for the forest reserve system in many parts of Africa. Obviously, the energy question has implications that produce economic, sociocultural, and biological impacts.

Other Causes

A number of related factors work to accelerate the rate at which agriculture, logging, and energy needs affect forest loss. They include population issues and political, social, and economic problems.

Figure 3.3 Women Are Primary End Users of Fuel Wood and Charcoal, as Is Demonstrated by This Vendor.

Population serves as a major driver for the abuse of forest space. In the poorer parts of Sub-Saharan Africa, where forests are a major source of food and services for daily living, population increases translate into higher demands for forest-related opportunities. Accordingly, the areas with the highest population growth rates also account for the highest rates of deforestation (UNDP 2001). The relationship between population numbers and deforestation is, however, not always linear. Brazil experiences high deforestation rates even though the density and rate of population increase is quite low. In effect, deforestation rates can be high without high population numbers.

Population can further affect forest stability as displaced people from wars, famine, and political instability seek refuge in forest regions. In the axis of the Congo, Rwanda, and Uganda borders, both primate populations and forests have become vulnerable because of the settlement of large numbers of refugees near pristine wildlife refuges and forests. Displaced people require food and other materials to survive and must depend on the forests. Rising concerns over the plundering of forest resources in the Democratic Republic of

Congo by victims of war have generated calls for an international probe and a novel idea: certifying the origin of global timber products.

Poverty is also considered a threat to the stability of forests because it forces people to overutilize the land for survival. The poor cannot easily access innovative management practices that are conducive to farm-level productivity and forest stability. The incidence of poverty in Africa has been exacerbated by national debt burdens and international monetary practices that are part of the World Bank's structural-adjustment program. (See Chapter 8 for more details.) The devaluation of national currencies, coupled with rigid fiscal policies, has in some cases led to a mass retrenchment and increased unemployment and poverty. Devaluation has also increased the value of imports, including basic foods such as rice and wheat. In light of the high prices and shortages of essential goods, the poor turn to the forest as an alternative source of livelihood. Saddling debt burdens make it impossible for governments to expand economic services and provide social safety programs fast enough. Any analysis of the causes of deforestation will be incomplete without establishing the link between poverty and the loss of forest resources.

Agricultural land use, logging, and fuel needs are considered major causes of forest loss in Sub-Saharan Africa, but, on closer examination, they appear to be mere outward expressions of the general weaknesses and failures of the human-economic and institutional frameworks of their affected environments. Small-scale subsistence agriculture, which produces the bulk of food needs in the continent, receives relatively little attention in terms of official research resources, investment, credit, and marketing opportunities compared with the export sector. Low productivity and income levels in the subsistence sector leave most rural dwellers overdependent on rural ecosystems for survival. This will not always favor optimal ecological functioning of local vegetation systems.

Policies in the energy sector have generally excluded marginal urban and rural populations. More worrisome is the fact that increased demand for fuel wood by the urban poor places additional burdens on the rural ecosystem. As the cost of electricity and other energy alternatives continue to be prohibitive for marginal urban and rural populations, the demand for fuel wood will continue to increase, thus threatening the stability of forest preserves.

Effects of Forest Degradation

The loss of forest land poses major planning and development challenges. African forests constitute the smallest and the most impoverished block of moist tropical forests worldwide. Depending upon the causal processes and on the degree of alteration in their respective environments, most forests are difficult to replace.

Forest loss has ecological and socioeconomic effects. The most significant ecological effects include the loss of biological diversity, accelerated erosion, and general degradation of ecological processes. Other than the Mediterranean region in the Cape province of South Africa, the forests of Sub-Saharan Africa hold the greatest diversity of plant species, as mentioned in Chapter 2. Rwanda and Burundi, which are noted for their populations of mountain gorillas, have lost much of their faunal and floral diversity. Significant habitat loss, due to large rural populations and agricultural encroachment, is among the major causes of habitat destruction in the region.

Similarly, the island of Madagascar, with its unique flora and fauna and a high degree of endemism, is likely to become more impoverished as habitats are seriously degraded.

The lemur population (about 30 species), which is unique in the world, is considered to be under the threat of extinction. The protected areas, which constitute only about 1.5% of the total land area, are carefully guarded. Also, the Sclater's Guenon, one of Africa's rarest and least known monkeys, found mainly in the forests of southern Nigeria, is also threatened. These species are now located in a few fragmented forest patches in the densely populated parts of the country (Oates, *et al.*, 1992).

Species retention is now highest only in protected vegetation systems that constitute about 3.9% of the total land area of Africa, an estimate below the world average of 4.8%, and higher only than for the Asian continent (World Resources Institute, 1992). The threat to biodiversity from deforestation also affects forest dwellers such as the Twa (pygmy people) of the eastern Congo Basin. An intact forest is a must for the Twa way of life, which has coexisted with the forest environment for millennia. They are increasingly resorting to a more sedentary life as decreasing forest coverage provides fewer opportunities for hunting and gathering. The threat to biodiversity is likely to affect the future genetic renewal of organisms. Resources, including pharmaceuticals, food, and protein-enriching substances, as well as religious and cultural symbols, are likely to be lost. The very basis of the ecological and economic sustainability of Africa could be lost through the loss of its species diversity.

The forests provide both direct and indirect benefits to its inhabitants. A recent government study in Cameroon concluded that more than half of local incomes earned by rural households were generated through subsistence extraction of food, fish, and meat from the local forests (Chege, 1994). Further evidence suggests that more than 1500 species of wild plants (fruits, nuts, seeds, roots, teas, herbs, and vegetables) are included in central and west African diets. Rapid forest loss and illegal hunting have collectively led to a decline in wildlife populations in many localities. As a result, bush meat is becoming a preserve for a smaller segment of the population, whereas it used to be a protein source for the majority of the population. Scarcity of game meat is common in countries where vegetation degradation and overhunting have destroyed wildlife populations.

Meat products from various species of wildlife, such as bush meat, are important in African diets. Hoskins (1990), citing FAO sources, reports that communities in Nigeria obtain 84% of their animal protein from bush meat, while 70% of Liberians and 60% of Batswana consume species of wildlife regularly. The degree to which wildlife survives as a vital ecological component is tied to the retention of appropriate vegetation systems. Uncontrolled and illegal hunting practices, and the seemingly high market prices for beef, fish, and other alternatives to bush meat, are likely to sustain the pressures on surviving wildlife, particularly in the nonreserved wildlife refuges. All species of mammals, with the exception of rodents, are reported to be overexploited in west Africa (Ntiamoah-Baidu, 1987).

Recent promotional campaigns in Zimbabwe and the rest of the southern African region have encouraged the development of private and communal land to maintain wildlife populations alongside nationally run park systems. (See Discussion Box 3.1.) Such partnerships are important, considering the scarcity of nationally owned habitats for wildlife sustenance.

Forests provide a wide range of goods and services that can be substituted for manufactured goods in rural subsistence economies. Fibers, poles, wild honey, medicinal herbs, and barks are all some of the benefits of the standing forest. With increased forest depletion, planners have to consider other sources of energy to complement wood to help minimize the possible effects of scarcity.

DISCUSSION BOX 3.1: LOCAL INVOLVEMENT IN FOREST ECOSYSTEM MANAGEMENT

Attempts to identify the root causes of deforestation in Sub-Saharan Africa must include a consideration of the dynamic changes in traditional African cultural values within a framework of changing socioeconomic and political conditions and their implications for the stability of forest ecosystems. Lalonde (1993) reiterated positive African cultural practices revolving around spiritual rituals, religious practices, social taboos, and sacred animal totems that served to protect ecological integrity. Ntiamoah-Baidu (1987) reported that the deciduous forest habitats of the black and white colobus (*Colobus polykomos*) and the mona monkey (*Cercopithecus mona*) continued to be protected around two villages in the Brong Ahafo region of Ghana (in otherwise degraded regional environments) because of local beliefs and traditions that considered the monkeys as the sons of the gods who protect the two communities. Mair (1976) cited the Mbuti pygmies as viewing the forests "as a being, especially benevolent, and a source of livelihood." Disturbances that offset such a harmonious relationship are considered abominations. Lalonde (1993) further expressed the reality that, in some of the most ecologically marginal areas of Africa, "knowledge of the local ecosystem simply means survival."

Over much of Africa, new postindependence frameworks for "modernizing" national economies and practices of modern scientific resource management rarely incorporated these traditional perceptions and practices toward natural resource utilization and retention. Such traditional attitudes are now deemed too slow or too insignificant to respond to some of the pressing human needs in many areas. Finally, a major constraint is the government control of traditional lands, which has effectively denied rural residents their traditional rights for land stewardship.

A strategy, for that matter, lies in the evolution of resource management practices that incorporate traditional values in ecosystem maintenance. Recent efforts by the Parks and Wildlife Management of Zimbabwe encourage the active participation of local populations. In 1986, the Communal Area Management Programme for Indigenous Resources (CAMPFIRE) was established to assist rural communities with the management of wildlife species. The net effect of this initiative was that, by the early 1990s, privately and communally owned lands accounted for nearly 50% of all areas allocated for wildlife, while national parks accounted for less than 30%. In the village of Masoka, in northern Zimbabwe, each household is reported to have received about $US450 from animal sales to a safari operator in a single year. In a low-income environment where average annual per capita income is only $US650, this represents significant earnings. Reports indicate that local and private involvement in wildlife management is growing in the entire southern African region.

Forest environments provide resources that sustain such vital sectors as the wood-products industry and agriculture, which are key national employers and revenue earners. According to UN statistics, Africa exported about $715.1 million worth of nonconiferous wood in 1987, the bulk of which came from Côte d'Ivoire, Gabon, Congo, Cameroon, and Liberia (UNDP, 1994). The forests provide optimal environments for the raising of such tree crops as cocoa, rubber, and coffee, which are the backbone of most of the economies in Sub-Saharan Africa. Efficient resource management strategies are needed to maximize the potential of forest regions.

From an ecological and socioeconomic standpoint, the forests are central to Africa's survival. The value of forests is immeasurable; they need to be managed efficiently to support Africa's drive toward sustainable development.

Response Strategies to Deforestation

Responses to forest decline have come from governmental, private, and nongovernmental agencies. Figure 3.4 illustrates some critical considerations for a balanced forest ecosystem. The complexity of responses to deforestation requires a set of strategies aimed at directly preserving the forest land base, at enhancing income-earning initiatives, and at the efficient management of forest resources.

Amongst the earliest and most effective responses to forest decline was the institution of forest reservations and protected areas by colonial governments. This policy was continued by postindependence governments. The underlying concept of forest protection was to

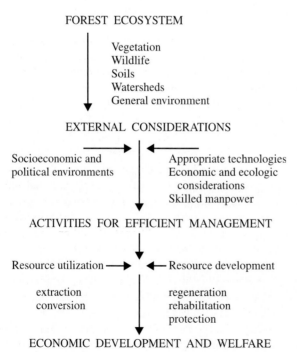

Figure 3.4 Critical Considerations for Balanced Forest Ecosystem Management, Adapted from Soerianegara, I (1982). Socioeconomic aspects of forest resources in Indonesia. In Socio-Economic Constraints in Tropical Management. New York: Wiley.

TABLE 3.4 Protected Areas and Conservation Budgets in Selected African Countries

Country	Number of Protected Areas	Protected Area Budget		
		Area (km^2)	Total Per km^2 US ($)	Region or Country US ($)
Cameroon	13	20,249	410,000	20.15
Central African Republic	12	58,374	480,000	8.22
Congo	10	13,338	40,000	3.00
Democratic Republic of Congo	8	85,842	1,000,000	11.65
Ethiopia	11	30,303	250,000	8.25
Gabon	6	10,439	420,000	40.23
Ghana	8	10,758	1,050,000	97.60
Kenya	36	34,726	18,200,000	524.10
Malawi	9	13,318	460,000	34.54
Nigeria	21	28,729	1,660,000	57.78
Senegal	10	21,758	620,000	28.50
Sudan	14	93,467	1,000,000	10.70
Tanzania	28	130,706	3,480,000	26.62

From Edwards. (1995). Conserving biodiversity: Resources for our future. In R. Bailey, ed., *The True State of the Planet*. New York: Free Press.

"fence off" a block of forests considered significant ecologically (and in some cases socioeconomically) to allow natural ecosystem processes to function without threats from exploitation. In most cases, regulations and sanctions ensured restricted access or barred access completely to nondesignated users (Osei, 1984). However, depending on the management plan for specific protected blocks, a level of access was accorded local inhabitants, such as for collecting fuel wood, seeds, herbs, and wild foods.

Forest improvement programs also ensured enrichment in the structure of the forests. In 1993, several countries, among them Kenya, Zimbabwe, and Ghana, devoted large proportions of their conservation budgets to protecting forested areas (Table 3.4). Limited access ensures the promotion of biodiversity and vegetation enhancement.

The history of reservation policies and the manner in which they were established in Africa, ironically, led to increased degradation in limited, unreserved areas. Notable examples occurred on the densely populated slopes of Mount Kenya, where the land factor provided a rallying cry during the struggle for Kenya's independence from colonial rule. In most parts of Africa, forest reserves were carved out with no consideration for local needs. Suspicion of colonial motives and the deprivation of traditional sources of need hindered local cooperation. Illegal occupation, trespassing, and "unauthorized" removal of forest resources by local populations tended to undermine the vitality of most reserves. Newer and more innovative programs, such as comanagement and local traditional integration, are now being employed to encourage a sense of participation and motivate community spirit.

Forest plantations have also been employed to stock poor and degraded forests. In most of Sub-Saharan Africa, the tree plantation system has been mainly a governmental effort, because local initiatives have been constrained by existing land-tenure systems and negative colonial government policies. However, reforestation programs are ongoing in the region.

Local forestry initiatives, usually with assistance from governmental and nongovernmental organizations, are particularly common in the Sahelian zone. Such programs have

> **DISCUSSION BOX 3.2: ECOSYSTEM
> REHABILITATION IN THE SAHELIAN ZONE:
> THE MAGGIA VALLEY PROJECT**
>
> The Maggia valley project in south-central Niger was established by CARE International in 1974. At the time, the region had been seriously denuded of its vegetation, which was a reliable source for food, fuel wood, and construction materials. The project's focus was, therefore, on assisting communities to plant shelterbelts and windbreaks to protect the vegetation and croplands. According to Ciesla (1993), neem (*Azadirachta indica*) was the principal species employed, because it adapts well to a wide range of climatic conditions, resists locust attacks, and provides shade for crops. It is a high-quality product for fuel wood, and its leaves have insecticidal and medicinal properties. The CARE project is reported to have established about 500 km of double-row windbreaks of neem in the Maggia valley, thus yielding increases in the millet and sorghum crop. In spite of its success, the Maggia valley project has confronted a few problems. Recently, there have been declines in the neem species due to site-related stress factors, such as low soil moisture, competition, intercropping, soil compaction, and such biotic factors as plant viruses. The implications are that program managers must work continuously to improve, adapt, and counter possible sources of constraints to ensure continuous success.

addressed multipurpose goals, such as countering soil erosion, improving water retention, and reducing wind erosion. Shaikh, *et al.* (1988) have documented a number of successful environmental programs involving agroforestry strategies in five Sahelian countries. Notable examples are the Maggia valley windbreak project in Niger (Discussion Box 3.2), which is being duplicated by the Koro project in Mali. Both projects were established by CARE International, a nongovernmental organization. The Kenyan Green Belt movement and other grassroots organizations all over Africa are responding positively to forest ecosystem maintenance.

Reforestation programs have been complemented by nonforest land-based programs. Such programs are aimed at providing income-building and service opportunities for rural populations. Definitive national development strategies that recognize the rural communities as vital links in sustainable development efforts are likely to succeed.

DESERTIFICATION IN SUB-SAHARAN AFRICA

As seen in Chapter 2, arid and semiarid climates dominate a large portion of Sub-Saharan Africa. Historical evidence, though, suggests that these semiarid and arid regions were once subject to humid climatic conditions. Evidence from fossil pollen analysis suggests that woodlands extended deeper into the deserts during wetter periods about 5000 years ago. Kunzig (2000) reports rock-wall paintings in the Egyptian Sahara that depict such scenes from ordinary daily lives as hunting, swimming, and cattle driving, from about 9000 to 7000 years ago. Today this area of the Sahara, including the Kebir highlands of southwestern Egypt, is hyperarid and is devoid of people and of plant and animal life. Cloudsley-Thompson (1977) also reports that lakes and rivers covered much of the Sahara south of the Tropic of Cancer, from Mauritania to the Sudan, between 8000 and 5000 years ago. Rainfall amounts were estimated to have been higher and less variable. Fossils of swamp animals such as the hippopotamus, dated to around 7000 years ago, have been found at Taoudini, about 650 km north of Timbuktu. Other evidence has located relic dunes well beyond the current margins of the Sahara Desert, suggesting a wider spread of desertlike conditions in the past (Obasi, 1985; Nichol, 1991; Lezine and Casanova, 1990).

Although desertification has been an ongoing environmental process, it was not perceived as a global threat until the second half of the 20th century. Desertification in the Sudano-Sahelian belt of Africa has, since the 1970s, received the special attention due to a series of disastrous droughts, such as the 1968–1973 drought, which caused an estimated 100,000 deaths and the loss of up to 80% of cattle stocks. These prompted such an impact that resolution 3202 (S-VI) of the UN General Assembly passed in May 1974 called the international community to undertake concrete and speedy measures to arrest desertification and assist in the economic development of affected areas (UNEP, 1992). As a result, the concept of desertification is now being reexamined and reevaluated to assess its meaning, its spatial and temporal scope, and its causes and effects (Rhodes, 1991).

Definition of Desertification

Desertification is a complex and often misunderstood concept; therefore, there is no consensus on its definition. Varied definitions equate desertification with the expansion of desertlike conditions, with forest and land degradation, and with the irreversible decline in the productive potential of an area (Binns, 1990). The term is often overstated, especially in those instances when it has been applied to all types of ecosystems besides arid, semiarid, and subhumid areas (Glantz and Orlovsky, 1983). Le Houerou (1992) suggested an alternative term—desertization—for the process of irreversible growth of new desert landscapes in arid regions that not long before had presented no such features. Dregne (1983) adds a human dimension when he defines desertification as the impoverishment of terrestrial ecosytems under the impact of man, leading to reduced productivity of desirable plants, undesirable alteration in the biomass and diversity of fauna and flora, accelerated erosion, and increased hazards of human occupancy.

The UNEP (1992) finally adopted a definition for desertification for the UN Conference on Environment and Development (UNCED or the Earth Summit) in 1991. Desertification was defined as land degradation in arid, semiarid and dry subhumid areas resulting mainly from adverse human impacts. Kerr (1998) added that people might be degrading drylands in the Sahel by changing the mix of plants without necessarily expanding the desert. Desertification obviously then involves the degradation of the capacity of land in arid areas, in terms of its biological and economic productivity, resulting from climatic effects and human impacts.

There is always the tendency for scholars to overstate and misinterpret the desertification process. Lamprey's (1975) study of the Sudan concluded that the Sahara was advancing at a rate of 3.42 miles a year (5.5 km), totaling 55.9 and 66.1 miles (90 to 100 km) in 17 years. Follow-up fieldwork by Hellden (1984) showed that there were no long-lasting desertlike conditions between 1962 and 1979 in the corresponding areas described by Lamprey. It is, therefore, necessary to make the appropriate distinctions between drought (short-term periods of below-average rainfall), degradation, and desertification (Binns, 1990). Drought conditions and forest or soil degradation can very well affect vegetative cover and productive capacity, but this might not necessarily lead to long-term, permanent damage amounting to the creation of new deserts. In making these distinctions, we must not lose sight of the potential damage that all three processes inflict on ecological systems and on human lives.

Spatial Coverage of Dry Lands in Sub-Saharan Africa

Africa contains 37% (6.68 million square miles or 17.3 million km^2) of the desertified land in the world. It is also estimated that 108 million Africans were affected by moderate desertification in 1980 (Grainger, 1982). While desertification is common in the arid areas of eastern and southern Africa, the southern margin of the Sahara Desert—the Sudano–Sahelian zone—has received substantial scholarly attention. This zone, which centers on latitude 15° N and is 124 to 248 miles (200 to 400 km) wide, ranges from Cape Verde in the west to Somalia in the east. Annual precipitation levels generally fall below 24" (600 mm), with a high degree of variability. Drought is a common occurrence here. Most of the ecosystems in the Sahel have adapted to periods of long drought and poor soils, and it is the lack of human adaptation that potentially creates a hazard. It is estimated that the desert fringes of the Sahel cover about 1.5 million hectares or 3.7 million acres of land. Worldwide, the potential annual cost of desertification in terms of lost agricultural opportunities has been estimated at $26 billion (Were, 1989). Problems in the Sudano–Sahel zone have captured international attention and are, therefore, examined next.

Causes of Desertification

Several causes, both physical and human, have been advanced to explain the desertification process in the Sudano–Sahelian region. The most frequently mentioned causes are deforestation (already discussed); changes in climatic conditions (resulting in droughts); overcultivation of marginal lands; overgrazing; inappropriate technology; weak institutional systems; and, to a lesser extent, political factors.

Drought, which reflects a major consequence of changes in climatic conditions, is considered to be among the most significant factors affecting desertification in the region. The term "drought" is generally used to describe a condition of lack of water. In some cases, drought is operationally defined as a certain number of days without rain; in others, it is described as a period when rainfall is significantly below normal (Wild, 1993). The UN Sudano–Sahelian Office (UNSO, 1992) refers to drought as a period of two years or more with rainfall well below average. The agency admitted a lack of consensus on definitions. Depending on the specific environment, a period characterized by the lack of water might be temporary (seasonally predictable or random) or long term (endemic). The critical factors are dryness that results in a period of drought and its effects on biotic life and surface processes.

Three major types of drought in Sub-Saharan Africa can be identified. These are meteorologic (or climatic), hydrologic, and agricultural. They are not mutually exclusive and can occur simultaneously. *Meteorologic drought* is the main cause of *hydrologic drought* (periods with low levels of stream flow) and agricultural drought (extended periods of soil dryness). Lack of rainfall leads to diminished surface stream flow; to loss of water in surface impoundments, such as freshwater lake systems and wetlands; and to reduced recharge of aquifers, which in turn affects underground inflow into springs and streams.

Agricultural drought results when soil lacks the necessary moisture for plant intake or for effective plant growth. It relates directly to rainfall and high rates of evapotranspiration. It could also relate to dryness of lakes, reduced stream flow, and the lowering of water tables (which affects irrigation water). In some cases, agricultural drought occurs even when rainfall seems adequate. This is because erosion, surface compaction, and steep slopes enhance surface runoff and limit water infiltration into the soil. Furthermore, some soils in valley bottoms and floodplains become so waterlogged that crops and soil organisms are damaged by diminished aeration. In this case, the waterlogged soil is as dry as a moisture-deficient soil.

Over all, meteorologic drought is considered to be the most widespread and most significant of all the potential drought types in the arid and semiarid areas of Sub-Saharan Africa. Meteorologic drought, for that matter, underlies much of the vulnerable environmental conditions that render the Sudano–Sahel belt susceptible to desertification.

Grainger (1990) outlines a number of theories that link climatic changes to drought in the Sahel. One theory attributes the Sahel's droughts to higher sea-surface temperatures in the tropical Atlantic. This has the effect of reducing the northward penetration of the southwest monsoon winds. Another attributes the prolonged droughts to the prevalence of dust storms caused by wind erosion. It is hypothesized that dust storms heat the atmosphere, thus inhibiting the upward flow of air that is needed for rainfall (Middleton, 1985). Another explanation is linked to the reflection of incoming solar radiation by desert surfaces—a process referred to as the *albedo effect*. Dry and bare surfaces have higher albedos than moist, covered surfaces. The argument here is that desert surfaces are cooler because they reflect more sunlight. Cooler air has less of a tendency to rise, thus limiting the probability of rainfall. While certain evidence lends credence to such a process, it has not been duplicated with much certainty in other environments with similar characteristics.

Since drought is mainly a climatic phenomenon, it becomes more of a threat when the human capacity to cope with it is lacking. As noted by Grainger (1990, cited in Wijkman and Timberlake, 1985, p. 12), "drought triggers a crisis but does not cause it. Overcultivation and grazing weaken the land, allowing no margin when drought arrives." The population of the Sahel is increasing while the carrying capacity of its land-resource base continues to diminish. Population pressure on land manifests itself in the form of overcultivation on marginal lands and overgrazing. Most of the already scarce arable land in Africa is monopolized by cash crops and by commercial agricultural activity. Subsistence farmers are usually relegated to marginal lands, which are intensively cultivated with little fertilizer use or proper crop management. These agricultural practices are compounded by the practices of nomadic pastoralists, who move from place to place in search of what little vegetation is left. In the culture of the Fulani pastoralist, social prestige and status come with owning more cattle. Overcultivation and overgrazing are leading causes of surface degradation. Every year, between 100 and 300 metric tons of wind erosion per hectare occurs in some arid zones of Africa (Le Houerou, 1992). Surface and soil degradation set the stage for vegetation loss and potential dune encroachment.

Furthermore, competition between grazing and sedentary agriculture has been a source of conflict across west Africa. Grazers from the Sahel, such as the cattle Fulani, often reach the forest and even coastal regions. Herders are usually accused of damaging crops and vegetation, and, lacking political clout, are forced to limit their activities to restricted and marginal areas, with potential damaging effects to the land. Improvements in veterinary services, basic infrastructure, and functional training for herders have boosted stock populations. Availability of modern wells and boreholes, for instance, now makes it possible for livestock to survive previously deadly droughts. Natural control of livestock populations, through such natural hazards as severe droughts, is being negated, while grazing grounds have failed to keep pace.

Inappropriate technologies and institutional systems also enhance the desertification process. In the arid regions of more developed societies such as Israel (the Negev Desert) and California, technological, institutional, and economic frameworks have been developed to cope with perennial aridity. Countered by superior infrastructure and efficient organizational structures, desertification there no longer generates the famines and forced migrations still besetting the Sahel.

Political factors accounted for the lethargic attitude towards desertification in the Sahel in the early 1960s. Most African governments were very slow to respond to the earlier droughts, and there were no efforts to mobilize communities or to create stockpiles of food in anticipation of future droughts. At times, governments disassociated themselves from any event or occurrence that had the potential to tarnish their self-image. Generally, governments have adopted a reactive rather than a proactive approach, allowing the problem to reach crisis proportions.

Strategies to Combat Desertification

Countering desertification requires a multifaceted planning and management approach. The UN Conference on Desertification (UNCOD) prescribed a set of comprehensive recommendations that provide a framework for African governments to develop action plans that tackle desertification problems (Discussion Box 3.3). In spite of its strong recommendations, the Action Plan was not very effective at the implementation stage and, therefore, has been further strengthened by new UN conventions and other global arrangements such as the Kyoto Protocol. The Kyoto Protocol on *greenhouse gases*, for instance, worked to allow credits to nations for afforestation and reforestation initiatives in recognition of forests as carbon sinks. Eventually, African countries will benefit from this facility within their national strategies to combat desertification. Beside the UN recommendations, a number of strategies have been initiated in the Sudano–Sahel region to counter the ecological and human effects of droughts and desertification. They include reforestation and revegetation, soil

Desertification in Sub-Saharan Africa

DISCUSSION BOX 3.3: HIGHLIGHTS OF THE UNITED NATIONS CONFERENCE ON DESERTIFICATION'S PLAN OF ACTION

Evaluation of Desertification and Improvement of Land Management

1. Countries should utilize appropriate methods, such as satellite imagery, to analyze the extent and magnitude of desertification.
2. Land-use planning and management techniques, based on sound methods, should be initiated to foster social and economic development.
3. Public participation should be an integral part of efforts to combat desertification.
4. Examine the effect of industrialization and agricultural development, and the process of urbanization on the ecology of arid areas.

Corrective Antidesertification Measures

5. Embark on environmentally sound management and development of water resources, using local materials and appropriate technologies, such as rainwater collecting from roofs, sand filters, water harvesting, low-cost reservoirs, solar and wind pumps, and composting.
6. Improved welfare for pastoral communities and better pastoral management (e.g., rotational livestock, managed water points, drought grazing reserves, use of crop residues, better land tenure and water rights, and price stabilization and marketing schemes).
7. Improve rain-fed farming by rational use of fertilizers, terracing, strip cropping, shelterbelts, sand stabilization, and diversification of farming systems.
8. Rehabilitate and improve existing irrigation schemes by combating waterlogging, salinization, and alkalinization.
9. Restore, maintain, and protect vegetative cover to stabilize and protect soils in degraded areas by establishing shelterbelts and other tree plantations.
10. Step up flora and fauna conservation efforts.
11. Develop national or intraregional systems to monitor climatic and ecological conditions of land, water, plants, and animals.

Socioeconomic Aspects

12. Examine social, economic, and political factors that have a bearing on desertification. Increase public awareness through education programs.
13. Adopt population and economic policies to control population growth.
14. Provide needed health and family planning services to people affected by desertification.

15. Human settlement in drylands should not conflict with land productivity and should take account of local climate, local building materials, and social habits.
16. Establish national systems to monitor the human conditions in areas affected by desertification.
17. Provide insurance schemes to cope with drought disaster and any other potential risks.

Strengthen Science and Technology at the National Level

18. Strengthen indigenous scientific and technical capabilities.
19. Explore alternative energy sources to fuel wood.
20. Training and education utilizing telecommunications technologies.
21. Establish a coordinated national machinery to combat desertification.
22. Integration of antidesertification progams into comprehensive development plans.

International Cooperation

23. UN agencies, international bodies, and nongovernmental organizations must cooperate to develop a uniform methodology to assess, monitor, and forecast desertification.

Source: Adapted from Grainger, A., *The threatening desert: Controlling desertification.* London: Earthscan in association with UNEP Nairobi (1990).

conservation and soil fertility improvement, water conservation, organizational initiatives, and policy initiatives (Shaikh, *et al.*, 1988).

Reforestation and revegetation is a logical step to counter forest degradation and desertification. Gritzner (1988) states that tree planting and shelterbelts can be established to increase precipitation interception, infiltration, and groundwater recharge; to stabilize dunes and soils; to protect against wind and wind-blown sand; and to yield fodder, fuel wood, and a broad range of minor forest products to support local communities. (Refer to Discussion Box 3.2.) Kenya's Green Belt Movement has been actively involved in educating communities about the advantages of tree planting. The movement is involved in tree planting projects and actively encourages local communities to plant trees, with the goal of halting deforestation and making fuel wood available. The founder, Wangari Maathai, works with a broad group of women coalitions and a range of participants, including school children. Responses to the campaigns and practical examples of this movement have taken root in various parts of the country and are being duplicated elsewhere in the region. Professor Maathai's environmental work in Kenya was recognized by the UNEP's *Global 500 award*. As a social activist and environmentalist, Dr. Maathai is likely to meet some opposition, but the long-term achievements of the organization will set a good example for local grass-roots organizations in devising and managing their own sustainable development initiatives.

The *Acacia albida* has been recommended as an appropriate species to spearhead reforestation efforts because of its reverse foliation properties. It is leafless during the agricultural season and, therefore, does not compete for light and moisture. During the dry season, it produces leaves that provide enough shade. Furthermore, the leaves it sheds contribute nitrogen and organic matter to farmers' fields. Wood for fuel, for local construction, for fencing material, and for medicinal substances is also derived from the acacia. These species, which are widely distributed in the Senegal area, could enhance environmental stability and agricultural productivity.

Vegetative cover and windbreaks are effective soil-conservation methods. Other soil-conservation methods include *strip cropping*, where close-growing vegetation, placed perpendicular to the flow of wind, provides protection to adjacent strips of land. The raised mounds also aid in the retention of water. Soil fertility can be improved through animal manure and nitrogen fertilizers. In the latter case, farmers will require the necessary finances and counseling on how to apply a judicious combination of fertilizer and water.

Water-conservation methods, such as water-retention dams, water catchments, and salt barriers, complement soil-conservation efforts. In the Sintet Foni Jarrol district of Gambia, diversion bunds, contour berms, and diking to retain freshwater and to keep saltwater out have been employed (Shaikh, *et al.*, 1988). On the slopes of mountains and plateaus, terracing is practiced to capture runoff and to retain soil moisture.

These efforts toward environmental rehabilitation cannot be successful without the proper organizational and policy initiatives. Arnould (1990) provides examples of such initiatives from the Sahelian communities of Gambia, Mali, Niger, and Senegal. In each case, successful grassroots initiatives at the village and district levels were initiated with the support of government and of international nongovernmental organizations that provided technical and financial assistance. In Niger, for example, a group of national and international technicians developed effective site-specific techniques to assist local village cooperatives with water catchments, windbreaks, and reseeding. In the small town of Koumpentoum, Senegal, forest and orchard projects have been succeeding through voluntary village labor, along with financial and technical assistance from Catholic Relief Services and from the US Agency for International Development.

CONCLUSION

Deforestation and desertification have been shown to provide significant environmental challenges in vulnerable ecological systems of Sub-Saharan Africa. Within such existing challenges is the looming threat of global warming trends. While conclusive evidence is not currently available to confirm persistent increases in temperature over the long term, any slight but persistent increase in temperatures in the arid and semiarid environments of Sub-Saharan Africa is likely to aggravate existing conditions and cause an expansion of arid lands. Most of the existing moist forests would degrade into poorer forests and woodlands, with significant consequences on biodiversity.

Although comprehensive governmental policies have not been forthcoming, there are some grassroots initiatives that address the environmental rehabilitation and the human development needs of affected regions. Such programs, while crucial, lack the necessary framework to decisively counter the challenges of deforestation and desertification, both in a wider regional sense and on an expedited time scale. International assistance, particularly through the United Nations, is necessary for severely impacted countries that have not been able to attain the requisite resources to implement past initiatives such as those outlined in Discussion Box 3.3.

The intergovernmental panel on forests, set up in 1995 by the Commission on Sustainable Development of the UN General Assembly, along with the provisions of "Agenda 21" of the UN Conference on Environment and Development, have specifically targeted solutions for desertification and drought that bode well for Sub-Saharan Africa (UN, 1995). The commission recently reported that there is a general decline in official development assistance and that there is a need to implement new and additional financial resources under "Agenda 21." Strategies aimed at deforestation and desertification need to include the critical physical and human factors behind these processes. Policies designed to create early warning systems; to improve job, income, and educational prospects; and to mobilize collaborative efforts at the local, national, and international level will likely minimize the impact of deforestation and desertification. Reforestation, revegetation, soil and water conservation, and biodiversity enhancement will also help. The increased use of Remote Sensing and Geographical Information Systems tools will improve information gathering, data analysis and accuracy, and the monitoring of the dynamics of deforestation and desertification in Sub-Saharan Africa. In the final analysis, our local environments are intricately linked with global ecological processes.

KEY TERMS

Agroforestry	Agricultural drought
Meteorologic drought	Hydrologic drought
Deforestation	Soil conservation
Desertification	Water conservation
Drought	Soil degradation
Strip cropping	Selective logging
Forest ecosystem management	Wood-fuel consumption
Forest degradation	

DISCUSSION QUESTIONS

1. Identify and evaluate the major causes and effects of deforestation in Sub-Saharan Africa.
2. To what extent is fuel-wood consumption a major threat in Sub-Saharan Africa?
3. Discuss and evaluate the magnitude and extent of desertification in Sub-Saharan Africa.

4. Discuss the role that human factors play in the degradation of forests and soils.
5. What, in your opinion, is the most effective way to combat desertification?
6. Are nongovernmental organizations effective partners?
7. Develop a descriptive model that outlines strategies for combating deforestation and desertification.

REFERENCES

ARNOULD, E. (1990). "Changing the Terms of Rural Development: Collaborative Research in Cultural Ecology in the Sahel," *Human Organization*, 49(4):339–350.

AVERY, D. (1995). Saving the Planet with Pesticides: Increasing Food Supplies while Preserving the Earth's Biodiversity, in *The True State of the Planet*, Bailey R., ed., New York: Free Press, 49–82.

BAILEY, R., ed. (1995). *The True State of the Planet*, New York: Free Press.

BATTERBURY, S. (1998). "Shifting Sands," *Geographical Magazine*, 70(5):40–45.

BINNS, T. (1990). "Is Desertification a Myth?" *Geography*, 75:106–113.

BONNIE, R., SCHWARTZMAN, S., OPPENHEIMER, M., & BLOOMFIELD, J. (2000). "Counting the Cost of Deforestation," *Science*, 288(5472):1763–1765.

CHEGE, N. (1994). "Africa's Non-Timber Forest Economy", *World Watch*, 7:19–23.

CIESLA, W. (1993). What Is Happening to the Neem in the Sahel, *Unasylva*, 41(160):14–19.

CLOUDSLEY-THOMPSON, J.L. (1977). *Man and the Biology of the Arid Zones*, Baltimore: University Park Press.

COLE, M. (1967). *The Ecology of the Alpine Zone of Mount Kenya*, The Hague: Junk Publishers.

DREGNE, H. (1983). *Desertification of Arid Lands*, Geneva, Switzerland: Harwood Academic Press.

ECKHOLM, E., et al. (1984). *Fuelwood, the Energy Crisis That Won't Go Away*, Washington, DC: International Institute for Environment and Development.

EDWARDS, S.R. (1995). "Conserving Biodiversity: Resources for Our Future," in *The True State of the Planet*, Bailey, R., ed., New York: Free Press, 211–266.

EL-FAROUK, A.E. (1996). "Economic and Social Impact of Environmental Degradation in Sudanese Forestry and Agriculture," *British Journal of Middle Eastern Studies*, 23(2):167–182.

FAIRHEAD, J. & LEACH, M. (1996b). "Reframing Forest History: A Radical Reappraisal of the Roles of People and Climate in West African Vegetation Change," in *Time Scales and Environment Change*, Thackwray S. Driver & Graham P. Chapman, eds., London: Routledge, 169–195.

FOOD AND AGRICULTURE ORGANIZATION OF THE UNITED NATIONS (FAO) (1993). *The Challenges of Sustainable Forest Management*, Rome: FAO.

—— (1999). *State of World's Forests*, Rome: FAO.

—— (2000). *State of World's Forests*, Rome: FAO.

GLANTZ, M. & ORLOVSKY, N. (1983). "Desertification: A Review of the Concept," *Desertification Control Bulletin*, 9:15–22.

GRAINGER, A. (1990). *The Threatening Desert: Controlling Desertification*, London: Earthscan in association with UNEP, Nairobi.

GRITZNER, J. (1988). *The West African Sahel*, Chicago: University of Chicago, Committee on Geographical Series.

HARRISON, P. (1987). *The Greening of Africa*, London: Paladin Grafton Books.

HELLDEN, U. (1984). *Drought Impact Monitoring, A Remote Sensing Study of Desertification in Kordofan, Sudan*, Khartoun, Sudan: Lunds Univesitets Naturgeografiska Institution in Co-operation with Institute of Environmental Studies, University of Khartoun, Sudan.

HOMER-DIXON, T. & NAZLI, C. (1993). *Global Accord: Environmental Challenges and International Responses*, Cambridge, MA: MIT Press.

HOSKINS, M. (1990). The contribution of forestry to food security, *Unasylva*, 41(160):3–13.

HOUSE, P. & WILLIAM, E. (1981). *The Carrying Capacity of a Nation*, Lexington, MA: DC Heath and Company.

IUCN (1990). *Biodiversity in Sub-Saharan Africa and Its Islands: Conservation, Management and Sustainable Use*, Geneva, Switzerland & Cambridge, UK: IUCN.

IUCN (1992). *The Conservation Atlas of Tropical Forests in Africa*, Singapore: Simon & Schuster.

KERR, R.A. (1998). "The Sahara Is Not Marching Southward," *Science*, 281(5377):633–634.

KUNZIG, R. (2000). "Exit from Eden," *Discover*, 21(1)(Jan):84–91.

LALONDE, A. (1993). "African Indigenous Knowledge and Its Relevance to Sustainable Development," in *Traditional Ecological Knowledge*, Inglis, J., ed., Ottawa, Canada: Canadian Museum of Nature and the International Development Research Center.

LAMPREY, H. (1975). *Report on the Desert Encroachment Reconnaissance in Northern Sudan*. UNESCO/UNEP Consultant Report.

LE HOUEROU, H. (1992). "Climatic Change and Desertization," *Impact of Science on Society*, 166:183–201.

LEZINE, A. & CASANOVA, J. (1990). "Across an Early Holocene Humid Phase in Western Sahara: Pollen and Isotope Stratigraphy," *Geology*, 18:264–267.

LEZINE, A. & CASANOVA, J. (1991). "Correlated Oceanic and Continental Records Demonstrate Past Climate and Hydrology of North Africa (0–140 ka)," *Geology*, 19:307–310.

MAIR, L. (1976). *African Societies*. Cambridge: Cambridge University Press.

MIDDLETON, N. (1985). "Effect of Drought on Dust Production in the Sahel," *Nature*, 316:431–4.

NICHOL, J. (1991). "The Extent of Desert Dunes in Northern Nigeria as Shown by Image Enhancement," *The Geographical Journal*, 157(1):13–24.

NTIAMOAH-BAIDU, Y. (1987). West African Wildlife: A Resource in Jeopardy, *Unasylva*, 39(2)(156):27–35.

NYERGES, A.E., & GREEN, G.M. (2000). "The Ethnography of Landscape: GIS and Remote Sensing in the Study of Forest Change in West African Guinea Savanna," *American Anthropologist*, 102(20):271–289.

OATES, J., ANADU, P., GADSBY, E., & WERRE, J. (1992). "Sclater's Guenon—A Rare Nigerian Monkey Threatened by Deforestation," *National Geographic Research & Exploration*, 8(4):476–491.

OBASI, G. (1985). "Understanding the Drought," *Bulletin of the Atomic Scientists*, 43–45.

OSEI, W.Y. (1984). *Forest Conservation in Ghana,* unpublished M.A. Thesis, Ottawa Canada: Carleton University.

OSEI, W.Y. (1987). "Woodfuel Research in Developing Countries: The State of the Art," *Ohio Geographers*, 15:31–43.

OSEI, W.Y. (1993). "Woodfuel and Deforestation—Answers for a Sustainable Environment," *Journal of Environmental Management*, 37:51–62.

References

PARDO, R. (1985). "Community Forestry: Building Success through People's Participation," *Unasylva*, 37(1)(147):36–43.

PERRINGS, C. and LOVETT, J. (1999). "Policies for Biodiversity Conservation: The Case of Sub-Saharan Africa," *International Affairs*, 75(2):281–305.

RAMAKRISHNAN, P.S. (1992). "Tropical Forests Exploitation, Conservation and Management," *Impact of Science on Society*, 166:149–162.

RHODES, S. (1991). Rethinking Desertification: What Do We Know and What Have We Learned? *World Development*, 19(9):1137–1143.

—— (1999). "Resources in Review," *Journal of Environmental Education*, 30(4)(Summer):43–44.

SHAIKH, A., ARNOULD, E., CHRISTOPHERSEN, K., HAGEN, R., TABOR, J., & WARSHALL, P. (1988). *Opportunities for Sustained Development: Successful Natural Resources Management in the Sahel*, Washington, DC: Energy/Development International/USAID.

SINN, H. (1988). "The Sahel Problem," *Kyklos*, 41:187–213.

UNITED NATIONS DEVELOPMENT PROGRAM (2001). *Human Development Report of 2001: Making New Technologies Work for Human Development*, New York: Oxford University Press.

UNITED NATIONS CONFERENCE ON ENVIRONMENT AND DEVELOPMENT (1993). Report of the United Nations Conference on Environment and Development: Rio de Janeiro, 3–14, June 1992. New York: United Nations.

UNITED NATIONS ENVIRONMENT PROGRAMME (UNEP) (1992). Status of Desertification aand Implementation of the United Nations Plan of Action to Combat Desertification. Nairobi, Kenya: UNEP.

UNITED NATIONS SUDANO–SAHELIAN OFFICE (UNSO) (1992). *Assessment of Desertification and Drought in the Sudano–Sahelian Region 1985–1991*, New York: United Nations.

UNITED NATIONS (1995). "New Panel to Combat Forest Degradation," *United Nations Chronicle*, 32(3):69.

WERE, M. (1989). "Ecological Upheavals with Special Reference to Desertification and Predicting Health Impact," *Social Science Medical*, 29(3):357–367.

WILD, A. (1993). *Soils and the Environment: An Introduction*, Cambridge, UK: Cambridge University Press.

WIJKMAN, A. & TIMBERLAKE, L. (1985). Is the African Drought an Act of God or of Man? *The Ecologist*, 15(1/2):9–18.

WORLD RESOURCES INSTITUTE (1992). *World Resources: 1992–1993*, New York: UNDP and World Bank.

—— (1998). *World Resources: 1998–1999*, New York: UNDP and World Bank.

PART II

Sociocultural Context

4

Historical Geography of Sub-Saharan Africa: Opportunities and Constraints

Ian E.A. Yeboah

INTRODUCTION

Historical geography involves the study of how past events influence the spatial organization and human geography of the present. It focuses on how societies evolve and how historical factors shape and mold human and cultural landscapes. The purpose of this chapter is to identify the importance and effects of the *triple heritage* on the present-day spatial organization and development of Sub-Saharan Africa. In chronological order, the triple heritage relates to the indigenous, Islamic, and Western influences on the physical and human landscape of Africa (Mazrui, 1986). Although the triple heritage constitutes the overriding theme, this chapter also examines some specific aspects of Africa's history: the achievements of ancient, medieval, and modern kingdoms; the impact of the Atlantic slave trade; the prejudices and misconceptions of European explorers; and the controversies surrounding the social, economic, and political impacts of colonialism. The chapter concludes with a brief assessment of the decolonization process.

INDIGENOUS HERITAGE OF SUB-SAHARAN AFRICA

The indigenous heritage comprises the ancient, medieval, and modern civilizations and their associated cultural value systems. This heritage has had a differential impact on the regions of Sub-Saharan Africa: The western and northeastern regions have experienced a long history of indigenous influences; the eastern and southern parts have had one only more recently. The history of the indigenous heritage is categorized into five major periods: Africa as the cradle of man; the era of ancient civilizations; the migratory patterns prior to European contact; the medieval period; and the era of modern kingdoms, which overlaps with the beginning of European contact. The first four periods have been referred to inappropriately as the "prehistory of the region" because their story was unwritten. This, however, does not mean that the region was without a history prior to European contact. What follows demonstrates that a rich history, characterized by well-established ancient and medieval civilizations, existed before the 15th century. The distinction between ancient, medieval, and modern attributes of the indigenous heritage has to do with the temporal and geographical location of civilizations that existed in each of these periods. In addition, the use of iron, the effect of Islam, and European contact are other issues that provide distinctions between these three periods of the indigenous heritage.

Sub-Saharan Africa: Cradle of Humans

The work of Leakey in the discovery of *Zinjanthropos* in the Olduvai Gorge and the discovery of *Lucy* in Ethiopia are strong indications that the African Rift Valley system was the first home of humanity (Ki-Zerbo, 1990; Graham, 1994). The first people of Africa, however, lived in small bands of hunter–gatherers. About 10,000 years ago, hunter–gatherers began to settle permanently along the Nile River to domesticate plants and animals and exchange goods. The permanence of settlement, the domestication of plants and animals, and the establishment of local trade networks marked the beginning of ancient civilizations in Africa, as they became quite distinct from the typical hunter–gatherer economy.

The first ancient civilizations were stateless, because political authority was often vested not in a particular ruler, but in the elders of a band or tribe. Associations were based on kinship and lineage (O'Toole, 1992). Thus, by 3500 B.C. such groups had developed structured political arrangements based on kinship and coresidency. It is the development of these political structures that led to the formation of state societies and civilizations such as ancient Egypt. Controversy exists about whether ancient Egypt was white or black. Earlier historians report ancient Egypt as being a white civilization, but the Afrocentric school, based on the work of Diop (1974), argues that it was black. Graham (1994) suggests that Egypt consisted of many colors; however, the important issue is that Egypt was an African civilization. Although Egypt is not a part of Sub-Saharan Africa, the connections between Egypt and the Sub-Saharan African civilizations are significant enough to warrant using Egypt as a starting point for Sub-Saharan African civilizations.

The Era of Ancient Civilization

The two earliest civilizations to emerge in the highlands of present-day Ethiopia were Kush, with its capital Meroë (c. 2000 B.C. to A.D. 300), and Axum (c. 200 B.C. to A.D. 700). The history of Kush was closely linked with ancient Egypt, and in fact, Kushites once ruled Egypt between 700 and 500 B.C. Kush eventually declined and was succeeded by Nubia (c. 500 to 700 A.D.), a Christian civilization that was later conquered by Muslims. Another

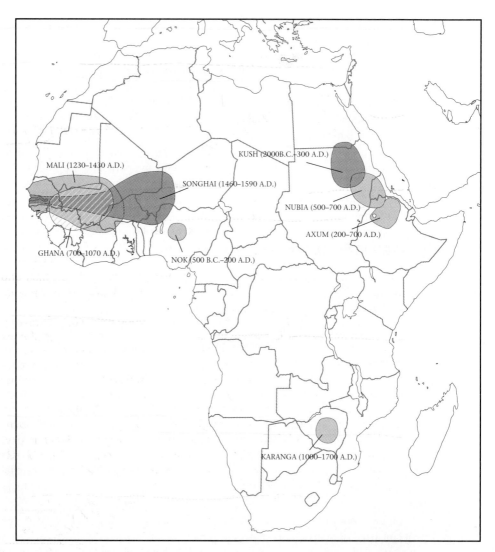

Figure 4.1 Ancient and Medieval Civilizations of Sub-Saharan Africa.

civilization, the Nok civilization (c. 500 B.C. to A.D. 200), emerged along the Niger–Benue confluence in west Africa (Figure 4.1). Nok was noted for its metal working technology and its trade links with Carthage in North Africa.

Although these ancient societies have not been researched as extensively as Egypt, they were civilizations in their own right (Discussion Box 4.1). All these civilizations were characterized by permanent settlement, domestication of plants (cereals and roots) and animals, and well-established political structures. Other distinguishing features were their use of iron in making tools and the long-distance trade networks they established. Kush, for example,

DISCUSSION BOX 4.1: INDIGENOUS SUB-SAHARAN AFRICA WAS CIVILIZED

Even though it is elusive to determine what the term "civilization" means, all the ancient, medieval, and modern civilizations of Sub-Saharan Africa were accomplished in science, technology, and social and political organization. Ancient trading routes between Nok and Carthage (Graham, 1994), of which the early rulers of Rome were envious, plus the trans-Saharan trade routes of the Ghana, Mali, and Songhai empires, show the extent of long-distance trade by medieval civilizations. The ruins of stone buildings in Kush–Meroë–Nubia, the sophisticated terra-cotta sculptures of Nok (Graham, 1994), the magnificent ruins of the city of Great Zimbabwe (Mazrui, 1986), and the significance of cities such as Gao and Kumbi–Saleh are testimony to the level of urban sophistication during the ancient and medieval periods. The establishment of Coptic Christianity in Axum by King Exana in A.D. 350 and the Islamic influence on Mali and Songhai are testimony to the nature of religious and social organizations that existed.

Meroëitic and Coptic scripts of Nubia are good examples of systems of writing that existed in ancient times. In addition, universities established at Timbuktu and Jenne show the extent of higher learning that existed in the Mali empire during the medieval period. The development of a pharaonic system in Nubia and the astuteness of the Islamic rulers of the Mali and Songhai confederacy (Khapoya, 1994) are classic cases of the kinds of political organization that existed in medieval civilizations. Mansa Musa, the famous ruler of Mali, on his pilgrimage to Mecca (1324–25) spread so much gold that, when he left Cairo, its price plummeted. This, in combination with the metal workings of Kush–Meroë–Nubia and of Mali, indicate that metallurgy was a major component of most of the ancient and medieval civilizations of Sub-Saharan Africa.

The modern and late-medieval kingdoms, such as the Ashanti, Mossi, and Zulu, were, and are still, characterized by elaborate and sophisticated systems of such political organization as chieftaincy. They continue to be characterized by systems of social organization that involve kinship, lineage, descent, and age-cohort association. Long-distance trade with north Africa and coastal Europe shows the level of economic sophistication achieved by the modern kingdoms in Sub-Saharan Africa. Specialized labor, tool making, trade, and religion were elements of these kingdoms. Culture traits of the modern kingdoms, such as the strong linkages between the living, the dead, and posterity and the significance of rituals, rites of passage, and the extended family, are all testimony to their advanced levels of civilization. The early European stereotypes of Sub-Saharan Africa and its people as a "dark" continent occupied by uncivilized savages are, therefore, inaccurate. Indigenous Sub-Saharan Africa was civilized.

was known for its impressive architecture (ruins of Meroe still exist today), its irrigation systems, its scripts, and its large metal industry.

Iron, Islam, and Medieval Civilizations

During the early medieval period, empires with both political structures and social orders developed as the ancient civilizations emerged in Sub-Saharan Africa (Davidson, 1966). In west Africa, three such empires evolved that occupied what in the present day are the savanna and Sahel belts. These civilizations (or so-called Sudanic empires)—Ghana (c. A.D. 700 to 1070), Mali (c. A.D. 1230 to 1430), and Songhai (c. A.D. 1460 to 1590)—overlapped each other spatially and temporally (Figure 4.1). There were, however, two features that distinguished the Sudanic empires from the ancient civilizations. First, Islam was an important organizing philosophy in the empires and had the strongest impact on Mali and Songhai (Olaniyan, 1982). Ghana, which was not Islamic, was eventually conquered by Muslims. Second, the Sudanic empires mined gold with iron implements, resulting in the creation of extensive trade networks across the Sahara Desert into northern Africa. Other commodities traded were salt and slaves.

Fewer civilizations emerged in southern Africa during the early medieval period. Around A.D. 1000, Karanga evolved into what is present-day Zimbabwe. Karanga was actively engaged in the production of gold for international trade. Its capital was Great Zimbabwe, a city built of stone without mortar. (See Figure 4.2.)

Hardly any major civilizations evolved in east Africa during the medieval period. Rather, trade contacts with people of Arabia and the Indian Ocean resulted in the creation of a number of city–states between A.D. 700 and 1500. Examples were the Swahili city–states

Figure 4.2 The huge walls and ruins of Great Zimbabwe, the capital of ancient Karanga, are considered to be the largest ancient structures south of the Pyramids.

of Mogadishu, Kilwa, Mombasa, and Sofala (Kwamena-Poh, *et al.*, 1982). The relative absence of civilizations in the interior of east Africa was attributed to its limited access, its sparse populations, and the lack of awareness of its resource potential.

The relative absence of empires in southern, eastern, and central Africa was bound to change because of a series of migration waves that swept across each region, resulting in more permanent settlement patterns. By the latter part of the medieval period, a fair number of empires (shown in Figure 4.3) had emerged not only in west Africa (Tekrur, Wolof, Mande, Wangara, Mossi, Borgu, Nupe, Kanem–Borno, and the Hausa city–states), but also in east and central Africa (Loango, Kongo, Luba, Kitara, Mwene–Mutapa, and Changamire).

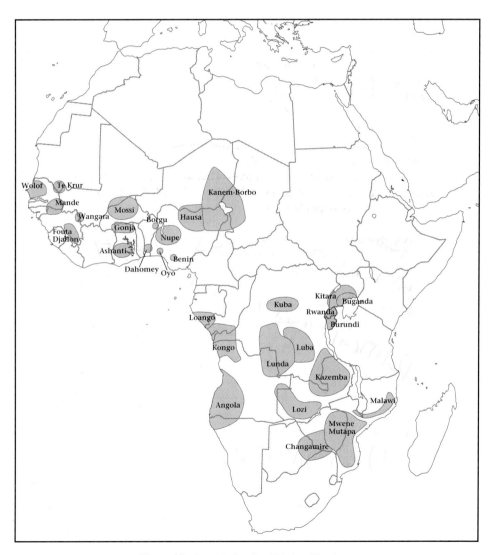

Figure 4.3 Late Medieval and Modern Kingdoms.

Some of these kingdoms still exist today. To understand how empires formed in the late medieval and subsequent periods (especially in east, central, and southern Africa), we need to examine the migration patterns of those who populated the region.

Bantu Migrations in Sub-Saharan Africa

The population densities of east, central, and southern Africa began to increase during the latter part of the medieval period as Bantus began to migrate from the Benue River, a tributary of the Niger River. The Bantu migrations occurred over a long period of time in a series of spurts and bursts, rather than in one major wave. Based on linguistic associations (the commonality of the suffix *ntu* in languages), cultural trait similarities, and archeological artifacts, historians and anthropologists have determined that, beginning in about 500 B.C., a series of migrations from the Benue River begun to fan out eastward, southwestward (Curtin, *et al.*, 1995), and even westward. Figure 4.4 shows these migratory paths. It is speculated that these migrations were caused by population pressure on land and by political forces of dissidence in the source regions. The eastward thrust of these migrations populated the Great Lakes District of east Africa. The southwestward thrust was through Cameroon into the rainforest and central African regions. By A.D. 1000, however, these migrations were over and had extended to the southernmost part of the continent.

The Bantu migrants encountered *autochthones* (aboriginal groups) who had not formed elaborate economic and political associations. The Bantu migrants introduced a number of innovations: iron smelting, a superior political and economic structure, the herding of cattle, and the cultivation of crops such as bananas. These innovations served as catalysts for the organization of civilizations like that at Karanga during the medieval period in southern Africa. Prior to these migrations, there is no evidence to support the existence of major civilizations in central, east-central, and southern Africa. Bantu migrations spurred further developments of such empires as the Loango, Kongo, Luba, Kitara, Mwene–Mutapa, and Changamire in east and central Africa.

The Era of Modern Kingdoms

Toward the end of the medieval period, especially after 1600, a number of modern kingdoms emerged, particularly in the forest belt and in areas where the Bantu migrations had spread (Kwamena-Poh, *et al.*, 1982). Most of these resilient kingdoms have persisted to the present time and form the basis of a number of cultural systems in the region. Examples of modern kingdoms include the Cayor, Baol, Futa Toro, Futa Djallon, Jolof, Kaarta, Segu, Macina, Ashanti, Gonja, Benin, Borgu, Nupe, Dahomey, and Oyo in west Africa (Figure 4.3). In central and east Africa, examples include Kuba, Lunda, Lozi, Kazembe, Malawi, Burundi, Rwanda, Buganda, Bunyaro, Nkore, Shona, and Angola (Kwamena-Poh, *et al.*, 1982). Modern kingdoms emerged upon the demise of the earlier medieval civilizations and were precipitated by trade in gold and slaves with north Africa and by the need to form political associations to protect both their trade routes and themselves.

Prior to European contact, these kingdoms engaged in the trans-Saharan trade in gold, ivory, ostrich feathers, salt, and slaves. The more powerful kingdoms controlled the trade routes and grew in importance. After European contact, the trade activities of most of these

Indigenous Heritage of Sub-Saharan Africa

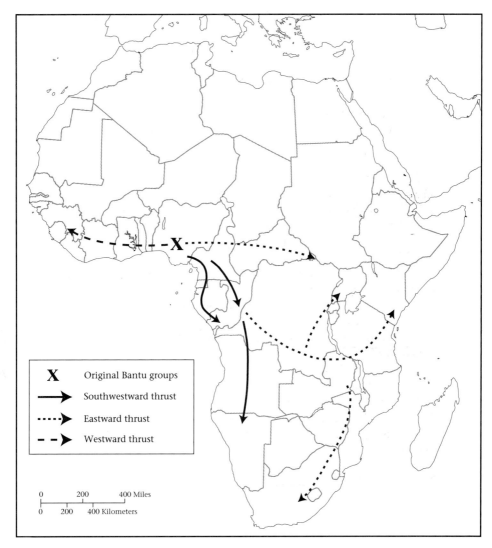

Figure 4.4 Bantu Migrations.

kingdoms turned toward the coast with trade in gold, ivory, and slaves (on a much larger scale) as the main commodities. In addition, a European weapon, the gun, was introduced to some groups, giving them a definite political and military advantage over their neighbors.

In spite of the wave of Bantu migrations into southern Africa, very few kingdoms were formed in this region during the early phases of the modern era. The San and the Khoikhoi, who were the two dominant groups in the region, were predominantly hunter–gathers. Around the middle of the 19th century, threats from overpopulation and overgrazing set in motion the shaking up of peoples in southern Africa, or the *Mfecane*, to create kingdoms such as the Zulu, Ndebele, and Sotho. O'Toole (1992) suggests that resistance to European control, coupled with the need to organize trade and defense, prompted these states' formations during the Mfecane in southern Africa.

Implications of the Indigenous Heritage of Sub-Saharan Africa

Early Europeans (such as David Livingstone, Henry Stanley, and Mungo Park) who explored the interior of Sub-Saharan Africa portrayed the region and its people as uncivilized and barbaric. The explorers and the missionaries who followed them perpetuated negative stereotypes through accounts and books that they wrote (Jarosz, 1992). Examples of such books are Henry Stanley's *In the Darkest Africa* and *Through the Dark Continent*. At the end of Stanley's 1887 journey on the Zambesi, he thanked God for his guidance and protection in piercing the dark continent. Some of these stereotypes exist today and have been the basis of racist attitudes towards Sub-Saharan Africans and Black people all over the world. The level of sophistication of ancient, medieval, and modern kingdoms in the region, however, shows that such stereotypes were misrepresentations of reality. (See Discussion Box 4.1.)

Contrary to early European accounts, indigenous Sub-Saharan Africa was civilized, with the region's indigenous heritage forming the foundation of its cultures and civilizations. Traditional forms of governance, social organizations based on lineage and kinship, and the traditional religious value systems embedded in all aspects of daily living (see Chapter 6) continue to be important guiding principles for all Africans. In addition, the indigenous influence has interacted with the Islamic and Western influences to produce the present state of civilization, development, and spatial organization. This three-way interactive process has also contributed to social unrest, political instability, and economic and cultural dependency, as later chapters on development and political geography illustrate.

It must be emphasized that, although the indigenous history of Sub-Saharan Africa was not written, this period was not void of civilizations. With time, ancient and medieval civilizations and some of the modern kingdoms in Sub-Saharan Africa declined. Karenga (1993) suggests a combination of internal and external factors that contributed to the decline of civilizations, empires, and kingdoms in the region. These factors include the prevalence of droughts, conflicts, and external threats. European contact was particularly responsible for the demise of modern kingdoms. The important issue is that, despite the decline of some empires and kingdoms, the indigenous influence constitutes the foundation of present-day culture and has influenced the spatial organization and development of Sub-Saharan Africa.

THE ISLAMIC INFLUENCE IN SUB-SAHARAN AFRICA

Islam has existed in Sub-Saharan Africa since at least A.D. 700. Islam has its roots in a series of revelations that the Prophet Muhammed received from Allah, beginning in A.D. 613. The religion brought to the Arab world a unifying religious faith; its precepts constitute a revision and embellishment of Judaic and Christian beliefs (de Blij and Muller, 1998). Prior to Islam, Judaism, Christianity and Zoroastrianism existed in southwest Asia. Islam was introduced from Arabia; it is not original to Sub-Saharan Africa. Yet, Islam has a major influence on the cultural, economic, and political systems of regions, especially in the Sahel and savanna belts of west Africa and along the coast of east Africa. The forest belt and interior of the eastern

The Islamic Influence in Sub-Saharan Africa

and southern parts of Africa have barely felt the impact of Islam, because the religion was spread, for the most part, through the trans-Saharan trade (Mazrui, 1986). The prevalence of the tsetse fly also explains the limited diffusion of the religion in the forest belt.

The Spread of Islam in Sub-Saharan Africa

In Sub-Saharan Africa, Islam spread along three specific paths (Figure 4.5). The first was by contact between Arabian traders and the people along the coast of east Africa and its surrounding islands (Zanzibar, Pemba, and the Comoros) (Mazrui, 1986). This first wave of

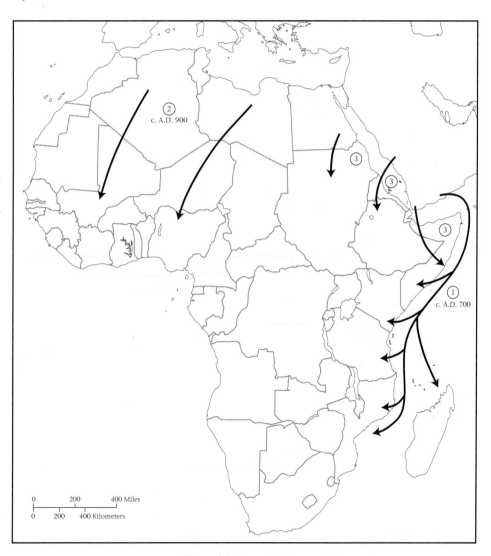

Figure 4.5 Spread of Islam.

Islamic influence began around A.D. 700. For reasons identified in previous paragraphs, these contacts and the subsequent spread of Islam were confined to the coast (Kwamena-Poh, *et al.*, 1982).

The second wave of Islamic diffusion began around A.D. 900 and continued until the 19th century. For the most part, the diffusion process was enhanced by trade via trans-Saharan routes. However, toward the end of this period, the religion was diffused through holy wars or jihads, as Fulani zealots, preachers, and warriors searched for grazing lands. These jihads influenced the political and social organization of the western Sudanic civilizations.

The third way in which Islam diffused to Sub-Saharan Africa was by trade between Egypt and Arabia and within the horn of Africa. By the 14th century, Islam was well entrenched in this part of Africa (Olaniyan, 1982). What makes this diffusion process different from the other two is that, because the region was in close proximity to Arabia, the spread of the religion was also accompanied by the spread of Arabian culture. The twin processes of Islamization and Arabization partly explain the political unrest that continues to plague both northern and southern Sudan.

Aside from these earlier migratory waves, there have been more recent movements, resulting in the presence of Islam in several urban areas. Lagos in Nigeria, Accra and Kumasi in Ghana, and Abidjan in Côte d'Ivoire, originally not part of the Muslim belt, are now home to large and influential Islamic communities.

Islam and the Indigenous Heritage

As Olaniyan (1982) suggests, Islam has influenced indigenous cultures in Sub-Saharan Africa in many ways. The host cultures, in turn, have contributed their cultural traits to the religion. Islam introduced a new religion and an Arabic language, helped in the formation of states, influenced the development of Kiswahili as a trade language (along the east coast), established the basis for an Arabic educational system, and influenced food choice and dress patterns. However, the relationship between Islam and indigenous cultures was more of a conversation than a domination. Adherents to Islam have had the choice of adapting the religion to their specific local circumstances, as is evidenced by the "Wolofization" of the religion in Senegal (Olaniyan, 1982; Mazrui, 1986). At the beginning of the second half of the 19th century, Ahmadou Bamba emphasized the need to make Islam relevant to Senegalese Muslims. His main objective was to provoke Senegalese Muslims to protest against French imperialism, which threatened the social foundations of Islam.

Thus, Skinner's (1986) argument that Islam adapted to the specific milieu of various regions in Sub-Saharan Africa is justified. This is not to say that the basic principles of the religion have changed. Sub-Saharan African Muslims still follow a sacred way of life that represents God's final revelation of His Prophet Mohammed. By the same token, Muslims in Africa, like their counterparts in Arabia, adhere to the five pillars of the religion: an affirmation of oneness in God; prayer five times a day facing Mecca; fasting during Ramadan; giving alms to the needy; and making a pilgrimage to Mecca and Medina at least once in a lifetime if it can be afforded.

Islam and the Western Heritage

Islam did not occur in isolation from Western and colonial influences. Even though colonialism was affiliated largely with Christianity, it facilitated the spread of Islam in Africa (Olaniyan, 1982). With colonialism came a breaking away from traditional life by youth in some territories, to take advantage of opportunities offered by the colonial educational system. Some of these youths converted to Islam in the process. Also, reaction against Christianity as the colonial religion led to Islamic conversions. The building of an infrastructure to extract resources facilitated the spread of Islam to other parts of Sub-Saharan Africa, such as the forest belt. Because of the higher literacy rates associated with Muslims at the beginning of the colonial period, they were relied upon by colonial administrators to facilitate trade and day-to-day government operations, as was the case of indirect rule in northern Nigeria. This elevated the status of Muslims.

Compared with the Western influence, Islam's effect on Sub-Saharan Africa was not as disruptive, partly because it seemed more accommodating to a larger cross section of Sub-Saharan Africans (Mazrui, 1986). For example, Islamic groups have adopted the use of the African drum in certain religious practices. The religion also allows polygamy, in agreement with a number of traditional African societies, and has been less critical of female circumcision. Yet, some of its negative consequences are apparent in present-day political, economic, and social life. An example is the civil war between Islamic northern Sudan and non-Islamic southern Sudan, where the predominantly Islamic north wishes to force Islamic law (*sharia*) on the southerners who practice African traditional religions and Christianity.

THE WESTERN INFLUENCE IN SUB-SAHARAN AFRICA

European contact of any consequence with Sub-Saharan Africa started in the middle of the 15th century (c. 1450). By this time, a number of medieval kingdoms still existed, and some modern ones had begun to emerge. Of the three influences on Sub-Saharan Africa, the European influence was the shortest, yet the strongest, and the most pervasive.

The Western influence on Sub-Saharan Africa can be divided into four main periods: the period of initial contact, the period of enslavement, the age of land exploration, and the colonial period. The first three periods spanned from the middle of the 15th century to the middle of the 18th century. The colonial period lasted from 1884 to the early 1960s. European activities during these four periods were dictated by the changing needs (especially economic) of Europe. During the first three periods, the emphasis was on trade in commodities and slaves to meet mercantile capitalist needs. The colonial period's emphasis was on the extraction of mineral and agricultural resources to meet Europe's industrial capitalist needs.

The Period of Initial Contact

The first meaningful contact between Europeans and Africans was, to a large extent, incidental rather than planned. Europe was searching for a sea route to India because Muslims, who controlled the land route, posed a threat to their trade in silk, porcelain, and spices

Figure 4.6 The Portuguese completed construction of Elmina Castle in 1482. It is one of several former trading and slave posts located along the coast of west Africa.

(Mazrui, 1986). Europeans originally used the coast of Sub-Saharan Africa for supplies on their journey to Asia. Their initial intention was not to establish a foothold in the region. Yet, Europeans were well aware of Sub-Saharan Africa's wealth in such commodities as gold, which had reached European markets via the trans-Saharan trade.

The first contacts were in west Africa and were limited to the coast, partly because established kingdoms prevented Europeans from penetrating the interior to disrupt their trade (Kwamena-Poh, *et al.*, 1982). Portugal was the first country to sponsor expeditions to Sub-Saharan Africa, followed by the Dutch, Danes, British, French, Germans, Italians, and Spaniards. When sea explorers such as Vasco da Gama arrived on the coast of west Africa, the best concession they could gain from local chiefs was the building of trading posts and forts along the coast. This led eventually to the establishment of 50 such forts along the west African coast, 43 of them in present-day Ghana (Figure 4.6). Coastal trade was legitimate, and it included the exchange of gold, ivory, cola nuts, and palm products from Africa for alcohol, guns, and sugar from Europe. The system of exchange was one of barter.

Because Europe's initial purpose was to find a sea route to India, the exploration of east Africa occurred much later. It was not until the 1530s that Portugal explored the east African coast. In fact, Portugal established its headquarters in Mombasa only in 1592 (Kwamena-Poh, *et al.,* 1982). This was because Portugal needed gold to pay for its trade in silk and spices with Asia, so it turned to the east African coast for its needs.

The Enslavement of Sub-Saharan Africans to the Americas

Trade between Europe and Sub-Saharan Africa continued until Europe "discovered" the Americas in 1492. The potential of the Americas to provide commodities such as tobacco, sugar cane, and cotton changed the nature of the relationship between Sub-Saharan Africa and Europe (Karenga, 1993). The sweet tooth of Europe, which manifested itself in its demand for tobacco and sugar, implied that plantations in the Americas would provide such commodities. The native Indian population either had been decimated by European diseases or could not be held as captive labor to work on plantations. Therefore, Europe turned to Africans as a source of workers for their plantations. This ushered in the transportation of Africans across the Atlantic to work as slaves (Figure 4.7).

Estimates of the number of Africans transported across the Atlantic between the late 1490s and the 1880s range between 6 million and 30 million. Most slaves came from the coast of west Africa as far south as Angola. The effect of this enslavement was devastating, especially in the middle belt of west Africa (Discussion Boxes 4.2 and 4.3). It must be borne in mind that this era coincided with the popularizing of the concept of race by Europe and its attendant Eurocentricity (Banton and Harwood, 1975). Because of the racist attitudes of Europeans, it was easy to justify the enslavement of people of a different skin color and culture. This racist attitude carried over into the age of land exploration of Sub-Saharan Africa.

The abolition of the shipment of slaves in 1808 (by Britain) reduced the rate at which slaves were transported across the Atlantic until about 1880. The abolition resulted from humanitarian efforts in the United States and Europe. The activities of the "Back to Africa Movement" and the "Maroon" communities, and the humanitarian efforts of Granville Sharpe and William Wilberforce, were instrumental in this effort. Western-educated and freed slaves such as Olaudah Equiano of Nigeria and Ottabah Cugoano of Ghana also played a major role in the abolition process (Kwamena-Poh, *et al.*, 1982). When slavery ended, the interests of Europe in Sub-Saharan Africa waned, and Britain, for example, almost pulled out of its holdings in west Africa, except for Sierra Leone. With time, alternative commodities for trade, such as groundnuts and palm oil, began to emerge on the west African coast. These commodities, in conjunction with the traditional trade in gold and ivory, were primarily responsible for the names given to such coastal territories as the Gold Coast (Ghana), Slave Coast (Nigeria), and the Ivory Coast (Côte d'Ivoire) (Kwamena-Poh, *et al.*, 1982).

The Age of Land Exploration in Sub-Saharan Africa

In spite of the fact that Europe's interests in Sub-Saharan Africa had waned after the era of slavery, the industrial revolution provided an impetus for Europe to maintain its presence in the region. Europe needed raw materials for its industries. This introduced an era of land exploration with the purpose of determining the resource potential of Africa's hinterland.

Scientific and geographic curiosity, the potential for resources in the interior, and the prestige associated with laying claim to territory led to land explorations on behalf of various

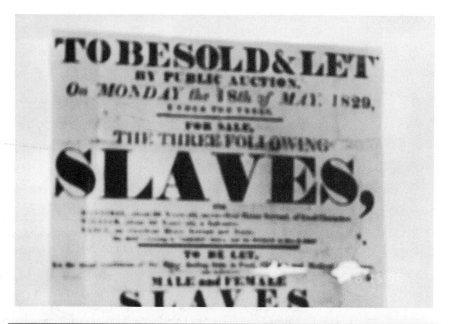

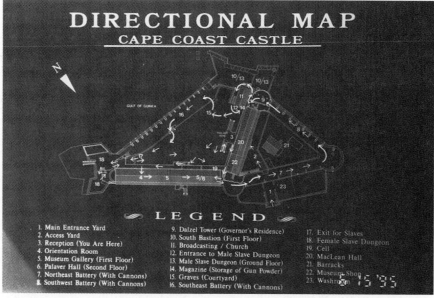

Figure 4.7 "Slaves for Sale" Sign Posted on Wall of Cape Coast Castle in Ghana.

DISCUSSION BOX 4.2: THE CAPTURE AND SHIPMENT OF AFRICANS TO THE AMERICAS: WHO WAS RESPONSIBLE?

From the late 1490s to the late 1880s, it is estimated that between 6 and 30 million Sub-Saharan Africans were forcibly captured and dehumanized, both in captivity and during the *middle passage* across the Atlantic to the Americas. Who is to blame for the capture and shipment of slaves to the Americas? Was this activity carried out solely by Europeans, who had relatively limited knowledge of the interior of Sub-Saharan Africa, or were Europeans aided by some Africans?

Both Rodney (1974) and Karenga (1993) argue that, because the capture and shipment of slaves across the Atlantic was against the will of the enslaved and because the actual people shipped across the Atlantic were not remunerated, the act was an enslavement of people rather than a trade in slaves between Africans and Europeans. In addition, the manner in which Europeans physically acquired slaves (by kidnapping, banditry, and trickery) suggests that this was an enslavement, rather than a trade.

Europeans justified the capture and shipment of slaves to the Americas on the premise that Africans were already engaged in slavery across the Sahara Desert. Differences, however, exist between slavery across the Sahara and slavery across the Atlantic. In the former case, slavery was less dehumanizing. Some slaves are known to have become important members of their new societies. For example, Juan Latino, who was sold into bondage to Spain from Ethiopia, became a professor; in 1556, he delivered the Latin Oration at the beginning of the school year. Also, Ibrahim Hannibal, who was captured in Eritrea by Turks and sold into slavery in Russia, served as a distinguished major general in the Czar's army. Slavery across the Atlantic dehumanized slaves to mere commodities and properties, and there were no opportunities for slaves to either buy their freedom or move up in their new societies.

There is no doubt that Sub-Saharan Africans share a responsibility in the activities of that period. They also physically captured slaves to exchange through ethnic wars and banditry (M'Bokolo, 1994). Yet, considering the power relations between Europeans and Sub-Saharan Africans and the economic motive presented to Sub-Saharan Africans, which group bears the brunt of the responsibility?

To ensure that slaves were readily available, Europeans introduced goods such as guns, sugar, wax prints, and alcohol to Sub-Saharan Africans. European guns were used by some Africans in ethnic wars and by bandits to capture prospective slaves. Commodities such as sugar, wax prints, and alcohol changed the tastes of Sub-Saharan Africans, thus creating a cultural dependency.

Not all Sub-Saharan Africans engaged in, or supported, the capture of slaves. For example, Queen Nzinghu of Matamba in Angola resisted the capture of slaves for over 30 years. Also, Tomba of Baga (present-day Guinea), King Nzega Meremba of Kongo, and King Agaja Trudo of Dahomey all resisted the capture of slaves by the Portuguese (Karenga, 1993).

DISCUSSION BOX 4.3: AN ORAL HISTORY OF SLAVERY IN THE INTERIOR OF SUB-SAHARAN AFRICA

Most of the writing about slavery has been on how slaves were taken from the coast of west Africa and shipped across the Atlantic Ocean to the Americas. Not much attention has been paid to the nature of slavery in source regions in the interior of Sub-Saharan Africa. Part of the reason for this is that no records were kept of how and where slaves were taken from before they arrived on the coast. Also, the capture of Africans as slaves was not a well-organized or regulated activity, so the source regions of slaves have not kept artifacts associated with slavery.

There is no doubt that the nature of slavery in the interior of Sub-Saharan Africa, especially the savanna and Sahel belts, provides a holistic picture of the nature of slavery as a whole. The town of Salaga, a Gonja town, in the northern region of Ghana was associated with the trade in slaves in the interior of Africa. Today, it is the backwater of Ghana and can be approached either by ferry from Yeji on the Volta or by road from Tamale (a mere 80 miles takes about 6 hours). Chief Haruna is a resident of Salaga and a retired education officer with 40 years of teaching experience. He narrates an interesting set of stories told to him by his grandparents. See what you think of it. Does it make you think of the experience of slaves in the interior of Africa and does it add to your understanding of slavery as a whole?

Paraphrase of Chief Haruna's Stories

Salaga is an old trading center in the historic long-distance trade between the forest, savanna, and Sahel belts of West Africa. It is located in the savanna belt and so had a trading advantage. The town got its origin as a place where large caravans met to exchange goods from the different ecological zones and even from Europe. Initially, this trade was legitimate, with kola and gold coming from the forest belt, manufactured goods like guns from Europe, and woven clothes, leather goods, and trinkets from the rest of the savanna and Sahel belts. The Juabeng from Ashanti were the main traders from the south. The Grushie and Mossi from present-day Burkina

Faso, the Hausa from Nigeria, and the Zamrama and Gruma from Niger and Mali were the traders from the savanna and Sahel belts. The indigenes of Salaga did not want to mix with these traders, so they set up an adjoining part of the settlement that was settled by foreign or ethnic traders. Each of these ethnic traders established their own quarters in Salaga and today there are 16 such quarters.

The transition from legitimate trade in goods to a trade in slaves came with the demand for slaves from European traders. Traders from Juabeng passed the signal for the demand for slaves on to their counterparts in Salaga, who then went into the savanna and the Sahel to capture slaves through raiding, trickery, banditry and kidnapping. No Europeans ever came to Salaga to buy or capture slaves; they could not ensure their safety from bandits or diseases such as malaria. It was the traders from Grushie, Mossi, Hauza, Zamrama, and Gruma who brought slaves to Salaga to be exchanged for kola and for manufactured goods from Europe (guns, alcohol, and cloth).

Slaves were mostly the physically fit who had been captured because of military, political, or economic vulnerability. Those who were captured out of ethnic wars, tricked, or owed a debt could also be taken as slaves. Also, notorious members of society such as thieves, troublemakers, and political dissidents were also sold as slaves. Females were hardly ever taken as slaves. The physical fitness of those taken as slaves was what made them attractive to slave traders. To ensure that they remained physically fit, they were often fed while shackled to trees and fixed locations in the 16 quarters or sections of Salaga. Before they were taken to market, they were asked to bathe and oil their skin with shea butter. The main market for slaves used to be centered on an old baobab tree with metal shackle hooks nailed into it. Unfortunately, this tree has been felled to make way for the construction of a gas station! Most of the slaves were brought to Salaga during the dry season when it was possible to travel long distances. Because streams could dry out during the dry season, a series of wells were dug just outside present-day Salaga where the slaves could take a bath. These wells still exist today.

Once the southern traders of Juabeng bought slaves, they were shackled together with chains on their legs, shoulders, and arms and were marched down to the coast where they were sold to other middlemen. Often the journey south was across the Volta River, which was much smaller because it had not been dammed at Akosombo. These trips could last anywhere between three weeks and two months and sometimes went through settlements such as Nkoranza, Yeji, and Kete Krachie. Sometimes, the southern traders were fortunate enough to sell their slaves to middlemen, who then continued the journey to the coast. Effectively, a series of middlemen worked in various chains of trade in slaves to get them from the savanna and Sahel belts of west Africa to the coastal ports of Cape Coast, Accra, and Elmina. It is in the forts and castles of these towns that most visitors to west Africa get to know about slavery.

Crowns in Europe (Khapoya, 1994). Between 1840 and 1890, land explorers such as Henry Stanley (an American employed by King Leopold of Belgium), David Livingstone (along the Zambezi), and Mungo Park (along the Niger River) began to explore hinterland regions, following a combination of existing trade routes and river courses. It was during this period that Europe allegedly "discovered" the lakes and rivers of Sub-Saharan Africa, even though Africans had lived in these parts for centuries.

For most parts of Sub-Saharan Africa, this was a period not of settlement, but of exploration. In southern Africa, however, the British and Dutch (Boers) had begun to colonize territory during this period, largely on account of the appeal of the Mediterranean climate and the abundance of mineral resources in the region. During this period, the British and Boers began to develop conflicting interests over the expansion and annexation of territory, leading to the Boer Trek (1836–38). The Boer Trek established, through force, a number of pioneering communities in the Natal region and the regions west of the Drakensberg Mountains. In later years, it was the established Boer presence in most of present-day South Africa that led the British, under Cecil Rhodes, to create Southern Rhodesia in 1890, to encircle what was then the South African Republic. In the process, Rhodes massacred the Matabele to take over their land.

The period after the cessation of the slave trade across the Atlantic was, therefore, characterized by Europe's land exploration of Sub-Saharan Africa. The reports of explorers about the resource potential of Sub-Saharan Africa led to intense competition for territory amongst European powers and set the stage for the scramble for territory, the Berlin conferences, the partitioning of the region, and European colonialism. Europe's desire to hold territory in Sub-Saharan Africa was precipitated by the shift from mercantile to industrial capitalism in Europe and by the attendant industrial revolution. Land explorers were followed by missionaries who began the conversion of Africans from their "darkened," "uncivilized," "heathen," and "barbaric" lives. The full impact of these missionaries became apparent during the colonial period.

The Colonial Period, 1884 to the 1960s

The colonial period in Sub-Saharan Africa lasted for only about 75 years for most countries, but it had the most pervasive impact. Competition among Europeans for territory in Sub-Saharan Africa was a result of the increased knowledge derived from land explorers and missionaries. In 1884–85, European nations met in Berlin to determine the ground rules for the partitioning of Sub-Saharan Africa into British, French, Portuguese, German, Belgian, Italian, and Spanish territories. This process occurred without the consent of Africans. The stage, therefore, was set for the formal colonization of Africa.

Germany lost its colonies after its defeat in World War I. Togo and Cameroon were split between the British and the French. Tanganyika was given to the British and Rwanda and Burundi went to the Belgians. Figure 4.8 shows the colonial map of Sub-Saharan Africa in 1956. The only countries in the region that were not colonized were Liberia and Ethiopia.

Kourouma (1992) suggests that differences in African and European culture, such as notions of equity and efficiency and Africa's emphasis on collective responsibility as opposed to Europe's emphasis on individual attainment, led to the colonization of Sub-Saharan

The Western Influence in Sub-Saharan Africa

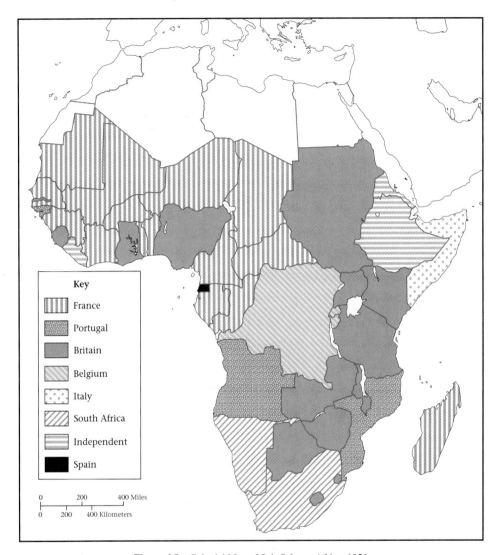

Figure 4.8 Colonial Map of Sub-Saharan Africa, 1956.

Africans. In a sense, the capitalistic ideology of Europe gave it the upper hand in its contact with Sub-Saharan Africa. Colonialism involved not just political subordination, but also economic, cultural, and political domination.

The Economic and Cultural Impacts of Colonialism

With the exception of southern Africa and parts of east Africa (Kenya) where settlement was a major motive for colonialism, the prime objective behind Europe's colonization of other Sub-Saharan regions was to extract agricultural and mineral resources to feed the burgeoning industries at home. Manufactured goods from Europe's industries found a ready market in Sub-Saharan Africa. To carry out its objectives, Europe had to link Sub-Saharan Africa to the world economy in a peripheral and subordinate manner to reap the benefits of its policy. The net effect of this policy of resource extraction was the creation of dual and dependent economic structures.

The dual economy consisted of a modern, cash-crop (cocoa, coffee, tea, rubber, timber, cotton) and mineral (gold, bauxite, copper) export sector, coexisting with a traditional, food-crop (millet, maize, yam, plantain) sector geared largely for local consumption. In eastern Africa, the presence of European settlements led to the appropriation of land for the purpose of establishing tea, coffee, and cotton plantations. In west Africa, workers on cocoa, palm-oil, and groundnut small holdings were relied on to produce for export. Europeans in mineral-rich regions established mines and exploited migrant Black labor to extract gold, diamonds, copper, and bauxite.

To ensure the relegation of Africans to mines, plantations, and small holdings, forced labor laws and poll and house taxes were enacted (Davidson, 1994). This had a detrimental effect on the traditional sector, as land and labor were siphoned off to the modern sector. In countries like present-day Democratic Republic of Congo, where the Belgians cultivated cotton, plantation workers were brutalized. For example, the failure to meet certain cotton quotas often resulted in the severing of a hand (Mazrui, 1986).

The siphoning off of land and labor from traditional to modern sectors implied that food production for local consumption in the traditional sector declined. This meant that Sub-Saharan Africans had to rely on European manufactured goods to meet their basic needs. For Europe, this was exactly what it intended to achieve. Sub-Saharan Africa had become a market for Europe's manufactured goods, and the tastes and preferences of Africans had shifted toward a preference for European, rather than African goods (Mazrui, 1986). Thus, the economic dependency of Africa was reinforced by a dependency on material culture.

The Eurocentric educational system and the spread of Christianity by missionaries emphasized the "superiority" of everything European and the "inferiority" of all that was African. It is, therefore, not surprising that Sub-Saharan Africans, even today, aspire to European values. In some cases, this cultural dependency was facilitated by European administrations enacting policies to stifle the development of local industry and inhibit the growth potential of traditional manufactured goods.

Such economic and cultural dependency led to the underdevelopment of Sub-Saharan Africa and the development of Europe, because limited capital and skill transfer occurred between Europe and its subordinate region (Davidson, 1994). Thus, the poverty that is characteristic of Sub-Saharan Africa today can be traced to its peripheral linking to the global capitalist economy (described in Chapter 8). What is remarkable about all this is that the dual economy and dependency relationship between Europe and Sub-Saharan Africa has been perpetuated by the multinational corporations (such as Unilever, Nestlé, and Firestone). In essence, the multinational corporation has assumed the role of the colonial state, and it still exerts economic and political control over independent countries of Sub-Saharan Africa. This is what *neocolonialism* means. Unlike in the colonial situation, where taxes and forced labor were relied on, in the contemporary neocolonial situation, the development of synthetics for Sub-Saharan Africa's cash crops and mineral resources and the prevalence of unfavorable terms of trade continue to marginalize African countries in the global economy (Rodney, 1974; Davidson, 1994). Not only did Europe disenfranchise Sub-Saharan Africa, but the very nature of the scramble for territory also rendered several countries spatially and economically unviable in terms both of their size and of their physical infrastructure. For example, the Gambia, a small country about 30 miles wide and 300 miles long, was maintained as a British colony only because the Gambia River gave Britain access to the interior of west Africa from the west coast. The French offered to swap the whole of present-day Côte d'Ivoire for the Gambia (because it was contiguous to Senegal), but Britain refused. So, at independence, the Gambia was created with limited resources (apart from groundnuts) to survive in the international economy (Mazrui, 1986).

Because Europe was geared primarily towards the extraction of mineral and agricultural resources from Africa's hinterland and the distributing manufactured goods from Europe to these same hinterlands, transportation lines (especially railways) were designed to be linear from resource locations to major ports. No effort was made to establish lateral connections between railways in different colonies (Davidson, 1994). A contemporary railway map of the region reveals the linearity and lack of lateral connections between countries. In addition, the colonial powers did not bother to synchronize their technology of railway development. Thus, in West Africa, railways in the former British and French colonies operate on different gauges.

The Political and Social Impacts of Colonialism

Colonialism has resulted in three major political dilemmas for present-day countries. The first relates to the arbitrary way in which the region was partitioned at the Berlin conference of 1884. Basically, geometric lines were drawn on a map to form colonies without any regard to the distribution of ethnic groups (Mazrui, 1986). At independence, these boundaries were more or less the basis for creating nation–states. The result is the agglomeration of diverse ethnic groups that are often at odds with each other in terms of territory, trade, religion, and customs. In Ghana, for example, 50 different ethnic groups speak 250 dialects in an area of 96,000 mi^2. In Nigeria, at least 400 ethnic groups exist in a country with a 2001 population estimate of 127 million. In addition, some ethnic groups were split between countries, such as the Ewe in Ghana and Togo and the Hutu and Tutsi in Rwanda and Burundi. The arbitrariness of the partition process and the balkanization that occurred with independence have resulted in a number of ethnic and border conflicts. Examples include the Biafran War in Nigeria (1967–70), the conflict in Sudan between the Islamic and Arabic north and the traditional south, and the continuous saga of violence and terror that people in Burundi and Rwanda have endured since their independence.

The second dilemma is that, to ensure the efficient extraction of resources from the interior of the colonial territories to the coast, colonial administrations had to set up an administrative structure to ensure law and order in their holdings. Each colonial power relied on its own style of administration. The British relied upon a system of *indirect rule* that utilized traditional political institutions as buffers between them and the local population. In stateless societies, the British relied upon a system of *direct rule*, which involved them in all aspects of the day-to-day activities of government. The French relied upon a system of *assimilation*, where the objective was to get people to adopt the French way of life. Because of the limited access to education, only a few people from the French colonies ever became totally assimilated. The Belgians were *paternalistic* and governed their territories, such as the Congo (Zaire), with an iron fist (Davidson, 1994).

These administrative philosophies were in keeping with the racist policies of the colonial powers. For example, a social ramification of British policy was that the Black African was an inferior species who could be civilized, yet without ever becoming as civilized as a Briton. For the French, the Black African was still inferior, but with the right kind of tutelage he or she could be converted to French culture. For the Belgians, the Black African was a resource to be exploited for personal gain, first by King Leopold and later by the tiny state of Belgium (Khapoya, 1994). These attitudes were revealed in the social policies of each colonial power. The Belgians limited the education of the African to only an elementary level. It is, therefore, not surprising that, in Democratic Republic of Congo at independence, only 16 people had received a university education and that in terms of manpower and skilled labor, the country was poorly prepared for independence.

The British established university colleges such as Legon (Ghana), Ibadan (Nigeria), and Makerere (Uganda) by 1948. In addition, they built hospitals and other social infrastructure to meet the basic needs of people in their territories. Thus, at independence, most countries were well prepared to take over from British colonial administration (Mazrui, 1986). The political impact of these administrative systems is that, irrespective of the differences in philosophy, they were all authoritarian, with all power vested in the hands of the European administrators (Davidson, 1994). The British district officer in colonial west Africa, for example, represented legislature, judiciary, and executive elements of district government and was thus very powerful. In east Africa, colonial administrators appointed chiefs in formerly stateless societies. Some of these chiefs were interested in perpetuating their own power. The authoritarian nature of government, the vesting of so much power in colonial administrators, and the desire of some chiefs to perpetuate their own power, in conjunction with the fact that chiefs (in indigenous Sub-Saharan Africa) were often appointed for life, resulted in a situation where many of Africa's heads of states, since independence, have reverted to a one-party system (M'Bokolo, 1992). Examples include Nkrumah of Ghana, Sekou Toure of Guinea, Kaunda of Zambia, and, more recently, Mugabe of Zimbabwe. As is discussed in Chapter 5 on political geography, the presence of one-party states, in the light of limited economic growth and development, has been a major cause of political instability since the independence of African states.

The third political dilemma of the colonial era is that political independence was accompanied by the adoption of Western models of government, which were characterized by the nation–state concept. Most countries adopted the Westminster system of democracy, where there is a separation of the executive, legislative, and judiciary branches, and where the military is assigned the responsibility of national defense. Such strictly European notion of government is foreign to the Sub-Saharan Africa (Mazrui, 1986). It is, therefore, not surprising that this form of government has failed in most countries. In its place, the military has defined itself as another arm of government, rather than a defense force. The implications of these legacies for the political instability and economic decline of Africa as well as alternatives to these systems of political organization are discussed in Chapter 5.

Independence from the West

Colonialism was largely detrimental to the social and economic well-being of Africans. The inherent unfairness of the colonial system resulted in Sub-Saharan Africans agitating for their independence after World War I (Davidson, 1994). In fact, with Italy's invasion of Ethiopia in 1936, a heightened nationalistic attitude emerged in Sub-Saharan Africa, with an emphasis on independence and self-determination. This signaled the beginning of the independence movement. Because of World War II, these demands were put on hold, but, with the cessation of hostilities in 1945, Sub-Saharan Africans continued their demands for a fairer shake in their relationship with Europeans on their own soil. Both Mazrui (1986) and Davidson (1994) identify a number of local and global events that converged to strengthen the demands for independence among Sub-Saharan Africans.

First, the subsequent attainment of independence by India (then known as the jewel of the British Empire), in 1947, stirred much optimism in Sub-Saharan colonies. Second, pressure groups who had witnessed the inhumanity of colonialism while fighting side by side with their African counterparts in World War II began to urge leaders in European capitals to end colonization. Third, after Italy invaded Ethiopia, several nationalist movements emerged, fueled further by the activities of African students who had studied in Europe and America. Activists such as Nkrumah, Diop, and Senghor returned home to organize political platforms. The involvement of the latter two in the Negritude movement propelled them to the forefront of the independence struggle in their countries. These nationalistic activities resulted, in some cases, in the creation of ungovernable situations (such as strikes and political disobedience) for the colonial governments. In Kenya, the Kikuyu-led Mau Mau movement resisted the colonial and missionary annihilation of local culture. Fourth, civil rights movements in the United States linked their cause to the independence struggles in Sub-Saharan Africa, giving them more recognition in the United States (the new world power after World War II). Fifth, Europe was demolished during World War II and had to turn to America for help in reconstructing its economy and infrastructure. This made the cost of administering colonies even more burdensome.

The cumulative effect of these factors resulted in Ghana being the first colony to attain its independence—in 1957, after six years of limited self-rule under the British. By 1965, almost all countries in Sub-Saharan Africa had obtained their independence. The hold-out colonialists were those who had come to settle, such as the Portuguese in Angola and Mozambique and the British in Zimbabwe. In most countries, independence was achieved relatively peacefully. Those countries that experienced a more rocky transition were those with settler colonialists (the Portuguese territories of Angola and Mozambique had to go through civil wars to attain their independence) and those that were inadequately prepared for independence, as was the case in Democratic Republic of Congo. Namibia was in a unique situation—it was a colony of South Africa; its independence was therefore late in coming (1990).

For most of Sub-Saharan Africa, therefore, the colonial era lasted no more than 75 years, but, by the end of this period, the economic, social, political, cultural, and, more importantly, spatial structure of these countries had changed drastically. The alleged modernizing influence of Europe had disrupted traditional ways of production and ways of life to suit self-serving interests of resource extraction. Sub-Saharan Africa was, therefore, marginalized from the global economy, and this peripheralization has been one of the main obstacles to its achieving any form of economic, social, political, cultural, and spatial development today (Mazrui, 1986; Davidson, 1994).

CONCLUSION

The history of Sub-Saharan Africa is the result of the combined effects and layering of three main influences: indigenous, Islamic, and Western (especially colonialism). Colonialism lasted the shortest time of these influences, yet it has had the most pervasive effect on the region.

Because several ancient, medieval, and modern civilizations existed in various parts of Sub-Saharan Africa, subtle differences exist in the effects of the indigenous heritage from region to region. Yet, there are some underlying commonalities that link the cultures of Africa. The Islamic influence is localized mostly in the northern savanna, Sahel, and desert regions of west Africa and in the coastal regions of east Africa. Other parts of the region are not affected in a significant way.

Today, the spatial economies of African countries have evolved differently, out of the variety of colonial legacies inherited. Yet, the colonial powers mostly had in common the single purpose of extracting resources from their colonies, thus relegating them to a state of economic and cultural dependency. To a large extent, the colonial period has created a number of conditions that ensure that economic underdevelopment and political instability will be the order in the region. Since independence, events in Sub-Saharan Africa and at the global level have exacerbated these depressing conditions.

After independence, most countries in Sub-Saharan Africa were saddled with the effects of the colonial period. Their economies are characterized by duality, dependency, and unfavorable terms of trade. All countries are still linked peripherally to the global economy as producers of primary resources and agricultural goods, and as importers of manufactured goods. Today, multinational corporations play a major role in this peripheral linkage. Hence, most countries exist in a state of a new or neocolonialism. This, in conjunction with the politics of conflict and social unrest, justifies Davidson's (1992) claim that the "Blackman has a burden." Sub-Saharan culture, economics, and politics, therefore, can best be appreciated when its history, along with its triple heritage, is understood.

KEY TERMS

The triple heritage
Indigenous heritage
Medieval civilization
Neocolonialism
Assimilation
Berlin conference
Western heritage

Cultural dependency
Ancient civilization
Colonialism
Indirect rule
Paternalism
Islamic heritage
Slavery

DISCUSSION QUESTIONS

1. How relevant is the concept of the triple heritage in our understanding of Sub-Saharan Africa's history?
2. Distinguish between ancient, medieval, and modern civilizations and kingdoms of Sub-Saharan Africa. What has been the contribution of the indigenous heritage to present-day Sub-Saharan Africa?
3. "It was an enslavement of Africans across the Atlantic, not a trade in slaves"—discuss. To what extent can Europeans alone be blamed for the slavery that Africans endured? Reason out the effects of slavery on Sub-Saharan Africa's underdevelopment.
4. What impact did the spread of Islam have on the social and cultural development of Sub-Saharan Africa?
5. Describe and evaluate the economic and political impacts of colonialism.
6. What specific opportunities and constraints does the triple heritage of Sub-Saharan Africa offer its people in their contemporary development effort?

REFERENCES

BANTON, M. & HARWOOD, J. (1975). *The Race Concept*, New York: Praeger.

CURTIN, P., FEIERMAN, S., THOMPSON, L., and VANSINA, J. (1995). *African History: From Earliest Times to Independence,* London: Longman.

DAVIDSON, B. (1966). *African Kingdoms*, New York: Time Incorporated.

DAVIDSON, B. (1994). *Modern Africa: A Social and Political History*, London: Longman.

DE BLIJ, H. & MULLER, P.O. (1998). *Geography: Realms Regions and Concepts*, New York: John Wiley and Sons.

DIOP, C.A. (1974). *The African Origins of Civilization: Myth or Reality*, Westport, CT: Lawrence Hill and Co.

GRAHAM, J.D. (1994). "Political Development in Historical Africa," in *The African Experience,* Khapoya, V.B., ed., Englewood Cliffs, NJ: Prentice Hall.

JAROSZ, L. (1992). "Constructing the Dark Continent: Metaphor as Geographic Representation of Africa," *Geografiska Annaler*, 74(B):2.

KARENGA, M. (1993). *Introduction to Black Studies*, Los Angeles: The University of Sankore Press.

KHAPOYA, V.B. (1994). *The African Experience*, Englewood Cliffs, NJ: Prentice Hall.

KI-ZERBO, J. (1990). *General History of Africa 1: Methodology and African Prehistory*, Berkeley, California: University of California Press.

KOUROUMA, A. (1992). "Africa's Long March," *The UNESCO Courier*, July–August.

KWAMENA-POH, M., TOSH, J., WALLER, R., and TIDY, M. (1982). *African History in Maps,* Essex, UK: Longman.

MAZRUI, A.A. (1986). *The Africans: A Triple Heritage*, Boston: Little Brown and Company.

M'BOKOLO, E. (1992). "Promise and Uncertainty," *The UNESCO Courier,* November.

M'BOKOLO, E. (1994). "Who Was Responsible," *The UNESCO Courier*, October.

OLANIYAN, R. (1982). "Islamic Penetration of Africa," *African History and Culture*, Olaniyan, R., ed., Lagos, Nigeria: Longman.

O'TOOLE, T. (1992). "The Historical Context," in April Gordon and Donald Gordon, eds., *Understanding Contemporary Africa*, Gordon, A. & Gordon D., eds., Boulder, CO: Lynne Rienner Publishers.

RODNEY, W. (1974). *How Europe Underdeveloped Africa*, Washington, DC: Howard University Press.

SKINNER, E.P. (1986). "The Triple Heritage of Lifestyles," in Mazrui, A.A. & Levine, T.K., eds., *The Africans: A Reader*, New York: Praeger.

5

Political Landscape of Sub-Saharan Africa: From Instability to Democratization?

Ian E.A. Yeboah

INTRODUCTION

One aspect of political geography deals with an analysis of the systems of class or group conflict over time and space (Dear, 1988). Another aspect deals with the structures of such political organization as systems of government. As a subfield, political geography attempts to identify conflicting interests between groups (such as political, ethnic, and community organizations) and to analyze the spatial manifestations and ramifications of such conflicts. Sub-Saharan Africa is a fertile region for the study of political geography. This is because the history of the region (its indigenous, Islamic, and Western influences), in conjunction with its economic poverty, cultural diversity, and status in the global community, has created a set of circumstances that manifest themselves in a landscape of political instability.

The level of political instability in Sub-Saharan Africa, however, is not entirely hopeless, in view of such recent developments as the democratic transition of the early 1990s and the end of political apartheid in South Africa. What is surprising, though, is that the indigenous heritage of the region, despite its potential for providing more enduring political structures, has often been ignored in analyses of political instability. There is a stark contrast between the political climate of much of Sub-Saharan Africa from the independence era in the 1960s to the early 1990s and that from the early 1990s onwards. From independence to the early 1990s, the political landscape of the region was relatively unstable. Since the early 1990s, African countries have embarked on multiparty elections in a variety of ways, and a democratic transition seems to be underway. Therefore, this chapter is divided into two parts. The first part describes and analyzes the political landscape of the region prior to the early 1990s. It begins with a characterization of the region as unstable and demonstrates the extent of despair and hopelessness that plagued the region. This is followed by an examination of the causes of political instability. The Democratic Republic of Congo (formerly Zaire) and Rwanda are used to demonstrate how political instability manifested itself in African countries. To transition to the second part, scenarios for political stability that have been tried in the past are assessed. The second part then assesses recent political events since the early 1990s and focuses on the emerging optimism by evaluating the new wave of democratization. Malawi is used as an example to demonstrate that there is hope for political stability in the region. In addition, the emergence of South Africa as a political and economic giant, the smooth transition of government in Ghana, and the relative success of Botswana in incorporating indigenous aspects of government provide other signs of hope. It must be stressed from the outset that, even though there are signs of change for the better, the old order of political instability and chaos still exists in some ways in many countries. Thus, the democratic transition in Sub-Saharan Africa is both an incomplete and a reversible process.

SUB-SAHARAN AFRICA PRIOR TO THE EARLY 1990s: A LANDSCAPE OF POLITICAL INSTABILITY AND CHAOS

An apt description of Sub-Saharan Africa's political landscape prior to the early 1990s is that it was characterized by instability and chaos. Political instability in most Sub-Saharan countries was grounded in a combination of the following four forms: (1) the preponderance of military coup d'états, (2) the presence of one-party states and dictatorships, (3) ethnic and civil unrest resulting in the devolution of state systems, and (4) border disputes between and within countries (Mazrui, 1986; Hayward, 1986; Gordon, 1992). The specific nature of political unrest differed from country to country.

Figure 5.1 highlights those countries that were under military rule or one-party states in 1992 (just after the democratization process began), and Table 5.1 shows the number of military coups each country endured from independence to the year 1995. There is a geographical pattern to the incidence of military coups and one-party states. Countries in west Africa, and to a lesser extent, central and east Africa seem to have experienced more coups than those in southern Africa. Countries in southern Africa and, to a lesser extent, east Africa, where settler colonialism was the norm, seem to have had fewer military coups and one-party states. In west Africa, for example, only Côte d'Ivoire, Senegal, and Cameroon had not experienced coups prior to 1995, whereas, in southern Africa, only Lesotho had experienced a military coup by 1995. Failed attempts at coups are not considered in this analysis. No country has had as many coups as Bolivia in South America, but most countries in Sub-Saharan Africa have experienced protracted periods of military rule. Political problems associated with southern Africa were related to the struggle for independence and to racist policies associated with white settlers.

Sub-Saharan Africa Prior to the Early 1990s

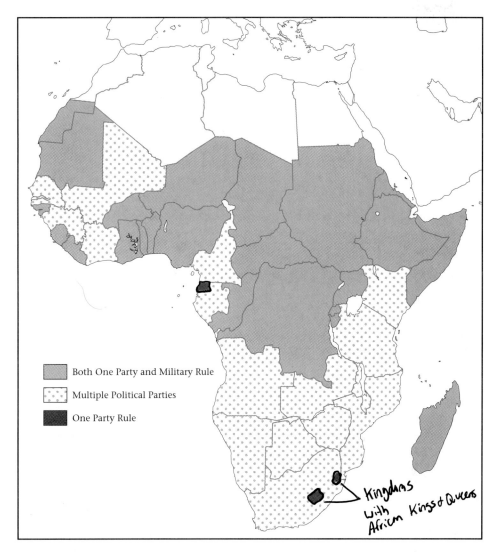

Figure 5.1 Sub-Saharan Countries with Military Rule and One-Party States, 1992. Compiled from Ramsay, F. J. (1995). *Global Studies: Africa*, Guilford, CT: Dushkin; Griffiths, I.L.L. (1993). *The Atlas of African Affairs*, London: Routledge.

By 1995, Ghana, the first country to attain independence in 1957, had experienced five military coups, and 25 of its 38 years of independence had been strictly under military regimes. Nigeria, the most populous country in the region, had spent 25 of its 35 years of independence under five different military governments. Mauritania had experienced three military coups; Uganda, Burkina Faso, and Sudan had each experienced five military coups. Even Madagascar, which is not spatially contiguous to Africa, experienced a coup in 1972. The problem, therefore, is not just a matter of military rule, but of its preponderance.

TABLE 5.1 Successful Military Coups by Country (1995)

Country	Number of coups	Years	Country	Number of coups	Years
West Africa					
Mauritania	3	1978, 1979, 1994	Mali	2	1968, 1991
Niger	1	1974	Chad	4	1975, 1979, 1982, 1990
The Gambia	1	1995	Guinea Bissau	1	1980
Guinea	1	1984	Sierra Leone	3	1967, 1968, 1992
Liberia	2	1980, 1990	Togo	2	1963, 1967
Burkina Faso	5	1966, 1980, 1982, 1983, 1987	Benin	5	1963, 1965, 1967, 1969, 1972
Ghana	5	1966, 1972, 1978, 1979, 1981	Nigeria	5	1966, 1966, 1975, 1983, 1985
Central Africa					
Equatorial Guinea	1	1979	Congo	3	1966, 1968, 1977
Central African Republic	3	1965, 1979, 1981	Democratic Republic of Congo	2	1960, 1965
East Africa					
Madagascar	1	1972	Rwanda	1	1973
Somalia	2	1969, 1991	Ethiopia	2	1974, 1991
Sudan	5	1958, 1964, 1969, 1985, 1989			
Southern Africa					
Lesotho	3	1970, 1986, 1991			

As of 1992, when the democratization process began in Sub-Saharan Africa, only 18 countries were not under some form of military dictatorship. The majority of these countries were in southern Africa (Figure 5.1). The few exceptions in other regions were Kenya and Tanzania in east Africa; Cameroon and Gabon in central Africa; and Gambia, Côte d'Ivoire, Mali, and Senegal in west Africa. Therefore, Griffiths (1994) was justified, at least before 1992, in suggesting that the region's penchant for coups continues unabated and that military coups accounted for the greatest number of changes in government. Since 1992, local and international activists and a new breed of leaders have stepped up efforts to advocate democratization in the form of multiparty governments. This has altered the political status of countries such as Ghana, Benin, Malawi, Zambia, Cape Verde, and Madagascar, where civilian elections have been held. Yet, other countries, such as Gambia and Sierra Leone, have reverted to military dictatorships, in spite of efforts to install civilian governments, proving that democratization in the region is neither universal nor irreversible.

For countries that had not experienced military coups prior to the early 1990s, one-party states, dictatorships, and perpetual leaders were a major attribute of their political landscape. One-party states tend to restrict basic individual liberties of speech and association. In Sub-Saharan Africa, there was a tendency for most leaders to monopolize political power by creating one-party states to stifle any type of initiative from opposing groups. In addition, the one-party state was often associated with one particular head of state as the ruler for life; therefore, these one-party states were also centered on one person. Often, the one-party state was achieved by the creation of a police state with strong lip service paid to socialist ideologies. All resources were targeted toward the perpetuation of the one-party state, resulting in the neglect of productive sectors of the national economy.

In some cases, one-party states began as military regimes, such as in the Democratic Republic of Congo, where Mobutu Sese Seko was its only president from 1965 till his overthrow in 1996. In other cases, one-party states began with democratically elected governments, which then degenerated into one-party states where all opposition parties were banned and individual freedoms suppressed. An example of someone in such a situation was Houphouet-Boingny, who was the leader for life of Côte d'Ivoire. Kenneth Kaunda, who led Zambia from 1964 to 1991, and Julius Nyerere, who ruled Tanzania from 1961 to 1985, when he voluntarily relinquished power, also fall into this category. Others include Kamuzu Banda, who was the only president of Malawi from 1964 to 1994, when he was voted out of office, and Robert Mugabe of Zimbabwe, who created a one-party state characterized by intimidation, the absence of an effective opposition, and the loss of individual liberties. To a certain extent, the recent political malaise in Liberia can be attributed to the desire of the former head of state, Samuel Doe, to monopolize power and remain leader for life. The Liberian crisis was partly caused by conflicts between Americo-Liberians and ethnic groups, but the role of a one-party dictatorship should not be downplayed.

Sub-Saharan Africa's political landscape prior to the early 1990s was also characterized by ethnic and civil unrest. The classic cases are the civil war in Chad between 1975 and 1983, the attempt by Katanga to secede from the Democratic Republic of Congo right after independence in 1960, and the secessionist Biafran war fought between the Igbos and the rest of Nigeria between 1967 and 1970. Estimates are that the Biafran war cost at least one million lives and disrupted many families. The recent saga of ethnic cleansing between the Hutu and Tutsi in both Rwanda and Burundi also has ethnic underpinnings. Often, these ethnic conflicts were reinforced by colonial policies that favored one group over another, as in the case of Rwanda. In some cases, such as between Ethiopia and Eritrea, the ethnic unrest led to devolution after a protracted war. Although conflicts existed in Somalia, which resulted in a civil war in 1990, it must be stressed that the war was a result of clan disputes rather than of ethnic differences.

Border and territorial disputes, for the most part, occur between, rather than within, countries. An example of a cross-country border dispute is the one between Ghana and Togo. Border disputes between Nigeria and Cameroon have been brewing since independence, but, fortunately, have not escalated into a full-scale war. Occasionally, there are boundary disputes between ethnic groups within a country's border, such as the one between the Kukombas and the Nanumbas in the northern part of Ghana in 1993 and 1994.

The combined effects of the preponderance of military coups, one-party states, ethnic unrest, and border disputes prior to the early 1990s prompted calls by local and international groups for the protection of human rights and the restoration of civil liberties, participatory democracies, and multiparty elections. This quest for democracy initially resulted in major confrontations between governments and citizens in several countries, among them Benin, Ghana, Malawi, Nigeria, Togo, the Democratic Republic of Congo, and Zambia. For the most part, these countries have since embarked on various elements of political change that can be summarized as democratization.

Causes of Political Instability in Sub-Saharan Africa

There are underlying factors that were, and are still, responsible for the plethora of problems in Sub-Saharan Africa's political landscape. These causes are related to specific historical, social, cultural, and economic circumstances and to the global status of the region. In discussing the political landscape of the region, such factors should not be ignored. According to Ake (1973), Mazrui (1986), and Khapoya (1994), eight factors have contributed to the destabilization of the political landscape in Sub-Saharan Africa described in the previous section:

1. ethnic differences and tensions in the region;
2. the triple heritage of the region and the conflicts between its branches;
3. political ineptness and incompetence of some leaders;
4. the legacy of the Western form of government and the poorly defined role of the military;
5. unrealistic expectations placed on state institutions;
6. undue international interference in the internal and sovereign affairs of states;
7. racial and discriminatory policies in some states; and
8. the struggle for independence in settler colonies.

Ethnic tensions in Sub-Saharan Africa are a direct result of the way in which the region was balkanized, without any regard to the ethnic affiliations in the newly created states at independence. As was explained in Chapter 4, the colonial authorities did not take cognizance of the diversity of ethnic groups and the kinds of tensions and hostilities that existed among them. The effects of these tensions and hostilities have already been documented in the preceding section. Discussion Box 5.1 also provides a detailed account of the Hutu and Tutsi debacle in Rwanda.

The region's triple heritage also contributed to the state of the political landscape in Sub-Saharan Africa. Of relevance here is the incompatibility between the traditional political structures, on the one hand, and Islamic (shariah law)- and Western (Westminster style)- imposed systems of government, on the other hand. The conflicting natures of these three systems make it difficult to implement appropriate administrative procedures. More often, split loyalties and parochial interests emerge. For example, the Ashanti in Ghana consider themselves, first and foremost, as Ashanti with loyalties to the Asantehene (king of the Ashantis) rather than as Ghanaians with loyalties to the head of state. In Uganda, the incompatibility between traditional authority and the Westminster system of government instituted at independence (among the Buganda) has destabilized the country since its independence in 1962. Because of the relative wealth, educational attainment, and political clout of the Buganda over the northern groups of Uganda, a coalition government that incorporated traditional authorities was created. Mutesa II, king of the Buganda, was made president, and Milton Obote (a Langi from the north) was designated prime minister. In 1966, Obote used the northerner-dominated army to overthrow Mutesa II; his basic reason was to abolish tribalism. This situation, along with Obote's dictatorial rule, set the stage for the downward spiral of Uganda into political chaos and instability under Idi Amin. Also, the

DISCUSSION BOX 5.1: RESTLESS POLITICAL LANDSCAPES: GENOCIDE AND POLITICAL CHAOS IN RWANDA

A massacre of people that is best described as a genocide occurred in Rwanda, beginning on April 6, 1994. Estimates reveal that up to 500,000 people were slaughtered in the civil war between the majority Hutu and the minority Tutsi. This was an attempt by the Hutu elite to cleanse the country of the Tutsis once and for all. In the process, those Hutus who were sympathetic toward Tutsi were also targeted by death squads. The massacre was often carried out, not by sophisticated armaments, but by hand-to-hand combat, using rudimentary weapons such as machetes and clubs. In addition, most of the killing done by Hutu death squads was perpetrated by a youth wing of the ruling National Revolutionary Movement for Democracy (MRND), known as *interahamie* (those who attack together), and the Coalition for Defense of Freedom (CDR), known as *impuzamugambi*. The killing also generated a massive wave of refugees, most of whom ended up in surrounding countries: the Democratic Republic of Congo, Uganda, and Tanzania. The killings in Rwanda destabilized the country politically. Inasmuch as genocide in any country is despicable, it is necessary to understand the underlying causes of such acts of brutality committed against other humans. Why have Hutu and Tutsi, who lived together peacefully before European contact, now become archenemies to the extent that one group wants to eliminate the other?

The trigger for the massacre was the shooting down of an airplane conveying the then president of Rwanda, Juvenal Habyarimana (a Hutu), and the then president of Burundi, Cyprien Ntaryamira, returning from peace talks in Dar es Salaam, Tanzania. Although the identity of the perpetrators was unknown, it was suspected that extremist Hutus in the Presidential Guard deliberately shot down the airplane to instigate the majority Hutu to massacre the Tutsi. The cause of the genocide, however, goes deeper than the mere shooting down of an airplane.

Rwanda, a country with a 2001 population estimated at 7.3 million and a total land area about the size of Maryland, is populated by three ethnic groups: the Hutu, the Tutsi, and the Twa. The Twa are hunter–gatherers and are the original settlers. The Hutu, who traditionally are sedentary agriculturists, are descendants of Bantu migrants from west Africa. The Tutsi are traditionally herders and are believed to be descendants of Nilotic migrants from Ethiopia. Although the Hutus are the larger group (about 89%), the Tutsi (about 10%) were the ruling class before European contact. During this era, both groups lived in symbiotic harmony. The hatred between the two groups evolved with European contact.

Rwanda was originally colonized by the Germans. After World War II, it became a protectorate of Belgium. Both the Germans and Belgians adopted a system of indirect rule. Tutsi rulers (known as *Mwami*) acted as intermediaries between Belgians and the locals. This system of indirect rule was associated with European notions of a racial hierarchy and differences between the Tutsi and Hutu. The Belgians, and the Germans before them, believed that the Tutsi were tall and slender "black Caucasians," conquerors from Ethiopia, and a superior, aristocratic race. The relatively stocky Hutu were believed to be peasants who were incapable of playing a role in civilized society. This attitude introduced a stratification between the Tutsi and Hutu during the colonial era. In fairness, the Tutsi believed that their rule over the Hutu was of divine origin, which however, did not imply superiority of the Tutsi over the Hutu. The colonial administration in Rwanda provided education to the sons of mostly Tutsi chiefs and subchiefs to fill positions in administration and the civil service. The Tutsi, therefore, received the best jobs during the colonial period. Thus, the period of colonial rule under both Germany and Belgium reinforced Tutsi domination and control. The colonial era did, however, foster a gradual political, social, and economic evolution of the Hutu, by imposing some restrictions on the Tutsi monarchy and giving some Hutu some access to education.

With such a social stratification, it is not surprising that, in 1959, an awakening of Hutu consciousness created a momentum that carried over into a collapse of the Tutsi monarchy. Hutu from the northwest region, which had escaped control by Mwami, led a rebellion against the Tutsi. These actions resulted in Rwanda's independence in 1962. Thus, the Tutsi domination during colonial rule had been reversed. Interethnic competition for power resulted in many Tutsi fleeing from Rwanda to Uganda, Tanzania, and Burundi. It was this group of exiles who formed the Rwanda Patriotic Front (RPF) in Uganda. The RPF demanded a total repatriation of all exiles to Rwanda, on the grounds that they had not been integrated into their host society. In 1973, Juvenal Habyarimana seized power in Rwanda by means of a military coup and embarked upon a dictatorial and authoritarian one-party state. With such repression, internal and RPF opposition grew against Habyarimana's government. The RPF invaded Rwanda in the early 1990s, and the French interceded to help broker a cease-fire between the government and the RPF. As previously stated, when Habyarimana's airplane was shot down, he was returning from a peace conference. The Presidential Guard, who had shot down the airplane to provoke a Hutu reaction, turned the tables by blaming the RPF for the incident. This set off the massacres in Rwanda and the massive refugee problem associated with it. The international community, especially nongovernmental agencies, came to the rescue of the refugees. Unfortunately, international response to the killings and the refugee problem were not forthright enough to stem the problem. The RPF then launched an attack on the ruling MRND and succeeded in taking over the reigns of government in Rwanda. In 1995, the United Nations brokered a cease-fire in Rwanda, and Paul Kagama, the leader of the RDF, was elected president in the subsequent election. Progress towards stability has been made, but the road toward resolving Rwanda's problems seems to be a long one.

> Rwanda's experience demonstrates that ethnic differences, and the way in which ethnic groups were plunged into a state system at independence, can be detrimental to the political viability of countries in Sub-Saharan Africa. What is revealing about Rwanda is that, even though there are only three ethnic groups, the way in which colonial administrations favored one group over the other created social stratification and aroused animosity between ethnic groups. Ethnicity does not necessarily have to give rise to violence, but colonial policies have manipulated ethnicity and reinforced ethnic differences by favoring one ethnic group over another. Rwanda is not in a unique situation; similar situations occurred in other parts of Sub-Saharan Africa, such as Burundi, Sudan, Uganda, and Nigeria. It is therefore not surprising that ethnic differences constitute a major cause of political instability in the region.
> Compiled from
> 1. de Waal, A. & Omaar, R. (1995). "The Genocide in Rwanda and International Response," *Current History*, 94(591):156.
> 2. Gibbs, N. (1994). "Why? The Killing Fields of Rwanda," *Time Magazine*, May 16.
> 3. Nyrop, R., et al. (1995). *Rwanda: A Country Survey.* Washington, DC: Foreign Areas Studies.

conflict in Sudan is attributed to the incompatibility of Islamic, or shariah, rule (and Arabic culture) in the north with the non-Islamic (non-Arabic) culture in the south.

Another factor that contributed to Sub-Saharan Africa's unstable political landscape previously described is the absence of competent, committed, and dedicated leadership. The region's political leadership was characterized by ineptness, corruption, greed, and naiveté. This problem seems to be endemic to most countries in the region. These traits, though, are not exclusive to Sub-Saharan Africa; however, the extreme forms of underdevelopment and poverty in the region have made its leaders and bureaucrats more vulnerable to these traits. Examples of sheer greed, corruption, and incompetence are associated with Mobutu Sese Seko of the Democratic Republic of Congo, Siad Barre of Somalia, Kenneth Kaunda of Zambia, Kamuzu Banda of Malawi, and "Emperor" Bokassa in Central African Republic. The case of Mobutu is discussed in detail in the next section of this chapter. In the case of Siad Barre of Somalia, his naiveté in playing off the Soviet Union against the West (the United States) during the Cold War set the stage for political instability in Somalia (Hamrick, 1993). The absence of effective leadership, the idea of ruler for life, and the authoritarian rule of colonial administrators inherited at independence helped perpetuate the existence of one-party states in Sub-Saharan Africa.

The legacy of the Western system of government—in particular, the poorly defined functions of a military institution under such a system—contributed to the unstable political climate in Sub-Saharan Africa. At independence, most countries in the region adopted Western systems of government associated with unicameral legislatures with only one level of lawmakers. Nigeria (partly because of the complexity of social, political, and economic issues) is one of a few countries to adopt a bicameral system where two levels of representatives act as checks and balances on each other. Currently, South Africa, Botswana, Namibia, and Swaziland have bicameral legislatures, as is demonstrated in Table 5.2. The concept of a military regime is not totally foreign to the region, because warrior groups, with specifically defined functions, existed in ancient and medieval Sub-Saharan Africa. However, since independence, the military (under the Western system that most countries inherited) has been regarded as a foreign institution with a poorly defined role in national development. With limited checks and balances on legislative and executive branches of government, the military has resorted primarily to organizing coups to intervene in the democratic process. Most military units in the region have the erroneous assumption that the military should serve as a corrective institution to which all branches of government are accountable.

Most military interventions were, and are still, predicated on the notion that the military seek to improve the circumstances of ordinary citizens. For the most part, most military regimes in Sub-Saharan Africa have not been able to achieve this purpose. This lack of success betrays the irresponsibility of military coups. Examples of such irresponsibility include Idi Amin of Uganda (a joke on the international scene), Valentine Strasser of Sierra Leone (a 27-year-old youth with limited government experience), and Samuel Doe of Liberia (of very dubious credentials). In fact, in most Sub-Saharan African countries, the military has always seen government as an avenue to amass personal wealth; this is particularly so in Nigeria (Suberu, 1994).

The economic underdevelopment and poverty that still characterize most of Sub-Saharan Africa have also been destabilizing factors. Citizens of each country have placed high and, often, unrealistic expectations on the state to eradicate poverty, despite the limited national resources available. This has resulted in the emergence of several interest groups, who often are at odds with each other. These interest groups can develop along ethnic lines, as was the case with the Biafran war in Nigeria. Unfortunately, most states in the region have been unable to meet the expectations placed on them and, in the process, have lost any semblance of legitimacy. Ninsin (1989) suggests that all democratically elected governments in Ghana since independence have lost their legitimacy, setting the stage for the military to take over power. This is typical for most Sub-Saharan countries.

Another factor explaining the unstable political situation is the undue interference of external forces in the economic and political sovereignty of Sub-Saharan states. During the Cold War, international interference was often within the realm of the strategic and political importance of a country and its leaders. In more recent times, the activities of both Libya and France in Chad contributed to a protracted civil war in that country between 1975 and 1983. Also, the activities of the Soviet Union, the United States, South Africa, the Democratic Republic of Congo, and Cuba in Angola contributed toward the protracted civil war (1975–91) between the ruling Popular Movement for the Liberation of Angola and the rebel National Union for the Total Independence of Angola. (See Figure 5.2.)

South Africa also interfered in the affairs of the states bordering it, to ensure the preservation of its apartheid system. Because Namibia was a colony of South Africa, the South African Defense Force waged a war of destabilization against the leading freedom

TABLE 5.2 Types of Governments from Selected African Countries

Country	Legislative Branch	Legal System	Comments
Angola	Unicameral National Assembly	Based on Portuguese civil law and local customary law	About 12 parties participated in the 1992 elections. The Popular Movement for the Liberation of Angola or MPLA has been the ruling party since 1975.
Benin	Unicameral National Assembly	Based on French civil law and customary law	Benin abandoned years of military coups along with Marxist–Leninist principles to engage in multiparty democratic reforms in 1990. It has as many as 20 political parties with Renaissance Party as the majority party.
Botswana	Bicameral parliament	Based on Roman–Dutch law and local customary law	Four major political parties with the Botswana Democratic Party as the ruling party. Bicameral Parliament includes a House of Chiefs: an advisory 15-member body of chiefs from eight principal tribes.
Democratic Republic of Congo	300-member transitional constituent Assembly established in August 2000	Based on Belgian civil law and tribal law	Democratic Republic of Congo has endured years of political dictatorships under Mobutu Sese Seko and Laurent Kabila. Joseph Kabila became president after his father was assassinated, and the country will hopefully transition into a more representative government.
Kenya	Unicameral National Assembly	Based on English common law, tribal law, and Islamic law	As many as 12 political parties were represented in the December 1997 elections. Daniel Arap Moi's ruling political party—Kenya African National Union (KANU)—has dominated the political scene for some time now.
Malawi	Unicameral National Assembly	Based on English common law and customary law	In 1999, President Muluzi's ruling party, the United Democratic Party, won the election with a majority vote of 93 out of 192 seats.
Namibia	Bicameral National Council and National Assembly	Based on Roman–Dutch law and 1990 constitution	Namibia gained its independence in 1990 and established a bicameral legislative system to accommodate representatives from the 13 regions. The ruling party is Sam Nujoma's Southwest African People's Organization or SWAPO.
Nigeria	Bicameral National Assembly and House of Representatives	English common law, Islamic shariah law (in some northern states), and traditional law	Nigeria has had its fair share of military regimes and since May of 1999 has elected a civilian government with President Obasanjo's People's Democratic Party (PDP) emerging victorious. The bicameral system is designed to accommodate the more than 500 ethnic groups and the 36 states.
South Africa	Bicameral National Assembly and National Council of Provinces	Roman–Dutch law and English common law	South Africa's bicameral legislature is designed to accommodate the provincial interests and to protect the rights associated with the different ethnic and racial populations.
Swaziland	Bicameral advisory body (Libandla) consisting of Senate and National Assembly	Based on South African Roman–Dutch law in statutory courts and Swazi traditional law and custom in traditional courts	Swaziland is a monarchy where the king (Mswati III) appoints the cabinet and judges based on the prime minister's recommendation. Political parties were banned by the 1978 constitution.
Tanzania	Unicameral National Assembly	Based on English common law	Tanzania's government provides opportunities for women to participate in the legislature. Thirty-seven of the 274 seats in the national assembly are allocated to women nominated by the president.

Source: Compiled from the CIA Factbook and the Political Reference Almanac.

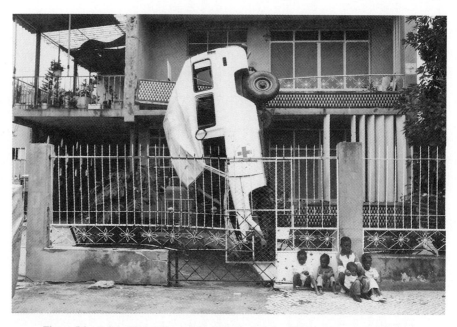

Figure 5.2 Political Instability and Civil War in Angola. Children in the city of Huambo sit in front of the home of a Red Cross official, whose vehicle was upended by a 500-kg bomb in September 1993. (Courtesy of Jeremy Harding.)

movement—the South West African People's Organization (SWAPO)—between 1963 and 1989. South Africa undertook similar destabilizing activities in Mozambique. The Soviet and American intervention in Ethiopia and Somalia, and the arming of these two countries during the Cold War, contributed to the proliferation of guns in Somalia. These very same guns were used by Mohammed Farah Aideed, Ali Mahdi Mohammed, and other clan leaders to intimidate residents of Mogadishu in the early 1990s (Hamrick, 1993). Ironically, the US military presence in Somalia in the early 1990s took place in order to clean up the mess it had helped create.

In countries where colonialism was associated with Caucasian settlement (especially in southern Africa), racial differences and racism were contributing factors to political instability. The apartheid system in South Africa denied blacks such fundamental human, political, social, and economic rights as the right to vote, the right to quality education, freedom of association, freedom of movement, and free choice of residence. Invariably, international and local pressure from both black and white South Africans has ended official apartheid.

The struggle for independence was a leading cause of political instability, particularly in the settler colonies of southern Africa. For example, between 1964 and 1974, the Mozambique Liberation Front undertook a guerrilla war against Portugal for the country's independence. Also, the Zimbabwean African People's Union under Robert Mugabe and the Zimbabwean African National Union waged guerrilla war against the racist white government in Salisbury (Harare) between 1966 and 1978. In Kenya, the Mau Mau movement staged by Kikuyus and other ethnic groups was a form of cultural and political resistance to settler colonial rule. Even though more than 12,000 Kikuyu were massacred between 1951 and 1955, the psychological victory of the Mau Mau set the stage for the nationalist movement that finally won the country's independence in 1963.

The eight factors listed earlier are by no means exhaustive; however, they constitute the major causes of political instability in Sub-Saharan Africa. The causes of instability differ from country to country; in most cases, however, a combination of factors has had a destabilizing effect. To appreciate specific ways in which the factors relate to specific countries, it is appropriate, at this juncture, to examine one case study, the Democratic Republic of Congo, which, in many respects, exhibits most of the causes of political instability so far examined. Perhaps racial differences and the struggle for independence are the only two causes of political instability that are not characteristic of the case of the Democratic Republic of Congo.

The Democratic Republic of Congo: a microcosm of political chaos and instability The political impasse in the then Zaire (now the Democratic Republic of Congo), where a struggle for power ensued between Mobutu Sese Seko and his opponents, stems from the corruption, greed, and incompetence of the country's president for life (Richburg, 1994). Since the overthrow of Mobutu by Laurent Kabila and the subsequent ascension to power by his son, Joseph (after Kabila's assassination), the situation in the Democratic Republic of Congo has not improved. In a historical sense, the country's political chaos can be traced to the effects of the diversity of ethnic groups, its colonial experience, foreign intervention (especially during the Cold War), and the leverage that Mobutu lost to the United States after the Cold War. But, above all, greed and incompetent leadership are demonstrated in this state of affairs.

At independence in 1960, Belgium had not established any institutions of higher learning, even though a well-established system of primary education existed. In addition, only 16 people, despite the Democratic Republic of Congo's size, had received a university education. Thus, the country, in terms of expertise, was not prepared for political independence (Kelly, 1993). Despite the lack of local personnel to administer the country, the brutality of both the Leopold era and the colonial period had translated into a hatred of whites. After independence, this hatred translated into violence against Belgians, who fled the country (Kelly, 1993). Hence, the expertise of Belgian administrators could not be utilized. In fact, the racist attitudes of Belgian officers in the army resulted in a mutiny of the Democratic Republic of Congo's troops on June 30, 1960 (within the first week after Joseph Kasavubu and Patrice Lumumba took office as president and prime minister, respectively). In addition, ethnic tensions that already existed in the country were heightened, resulting in the desire of the Katanga province (now Shaba) to secede.

With such a volatile situation, Patrice Lumumba called for assistance from the international community to ensure peace in the country. Yet, divisions between Lumumba and Kasavubu worsened. The secessionist war of Katanga and ethnic unrest wore on. In the end, Lumumba's confidant in the army, Mobutu Sese Seko, took over power in a military coup in 1965 and handed Lumumba to Katanga rebels, who executed him. Mobutu's coup was supported by the CIA of the United States under the pretext that he was the only person who could hold the then Zaire together. The argument was that it was either Mobutu or chaos; eventually, it became both (Aronson, 1993). The fact of the matter, though, is that the problem of ethnic unrest was almost solved by the time of Mobutu's coup. At that time, only one ethnic group had not signed a peace accord brokered by the United Nations.

Mobutu ruled the Democratic Republic of Congo from 1965 to 1996 as a one-party dictator. His government continued to exhibit high levels of corruption, incompetence, and greed, which crumbled the country's social, economic, and political infrastructure. Richburg (1994) provides the following examples of the extent of this decay: The per capita income of the county in 1990 was $US170, which is equivalent to only 10% of its 1960 level. In his 30 years in office, Mobutu did not build a single hospital in the whole country. Only 3% of the national budget went to health and education, 23% went to the military, and 50% of it lined the pockets of Mobutu and his peons. Richburg (1994) also describes the extent to which Kinshasa, once an elegant city with European-style boulevards, had been overrun by debris, and how elephant grass and overgrown trees obscure abandoned government buildings. In Richburg's words, "the jungle is claiming the capital."

During that period of decline in the Democratic Republic of Congo, Mobutu's personal wealth, through kleptocracy, became enormous; it included 12 French and Belgian chateaus, a Spanish castle, and a 32-bedroom Swiss villa. His net worth was estimated to be between $US3 billion and $US7 billion. Yet, the child mortality rate in the country was 50%. This was the extent of the corruption, greed, and incompetence of Mobutu. Unfortunately, the West was aware of Mobutu's pillage of the Democratic Republic of Congo, but hardly dissented, because Mobutu was a friend during the Cold War (Aronson, 1993). The Democratic Republic of Congo was an important supplier of industrial diamonds, copper, cobalt, and uranium, which are vital for aircraft and defense industries. In addition, Mobutu allowed the Democratic Republic of Congo to be used as a CIA base for both overt and covert operations into Angola in opposition to the Cubans (Kelly, 1993). Mobutu essentially utilized the Democratic Republic of Congo's geostrategic and geopolitical advantages to gain leverage in international negotiations.

With the end of the Cold War and the disintegration of the Soviet Union in 1991, Mobutu ceased to be relevant to the United States. American financial and moral support for Mobutu both ended, and Western pressure to democratize increased (Kelly, 1993). Although the United States abandoned Mobutu, his opponents were initially too weak to overthrow him. This situation resulted in a political power struggle or stalemate in the country (Richburg, 1994). In 1992, politicians, civic leaders, intellectuals, and clerics in the Democratic Republic of Congo organized to strip Mobutu of his power by appointing Etienne Tshisekedi as prime minister. Mobutu, however, ignored these efforts at pluralism and continued to rule as he pleased. He fired Tshisekedi and appointed his staunch supporter, Faustin Birindwa, as prime minister. Tshisekedi refused to resign, resulting in a country with two prime ministers, neither of whom was capable of governing. This explained the stagnation of government and civil society in the Democratic Republic of Congo after 1992, which was characterized by a *de facto* decentralization of administrative, legal, and political functions in the country.

In July 1994, national conference delegates in the Democratic Republic of Congo, in an attempt break the political stalemate, elected an interim prime minister relatively acceptable to both sides of the impasse: Kengo wa Dondo. Elections were scheduled for July 9, 1995, with Mobutu's Popular Movement for the Revolution as one of the leading contenders. Mobutu's supporters expected him to win because he had kept the country from civil and ethnic war and because he was the only candidate with the credentials of a chief or strong man (Shiner, 1994; Press, 1995). Issues yet to be resolved that could hold the elections back were the need for a census, a proper registration of voters, and a referendum on the transitional constitution. By December, 1995, these elections had not taken place, and Mobutu was still in power despite the instability in the country.

In 1997, a relatively unknown revolutionary, Laurent Desire Kabila, along with his Alliance of Democratic Forces for the Liberation of Congo (ADFL), took advantage of three things to overthrow Mobutu (Schatzberg, 1997). First, the Tutsi (Banyamulenge) from the east, who had been stripped of their Zairian citizenship by Mobutu in 1981, started a revolt backed by Rwanda and Uganda. Second, Mobutu's diagnosis of prostate cancer meant that he was out of the country for treatment and thus could not be the intimidating leader that he had to be. Third, Mobutu had pillaged state institutions so much that most people were glad to see him overthrown (Aronson, 1998). It is not surprising that, because Mobutu had not paid his army in months, most of them fled upon the advance of ADFL forces (a reflection of the decay of state institutions); the residents of Kinshasa welcomed the rebels into the city.

Kabila's rule, from 1997 to January 2001, was no different from Mobutu's. He abandoned his initial promise to organize multiparty elections. Kabila "regularly and ruthlessly violated the human rights of the Congolese people, killing, torturing, imprisoning, and causing 'disappearance' of any who he thought threatened him or his regime. Among those who suffered most were political opponents, leaders of civil society, human rights activists, and journalists" (Human Rights Watch, 2001). Some allegations were that Kabila followed Mobutu's example of kleptocracy and lined his personal bank accounts with state money. In the end, those who supported Kabila's taking over power (Rwanda and Uganda) waged a war against him. During this war Angola, Zimbabwe, and Namibia were Kabila's supporters. Kabila's unwillingness to allow the United Nations to investigate allegations of genocide by the Hutus who had brought him to power also curtailed the flow of badly needed international aid that was to be the foundation of his $US4 billion trust fund for national reconstruction. The turmoil that surrounded his rule culminated in his assasination by one of his bodyguards in January 2001. Interestingly, his unknown son Joseph, at somewhere between 28 and 31 years old, was appointed head of state.

The history and present situation of the Democratic Republic of Congo make it appropriate to say that the nation reflects a microcosm not only of economic decay, but also of the identified causes of political instability in Sub-Saharan Africa. It is apparent that the country's political instability was mainly due to Mobutu's corruption, greed, and incompetence, in conjunction with international interference during the Cold War. Yet, historically, the diversity of ethnic groups, the Belgian policy of not preparing the country for independence, the racist attitudes of colonial administrators, the struggle for independence, the incompatibility of Western and traditional forms of political organization, the irresponsible role of the military, and the poverty and resultant corruption in the country have all contributed to the Democratic Republic of Congo's political instability. Even though the Democratic Republic of Congo exhibits most of the factors that cause political instability in Sub-Saharan Africa, it must be emphasized that the combined effect of these factors differs from country to country.

SCENARIOS FOR POLITICAL STABILITY IN SUB-SAHARAN AFRICA

After outlining the nature and causes of political instability in Sub-Saharan Africa, it is logical at this juncture to consider the following question: What alternative approaches have been adopted to solve problems of political instability in individual countries, and how successful have they been? Based on the work of Hayward (1986) and Khapoya (1994), six scenarios are presented next.

The first is the legacy of the *Western system of government*—the *Westminster parliamentary* and *the Federal systems*. This approach has been adopted by almost all countries that gained independence peacefully, such as Ghana, Nigeria, Kenya, Zimbabwe, and Botswana. The foreignness of the nation–state (a European concept) system of government has, to a large extent, contributed to military interventions (Hayward, 1986). Yet, because it is the system that colonial administrations introduced to Sub-Saharan Africa at independence, most countries seem to identify with it as the only viable form of national government. In today's Sub-Saharan Africa, demands for democratization are synonymous with these forms of government. The broader question, though, is whether there are any alternatives to such a form of government in terms of democracy.

The second approach is *African socialism*. Khapoya (1994) suggests that, even though some countries declared that they were adopting socialist policies, in reality they were not socialist. Examples of these are Kenya under Kenyatta and Ghana under Rawlings. In the latter case, the Rawlings regime completely turned around to embrace privatization, which was counter to its original intentions of creating a socialist state based on "popularist principles." The classic case of genuine African socialism is Tanzania's *Ujamaa* or "familyhood" under Nyerere (referred to in Chapter 8). Other countries, such as Senegal under Senghor, Guinea under Sekou Toure, and Ethiopia under Mengistu, have tried African socialism, but not to the extent and intensity of Tanzania. In the case of Tanzania, for example, Ujamaa as an approach to political development involved a territorial and economic reorganization of space, production, and social relations (Sendaro, 1987). Yet, the fact that it implied collectivization and socialism meant that the Western world and its financial arms, the World Bank and the International Monetary Fund (during the Cold War), were not willing to support it.

A third approach that has been tried in the region is what Hayward (1986) refers to as a *controlled political development with an open economy*, or what Khapoya (1994) terms *African capitalism* (with specific reference to Kenya under Kenyatta). Côte d'Ivoire under Felix Houphouet-Boigny is the best example (refer to Chapter 8). At independence from France in 1960, Houphouet-Boigny encouraged free enterprise with the full participation of citizens in the parliamentary system. Yet, political power was largely confined to a one-party state under the paternalistic autocracy of Houphouet-Boigny. With support from the French, a large expatriate Lebanese community, and the linking of the CFA franc currency to the French franc, the country was able to establish an industrial base, albeit one that was geared toward the production of consumer goods. The distribution of the gains of this development were, however, highly concentrated in the hands of a few political and economic elites. With a downturn in its economy in the late 1980s and early 1990s, political opposition against Houphouet-Boigny emerged. Upon his death in 1994, the current president, Henri Bedie, instituted reforms to ensure a fairer participation of citizens in the political and economic arena.

Although Houphouet-Boigny has often been cited as a leader who relied on a controlled political environment amidst a free-enterprise economy, his personality exemplifies the fourth alternative solution to political problems in Sub-Saharan Africa. This involves having a *charismatic individual leader as a guide*. Julius Nyerere in Tanzania, Jomo Kenyatta in Kenya, and Kwame Nkrumah in Ghana are also good examples of this type of leader. Nkrumah was known as *Osagyefo*, or The Savior; Nyerere was viewed as *Mwalimu*, or The Teacher; Houphouet-Boigny was viewed as *Le Vieux* (The Old Man) in Côte d'Ivoire; Kenyatta was referred to as *Mzee*, or The Old Man; Mobutu Sese Seko of the Democratic Republic of Congo was viewed as *Le Guide*. By the way, his original name is Joseph-Desire Mobutu, but his full adopted name is Mobutu Sese Seko Kuku Mgbendu wa za Banga, which roughly translates to "*the all-powerful warrior who, by his endurance and will to win, goes from contest to contest, leaving fire in his wake*" (Lamb, 1982). The problem with this approach to political stability is that too much stock is placed in the leader (Lamb, 1982), to the extent that, once his performance begins to decline, either on the economic or social level, his popularity declines, leading to a vacuum in leadership. It is not surprising that Nkrumah of Ghana was overthrown in 1966, despite his prominent role in the independence struggle of his and other African countries. Even Mobutu was overthrown, despite his tight grip on power for over 30 years. In cases where such leaders maintain power, their appeal declines to the extent that either opposition groups develop or they are pressured to hand over power, as was the case with *Le Vieux* and *Le Guide*, respectively.

A fifth approach suggested for maintaining political stability in Sub-Saharan Africa is to view an *individual leader* as a *godlike figure*. Haile Selassie in Ethiopia and Nelson Mandela in South Africa are examples of this type of leader in both a historical and contemporary sense. Haile Selassie was referred to as the *Conquering Lion of Judah, the Elect of God, and the King of Kings*. Apart from reestablishing the monarchy of ancient Ethiopia, he orchestrated the war of attrition against Italian occupation in the 1930s and, during the period of independence struggle in Africa, was an important symbol of strength. His dignity and personal presence during, and after, Italian occupation added to his stature in Ethiopia and on the continent. Mandela's struggle against apartheid and his 27-year imprisonment provided him the legitimacy of a political and spiritual leader amongst both black and white South Africans. Gevisser (1994) suggests that Mandela's near-seamless statesmanship accounted for the smooth transition to majority rule in South Africa. The problem with such an approach to political stability is similar to that arising from dependence on a charismatic leader: Can South Africa continue with the peace that existed under Mandela's government? Can anybody achieve his level of statemanship? So far, South Africa seems to have survived fairly well under the leadership of Thabo M'beke. There is no doubt that the statemanship of Mandela cannot be matched. These are questions that plagued Ethiopia after the overthrow of Haile Selassie and that are bound to plague South Africa in the future. The other problem with this approach is the fact that such leaders begin to develop delusions of grandeur and a sense of invincibility that, history shows, can be very harmful.

Military "revolutions," the sixth approach, have been waged in a number of countries in an attempt to restore some semblance of order. Ironically, the presence of the military is a component of Africa's problematic political landscape. This approach involves the transformation of, at least, the top hierarchy of the military into a political governing body. Jerry Rawlings in Ghana, Ibrahim Babanguida and Sani Abacha in Nigeria, Mengistu Haile Mariam in Ethiopia, and Blaisse Campore in Burkina Faso are examples of revolutionaries and of the problems that revolts entail. Military regimes have not been able to weed out corrupt politicians and safeguard the interests of the ordinary people on the street, as they claim to do at the outset. They are usually accompanied by their own set of problems. As early as 1962, Finer argued that the military lacks administrative skills and legitimacy in governing a country. In addition, Suberu (1994) argues that the democratic recession plaguing Nigeria as of 1995 is due to the belief that the military can be used as a vehicle to control political power and amass wealth. Thus, the military in Nigeria seeks political power at all costs. It is not surprising that frustrated countries continue to seek an end to military regimes.

These six suggestions about how to solve Sub-Saharan Africa's political dilemma have met with mixed results. In the light of the dismal performance of most of these approaches, the question looming on the Sub-Saharan African political horizon is whether there is hope for countries to achieve some meaningful semblance of political stability. We turn our attention to this question in the next half of the chapter.

IS THERE HOPE FOR SUB-SAHARAN AFRICA?

Given the eight factors identified as causing political instability in Sub-Saharan Africa, it can be argued that the first line of attack for solving the political dilemma is to minimize ethnic tensions; synchronize traditional, Islamic, and Western political structures; seek and appoint committed and honest leaders; restructure the military as an institution of national development rather than of destruction; educate citizens of each country to be realistic in their expectations of the state; emphasize to the international community the need to cooperate rather than interfere in the activities of countries; and adopt policies which ensure that race is not the basis of differences in countries such as South Africa and Zimbabwe. Although this seems like a wish list of sorts, these are logical steps toward achieving some degree of political stability.

What will be needed as a catalyst, though, is a combination of a cadre of committed and honest leaders in conjunction with viable economic, social, and political development policies in each country. As Chege (1994) suggests, there is a need for African responses to African problems, ones in which honest statesmen and civic and religious leaders take it upon themselves to work as catalysts to bring about these changes. It is important to recognize in this effort that political development cannot be achieved in isolation from economic, social, and cultural development. The political landscape of the region can be changed once the roots of these problems are addressed. For Sub-Saharan Africa, this is an enormous task.

The enormity of the task, however, does not mean that there is no hope for the region. Four processes in operation in Sub-Saharan Africa point to strong options for solving political problems. The first process is the *democratization trend* that started in the early 1990s (Gros, 1998; Diamond, 1995; Chege, 1994). The second is the transition of South Africa from apartheid to majority rule since 1994 (Gevisser, 1994; Nelan, 1994). The third is the relative success stories of countries such as Benin, Uganda (Berkerley, 1994), and Botswana (Holm, 1994). A fourth factor that augurs for a silver lining in the region's political stability is its triple heritage and the potential advantages that the indigenous heritage can offer. What makes it ironic is that conflicts between the triple heritage have been identified as one of the causes of political instability.

The Democratization Process in Sub-Saharan Africa

Since 1989, most countries in Sub-Saharan Africa have embarked on a democratization trend, among them Ghana, Togo, Benin, Kenya, Ethiopia, Namibia, Zambia, Niger, Madagascar, Central African Republic, Malawi, Cape Verde, São Tomé and Príncipe, and Seychelles. Three reasons have been suggested for the democratization trend in the region. The first is the emergence of relatively enlightened leaders. The second involves societal pressures from civil societies, churches, student unions, the media, and professional bodies within African countries. The third relates to international pressures from donor agencies such as the International Monetary Fund and the World Bank (Gros, 1998). Before examining the specific effects of these factors on the democratization process, it is worth noting Diamond's (1995) observation that, in most African countries, the *second beginning* of democracy has been marred by rigged and unfair elections. This is a reflection of the fact that the new democratization in Sub-Saharan Africa is fraught with uncertainty. Such a state of affairs calls for an understanding of the nature of democratization in Sub-Saharan Africa.

Democratization, in this context, refers to "a transitional phenomena [sic] involving a gradual, mainly elite-driven transformation of the formal rules that govern a political system" (Gros, 1998). It is a process by which a country tries to become democratic. Gros divides democratization into two distinct phases. The first is political liberation, where political leaders of a country open the political system to competition. This phase involves the amendment of constitutions, to permit opposition parties to compete legally, and the establishment of multiparty elections. Most Sub-Saharan African countries are at this stage of democratization. The second phase, which is more difficult to achieve, involves creating conditions that will lead to the rule of law. This involves the separation of power between the executive, legistative and judicial branches of government; administrative accountability; the freedom of speech, assembly, and the press; and a hegemony of the civilian over the military. Almost all Sub-Saharan African countries that have made an effort to democratize have not reached this phase. Since democratization is a process, it is possible for a country to revert to undemocratic ways of governance, even though it may have achieved the first phase of the process. Democratization is therefore reversible.

Democratization in Malawi: societal pressures from the church and civil rights activists The case of Malawi illustrates the extent to which democratization is occurring in Sub-Saharan Africa and the kinds of challenges and uncertainties that are associated with the process. Mchombo's (1998) analysis of Malawi's democratization process provides an interesting backdrop for what is presented here. Dr. Hastings Kamuzu Banda became president of Malawi at independence in 1964. How did his regime survive over a 30-year period until his ouster from office in 1994? Inept leadership and societal and international pressures contributed to his downfall and the emergence of a new democracy in Malawi.

Banda's medical education in both the United States and England seemed to have prompted him to opt for a capitalist model of development for his newly independent country. However, his policies were tainted with a certain degree of authoritarianism in government (Mchombo, 1998). When nationalists such as Henry Chipembere, Kanyama Chimue, Augustine Bwaanausi, Orton Chirwa, and Willie Chokani sought more democratic participation in government, Banda cracked down and exiled them from the country. The threat from such exiles gave Banda the opportunity to use national security concerns to repress his people and their freedoms. Even student opposition to the government was clamped down. Maintaining the Banda regime in power was done in association with two cronies, Cecilia Tamanda Kadzmire (the state hostess) and John Tembo. In addition, Banda created the Malawi Young Pioneers (MYP) as a tool of intimidation. This was despite the fact that the country had a standing army. Over the course of his 30 years of rule, Malawi became not just a one-party state, but effectively a one-man state, with an ideology of *Kamuzuism*. As Mchombo (1998) argues, the brutality of the Banda regime was reflected in the political disappearance of people, economic malaise, inequality in the distribution of resources, and the severe curtailment of the press, academic community, and individual freedoms of speech and association. That is how the regime survived for so long.

Part of Banda's foreign policy focused on regimes that were under sanctions in Africa (South Africa, Zimbabwe, and Portugal), which provided him with an opportunity to get both financial and moral support from these sources. In fact, South Africa funded the movement of the capital from Zomba to Lilongwe. The move of the capital to Lilongwe demonstrates how Banda favored his Chichewa ethnic group (in the central part of the country) over other ethnic groups.

The first social group to exert pressure on Banda was religious. Even though the churches had stayed neutral about politics, in 1992 the Episcopal Conference of Malawi issued an 11-page document titled *Living Our Faith*. This letter detailed the ills of the Banda regime and the need for change. Sixteen thousand copies of the document were printed and distributed, with the aid of advances in communication technologies, to the remotest parts of the country. Malawi did not have a television station, but the availability of photocopy machines, fax machines, computers, and telephones made it possible to disseminate the first major criticism of the regime to almost all Malawians. For some reason, Banda's regime did not have a concerted reaction against the Bishops in the form of disappearances or major harassments.

Out of this criticism, and the lack of the regime's response, emerged the first opposition leader to Banda in the form of Chakufwa Chihana (of the northern Tumbuka ethnic group), an international civil rights activist. Chihana, who had been in exile in Lusaka, Zambia, referred to Banda's Malawi as "twentieth-century feudalism" (Mchombo, 1998). Chihana returned to Malawi after forming the Committee for a Democratic Alliance (ICDA). On his arrival, he was arrested by the state—an action that immediately made him a rallying point for democratic forces in Malawi. Subsequently, other activists formed another opposition group, the Alliance for Democracy (AFORD), in the north and made the jailed Chihana its chairman. The activities of AFORD encouraged the formation of a southern-based opposition group, the United Democratic Front (UDF), with Bakili Muluzi, a Muslim from the Yao ethnic group, as its leader. The emergence of these political groups and their leaders was significant, but the fact that they were affiliated with ethnic groups and specific geographic territories would have implications for future elections.

The initial agenda items of the opposition groups were to call for a multiparty democracy and later the disbanding of the MYP. Banda agreed to a referendum on the issue, which resulted in a vote for multiparty elections. Banda's interpretation of the vote was that although Malawians wanted multiparty elections they had not rejected his presidency. As fate would have it, Banda suffered a heart attack that resulted in his being flown to South Africa for treatment. The power vacuum created by his absence gave the army the chance to finally disarm and disband the MYP through Operation Bwezani. Elections were finally held on May 17, 1994, with Dr. Bakili Muluzi appointed as president and his United Democratic Front as the ruling party.

The case of Malawi demonstrates how a country arrived at the first stage of democratization (political liberation) and how the emergence of leaders and societal forces such as the church contributed toward the process. Banda was effectively forced to democratize and can be considered as an unwilling democrat (Gros, 1998). Malawi's democratization was achieved by opposing ethnic groups seeking political liberation, which may result in a different set of problems for the country in the future.

Ghana's democratization process: international influences The specific example of Malawi does not illustrate the influence of international pressure in democratization. The country that best illustrates this is Ghana. Ghana's recent democratization under Jerry John Rawlings is closely tied to pressures from international organizations. Although Rawlings claimed in 1981 to be a socialist, the reality of the declining economy he inherited made it apparent that there was no alternative to relying upon international donor agencies for economic survival (Ninsin, 1991; Panford, 1998). Rawlings first embarked upon an economic program sponsored by the International Monetary Fund and World Bank and then followed that with a structural adjustment program in the early 1980s. Ghana's chances of receiving international aid were based partly on the condition that it embark upon a process of democratization, which illustrates how pressures from the international scene compelled Rawlings to democratize. This decision was made within the context of the end of the Cold War; the West no longer felt that it had to tolerate excesses in strategic Third World allies to retain their support. Effectively, Rawlings was seen as an unwilling democrat, and, until he handed over power to John Kuffor in 2001, his commitment was viewed with suspicion. It must be stressed that the media (especially the proliferation of FM radio stations in the country) and the ubiquitous nature of cellular telephones had a lot to do with the fair elections of 2000. (See Discussion Box 5.2 for anecdotal evidence of this.)

DISCUSSION BOX 5.2: ANECDOTAL EVIDENCE OF THE ROLE OF THE MEDIA IN FREE AND FAIR ELECTIONS IN SUB-SAHARAN AFRICA

I flew into Accra (the capital of Ghana), from Amsterdam, on the evening of December 28, 2000. This was the day on which a run-off election for the presidency was being held between Mr. John Agyekum Kuffour of the New Patriotic Party (NPP) and Prof. John Atta-Mills of the National Democratic Congress (NDC). The NDC is the party of Jerry Rawlings, and there was anxiety in the general population about the possibility that the incumbent party would rig the elections. By 9:30 P.M., I was glued to both a television set and an FM radio station in Accra, listening to election results coming in from the regions. I was amazed at the speed with which both radio and television stations were reporting electoral returns. The last election I experienced in Ghana was a referendum on General I. K. Acheampong's controversial Union Government idea in the 1970s. It took almost a week to get any sense of what the national referendum results were. By 5:00 A.M. on the morning of December 29, it was clear that John Kuffour was the president elect of Ghana.

What created the difference in this election? Commentators and pundits believe the proliferation of FM radio stations and the liberalization of the airwaves in Ghana contributed to not only the speed of reporting election results but also the almost total absence of election rigging. Because FM radio stations and cellular telephones have become ubiquitous in almost all of southern Ghana, the results from most polling stations and districts were reported to radio stations and subsequently broadcast to the general public before they were even reported to the Electoral Commission. All political parties involved in the elections had representatives at polling stations. Once votes were counted, these representatives pulled out their cellular phones and called in the results to party officials in Accra, who then relayed the information to the FM and television stations. For a country that had historically been plagued by election rigging, the free press, the availability of cellular phones, and the ubiquitous nature of privately owned FM stations minimized any potential for election tampering.

The morale of this anecdotal evidence is that a free press (societal pressures) and the availability of technology promote fair elections and facilitate the democratization process. The irony of Ghana's presidential elections is that the one individual who fought for the liberalization of the airwaves in the late 1980s (Dr. Charles Wireko Brobbey) fared poorly in the presidential primaries and did not make it to the second round. Dr. Brobbey prepared the ground for Mr. Kuffour to win the presidency fairly. This is democracy at work!

The example and potential role of South Africa The remarkable demise of apartheid and the establishment of majority rule in South Africa since 1994 is another beacon of hope for countries in Sub-Saharan Africa. As Chege (1994) points out, Nelson Mandela's noble gesture of extending an olive branch to white South Africa after 27 years of incarceration is one of the positive indicators of political stability in Sub-Saharan Africa. Browne (1994) suggests that, with these changes, South Africa has become a powerful new player in the economic development and political stability of the region. It is true that South Africa's transition is plagued with problems, such as land redistribution and the speed at which the economy can be transformed (Gevisser, 1994; Nelan, 1994). This, however, does not diminish its potential in bringing about stability in Sub-Saharan Africa, especially in southern Africa.

For years, this multinational state of South Africa was torn apart by the inhuman policy of apartheid, a raison d'être designed to keep races apart through a set of rigid, hierarchical laws and by intimidation. Official apartheid began in 1948, but elements of racism existed in South Africa before then. A series of acts was instituted to discourage races from utilizing similar amenities (the Separate Amenities Act of 1953), to prevent mixed marriages and social contact between races (the Mixed Marriages Act and Immorality Act), and to segregate residential areas on the basis of race (the Group Areas Act of 1950). Superimposed on these was the policy of grand apartheid, whose ultimate purpose was to create independent black "homelands," based on ethnicity. Examples of these homelands were Bophutatswana, Venda, Ciskei, and Transkei.

These homelands, or *bantustans*, never stood a chance at gaining some semblance of statehood, because they lacked the economic and human resources necessary to become self-sufficient. About 75% of Africans were relegated to tiny, inhospitable living environments that accounted for only 13% of the total land area of South Africa (Smith, 1990). Also, these homelands had virtually no significant mineral reserves and lacked the potential for cultivation. The boundaries of these fragmented and landlocked territories were also drawn in such a way that they conveniently avoided the major core regions in South Africa, such as Witwatersrand, Durban, Port Elizabeth, and Cape Town. Consequently, there were severe backwash effects created, as able-bodied males frequently left the marginal homelands to work in the core regions (as migrant laborers), leaving the homelands with meager revenues to provide adequate educational and health facilities and an appropriate infrastructure. Given this scenario, the "homelands" were not states in the true sense of the word; they lacked the resource base to become economically and politically sovereign and self-sufficient. Furthermore, the international community recognized none of the homelands.

The African National Congress (ANC), Pan-African Congress, and other nationalist groups were at the forefront of the internal fight against apartheid. Even though its leaders, such as Nelson Mandela, were imprisoned in Robin Island for over 27 years, internal agitation through revolts, like the Sharpville Massacre and the Soweto Uprising of the 1960s and 1970s, are the best known internal challenges to the system of apartheid. International pressures on transnational corporations and the imposition of sanctions against the country also worked to dismantle apartheid. Apartheid effectively ended with the first free elections held for South Africans, in April of 1994.

South Africa is a major regional power and also dominates the rest of Sub-Saharan Africa in terms of industrial output, mining production, steel production, installed electrical capacity, telecommunications, and biotechnology. It has the densest road, rail, and air networks in Sub-Saharan Africa. Its rail and harbor network is the only reliable trade link with the outside world for land-locked countries such as Botswana, Lesotho, Swaziland, Zimbabwe, Zambia, Malawi, and even the Democratic Republic of Congo, and each year it draws millions of migrant workers from these countries. Its relative economic and industrial development could set an example for other Sub-Sahran African countries to regularize their economies.

South Africa has not just had its first free elections, its presidency has also successfully changed hands from Nelson Mandela to Thambo M'beke. This is a sure sign of political stability! South Africa is now at the center stage as political geographers grapple with the question, What next? A new Bill of Rights ensures equality regardless of race, gender, sexual orientation, physical disability, or age; free speech, movement, and residence; freedom of religion, politics, and conscience; fair trial, with no torture or forced labor; freedom to use any of 11 languages; economic, social, and cultural rights; the right to work; and the right to an education. The burning questions that remain are, Can the government develop a strong sense of nationalism and build on diversity? and How will the new government deal with the land question and the Truth and Reconciliation Commission? These are certainly issues that will take some time to resolve.

Some elements of apartheid still remain intact as blacks, whites, coloreds, and Asians adjust to the new social, economic, and political order. With respect to the contentious land question, the Restitution of Land Rights Act was enacted in November, 1994 to investigate land claims and restore ownership to those who lost land unjustly. Perhaps the major issues that South Africa has to address relate to unemployment and its associated urban crime and to the issue of HIV/AIDS and its potential to wipe out the most active populations of the country. The unequal development between blacks and whites, especially in terms of housing, access to health, and access to education, still presents problems. If South Africa can deal with these problems, it has the potential of being the beacon of hope and leadership in Sub-Saharan Africa.

Political stability of Botswana: indigenous and modern values in government The political history of Botswana provides another avenue of hope for political stability in Sub-Saharan Africa; it serves as a model for economic growth and the emergence of democracy in Sub-Saharan Africa. Holm (1994) argues that, even though Botswana is by no means a complete liberal democracy, the incorporation of traditional elements of political organization into the inherited Western system has provided a climate of relative stability since independence in 1966. For example, a House of Chiefs has an advisory role in matters of national interest. Because of this situation, Chiefs cannot stand for elections unless they resign their positions. In addition, "freedom squares" that offer citizens, especially illiterates, an opportunity to listen to, and question, politicians on local and national issues are well entrenched in urban and rural areas. The Western model of parliamentary representation has been refashioned to meet the specific needs of Botswana. Thirty-four members of parliament are elected to the National

Assembly, and they are responsible for selecting the president. The president, though, has more power than the assembly. He appoints his cabinet, and it is at his discretion to consult with the National Assembly on such matters. Over all, political rights and liberties of the population, as well as freedom of the press, are guaranteed. This is not to say that Botswana has no political problems. The ruling Botswana Democratic Party (BDP) has limited the actions of unions and has been the only party in power since independence, even though its leadership has changed. Yet, Botswana's system is an intermediate step for countries seeking political stability.

A factor that has often been neglected, yet has great potential, in establishing more enduring political structures is the attributes and cultural traits of the indigenous heritage. For example, the system of chieftaincy amongst the Akan and other ethnic groups of Ghana has survived despite Western influence. In addition, the notion of the extended family with its attendant system of kinship, lineage, and age cohorts demonstrates that the family is alive and well in Sub-Saharan Africa (Mazrui, 1986). The question is whether these traditional structures of political and social organization can be incorporated into the design of present-day political structures to make them more durable (Ayittey, 1991). As already pointed out in the case of Botswana, the House of Chiefs serves as an advisory committee to the president and National Assembly. In addition, the use of "freedom squares" has ensured popular participation. This is similar to the participation of citizens at both the national and local levels under Tanzania's Ujamaa system. It is high time that each country looked within itself to realize that it has an indigenous heritage that it can use to its benefit. To ensure political stability, therefore, the indigenous heritage that has served Sub-Saharan Africans for so long should not be ignored. Discussion Box 5.3 further illustrates the notion of adhering to employing the traditional system of indigenous goverment for political stability.

DISCUSSION BOX 5.3: THE POTENTIAL OF TRADITIONAL SYSTEMS OF GOVERNMENT FOR POLITICAL STABILITY

Prior to European contact with Sub-Saharan Africa, two types of political societies existed: state and stateless societies. State societies, such as the Ashanti and other Akan groups of Ghana, were characterized by elaborate political structures that included a hierarchy of chiefs. Stateless societies, such as the Masaai of Kenya and Tanzania, the Boran of Kenya, and the Fulani of West Africa, did not have chieftaincy systems, but instead were characterized by systems of elders and religious leaders. The political structures of both state and stateless societies ensured the smooth running of economic, social, religious, and cultural life. They helped preserve the internal coherence of societies.

The establishment of the Ashanti kingdom was primarily by the formation of a confederacy of people, but it had strong religious underpinnings.

A high priest, Okomfo Anokye, provided the confederacy with two symbols of strength and unity: the Golden Stool and the Seat of Government. The Golden Stool symbolizes the external link between the Ashanti and their ancestors—it is believed that the stool was summoned from the heavens. It represents a powerful focus at religious meetings that sanction the authority of the Ashanti king (Asantehene). It therefore represents the strength of the kingdom and provides a strong justification for the monarchy. Kumasi was selected as, and remains, the spiritual and political capital of the Ashanti. From this strategic location, the Ashanti could control their empire and trade routes to preserve their economic prosperity.

The Ashanti kingdom is still administered through a hierarchy of chiefs, with the *Asantehene* at the top. Beneath the Asantehene are the Paramount chiefs (*Omanhene*), located in towns such as Mampon (*Mamponhene*) and Effiduase (*Effiduasehene*). The next levels below are the *Ohene*, followed by the *Odikro* (subchiefs of villages). Each chief is given responsibilities, ranging from war to managing the treasury. This hierarchical arrangement is tied to Akan social organization. Ashantis believe that the ancestors of each town or community constitute seven clans. Irrespective of the level of hierarchy, Ashanti chiefs are chosen from the Oyoko clan. The queen mother (*Ohemaa*) initiates the choice of chief, and the council of elders (*Mpeninfo*), who represent the seven clans of the town, as well as the members of the community, have to approve the Ohemaa's choice before the new chief is enstooled. The new chief swears allegiance to the successive chiefs in the hierarchy. The day-to-day responsibility of the chiefs is to serve as a custodian of land, as a spiritual mediator between the Ashanti and their ancestors, and as an enforcer of law and order. The chief rules with the assistance of his Mpeninfo and the Ohemaa; therefore, his power is not absolute.

In spite of attempts by the British to subdue the Ashanti during the colonial era, this political structure ensured the survival of the kingdom, and it still thrives today. To what extent, therefore, can indigenous structures like this provide lessons for Sub-Saharan countries? Even though Botswana demonstrates that such indigenous structures can be used in establishing stable political regimes, in most countries their contributions are ignored. Western structures of governance have not fared well in Sub-Saharan Africa, yet indigenous structures such as that of the Ashanti are alive and well. Certainly, it is worth investigating the potential contributions of indigenous institutions; however, this must be done with caution, to ensure that modern government does not become tribalized.

Like other African countries, the democratization process in Malawi, Ghana, South Africa, and Botswana reveals that they have not passed beyond the political liberation stage and into the second stage of creating conditions for the establishment of the rule of law. No country in Sub-Saharan Africa (even South Africa) has reached this second stage of the democratization process. The emerging democratization trend in the region is fraught with uncertainty and could easily be reversed. However, the increasing accountability of governments in Benin, the relatively independent press in countries such as Kenya and Nigeria, the widening horizons of freedom of speech, and the competitive multiparty elections in Zambia, Cape Verde, Niger, Mali, Ethiopia, and Central African Republic suggest that there is something right with the region after all (Chege, 1994). It must be emphasized that, despite these positive developments, the disappointments in countries like Sudan, Nigeria under Abacha, the Democratic Republic of Congo, and Gambia are not encouraging. These examples call for the development of institutional capacity and the appointment of highly principled leaders in all countries to ensure true democracy.

CONCLUSION

It has been about 40 years since most countries in Sub-Saharan Africa attained independence from former colonialists. At independence, expectations were high for economic, social, and political advancement. Yet, until the early 1990s, the political landscape of the region was characterized by instability and chaos. A question that emerges from this situation is whether Sub-Saharan Africans are capable of governing themselves. A consideration of the causes of this unstable political landscape reveals that conflicts within the triple heritage of the region have been a major contributor. In addition, foreign intervention and the colonial legacy have had a detrimental effect. This, however, should not excuse the excesses of greed, corruption, and incompetence associated with the region's leadership. Political instability in Sub-Saharan Africa should not be attributed to colonial influences only, but also to Africans themselves and their actions.

There are promising trends emerging in the region that auger well for political stability. First, a wave of democratization has been occurring in some countries since the early 1990s. This democratization is a reversible process that is only in its formative stages. Initial impacts of such a trend provide a glimmer of hope for countries. Second, the emergence of South Africa as a beacon of both economic and political hope for African countries to emulate should not be downplayed. Third, the relative success of countries such as Botswana in devising political systems that incorporate both indigenous and modern elements of government suggests that there is hope for Africa.

Sub-Saharan Africans have to solve their problems by assuming the responsibility for electing, appointing, and supporting competent, committed, and honest leaders. These leaders must put their national interests ahead of their personal ones. In addition, indigenous elements of the triple heritage should be stressed in finding African solutions to African problems. This should be done within the framework of recognizing that political development cannot be separated from cultural, economic, and social development. The task of solving Africa's political problems are daunting, but examples from Botswana, Benin, Ghana, Malawi, and Ethiopia (in the last 10 years) and recent developments in Nigeria (under the new Obasanjo regime), Sierra Leone, and Liberia with respect to either a democratic transition or a cessation of civil wars indicate that the political climate in the region is not entirely hopeless.

KEY TERMS

One-party states
Nation–states
Coups d'état
African socialism
African capitalism
Apartheid

Secessionism
Osagyefo
Mwalimu
Devolution
Chieftaincy
Democratization

DISCUSSION QUESTIONS

1. Political instability in Sub-Saharan Africa prior to the early 1990s was caused solely by external influences. Discuss this statement.
2. Compare and contrast Western political structures with those in Sub-Saharan Africa.
3. Looking at the genocide in Rwanda, a conclusion that can be drawn is that Africans have not changed from the barbaric savages that Europeans portrayed them to be in the late 18th century. To what extent do you agree or disagree with this statement?
4. To what extent does the indigenous heritage of Sub-Saharan Africa contribute to both the cause and solution of problems of political instability?
5. What, in your opinion, are some viable solutions to political stability in Sub-Saharan Africa?
6. Is the Botswana model of political and economic development a suitable model for other Sub-Saharan countries?
7. Looking at the examples of countries such as Malawi and Ghana, it can be deduced that a democratization trend in Sub-Saharan Africa is occurring. To what extent is this statement a reflection of reality of the region?

REFERENCES

Aronson, D. (1993). "Why Africa Stays Poor and Why It Doesn't Have To," *The Humanist*, March–April.

Aronson, D. (1998). "Mobutu Redux?" *Dissent* (Spring): 20–24.

Ake, C. (1973). "Explaining Political Instability in New States," *The Journal of Modern African Studies*, 11(3):347–360.

Ayittey, G.N.N. (1991). *Indigenous African Institutions*, New York: Transnational Publishers.

Berkerley, B. (1994). "An African Success Story?" *The Atlantic Monthly*, 274(3):20–30.

Browne, R.S. (1994). "How Can Africa Prosper?" *World Policy Journal*, 11(3):29–39.

Chege, M. (1994). "What Is Right with Africa?" *Current History*, 93(583):193–197.

Dear, M. (1988). "The Postmodern Challenge: Reconstructing Human Geography," *Transactions of the Institute of British Geographers*, No. 13.

DIAMOND, L. (1995). "The New Wind," *Africa Report*, 39(5):50–54.
GEVISSER, M. (1994). "South Africa in Transition: Democracy in Living Color," *The Nation*, December 26.
GIBBS, N. (1994). "Why? The Killing Fields of Rwanda," *Time Magazine*, May 16.
GRIFFITHS, I.L.L. (1994). *The Atlas of African Affairs*, London: Routledge.
GROS, J. (1998). "Introduction: Understanding Democracy," *Democratization in Late Twentieth-Century Africa. Coping with Uncertainty*, in Gros, J.-G., ed., Westport, CT: Greenwood.
HAMRICK, S.J. (1993). "How Somalia Was Left in the Cold," *The New York Times*, February 6.
HAYWARD, F.M. (1986). "In Search of Stability: Independence and Experimentation," in *The Africans: A Reader*, Mazrui A.A., & Levine, T.K., eds., New York: Praeger.
HOLM, J.D. (1994). "Botswana: One African Success Story," *Current History*, 93(583):198–202.
HUMAN RIGHTS WATCH (2001). "Congo: The Kabila Legacy," *Human Right News*, January 19. (http://www.hrw.org/press/2001/01/kabila0119.htm).
KELLY, S. (1993). *America's Tyrant: The CIA and Mobutu of Zaire*, Washington, DC: The American University Press.
KHAPOYA, V.B. (1994). *The African Experience: An Introduction*, Englewood Cliffs, NJ: Prentice Hall.
LAMB, D. (1982). *The Africans*, New York: Random House.
MAZRUI, A.A. (1986). *The Africans: A Triple Heritage,* Boston: Little, Brown and Company.
MCHOMBO, S.A. (1998). "Democratization in Malawi: Its Roots and Prospects," in *Democratization in Late Twentieth-Century Africa. Coping with Uncertainty*, Gros, J.-G., ed., Westport CT: Greenwood.
NELAN, B.W. (1994). "Making Their Own Miracles," *Time*, 144(25) (December 19):52–53.
NINSIN, K.A. (1989). "Introduction: Thirty-Seven Years of Development Experience," in *The State, Development and Politics in Ghana*, Hansen E. & Ninsin, K., eds., London: Codesria.
NINSIN, K.A. (1991). "The PNDC and the Problem of Legitimacy," in *Ghana, The Political Economy of Recovery*, Rothchild, D., (ed.), Boulder, CO: Lynne Rienner.
NYROP, R., *et al.* (1985). *Rwanda: A Country Survey*, Washington, DC: Foreign Areas Studies.
PANFORD, K. (1998). "Elections and Democratic Transition in Ghana 1991–96," in *Democratization in Late Twentieth-Century Africa. Coping with Uncertainty*, Gros, J.-G., ed., Westport, CT: Greenwood.
PRESS, R.M. (1995). "Forgotten Zaire Struggles to Break Free of Past Shackles," *The Christian Science Monitor*, January 19.
RAMSAY, F.J. (1995). *Global Studies: Africa*, Guilford, CT: Dushkin.
RICHBURG, K.B. (1994). "A Microcosm of Disintegration," *Washington Post National Weekly Edition*, September 12–18.
SCHATZBERG, M.G. (1997). "Beyond Mobutu: Kabila and the Congo," *Journal of Democracy*, 8(4):70–84.
SENDARO, A. (1987). "Reflections on Implementation of Ujamaa Vijini Policy in Tanzania 1967–1986," *African Review*, 14, (2).
SHINER, C. (1994). "Mobutu Ascendant," *Africa Report*, 39(3):42–46.
SMITH, D. (1990). *Apartheid in South Africa*, 3d ed., Cambridge, UK: Cambridge University Press.
SUBERU, R.T. (1994). "The Democratic Recession in Nigeria," *Current History*, 93(583):213–218.
——— (1994). "Who Owns South Africa?" *The Economist*, 331(7865):35.

6

Culture, Conflict, and Change in Sub-Saharan Africa
Joseph Ransford Oppong

INTRODUCTION

It is impossible to understand the current cultural malaise and the multiplicity of ethnic conflicts in Sub-Saharan Africa without an appreciation of two fundamental contributory factors: the rich tapestry of cultural values and their expressions in precolonial Africa, and the colossal head-on collision with European culture during and after the colonial era. Almost every major problem of Sub-Saharan Africa—clan warfare in Somalia, ethnic conflict in Rwanda and Liberia, the land crisis, family breakdown, and rampant bribery and corruption—can be explained, at least in part, by these two factors. The cultural geography of Sub-Saharan Africa is, thus, fundamental to understanding the geography of the region—its political situation, its medical geography, its population dilemma, and the current development crisis.

In this chapter, we will explore some major tenets of precolonial Africa's varied cultural geography, its conflict with Western culture and the resulting changes, and how all these contribute to the current African situation. We will begin by examining the fundamentals of cultural geography and the process of culture change. We then examine selected elements of the rich tapestry of African culture: religion, family, and kin relationships; language and society; land tenure arrangements; and cultural expressions and symbolism. Afterward, we will examine the diffusion of European culture and its impact on African culture.

CULTURAL GEOGRAPHY

Cultural geographers study the culturally endowed elements of physical and human landscapes and the spatial variations resulting from human behavior and activity. Studying the culture of a group of people involves evaluating their way of life—how they live; what clothes they wear; what foods they eat; their customary habits, belief systems, speech patterns, and value systems. It also involves learning how these cultural traits manifest themselves in the physical landscape. For example, the arrangement of homes in traditional family compounds is dictated by traits that are inherent in patrilineal or matrilineal societies. Also, the spatial patterns of crops produced cannot be explained adequately by climate and soil characteristics alone. Food preferences, taboos (frequently based on religious belief), technology, and farming practices influence the choice of crops. Of course, farming practices are influenced by the system of land ownership and institutionally accepted arrangements regarding land acquisition and use—land tenure. Such imprints on the *cultural landscape* mirror the cultures that created them and involve the cumulative influences of successive occupants. Jackson (1984) sees the cultural landscape as a composition of human-modified spaces that serve as infrastructure or background for our collective existence. Thus, the cultural landscape includes sounds, smells, and all the different elements—the visual reflection of the culture—that give any place its unique flavor. It also reflects different cultural attitudes and tastes, for example, toward littering and human waste.

Cultural geographers also incorporate the theme of *cultural integration* into their analyses, recognizing that cultures are complex wholes (Jordan and Rowntree, 1982). As a result, they try to identify the functional relationships that exist between cultural traits—in other words, how the different parts fit together. For example, communal land ownership patterns in Sub-Saharan Africa are functionally related to the belief in ancestral spirits as active participants in society, worthy of reverence and worship because of their ability to rain curses or blessings on the living. Within the family or kin group, cooperation is the desirable value rather than competition. Family members pool their resources and help each other while competing against outsiders and nonfamily members. Abuse of the land by outright sale, for example, could provoke serious punishment, such as sickness or even death, from the ancestors. Sicknesses caused by angry ancestral spirits can be cured only after appropriate remedial actions, including pacifying sacrifices. Thus, health-seeking behavior is also related to the land-tenure system and religious beliefs.

Cultures can also be studied at different spatial scales. Any unit of geographic space occupied by people with one or more common culture traits is a *culture region* or *culture area* (Jordan and Rowntree, 1982). For example, we can distinguish the matrilineal inheritance culture region in Democratic Republic of Congo, Zambia, or Mozambique. More commonly, culture regions are defined by several related traits, such as religion, language, and social organization.

In effect, cultures do not develop or exist in isolation; they interact with the physical environment they occupy. *Cultural ecologists* study these interrelationships between cultures and their environments. Building styles and technology can be influenced as much by cultural perceptions of the appropriate way to build as they are by the materials available for building. Moreover, cultures differ in their recognition and use of materials occurring naturally in the physical environment. Naturally occurring substances perceived to be directly or indirectly beneficial to human beings are called *resources*. Because natural resources are culturally appraised, what is a resource to one culture might not necessarily be a resource to another culture. A goat's blood or locusts could be a delicacy in parts of Africa, but a nuisance in other parts, where cats might be the delicacy. In essence, traditional food habits reflect the surroundings of the physical environment.

The relationship between cultures and the physical environment they occupy has been the focus of intense study and debate. Some geographers hold that the physical environment, specifically climate and terrain, are the critical determinants of cultural development. This unidimensional argument assumes that human culture is the product of the physical environment and that similar environmental circumstances will likely produce similar cultures. This viewpoint, *environmental determinism*, has been used to explain underdevelopment in Africa. Nature's abundance in year-round warm climates, including abundant wildlife and naturally occurring foods such as berries, are seen as inhibiting technological advancement. On the other hand, extreme cold weather forced Canadians and other residents of high-latitude countries to develop not just a means of keeping warm, but also a technology of housing, agriculture, and food storage that is missing in Africa. Obviously, environmental determinism overemphasizes the role of the environment in cultural development. The environment is doubtless an important factor in cultural development, but it is seldom the sole determinant of human, political, or economic organization.

In contrast to environmental determinism, *possibilism* is the contention that humans, rather than their environments, are the primary determinants of cultural development. The physical environment offers opportunities for humans to make choices based on their perceived needs. Thus, cultural traits are simply the result of human decisions made within the range of possibilities provided by the physical environment. Again, this viewpoint could be a bit extreme. Technological advancement might moderate, but can never eliminate, the influence of the physical environment. What we considered technologically advanced solutions to human problems some time ago, such as DDT for eliminating pests, have come back to haunt us as environmental disasters. Consequently, the cultural ecological viewpoint, where a culture influences and is influenced by its physical surroundings, is more plausible.

Culture Change

Culture is dynamic and is subject to continual change over time. Culture changes can result from internal changes, such as improvements in technology, or from interaction with other cultures. The extent to which one culture is influenced by another can depend on the intensity and magnitude of the *cultural diffusion* process. This could occur as one group is acculturated or assimilated into another culture by adopting its language, or it could occur through the adoption of a new technology, idea, or innovation. Chapter 4 illustrates the dilemma that Sub-Saharan African countries face in dealing with the triple heritage of indigenous value systems and Islamic and European values.

Throughout history, ideas, civilizations, and inventions have diffused from major *culture hearths*, which are areas where distinctive cultures originate (Bergman and McKnight, 1993).

Cultural Geography

Precolonial African history demonstrates how technologies and innovations from the ancient civilizations of Kush, Nubia, and Axum diffused to the medieval kingdoms of the savanna states. Both Islam and Christianity spread in Sub-Saharan Africa through the process of cultural diffusion.

Diffusion can occur through a relocation and expansion process. *Relocation diffusion* involves the physical movement of people from one place to another. The Bantu language, which is a subfamily of the Niger–Congo language family, is currently confined to the eastern and southern parts of Sub-Saharan Africa. However, some historical and linguistic evidence traces it to the west African forest and savanna regions. *Expansion diffusion* is intensified at the source region, or cultural hearth, and has a snowball effect as it spreads from this region (Renwick and Rubenstein, 1995). Islam, for example, expanded from the Arabian peninsular hearth across the northern sections of Africa.

Two forms of expansion diffusion are *contagious* and *hierarchical* diffusion. When an idea or innovation spreads through direct contact, as is the case with the infectious disease AIDS, it is called *contagious diffusion*. The spread of a language, for example, is usually facilitated through direct contact. Usually, acceptance of an innovation decreases with time and distance. This *time–distance decay effect* occurs when areas closest to a cultural hearth accept the innovation earlier and more thoroughly than areas that are farther away in space and time. The distribution of the Islamic religion in Africa demonstrates this effect: its influence declines with increasing distance from the Middle East culture hearth and from northern Africa. Technological advances in the media have significantly diminished the influence of time–distance decay. Sometimes, however, proximity is less important than the level of interaction between major centers or people. For example, new fashion spreads internationally between major cities first, then later filters down the hierarchy or *cascades* to smaller centers and rural towns.

Obstacles to Cultural Diffusion

The cultural-diffusion process can be retarded or halted by *diffusion barriers*. In this day and age of telecommunications and time–space convergence, there are hardly any barriers that completely halt the diffusion process, as cultures have an opportunity to interact through news media and educational and cultural exchange programs. Even physical landscapes like the Sahara Desert slowed down, but did not prevent, the interaction between Arabian and African scholars during medieval times. Yet, there are still cultures, such as the Khoisans in southwest Africa and the pygmies of the rain forest, who remain isolated and zealously protect their cultural heritage.

Certainly, a number of political ideologies have hindered cultural advancement. We have all witnessed the detrimental effect that the South African policy of apartheid had on the economic and cultural advancement of blacks. Genocide in Rwanda has also had a debilitating effect on Hutus and Tutsis alike.

In most instances, we do not witness a complete assimilation or acculturation of a culture. What really occurs is a situation where cultures modify and blend certain aspects of one culture with their own to produce something different. For example, Pidgin English

Figure 6.1 Blending Traditional African and Western Music.

in Africa is a blend of English with the local language. Also, African music and dance forms have been modified by mixing with Western music and instruments like the guitar to produce a new form that is unique (Figure 6.1).

ELEMENTS OF AFRICAN CULTURE

Widespread diffusion of European culture makes it extremely difficult to identify an authentic African culture. Besides, although Africans share a remarkable cultural unity, they also have significant ethnic, linguistic, religious, and other differences. Shared characteristics, experiences, and worldviews give Africans a common identity that transcends differences arising from the great diversity of languages, ethnic identities, and religions. For example, Africans see the relation between human beings and nature in a way very different from the way that Western societies do. Christian theology sees God as separate from nature and humans as outside and above nature. Nature exists for the well-being of humans. In contrast, African traditional religions emphasize that spiritual forces are manifest everywhere in the environment. Gods and spirits are both associated with the major elements of the physical environment, including rivers, mountains, and such landmarks as rock outcrops and tree groves. This is different from *pantheism*, however. A human being does not consider himself or herself a god, as some Eastern religions, and lately the New Age religious movement espoused quite powerfully by Shirley Maclaine, claim. Rather, gods and spirits are concerned about and are actively involved in the daily lives of humans (Figure 6.2), bringing blessings

Elements of African Culture

Figure 6.2 Religious Influence in the Daily Lives of Africans. Religion influences every aspect of African life. Names of businesses and even passenger vehicles reflect religious beliefs and values, as illustrated by the "Holy Bar."

when pleased and curses when angered. Drought, crop failure, and personal accidents, for example, may be attributed to the wrath of gods or ancestral spirits denied due respect. Let us approach the geography of religion a little bit more formally.

Religion in Sub-Saharan Africa

Christian churches, Islamic mosques, and many small shrines for traditional religions constitute the religious landscape of Sub-Saharan Africa. Christianity is the oldest of the universalizing religions in Sub-Saharan Africa, having been in Ethiopia since the fourth century, well before the arrival of Islam (Parrinder, 1983). Since then, Christianity in the form of the Coptic Church has flourished in Ethiopia and actually served as a diffusion barrier, preventing the southward diffusion of Islam. Today, the country is an island of Christianity in a sea of Muslims. Nearly every village has a church, and there are many fine ancient churches and monasteries.

Christianity With colonialism came a more widespread extension of Christianity, particularly in tropical and southern Africa and in most ports and coastal towns (Figure 6.3). There is a strong Roman Catholic presence in the former Belgian territories, namely, Rwanda, Burundi, and Democratic Republic of Congo; a strong Anglican presence in former British colonies; and a prevalence of Presbyterians in Ghana, Malawi, Kenya, and Gabon (Murray, 1981). Like the pattern in Europe, Christianity in Africa comprises different denominations: Catholics;

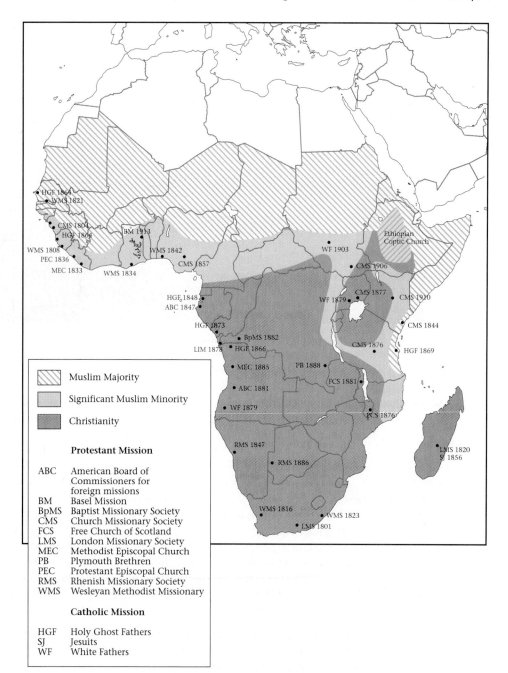

Figure 6.3 Religious Map of Sub-Saharan Africa.

Elements of African Culture

Figure 6.4 Religious Landscape: Presbyterian Church.

Protestant denominations, such as Methodist, Presbyterian, and Seventh-Day Adventist; Pentecostal denominations, such as Assemblies of God, Church of Pentecost; and, lately, independent churches. Church buildings generally follow European models in concrete Gothic and village chapel styles, but modern architectural styles are also common (Figure 6.4).

Forms of worship vary according to denomination, with one major difference—significant Africanization. For example, African carvings have been substituted for church decorations and drums for church bells in Catholic Masses, particularly in Uganda and Cameroon, and Pentecostal churches now feature a range of African music, including drumming and dancing. Worship in these churches is vibrant and lively, with a lot of dancing, reminiscent of traditional celebrations. Many churches are organized into dioceses and parishes.

Independent churches tend to be syncretistic. Some permit polygamy, most use the Bible and some Christian doctrines, and many are usually founded around a prophetic figure, such as Shembe of Natal, Kimbangu of Congo, and Harris of Côte d'Ivoire (Parrinder, 1983). Many stress faith-healing and baptism by immersion. On Sundays, processions of church groups accompanied by drumming and dancing can be seen in most African towns and villages.

Islam Islam dominates in northern Africa, but, in many large towns and cities in Sub-Saharan Africa, evidence of Islam—mosques and minarets—may also be found (Figure 6.5). From its hearth in Saudi Arabia, Islam diffused overland into Africa and, through conquest, spread through the savanna into the coastal regions, especially in Sierra Leone and parts of western Nigeria. Northeastern Africa is predominantly Islamic, except for Ethiopia. Somalia is almost wholly Islamic.

Figure 6.5 The Influence of Islam on the Religious Landscape.

Islam in Sub-Saharan Africa is predominantly Sunni. The Koran, the infallible word of God "recited" to Mohammed, is the final authority and standard for life and truth. All Muslims learn some verses in Arabic, and, throughout Africa, Islamic children can be heard in schools and mosques reciting Arabic verses from the Koran. In addition to repeated sayings of the basic creed, praying five times daily facing Mecca, fasting for one month each year, and giving alms, Muslims are encouraged to visit Mecca at least once during their lifetime. Those who have been to Mecca assume the title of hajji, and are usually called *alhaji*.

Traditional religions Deities and ancestral spirits are honored in sacrifices and ceremonies, usually undertaken at specific times of the year and conducted by priests devoted to their worship or other qualified persons, such as clan heads and group leaders. The power of gods and spirits is also reflected in the daily behavior of individuals—for example, in the avoidance of places, such as sacred groves, believed to harbor spirits. Because sickness might be caused by the gods, health-seeking behavior includes consultation with the traditional healer and offering of appropriate sacrifices.

In some places, stools are regarded as shrines for departed souls, and offerings are made at regular intervals. In addition, small temples for ceremonies in honor of *nature gods* dwelling in rivers and hills abound in west Africa, but less so in east Africa. Wooden carvings found in temples and household shrines usually indicate the designated god or his attendants. The Yoruba traditional religion in Nigeria has a four-tiered structure of spiritual or quasi-spiritual beings. At the top of the hierarchy is the Supreme Being *Olodumare*, and his subordinate ministers *Orisha* constitute the second tier. The third tier consists of deified ancestors, *Shango*, and this is followed by spirits associated with natural phenomena, such as rivers, lakes, mountains, and trees (Murray, 1981).

Divination and fortune-telling are popular activities in traditional religion. The Ifa oracle system of Nigeria is especially renowned (Parrinder, 1983). A primary activity of traditional priests is exposing those involved in witchcraft activity, protecting people from witchcraft curses, and curing those who are bewitched. Witches, supposedly, leave their bodies asleep at home at night to inflict harm on the souls of victims, harm leading to sickness and, eventually, death. Consequently, a primary component of health-seeking behavior includes the neutralizing of the effect of witchcraft activity and the invoking of spirit beings to defend the victim against malevolent witches.

Of course, all this was before the diffusion and widespread acceptance of Christianity and Islam, which are more prevalent amongst the urban and more educated populations in Sub-Saharan Africa. Countries caught in the transition zone between Islam in the north and Christianity in the south, such as Nigeria and Sudan, are rife with religious conflict. In Nigeria, sporadic outbreaks of violence between Islam and Christianity are quite widespread. In the Islam-dominated north, people are calling for Nigeria to be declared an Islamic State; in its predominantly Christian and animist south, some are calling for secession. Nigeria's future political stability depends on the peaceful resolution of this conflict.

Traditional religious beliefs have undergone considerable changes to become syncretistic, mainly through the influence of non-African religions, such as Christianity and Islam. Perhaps the best expression of blending the old religion with the new is the spiritist church, which combines elements of traditional ancestral worship with elements of Christianity and Islam. Angels and saints, like the ancestral spirits, are simply mediators between humans and the supreme, omnipotent deity, God. Traditional religious practices are widespread, with many people practicing traditional religion, while remaining active church members or Muslims.

Family and Kinship Relations

In its simplest form, the traditional African family consists of a husband and wife (or wives) and their children. They form an economic unit, because they cooperate in the maintenance of their common household. Frequently, a family consists of several households, including the head of family, his wife or wives, their unmarried children, and the husbands and children of their married daughters (Moore and Dunbar, 1969). All members of this family have obligations and ties to the other members. Moreover, the family includes both the living and the dead. *Kinship* comprises family ties or relations.

In a *bilateral kinship system*, descent and family ties are traced through both the mother's and the father's side. In contrast, kinship is traced through either the father's or the mother's line of descent exclusively in a *unilineal descent system*. North Americans tend to have a bilateral kinship system; Africans usually follow a unilineal system of descent. When kinship and descent are traced through the line of the male parent, it is called *patriliny* or a *patrilineal system*. In a patrilineal system, lines of descent and authority are linked to the father or husband, and the wife is gradually incorporated into her husband's descent group. Examples include pastoral societies in the savannas of west and east Africa, including the

Fulani, the Nuer in southern Sudan, the Masai (west Kenya and Tanzania), the Tiv (east central Nigeria), the Kikuyu (central Kenya), the Yoruba (southwest Nigeria), and the Ganda of Uganda.

Descent through the mother is called *matriliny*, or matrilineage. Links to the father's family are secondary when it comes to the inheritance of property. Resources are inherited from the mother's brother. Thus, the mother's brother, and not the father, is the matrilineal authority. This system has left some wives and children little or no property after the death of the husband and father. Countries like Kenya, Zambia, and Ghana have considered reforming inheritance rules to ensure that wives and children have access to land and other family property upon the father's death, although these "civil" laws often conflict with customary law. Examples of matrilineal societies include the Akan of central and southern Ghana and the Lamba, Bemba, and Tonga of Zambia.

Cultural groups that practice patriliny also practice *patrilocality*. After marriage, the wife leaves her home and family to live with or near her husband's family. Similarly, matrilineal groups practice *matrilocality*. The groom leaves his family to live with or near the wife's matrikin. After several years of staying with the wife's family and proving his ability to provide and care for the wife and children, the husband may request permission to move to his mother's brother's village to establish residence. This creates an *avuncular* residence, and the practice is called *avunculocality*.

In traditional African societies, marriage is a union, not of two individuals, but of two extended families. Over all, marriage is perceived as a civil contract between two families. The contract calls for the transfer of goods or money, or both, from the bridegroom's family to the bride's family in the form of *bride wealth*. This is not the "price" of a bride. Wives are not bought. The bride wealth may be seen as a recompense to the bride's parents for the loss of her services, a validation of the legality of the union, or a gift to seal the contract (Moore and Dunbar, 1969). The mode of payment of bride wealth varies. Among pastoral people, it could be the transfer of livestock—cattle, sheep, or goats; elsewhere it could be money, foodstuffs, or something else.

Several culture groups permit polygamy. While provoking jealousies, especially in homes where one wife is barren, polygamy has some benefits (Moore and Dunbar, 1969). The sick are seldom left uncared for, and the young or orphaned are often disciplined. Single parenting is not an issue, and prostitution is rare. Divorce is infrequent; it can be granted for a number of reasons—adultery, barrenness, unharmonious relationship with a mother-in-law, impotence—but only after counseling and efforts at family intervention and conflict resolution have failed.

Language and Society

One of the most intriguing aspects of Sub-Saharan Africa's cultural geography is its geography of languages. It is estimated that more than 1000 languages exist in the region. Most of these languages do not have a written tradition, and approximately 40 are spoken by 1 million or more people (Bergman and McKnight, 1993). Linguistic scholars, such as Greenberg (1966), have conducted extensive analyses in Africa; they identify four major language families in Sub-Saharan Africa: The Niger–Kordofanian, Nilo–Saharan, Khoisan, and

Afro–Asiatic (Semitic–Hamitic) families. A *language family* is a collection of individual languages related to each other through a common ancestral language before recorded history and, therefore, having shared, but distant origins (Renwick and Rubenstein, 1995). These families have *language branches* or *subfamilies* that constitute a collection of languages that share a common origin, but have evolved into individual languages.

Niger–Kordofanian family This is the largest language family in Sub-Saharan Africa, and it falls into two related, but distinct categories, the Kordofanian language branch and the Niger–Congo. The Kordofanian branch encompasses a small area in the Nuba hills of Sudan and consists of about 20 languages (Murray, 1981). The Niger–Congo branch, on the other hand, is spoken by more than 150 million people and stretches across half of Sub-Saharan Africa, extending from west Africa to the equatorial and southern regions. It includes the west Atlantic languages Wolof and Fula (in Senegambia) and Temne and Gola (in Sierra Leone and Liberia) (Figure 6.6); the Mande languages Malinke, Bambara (Guinea), and Soninke (Mali); the Central Bantoid (or voltaic) languages Mossi (Burkina Faso) and Dagomba (northern Ghana); the Guinean (or Kwa) languages Kru (Liberia), Akan (Ghana), Yoruba (western Nigeria), and Ibo (eastern Nigeria); the Hausa language; the central and eastern Sudanese languages Azande (Central African Republic) and Banda; and the largest subfamily, the Bantu.

Linguistic studies trace the origins of the Bantu language to the southeast margins of the Congo rain forest, although some evidence suggests some linkages to the west African forest and savanna regions. This is the largest language branch, stretching from the equatorial region to South Africa (Figure 6.6). The major Bantu languages spoken by 10 million or more people include Lingala (Democratic Republic of Congo), Swahili (Kenya, Tanzania), Ruanda-Rundi, Tswana-Sotho (Botswana and South Africa), and Zulu. Other languages spoken by more than 1 million include Bemba (Zambia), Luba (southern Democratic Republic of Congo), Shona (Zimbabwe), and Ganda (Uganda).

Nilo–Saharan The Nilo–Saharan language branch stretches in a west-to-east direction from the Songhai language in southwest Nigèr to the Nilotic languages Nuer and Dinka (in southern Sudan) and Luo and Masai (in southwestern Kenya). It also includes the Saharan languages Kanuri (which can be traced to the Kanem and Bornu Kingdoms of Lake Chad), Kanembu, and Teda.

Khoisan The Khoisan family is confined to the Kalahari Desert region in Namibia, Botswana, Zimbabwe, and South Africa. It dates back several millennia, having occupied the entire territory of southern and eastern Africa from Somalia to the Cape of Good Hope. Over the centuries, the Khoisans have been ravaged by diseases, Bantu invasions, and expropriation of land by European settlers, thus depleting their economic sustenance. The Khoisan language is affiliated with the Nama, Hottentots, and Bushmen.

Afro–Asiatic (Semitic–Hamitic) This family covers much of North Africa. In Sub-Saharan Africa, it is concentrated in Mauritania and the Horn, where such Semitic languages as Amharic, Tigre, and Tigrinya are spoken. It also includes Kushite languages, such as Somali and Mbugu (Tanzania), and the Chadic and Berber languages.

Besides these four major groups, there are two other language groups in the continent with non-African origins. The *Malay–Polynesian family* was introduced to Madagascar

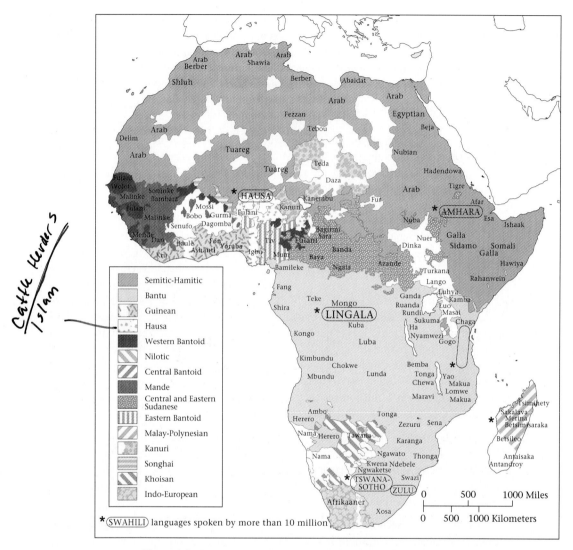

Figure 6.6 Distribution of Languages in Sub-Saharan Africa. (From Bergman, E. & McKnight, T. (1993). *Introduction to Geography*, Englewood Cliffs, NJ: Prentice Hall, p. 249.)

about 2000 years ago; while the Afrikaans language, a derivative of Dutch, has Indo-European origins dating back to 1602, when the Boers arrived in South Africa.

In many regions throughout the continent, people whose native languages are mutually incomprehensible use a common language, a *lingua franca*, for communication. Swahili, which developed along the east African coast from a fusion of Arabic with local Bantu, is now spoken widely in east Africa. It is also becoming increasingly popular in neighboring central Africa. In west Africa, Hausa, spoken by more than 50 million people, is the most

Elements of African Culture

important lingua franca, particularly in Nigeria and Niger. Pidgin languages, a mixture of African and European languages, are also becoming increasingly popular. In English-speaking West Africa, particularly Nigeria and, to some extent, Ghana, Pidgin English is popular. In Cameroon, with 200 dialects, pidgin is widespread (Discussion Box 6.1).

Language is a very important unifying (*centripetal*) as well as divisive (*centrifugal*) factor in Sub-Saharan Africa. Many people identify with an ethnic group, defined usually by language, rather than with the state. Most people in Kenya feel themselves to be Kikuyu, Luo, or Kamba, at least as much as, if not more than, they feel themselves to be Kenyan (O'Connor, 1992). Language is a key factor in maintaining a sense of identity. In Tanzania, in order to unify the people, Swahili has been promoted as the national language. Thus, tribal groupings are less important there than they are in Uganda or Kenya. Increased attachment to the ethnic group is more visible today in countries with traumatic ethnic-based political conflicts, such as Rwanda, Liberia, and Uganda.

DISCUSSION BOX 6.1: WEST AFRICAN PIDGIN ENGLISH

Pidgin English is popular in English-speaking west Africa. It is syncretistic, in that English vocabulary is mixed with local vocabulary to produce something unique, a mixture, as illustrated by the lyrics of a very popular high-life song in Ghana during the early 1980s. See if you can understand it. Akpeteshie is a local gin with high alcohol content, half and quarter are measures, and to booze is to get drunk.

Akpeteshie seller give me quarter; I go pay you tomorrow. *Credit.*

Akpeteshie seller give me half; I go pay you moon die. *— end of month.*

Too much problems on my head, too much wahala [*stress*] in my house.

I don't know whey thing I go do, I go pay you tomorrow. [*don't know what to do*]

My wife deh shout for money, my children deh cry for food.

So so wahala in my house, I go pay you tomorrow.

If I booze, I go forget my problems.

If I booze, I go forget my sorrow.

Na life e hard for me, I go pay you tomorrow.

I want to forget my problems, I want to forget my sorrow.

Na life e hard for me, I go pay you tomorrow.

I want to forget my problems, I want to forget my sorrow.

Na I don't get money, I go pay you tomorrow.

LAND TENURE

The multiplicity of land-tenure arrangements in Sub-Saharan Africa reflects not only the region's rich cultural diversity, but also the effects of European culture. It also helps to explain the fairly widespread land crises associated with land fragmentation, land alienation, and land-registration procedures. *Land tenure* refers to the fabric of rights and obligations that compose the relationship between human, customs, land, and society. Land-tenure systems can be considered a set of prescribed customary or procedural rules concerning people's rights to land, along with the institutions (social, political) that administer these rights, and the ways in which people hold that land (Downs and Reyna, 1988). Throughout most of traditional Sub-Saharan Africa, land is held communally, rather than individually. In fact, the land belongs to the living, to the dead (ancestors), and those yet to be born (future generations). Individuals have essentially custodial rights to land and to land-based resources, granted by the gods and ancestral spirits. In this strict sense, the land can be neither bought nor sold. Colonialism and modernization have changed the practice of communal ownership to some extent, and private land ownership is becoming more common. Some of the more common forms of land-tenure arrangements in Sub-Saharan Africa include family land, communal land, stool land, state ownership, and privately owned land.

Family Land

Family land is simply passed on through the lineage matrilineally or patrilineally, with rights of the land held jointly by a number of heirs. Usually, no monetary transactions can take place with such land. However, Dickerman (1988) points out that high rates of urbanization and growing land speculation have promoted the sale of family land holdings by individuals, sometimes without consulting with or gaining the consent of other family members. Two other problems associated with family land are (1) the increased fragmentation of land resulting from rapid population growth and further subdivision of family land among heirs and (2) the inequities associated with matrilineal inheritance. New statutes are being designed to ensure that spouses and children have legal control over the land property upon an intestate death.

Communal Land

Communal land means that the land belongs to the lineage, village, or community, and, under ideal circumstances, every member of the community has equal right to as much land as possible. Thus, there is no landless class. Access to land is pure *usufructuary*, or user rights, not ownership rights, and outright sale is prohibited. When the land is no longer being used, it reverts to the community. In almost all cases, the head of the village or clan has charge over land and its disposition. Decisions concerning usufructuary rights and land allocation are rendered in consultation with village or community elders. Problems with this type of arrangement include (1) a shortage of land, resulting from increasing population densities, and (2) the fact that aliens have no access to land, except through special arrangements or permission.

Stool Land

Stool land is land vested in the stool or skin, which is the symbol of kingship among the Yoruba in Nigeria, the Mossi in Burkina Faso, the Baganda in Uganda, and the Ga and Ashanti in Ghana. The traditional leader, chief, king, or skin has a sacred duty to hold the land in trust for the people. Subjects of the stool can access land for farming and shelter purposes, but in return they must provide customary services and pay homage.

State Land

State land can be viewed from a number of perspectives. It has been equated with public land, where land is acquired for a public purpose. It has also been associated with unoccupied land, in which case the state assumes the right to allocate it through the private market or through public domain, where it can be used for reforestation programs or game parks.

Mabogunje (1992) points out, however, that state ownership involves the transfer of all existing rights to the state, with the state then becoming the sole allocator of land. He estimates that there are about 20 countries in Sub-Saharan Africa that have nationalized all lands and done away with private freehold ownership. Some countries, such as Lesotho (1979) and Cameroon (1980), nationalized on the premise that they were continuing an African tradition in which ownership resided in the community and not the individual. The Congo (1979) and Senegal (1964) preferred to maintain the French legacy of ignoring customary practices and treating all land as belonging to the state. Zaire (1973) and Nigeria (1978) opted for nationalization on the grounds that it was the most efficient way of allocating land for public purposes. Most of the other countries, such as Somalia (1975), Ethiopia (1975), Mozambique (1979), and Zambia (1975), nationalized land on purely ideological grounds. Examples of countries that have not nationalized land include Botswana, Kenya, Uganda, Zimbabwe, and Mauritania.

Individual Private Ownership

With modernization, private ownership of land is gradually on the rise. Sometimes, a piece of family land is sold to obtain cash to support a needy family or clan member—for example, during a serious illness or for a costly education or any other event deemed necessary by the majority of the family. The decision belongs to the family or clan, and violators can expect to incur the wrath of ancestral spirits and other deities.

Individuals are allowed *freehold rights*—outright ownership—or leasehold rights that, in most cases, allow tenure from 49 to 99 years. Mabogunje (1992) states that freeholding is shunned by African governments because it is associated with the exclusionary policies of former colonial administrators that tended to create European reserves in cities and rural areas, particularly in east and southern Africa.

ADORNMENT, DRESS FORMS, AND SYMBOLISM IN AFRICAN CULTURE

Adornments and modes of dressing are cultural expressions and vary throughout Sub-Saharan Africa. Clothing is a status symbol. A person's status, role, age, and mood can be fairly well predicted from his or her grooming. Among the Akan of Ghana, red, black, and dark-brown colors are worn during mourning and grief (Figure 6.7). In contrast, white symbolizes joy, victory, and success. An Akan chief portrays his authority by his regalia; attendants' clothing, similarly, portrays subordination.

Moreover, the symbols and fabrics in African textile design make a clear statement. For example, the kente and adinkra cloths, both of which originated among the Akan people of Ghana, are full of cultural meaning. Adinkra is a highly valued, hand-printed and embroidered cloth. Originally, it was made to be used exclusively by the royalty and spiritual leaders for important sacred ceremonies and rituals. Each of the motifs that compose the body of adinkra symbolism has a name and meaning derived from a variety of sources: proverbs, historical events, human attitudes or behaviors, and elements from the surrounding physical environment (Figure 6.8). Thus, adinkra symbolism is a visual representation of the social thought, history, philosophy, and religious beliefs of the Akan people (Discussion Box 6.2).

Figure 6.7 Symbolism of Clothing: A Funeral Ceremony.

Adornment, Dress Forms, and Symbolism in African Culture

Figure 6.8 Meanings of Symbols in Adinkra Cloth. Adinkra is a highly valued hand-printed and embroidered cloth. Its origin is traced to the Asante.

DISCUSSION BOX 6.2:
ADINKRA SYMBOLISM

Meanings of Symbols in Adinkra Cloth

Adinkra is a highly valued hand-printed and embroidered cloth. Its origin is traced to the Asante people of Ghana and the Gyaman people of Côte d'Ivoire. However, the production and use of Adinkra have come to be associated more with the Asante people than with any other group of people. Adinkra cloths were made, and used exclusively by the royalty and spiritual leaders, for various important and sacred ceremonies and rituals. Today, adinkra cloths are used for a wide range of social activities, including festivals, church going, weddings, naming ceremonies, and initiation rites. Designers use adinkra symbols for a wide range of products.

Each of the motifs that make up the corpus of adinkra symbolism has a name and meaning derived either from a proverb, a historical event, a human attitude, an animal behavior, plant life, or the forms and shapes of inanimate and human-made objects. These are graphically rendered in striking geometric shapes. Meaning of motifs can be categorized as follows: aesthetics, ethics, human relations, and religious concepts. In its totality, adinkra symbolism is a visual representation of social thought relating to the history, philosophy, and religious beliefs of the Akan peoples of Ghana and Côte d'Ivoire.

Commonly used symbols, their names, their sources of derivation, their literal translations, and their symbolic meanings are presented in the following list, with names and meanings of the symbols presented in Twi (the language of the Akan peoples), then translated into English:

1. Gye Nyame (except God): Symbol of the omnipotence of God. Proverb: Abode santan yi firi tete: obt nte ase a onim n'ahyase, na obi ntena ase nkosi n'awie, Gye Nyame. (This great panorama of creation dates back to time immemorial; no one lives who saw its beginning, no one will live to see its end, except God.)

2. Kuronti ne akwamu (the two complementary branches of the state): Symbol of democracy, duality of the essence of life (interdependency and complementarity). Proverb: Tikoro nko agyina. (One person does not constitute a Council; or, one head does not constitute a jury.)

3. Odo nyera fie kwan (love does not get lost on its way home): Symbol of love, devotion, and faithfulness. Proverb: Odo nyera ne

fie kwan. (Love lights its own path; it never gets lost on its way home.)

4. Sankofa (go back to fetch it): Symbol of the wisdom in learning from the past in building for the future. Proverb: Se wo were fi na wo sankofa a yenkyi. (It is not a taboo to go back and retrieve if you forget.)

5. Osram ne nsromma (the moon and the star): Symbol of faithfulness, love, devotion, fondness, loyalty, benevolence, and the feminine essence of life. Proverb: Kyekye pe aware. (Kyekye—the North Pole Star—has a deep love for marriage. She is always in the sky waiting for the return of the moon, her husband.)

6. Another version of Sankofa.

7. Akoma (the Heart): Symbol of love, patience, goodwill, faithfulness, and endurance. A saying: Nya akoma. (Get a heart; or, be patient.)

8. Another version of Sankofa.

9. Nyame nwu na mawu (God does not die, and so I don't die): Symbol of the eternity of God and the human soul.

10. Dwenini mmen (ram's horns): Symbol of strength (of mind, body, and soul) and humility. Proverb: Dwenini ye asisie a ode nakoma na ennye ne mmen. (The ram may bully, not with its horns but with its heart.)

11. Hwenhwemudua (searching rod or measuring rod): Symbol of excellence, perfection.

12. Another version of Dwenini mmen.

Courtesy of Dr. Kwaku Ofori-Ansah

Similarly, the kente, a ceremonial cloth handwoven on treadle loom, is full of symbolism and meaning (Figure 6.9). Strips measuring about 4 inches wide are sewn together into larger pieces. Kente cloth comes in various colors, sizes, and designs and is worn during important social and religious occasions (Discussion Box 6.3).

(1) EMAA DA

(2) OBAAKOFO MMU MAN

(3) WOFRO DUA PA A NA YEPIA WO

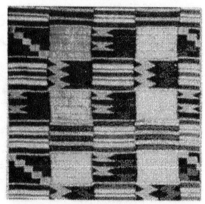
(4) ABUSUA YE DOM

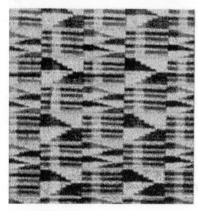
(5) ADWINASA

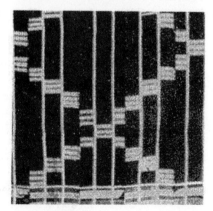
(4) AKYEMPEM

Figure 6.9 Kente Cloth Symbology.

Adornment, Dress Forms, and Symbolism in African Culture

(7) SIKA FUTURO

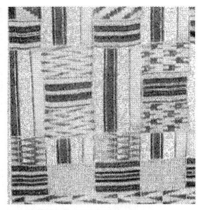

(8) TOKU KRA TOMA

(9) KYERETWIE

(10) NYANKONTON

Figure 6.9 (*Continued*)

DISCUSSION BOX 6.3: SYMBOLOGY AND MEANING IN GHANA'S KENTE CLOTH

Kente is an Asante ceremonial cloth that is handwoven on a horizontal treadle loom. Strips measuring about 4 inches wide are sewn together into larger pieces of cloths (Figure 6.10). Cloths come in various colors, sizes, and designs and are worn during very important social and religious occasions. In a total cultural context, kente is more important than just a cloth. It is a visual representation of history, philosophy, ethics, oral literature,

Figure 6.10 Kente Weavers.

moral values, a social code of conduct, religious beliefs, political thought, and aesthetic principles.

The term *kente* has its roots in the word kenten, which means *basket*. The first kente weavers used raffia fibers to weave cloths that looked like kenten (a basket) and thus were referred to as kenten ntoma, meaning basket cloth. The original Asante name of the cloth was nsaduaso or nwontoma, meaning "a cloth handwoven on a loom," and that name is still used today by Asante weavers and elders. However, the term kente is more popular today, within and outside Ghana. Many variations of narrow-strip cloths, similar to kente, are woven by various ethnic groups in Ghana and elsewhere in Africa. The focus here is on the Asante kente cloth. The Asante are one of the Akan peoples who live in parts of Ghana and Côte d'Ivoire.

1. Emaa da—Literally means "it has not happened before" or "it has no precedence." According to Nana Kwasi Afranie of Bonwire, the Asantehene's chief weaver, the cloth was designed and so named by one of the Asante kings who was so awed by the uniqueness of the pattern that he remarked "Eyi de emmaa da," meaning "this one has no precedence." The cloth was therefore reserved for the exclusive use of the king, but its use was later extended to people of high ranks. It is a symbol of ingenuity, innovation, uniqueness, perfection, and exceptional achievement.

2. Obaakofo mmu man—Literally means "one person does not rule a nation." It expresses the Akan system of governance, based on participatory democracy. The nine squares represent Mpuankron (nine tufts of hair), a ceremonial hair cut of some royal functionaries who help rulers make decisions. Originally the cloth was named Fathia fata nkruma. "Fathia is a suitable wife for Nkrumah." After the military overthrow of Nkrumah, the original significance of Mpuankron (participatory democracy) was applied to reflect the prevailing political atmosphere. The cloth symbolizes participatory democracy and warning against autocratic rule.
3. Wofro dua pa a na yepia wo—Literally means "one who climbs a tree worth climbing gets the help deserved." The cloth was designed to express the Akan social thought that maintains that any good individual effort deserves to be supported by the community. When someone climbs a good tree that has fruits on it, people around will give him a push, since they know they will enjoy the fruits of his labor. It is a notion that reinforces the importance of aspiring towards a worthy course. It symbolizes aspiration, hope, mutual benefits, sharing, and noble deeds.
4. Abusua ye dom—Literally means "the extended family is a force." Among the Akan peoples, the extended family is the foundation of society. Like a military force, members of the family are collectively responsible for the material and spiritual well-being, the physical protection, and the social security of all the members. The cloth was designed to celebrate and reinforce such positive attributes of the extended family system. In its many variations and background colors, the cloth symbolizes a strong family bond and the value of family unity, collective work and responsibility, and cooperation.
5. Adwinasa—Literally means "all motifs are used up." According to the elders, the designer of this cloth attempted to weave a unique cloth to please the Asantehene. In his effort, he used all the motifs then known to weavers in weaving one cloth. In the end he remarked that he had exhausted the repertoire of motifs known to Asante weavers. The cloth was, therefore, viewed as one of top quality and the most prestigious of kente cloths, besides those woven exclusively for Asante kings. It was, in the past, worn by kings and people of high status and wealth. It symbolizes royalty, elegance, creative ingenuity, excellence, wealth, perfection, and superior craftsmanship.
6. Akyempem—Literally means "thousands of shields." This is a reference to shields used by highly organized militia consisting of thousands of men and women who defended the Asante kingdom against external aggression. According to the military strategy of the Asante kingdom, the chief of the shield bearers, the Akyempemhene, and the rear guards of the king are his own sons. Shields once used as military weapons are now used in royal ceremonies to symbolize and commemorate the military prowess of the Asante kingdom. The cloth symbolizes military prowess, unity through military strength, bravery, political vigilance, and spiritual defense power.

7. Sika futoro—Literally means "gold dust." Before the use of coins and paper as money, gold dust was used as a medium of exchange among the Akan peoples and was therefore considered as a symbol of wealth and prosperity. The predominant use of intricately textured patterns in yellows, oranges, and reds replicates the visual characteristics of gold dust. The cloth symbolizes wealth, royalty, elegance, spiritual purity, and honorable achievement.
8. Toku kra toma—Literally means "Toku's soul cloth." The cloth is designed and named to commemorate the soul of a warrior queen mother of that name, who, though defeated and executed in a battle with Nana Opoku Ware I, the king of the Asante kingdom (1731–42), was viewed as a courageous woman. It commemorates that historic event and honors the soul of that queen mother for her bravery. In the past, such a cloth would be worn only by the royalty and people of high rank during very sacred ceremonies in which the spirits of the ancestors are venerated. The cloth symbolizes courageous leadership, heroic deeds, self-sacrifice, and spiritual vitality and rebirth.
9. Kyeretwie—Literally means "the lion catcher." The cloth was designed to commemorate an incident during the reign of King Kwaku Dua (1838–67), who tested the courage of his warriors by ordering them to catch a leopard alive. The appellation "Kyeretwie" was later appended to the names of some of the Asante kings whose bravery and leadership qualities were comparable to the courage needed to catch a leopard alive. The black vertical warp stripes represent the black spots in a leopard's fur. In the past, the cloth was worn only by the Asantehene (or by other chiefs, with his permission). The cloth symbolizes courage, valor, exceptional achievement, and inspiring leadership.
10. Nyankonton—Literally means "God's eyebrow" (the rainbow). It was created in exaltation of the beauty and mystery of the rainbow phenomenon. The arrangement of warp threads mimics the visual characteristics of the rainbow. This cloth symbolizes divine beauty, gracefulness, divine creativity, uniqueness, and good omen.

Courtesy of Dr. Kwaku Ofori-Ansah, Sankofa Publications

COLONIALISM AND DIFFUSION OF NON-AFRICAN CULTURE

Colonialism is the one factor that has most profoundly influenced African cultural development. European culture—language, religion, social organization, values—was simply imposed on Africans. For example, in place of indigenous African languages, French, English, and Portuguese became official languages in areas colonized by the French, the British, and the Portuguese, respectively (refer to Chapter 4). Also, the two leading religions in Africa today, Christianity and Islam, are both imports. Moreover, colonial administrations "condemned everything African in culture—African names, music, dance, art, religion, marriage, systems of inheritance—and completely discouraged the teaching of all these facets of African life in schools and colleges. Even the wearing of African clothes to work or school was banned." (Boahen, 1987) Colonial education fostered a disdain for "shameful" and "backward" African cultural practices. The outcome of this has been the loss of dignity and respect for African culture. Perhaps the result of such a cultural collision would not have been as traumatic if the differences between European and African culture had not been so drastic. (Refer to Table 6.1 for a sample of European and African cultural differences.)

TABLE 6.1 Selected Differences Between African and European Cultures

Culture Trait	African Culture	European Culture
Marriage	Union between families	Union between individuals
Bride wealth	Groom to bride's parents and extended family	Bride to groom
Polygamy	Permitted	Essentially monogamy
Annulment	Not easily dissolved	Easily dissolved
Values	Cooperation within family and community	Competition, even within family
	Emphasize community	Emphasize individuality
Money and relationships	Does not necessarily put a monetary value to everything, may freely share	Puts a monetary value to everything, including relationships
How wealth is demonstrated	Sharing, giving; provide for family members; take care of orphans	Accumulation, hoarding; orphanages
Attitude toward the poor	Immediate family members and community are responsible; poverty usually a sign of misfortune; sympathy and help; it could happen to you	Responsibility primarily of the state, welfare system; poverty usually a sign of laziness
Land	Community owned, not to be sold	Individual owned, may be sold anytime
Aging and the aged	Sign of maturity, experience, wisdom	Senility
	Supported by extended family, particularly children	Supported by social security, rarely children
	Live with children as essential, indispensable members of household and community	Senior citizen's homes utilitarian thinking, not revenue generating
Political	Allegiance primarily to the family unit, then community, ethnic, tribal group	Family relationships not as strong; allegiance primarily to the state
Children and child care	Childbirth, a sign of fertility, social respectability and status	Children more a liability than an asset
Child care	Child care not solely a parental responsibility, but family or community responsibility	Primarily parental responsibility or state
Child punishment	Spanking freely and commonly used	Spanking frowned upon, rarely used
Social security	Family support as social security, especially children and family as social security and insurance	Primarily individual income, employment or state based

Traditional African culture (like traditional Asian culture) emphasized cooperation and community; European culture emphasized competition and individuality. Africans emphasized sharing and giving; Europeans emphasized accumulation and hoarding. Whereas, in most African cultures, land could not be sold because it belonged to the living, the dead, and those yet to be born, European culture saw land as capital to be traded and commodified to accumulate more capital, where possible. Moreover, in areas where Europeans found prime land, such as in Zimbabwe, Namibia, and South Africa, Africans were pushed into reserves or had their land confiscated, thus creating serious artificial land shortages. Furthermore, the imposition of European land-tenure systems translated into huge individual landholdings and many landless people, instead of small community landholdings (O'Connor, 1992).

So all pervasive was the impact of European culture on African culture that few culture traits emerged unscathed. Many African countries are striving to undo the negative impacts of colonialism on African culture, with little success. Nonetheless, the search for, and reconstruction of, authentic and indigenous African culture is still being pursued. In Democratic Republic of Congo, the late President Mobutu promoted a return to African cultural origins by, among other things, banning European personal and geographical names. He himself changed his name from Joseph Désiré Mobutu to Mobutu Sese Seko (Atsimadja, 1992). Various countries, on attaining independence, changed to more African names. Three examples are Gold Coast, which became Ghana; Rhodesia, which became Zimbabwe; and Upper Volta, which is now Burkina Faso. A formal transnational attempt at preserving and protecting Bantu culture led to the formation of the Centre International de Civilisation Bantu (CICiBA) in 1983. Its 10 members are Angola, Central African Republic, Comoros, Congo, Gabon, Equatorial Guinea, Rwanda, São Tomé and Príncipe, Democratic Republic of Congo, and Zambia.

MODERNIZATION AND CULTURAL CONFLICT

African culture is under tremendous stress due to the influences of Westernization and the practical realities of living in a global village. Traditional African values are being questioned, revised, or even abandoned. Policy options to deal with increasingly difficult questions are scarce. Let us look at the problem of aging and at the culture of caring for the aged in Sub-Saharan Africa, to illustrate the problem.

Modernization and the Aged in Sub-Saharan Africa

Africans view the aged with dignity and respect. Old people are the foundation of village life, and a village without old people is like a hut eaten away by termites (Kabwasa, 1982). Old age is a desirable stage in life, a time of wisdom, honor, and teaching (Carothers, 1953). Thus, an older person is considered a wiser person, worthy of honor and respect. In traditional African societies, the extended family plays a central role in providing and caring for the elderly (United Nations, 1985b). It provides for the elderly in an integrated system: In exchange for providing child-care services and help with various miscellanies, the elderly receive food, clothing, accommodation, and love. They belong, have security, and play a functional role in society. It is generally considered an honor for a family to take good care of the aged; therefore, those who shirk this responsibility are disdained.

Modernization, urbanization, and migration are rapidly changing traditional African values, putting the elderly increasingly at risk of isolation. Younger members of the family, once they obtain jobs in urban centers, return to the village less often, maybe only once or twice a year. While they still feel a sense of responsibility for the elderly, they are thwarted by geography and changing times. The elderly are, thus, left with little or no means of income, poor housing, and frequently poor health. Throughout the continent, policies to provide for such elderly have been shrouded in controversy.

Institutional care is not as well accepted in Sub-Saharan Africa as it is in other parts of the world. An African representative to the World Assembly on Aging in Vienna in 1982 made a strong policy statement rejecting institutionalization as a mode of caring for the elderly (United Nations, 1985b). Droskie (1977) argues that admission to an old-age home should be a last resort for the elderly. A representative of Cameroon at the World Assembly on Aging argued that Africans must resist making the mistakes of the industrialized countries.

Nevertheless, institutional care is available in parts of Sub-Saharan Africa for those with no alternatives. According to the United Nations (1985b), Ethiopia had 4 homes, with plans for more; Malawi had a home for elderly veterans; Mauritius had 12 residential homes; Rwanda had 5; and Zambia cared for aging persons in government institutions. Kenya, trying to avoid duplication of homes in the Western world, has planned integrated community facilities. In these facilities, the elderly will live with persons from other age groups.

Creating alternative support mechanisms, such as nursing homes, would cause further social and economic burdens. Consequently, African policy makers are urging the development at the community level of strategies for strengthening the family to provide social protection for the elderly. Suggested mechanisms for strengthening the family include financial support to families caring for the elderly, educational programs to the whole community on how to care for the elderly, legal action that obliges children to provide for the elderly, and the provision of economic and service infrastructure that encourages younger people to remain in rural areas and continue to care for the elderly. The goal is to preserve the traditional support system and create an integrated rural society where the young and old may live in symbiosis. Family and community resources are great sources of strength; the reinforcing of these foundations would allow other needs to be addressed, such as health care, housing, social welfare, income security, employment, and education.

Values and Expectations in African Society

Another area of stressful cultural conflict is in the domain of nepotism, bribery, and corruption. In traditional African society, kinship obligations require the privileged to extend favors and assistance to kin less privileged. Societal pressure bears on those in positions of authority to help younger or less well-placed kinsmen to obtain similar or higher positions. With modernization, however, this practice can no longer be feasible, since it is the most qualified person who deserves to be hired, not the closest relative. Ethnic solidarity and kinship obligations frequently clash with expectations of Western institutional approaches and hiring practices. Hence, people from the clan or extended family ask how well-placed politicians or academics might help them to obtain a job or scholarship. The ordinary people who exert these pressures on more successful ethnic kin are often oblivious to the moral inconsistencies of the situation. Helping kinsmen was traditionally acceptable—in fact, desirable, creating what Mazrui (1986) has described as cultural confusion. African leaders are torn between the Western value of individual merit and the collective solidarity of the traditional culture.

Another manifestation of this cultural confusion is in bribery and corruption. The traditional value of reciprocity in the act of serving is a source of confusion. Dignity often demanded that the beneficiary of a favor should give something in return. Giving a favor to a person in anticipation of asking them for one later was also perfectly acceptable. From the Western perspective, these constitute bribery and corruption.

A final manifestation of the cultural confusion that assails Africans is the misappropriation of government funds. Because the colonial regime lacked legitimacy, colonial government property lacked respect. Thus, it was almost a patriotic duty to misappropriate the resources of the colonial government. Stealing from a foreign "thief" was simply an act of heroic restoration (Mazrui, 1986). Postcolonial Africa still suffers from the cynical attitudes to government generated by the colonial experience. Government has changed hands, but conceptions about government have not changed as radically. Therefore, some citizens still consider it heroic to steal from their African governments.

CONCLUSION

Understanding Sub-Saharan Africa without reference to the rich culture and its collision with colonialism is impossible. Nevertheless, the integrity of this rich culture is under attack, particularly from the impact of colonialism and the diffusion of European culture. Strong kinship ties are gradually being weakened, and extended family ties are crumbling under the impact of urbanization influences. The clash of Western values with traditional African values creates considerable cultural confusion for Africans. Also, Cold-War superpower rivalry, coupled with greed and ethnic favoritism among African political leaders, has had disastrous effects on the cultural integrity of the continent. Rapid population growth facilitates changes in land tenure and a loss of traditional support mechanisms. How Sub-Saharan Africa responds to these challenges now will determine whether an authentic African culture will remain.

KEY TERMS

Culture trait
Cultural integration
Cultural landscape

Family and kinship
Bilateral kinship
Patrilineal system

References

Cultural ecology
Cultural diffusion
Coptic Christians
Syncretistic religions
Traditional religions
Spirit shrines
Stool land
State land
Land tenure
Usufructuary rights

Matrilineal system
Language family
Language branch
Pidgin English
Family land
Communal land
Adinkra Cloth
Ancestral spirits
Lingua franca

DISCUSSION QUESTIONS

1. Are traditional African land-tenure arrangements an obstacle to economic development? Is nationalizing all land a viable solution?
2. Cultural differences, including language, religion, and ethnicity, are the root cause of all of Sub-Saharan Africa's problems. Discuss this statement.
3. "Somalia and Liberia indicate that peace and stability are impossible in Sub-Saharan Africa." Consequently, Somalia and Liberia reflect the true future of the region. Explain why you agree or disagree.
4. Compare and contrast cultural value systems in the West with those in Sub-Saharan Africa.
5. Identify the major languages spoken by 10 million or more people in west, east, central, and southern Africa. To what extent can these languages be centripetal forces?
6. Describe the major attributes of traditional belief systems in Sub-Saharan Africa.
7. To what extent has colonialism been a detriment to the cultural integrity of Sub-Saharan Africa?
8. What Westerners call "bribery and corruption" in Africa are simply a Western misunderstanding of traditional African ways of doing business. To what extent do you see these as problems? Should, and can, they be eliminated?

REFERENCES

ATSIMADJA, F. (1992). "The Changing Geography of Central Africa," in Chapman, G. P. & Baker, K.M., eds., *The Changing Geography of Africa and the Middle East*, New York: Routledge.

BERGMAN, E. & MCKNIGHT, T. (1993). *Introduction to Geography*, Englewood Cliffs, N.J: Prentice Hall.

BOAHEN, A. (1987). *African Perspectives on Colonialism*, Baltimore: Johns Hopkins University Press.

CAROTHERS, J. (1953). *The African Mind in Health and Disease*, Geneva, Switzerland: World Health Organization.

DICKERMAN, C. (1988). "Urban Land Concentration," in *Land and Society in Contemporary Africa*, Downs, R. & Reyna, S. eds., Hanover, NH: The University Press of New England.

DOWNS, R. & REYNA, S., eds. (1988). *Land and Society in Contemporary Africa*, Hanover, NH: The University Press of New England.

DROSKIE, Z. (1977). "The Role of the Voluntary Welfare Agency in the Care of the Aged," *South African Medical Journal*, 51:436–437.

GREENBERG, J. (1966). *Languages of Africa*, 2d ed., Bloomington IN: University of Indiana Press.

JACKSON, J. (1984). *Discovering the Vernacular Landscape*, New Haven, CT: Yale University Press.

JORDAN, T. & ROWNTREE, L. (1982). *The Human Mosaic*, New York: Harper Collins.

KABWASA, N. (1982). "The Eternal Return," *UNESCO Courier*, 14–15.

MABOGUNJE, A. (1992). *Perspectives on Urban Land and Urban Management Policies in Sub-Saharan Africa*, Washington, DC: World Bank.

MAZRUI, A. (1986). *The Africans: A Triple Heritage*, Boston: Little, Brown and Company.

MOORE, C. & DUNBAR, A. (1969). *Africa Yesterday and Today*, New York: Praeger.

MURRAY, J. (1981). *Cultural Atlas of Africa*, Amsterdam: Elsevier International Projects.

O'CONNOR, A. (1992). "The Changing Geography of Eastern Africa," in *The Changing Geography of Africa and the Middle East*, Chapman, G.P. & Baker, K.M. eds., New York: Routledge.

PARRINDER, G. (1983). "The Religions of Africa," in *Africa South of the Sahara*, London, UK: Europa Publications Limited.

RENWICK, W. & RUBENSTEIN, J. (1995). *People, Places, and Environment: An Introduction to Geography*, Upper Saddle River, NJ: Prentice Hall.

UNITED NATIONS (1985a). *Demographic Indicators of Countries: Estimates and Projections as Assessed in 1980*, New York: United Nations.

UNITED NATIONS (1985b). *The World Aging Situation: Strategies and Policies*, New York: United Nations.

7

Population Geography of Sub-Saharan Africa
Samuel Aryeetey-Attoh

INTRODUCTION

Population geography is the study of the temporal dynamics and spatial variations associated with the components of population change. Population geographers analyze the following:

1. How populations are distributed and arranged in space.
2. The dynamics of human populations with respect to changes in birth, fertility, and mortality rates.
3. Changes in the age, sex, and ethnic composition of human populations.
4. The spatial interaction that occurs between and within regions as a result of human migration.

This chapter addresses these issues in the context of Sub-Saharan Africa. Rural–urban migration is dealt with in more detail in Chapter 8, and the issue of refugee migration is given special consideration. The chapter concludes with a review of government approaches to family planning.

GENERAL POPULATION TRENDS

Sub-Saharan Africa's population was estimated at 673 million people in 2001 and continues to grow at a fairly rapid rate, 2.5% a year. With a doubling time of approximately 28 years, the population is projected to increase to about 1.56 billion by the year 2050 (Population Reference Bureau, 2001). By then, the Sub-Saharan region will have three and one-half times as many people as North America, twice as many as Europe, and more than 700 million more people than Latin America and the Caribbean's projected 814 million people. Sub-Saharan Africa is second only to Asia, which will continue to increase its lead in population over all other regions of the world.

Sub-Saharan Africa continues to have the highest fertility and mortality rates in the world, along with the highest proportion of young dependents, although the evidence that follows reveals considerable subregional variations in these trends. Further evidence suggests a gradual decline in these trends.

In spite of rapidly growing populations, almost half of the 48 Sub-Saharan countries have populations under 10 million (Table 7.1). Only 7 countries have populations over 25 million; they include, in descending order of population, Nigeria, Ethiopia, The Democratic Republic of Congo, South Africa, Tanzania, Sudan, and Kenya.

As a cautionary note, it should be mentioned that these population figures are estimates or indirect measures of what should be actual, observable data sets. Sub-Saharan Africa has been plagued with several population enumeration and survey problems; this was especially problematic prior to 1974. Some countries have not conducted comprehensive surveys over the last 20 years. Ethiopia, for example, has never had a complete population census, although the one conducted in May 1984, was estimated to be 85% complete. Also, Ethiopia conducted its first demographic and health sample survey in 2000. As a result, the current population estimate, 65.4 million, is based on the 1984 count (42.2 million). Other countries have encountered similar problems related to undercounting (Sierra Leone, Liberia, and Central African Republic) and overinflated numbers (Nigeria and Zambia). Recently, with assistance from nongovernmental organizations and the United Nations, census and survey enumeration procedures have improved. Most Sub-Saharan countries conducted national or partial census surveys between the World Population Conferences of 1974 (in Bucharest) and 1984 (in Mexico). These have been supplemented with World Fertility Surveys and Demographic and Health Surveys, which provide detailed accounts of the behavioral and attitudinal attributes of African households. Given these improvements in information technology, the world is becoming more aware and better informed about population issues in Africa.

Population Density and Distribution

Population density is measured in a number of ways. A standard approach is the *crude* or *arithmetic density*, expressed as a ratio of the number of people per unit of land area (square miles or square kilometers). A more revealing index, which incorporates cropland or arable land, is the *physiological density*. It is useful in assessing a country's ability to sustain itself in terms of food production or in determining the *carrying capacity* of a region—the

General Population Trends

TABLE 7.1 Population Data for Sub-Saharan Africa, 2001

Country	Population Mid-2001 millions	Birth Rate Per 1000	Death Rate Per 1000	Natural Increase Annual (%)	Infant Mortality Rate	Total Fertility 1970–75	Total Fertility 2001	Dependency Ratio % Age <15	Dependency Ratio %Age +65
WORLD	6137	22	9	1.3	56		2.8	30	7
Developing Countries	4944	25	8	1.6	61		3.2	33	5
Sub-Saharan Africa	673	41	15	2.5	94		5.6	44	3
Western Africa	245.7				88		5.8	45	3
Benin	6.6	45	15	3.0	94	7.1	6.3	48	2
Burkina Faso	12.3	47	17	3.0	105	7.8	6.8	48	3
Côte d'Ivoire	16.4	36	16	2.0	112	7.4	5.2	42	2
Chad	8.7	49	16	3.3	103	6.7	6.6	48	
Ghana	19.9	32	10	2.2	56	5.8	4.3	43	3
Guinea	7.6	41	19	2.3	98	7.0	5.5	44	3
Liberia	3.2	49	17	3.1	139		6.6	43	3
Mali	11	50	20	3.0	123	7.1	7.0	47	3
Mauritania	2.7	43	15	2.8	106	6.5	6.0	44	2
Niger	10.4	53	24	2.9	123	8.1	7.5	50	2
Nigeria	126.6	41	14	2.8	75	6.9	5.8	44	3
Senegal	9.7	41	13	2.8	68	7.0	5.7	44	3
Sierra Leone	5.4	47	20	2.6	153	6.5	6.3	45	3
Togo	5.2	40	11	2.9	80	7.1	5.8	47	2
Eastern Africa	243.6				97		5.7	45	3
Burundi	6.2	42	17	2.5	75	6.8	6.5	48	3
Eritrea	4.3	43	13	3.0	80	6.5	6.0	43	3
Ethiopia	65.4	44	15	2.9	97	6.8	5.9	44	3
Kenya	29.8	34	14	2	74	8.1	4.4	44	3
Malawi	10.5	46	23	2.3	104	7.4	6.4	47	3
Mauritius	1.2	17	7	1	15.6	3.2	2.0	26	6
Mozambique	19.4	43	22	2.1	135	6.6	5.6	44	3
Rwanda	7.3	39	21	1.8	107	8.3	5.8	44	3
Somalia	7.5	48	19	3.0	126		7.3	44	3
Sudan	31.8	34	11	2.4	74	6.7	4.9	43	2
Tanzania	36.2	41	13	2.8	99	6.8	5.6	45	3
Uganda	24	48	19	2.9	97	7.1	6.9	51	2
Central Africa	77.3				88.6		6.6	47	3
Cameroon	15.8	39	12	2.7	77	6.3	5.2	43	3
Central Africa Rep.	3.6	38	18	2.0	98	5.7	5.1	44	4
Congo Rep.	3.1	46	16	3.0	105	6.3	6.3	43	3
Democratic Republic of Congo	53.6	47	16	3.1	106	6.3	7.0	48	3
Gabon	1.2	32	16	1.6	57	4.3	4.3	40	6
Southern Africa	83.8				92	6.5	4.7	42	5
Angola	12.3	50	25	2.4	198	6.6	6.9	48	3
Botswana	1.6	31	20	1.0	60	6.6	3.9	41	4
Lesotho	2.2	33	13	2.0	84	5.7	4.3	40	5
Namibia	1.8	36	17	1.9	68	6.5	5.0	43	4
South Africa	43.6	25	14	1.2	57	5.4	2.9	34	5
Swaziland	1.1	41	20	2.0	109	6.5	5.9	46	3
Zambia	9.8	45	22	2.3	95	7.4	6.1	45	2
Zimbabwe	11.4	29	20	0.9	65	7.4	4.0	44	3

Source: Population Reference Bureau. (2001). *Human Development Report, 2001*, Washington, DC: Population Reference Bureau, Inc.

technological capacity of a region to sustain itself, given its human and natural resource base. The *agricultural density* is a slight modification of the physiological density; it is a measure of the ratio of the farm population to arable land.

A look at the crude population densities in Table 7.2 reveals six categories of countries with varying densities. The small-sized countries with high densities include Rwanda, Burundi, and the island country of Mauritius; those with sparse populations include Equatorial Guinea, Liberia, and Guinea Bisseau. Swaziland and Lesotho have medium densities. Large-sized countries with above-average densities include Nigeria, Ethiopia, and Kenya; those with below-average densities include Democratic Republic of Congo, Angola, Namibia, and the Sahel states Mali, Niger, Mauritania, and Chad. South Africa, Zimbabwe, Cameroon, and Tanzania have medium densities.

A look at the physiological densities provides us with an even more revealing picture about the intense use of arable land. It also explains the important role the environment plays in the arrangement of people over space. For example, countries located in arid to semiarid zones have physiological densities that are at least 30 times as high as their crude densities. Countries like Mali, Chad, Niger, and Mauritania (all of which lie within the Sahel) and Namibia and Botswana (which have part of their territories in the Kalahari desert) lend credence to this fact. Also, a comparison of Angola and Benin shows that, although Angola is larger and has a much lower arithmetic density, its physiological density is higher than Benin's, an effect of the presence of the Namib Desert and of a hilly environment covered by an extensive plateau.

Figure 7.1 reveals a density pattern consistent with the characteristics just described. Major zones of dense settlement include the west African coastal belt stretching from Dakar, Senegal to Libreville, Gabon, and a north–south belt stretching from the Ethiopian Highlands down through Lake Victoria and the copper belt of the Democratic Republic of Congo and Zambia and ending in the Witwatersrand district of South Africa. Three broad zones of sparsely populated regions include, from north to south, the Sahel region extending from Dakar in the west to Mogadishu in the east, the west central forest regions of the Democratic Republic of Congo and Gabon, and the arid–semiarid region of southwest Africa.

These spatial distributions coincide with a number of environmental (vegetation, soil, climate, topography), developmental (levels of urbanization, industrialization, and agricultural development), and sociopolitical characteristics (oppressive regimes, ethnic disputes, resettlement schemes). For example, the west African coastal strip contains most of the region's urban, economic, and political centers. (See Chapters 8 and 10.) Other economic centers, such as the copper belt of the Democratic Republic of Congo and Zambia, the diamond and gold mining centers of South Africa's Witwatersrand region, and the tourist centers of the Lake Victoria borderlands attract large population clusters. Ethnic disputes and political persecutions have also triggered a massive flow of refugees to a number of African countries, as described later on in the chapter.

These patterns of uneven distribution are also evident on a microscale within Sub-Saharan countries. In Tanzania, the sparsely settled, environmentally hazardous central region is surrounded by dense population clusters on the fertile slopes of Mount Kilimanjaro in the north, on the shores of Lake Victoria in the northwest, on the shores of Lake Malawi in the southwest, and in the economic center, Dar-es-Salaam on the east coast. Government policy on resettling people into cooperative, nucleated settlements (Ujamaa villages) has also had

TABLE 7.2 Arithmetic and Physiological Population Densities in Sub-Saharan Africa

	Land Area (mi^2)	Population per mi^2	Crop Land (mi^2)	Density per mi^2 arable land
Sub-Saharan Africa	9,378,573	72	618,494	1088
Angola	481,350	26	13,834	889
Benin	43,483	152	7154	923
Botswana	224,606	7	5301	302
Burkina Faso	105,792	116	13,216	931
Burundi	10,745	577	5150	1204
Cameroon	183,568	86	27,042	584
Cape Verde	1560	256	151	2649
Central African Republic	240,530	15	7745	465
Chad	495,753	18	12,375	703
Congo Republic	132,046	406	645	83,101
Côte d'Ivoire	124,502	132	14,104	1163
Democratic Republic of Congo	905,351	59	30,309	1768
Equatorial Guinea	10,830	46	888	563
Ethiopia	426,371	153	53,784	1216
Gabon	103,347	12	1745	688
Gambia	4363	321	672	2083
Ghana	92,100	216	10,425	1909
Guinea	94,927	80	2807	2708
Guinea–Bissau	13,946	86	1293	928
Kenya	22,4081	133	9359	3184
Lesotho	11,718	188	1236	1780
Liberia	43,000	74	1436	2228
Madagascar	226,656	72	11,888	1380
Malawi	45,745	230	9232	1137
Mali	478,838	23	8058	1365
Mauritania	395,954	7	768	3516
Mauritius	788	1523	409	2934
Mozambique	309,494	63	11,958	1622
Namibia	318,259	6	2556	704
Niger	489,189	21	13,896	748
Nigeria	356,668	355	120,985	1046
Rwanda	10,170	718	4436	1646
Senegal	75,954	128	20,178	481
Sierra Leone	27,699	195	6954	777
Somalia	246,201	30	4008	1871
South Africa	471,444	92	50,857	857
Sudan	967,494	33	48,259	659
Swaziland	6703	164	633	1738
Tanzania	364,900	99	20,232	1789
Togo	21,927	237	5552	937
Uganda	93,066	258	25,888	927
Zambia	290,583	34	20,224	485
Zimbabwe	150,873	76	10,795	1056

Source: Compiled from the Population Reference Bureau, 2001, and from the World Resource Institute, 1998.

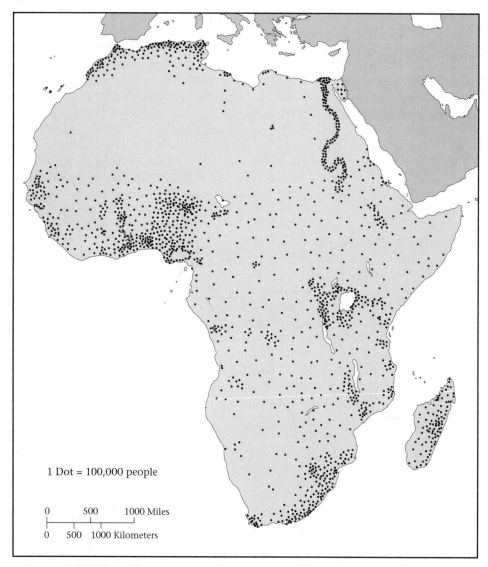

Figure 7.1 Sub-Saharan Africa: Population Density. (Data Compiled from Population Reference Bureau, 2001.)

its impact. In Botswana, the majority of its 1.4 million people are located in the well-endowed and best developed area, the eastern strip, far removed from the desolate Kalahari in the south and west. In the Democratic Republic of Congo, the population density in the great forest region is about half the national average of 49 people per square mile (19 per square kilometer). Chad's southern region is its agricultural heartland—the center for cotton and groundnut cultivation. This region has twice the annual rainfall of the northern Sahel.

THE DYNAMICS OF POPULATION CHANGE

Population growth is a function of birth (or fertility), mortality, and net migration. Data on migration are often difficult to track, and most African countries do not provide accurate estimates; therefore, birth, fertility, and mortality rates are primarily relied on to gauge net changes in population. Several estimates on fertility and mortality have improved as more and more countries participate in the World Fertility Survey and the Demographic and Health Survey.

The evidence from Table 7.1 suggests that the *rates of natural increase* (the difference between birth rates and death rates) in Sub-Saharan Africa are relatively high. The average rate, 2.5%, is almost double the world average, 1.3%. (A 2.5% annual rate of growth can result in a country doubling its population in 28 years.) These high rates are attributed to the fact that fertility rates remain high, while mortality levels have experienced dramatic declines in the past 30 years. On a regional scale, west and east Africa have above-average rates, with countries like Togo, Côte d'Ivoire, Niger, and Tanzania leading the way; southern Africa has below-average rates. The consensus among population geographers is that rates of natural increase close to 2.0% are indicative of stabilizing populations.

Fertility Levels and Trends

Sub-Saharan Africa continues to maintain the highest total fertility rates in the world. The *total fertility rate* is the average number of children a woman will bear during her reproductive years, usually between 15 and 49 years old; some analysts however have expanded this range to include from 10 to 55 year olds. Families in the region average an estimated 5.6 children, although there is considerable variation by region, socioeconomic status, and place of residence (rural vs. urban).

While the majority of countries in east and west Africa have fertility rates above the 5.6 average, several countries in middle and southern Africa have below-average rates. (See Table 7.1.) Doenges and Newman (1989) revealed a belt of markedly lower fertility rates in middle Africa stretching from southwest Sudan through the Central African Republic into the northern sections of the Democratic Republic of Congo. More evidence of low fertility was uncovered in parts of Cameroon, Gabon, the Lake Victoria region, the savanna-Sahel belt, highland Ethiopia and central Tanzania and among the San peoples of Botswana and Namibia. These zones of "impaired fertility" had high incidences of gynecological disorders and pelvic inflammatory diseases. Gonorrhea, for example, is mentioned as a major contributor to infertility among the Azande and Bakweri in the equatorial region, the Baganda and Teso in Uganda, and some Fulani groups in west Africa. Syphilis has also lowered the fertility levels of the Nzakara of the Central African Republic, and schistosomiasis is widespread around Lake Victoria. Although Table 7.1 provides evidence of lower fertility rates in Central African Republic and Gabon, Doenges and Newman found even lower rates, in the range from 2.0 to 5.0, among specific groups.

Disease vectors are not solely responsible for low fertility rates in Sub-Saharan Africa. Countries such as Botswana, Ghana, Kenya, and Zimbabwe have made significant inroads in their family-planning efforts to reduce fertility, as will be seen later. South Africa also has low fertility rates; however, there are substantial differences among its races.

Further evidence shows fertility differentials between rural and urban areas and by socioeconomic status. Malawi's Demographic and Health Survey in 2000 revealed a rural–urban differential between 6.7 and 4.5. Zambia and Namibia's 1992 Demographic and Health Survey revealed differentials of from 7.1 to 5.8 and from 6.3 to 4.0, respectively. Fertility levels also varied by educational levels in each of these countries, with declines associated primarily with those with a postsecondary-level education. In Malawi, fertility rates were 7.3 for those with no education, 6.7 for those with a primary-level education, 6.0 for those with a secondary-level education, and 3.0 for those with a postsecondary-level education.

Causes of high fertility Population geographers identify two types of determinants associated with fertility rates: The first are the direct or "proximate determinants" that relate to the behavioral and biological aspects of fertility; the second are indirect factors, such as socioeconomic (one's income, education, and residential status, as described in the previous section), cultural, historical, environmental, and politicoinstitutional (government policy, political will, effectiveness of medical institutions) factors. Examples of proximate determinants include marriage or nuptiality, postpartum abstinence, breast-feeding, lactational amenorrhea, and natural and pathological sterility (Sonko, 1994; Jolly and Gribble, 1993; Bongaarts, *et al.*, 1990).

Marriage patterns in Sub-Saharan Africa have a number of features that are quite distinct from those in North America and Europe. Most marriages, particularly in traditional societies, are universal and occur at an early age. Also, high rates of remarriage and polygyny negate any potential effects that divorce or widowhood might have. Recent Demographic and Health Surveys indicate that the median age at first union among women ranges from 15.7 in Mali to 19.7 in Ondo State, Nigeria (Jolly and Gribble, 1993). Also, the proportion of women who ever married in the 20-to-24 age group was 98% in Mali in 1987, 83% in Uganda in 1988 to 1989, 75.8% in Togo in 1988, and 30.3% in Botswana in 1988 (van de Walle, 1993). Bongaarts, *et al.* (1990) further estimate that, after entering their first union, 90% of Sub-Saharan African women spend their remaining reproductive life in some kind of conjugal union, when remarriage and other forms of conjugal partnership are taken into account.

Polygyny is fairly widespread in the region. Pebley and Mbugua (1989) report that 20% to 50% of married men in southern west Africa and west central Africa have more than one wife. Polygyny might enhance fertility over all, but there is no consensus to that effect; women in such unions have lower fertility levels than those in monogamous marriages, because less time is shared with the husband.

Postpartum abstinence, abstaining from intercourse following birth, is one of the traditional methods of fertility control widely practiced in Sub-Saharan Africa. It is often associated with breast-feeding and lactational amenorrhea (the length of time during which ovulation is blocked by lactation). Breast-feeding delays the normal pattern of ovulation and protracts amenorrhea, thus having an effect on the spacing of children. Breast milk contains nutrients that protect against childhood diseases such as viral infections, kwashiorkor, marasmus, and diarrhea. Sonko (1994) estimates that, on average, women in Africa breast-feed for up to 18 months. Jolly and Gribble (1993) estimate that the mean duration ranges from 17.5 months in Liberia to 23.9 months in Burundi. Mean duration of abstinence ranges from 2.4 months in Burundi to 22.7 months in Ondo State, Nigeria.

Primary sterility (the inability to conceive) and *secondary sterility* (the inability to conceive after one or successive births) have both natural and pathological causes. Larsen (1994), who measured primary sterility as the proportion of childless women married before age 20, estimated it to be less than 3% in seven countries (Botswana, Burundi, Ghana, Kenya, Sudan, Togo, and Zimbabwe) and between 3% and 4% in nine countries (Benin, Côte d'Ivoire, Lesotho, Liberia, Mali, Mauritania, Nigeria, Senegal, and Uganda). Cameroon and Sudan had a rate of 5%. An age-specific sterility survey of 17 countries revealed that only 4% of Burundi women (1987) aged 20 to 24 were sterile, compared with 14% in Cameroon (1991). For those aged 40 to 44, the proportions were 39% and 70%, respectively.

In addition to Cameroon, Lesotho (72%) and Sudan (70%) have the highest proportion of sterility rates among women aged 40 to 44. In Sudan, female circumcision is said to increase the incidence of sterility. The consensus among those who have analyzed the effect of proximate variables on fertility is that they are strongly influenced by the cultural and traditional belief systems, primarily in traditional African society (Caldwell and Caldwell, 1987; Acsadi, *et al.*, 1990; Sonko, 1994).

Cultural determinants of fertility The belief systems, customs, traditions, and values of Sub-Saharan Africans have a significant impact on fertility levels. As Caldwell and Caldwell (1987) point out, the lineage-based systems in Africa (usually patrilineal) are so coherent that they offer greater resistance to the success of family-planning efforts than their counterparts in any other region in the world. The family lineage is seen as an extension of the past and a link to the future; therefore, there is a high premium placed on individual spiritual survival and lineage survival. It is common practice to name children after ancestors to celebrate their rebirth. Among the Kikuyu of Kenya, a child who dies is regarded as a temporary visitor, and the next child is regarded as the reincarnation of the same spirit and given the same name. Among the Yoruba of Nigeria, after a father's death, his children compete to reproduce a son to return him to the living again.

From a religious perspective, fertility is equated with virtue and spiritual approval; it is associated with joy, the right life, divine approval, and approbation by both living and dead ancestors (Caldwell and Caldwell, 1987). Conversely, barren women are treated with disdain and ostracized from society. Often, the perception is that barren women have entered into a pact with evil spirits. In some societies, a child's death during birth or a woman's suffering through a difficult birth is cynically viewed as a sign of evil or sin. Mwambia (1973) points out that the Meru of Kenya consider it a risk to keep a barren woman in the homestead, for fear that she would use sorcery to kill other wives' children. It should be pointed out that several horror stories concerning barrenness in Africa were published in anthropological accounts in the 1950s and 1960s. Recent evidence suggests a more sympathetic approach towards barren women.

The African family structure is male dominated, and decisions about reproduction and family size are usually deferred to the husband. This explains the low prevalence of contraceptive rates in Sub-Saharan Africa. A study of six countries revealed that, although the use of modern contraceptive devices was low (lower than 5%), 86.5% of the women surveyed in Burkina Faso and 78.6% of those in Mali indicated a desire to use family planning (Wawer, *et al.*, 1991). The survey further indicated that women had difficulty talking to their husbands about family planning.

Another factor influencing fertility is that children are regarded as economic assets—a source of wealth and prestige, as well as a labor reservoir for household chores and agricultural production. Children are also required to offer tribute to their parents. Caldwell and Caldwell's survey of a Nigerian professional group in 1987 found that between one-third and one-half of their incomes went to parents, grandparents, and members of the extended family. This flow of wealth to the elderly is socially sanctioned and a religiously expected tribute. Children fear retribution when they do not comply with such social regulations.

A high premium is placed on children, so African women, aspiring to elevate their status, comply with their husband's request to have more children. In Haya, Tanzania, women are recognized for high fertility to the extent of being consulted when male elders gather to discuss pertinent matters (Kamuzora, 1987; cited in Sonko, 1994). Unfortunately, the subservence of women reduces their role to childbearing and nurturing. In addition, they are responsible for discipline, home education, and general welfare and are, essentially, relegated to a status of structural and social dependency. (See Chapter 9.) This dependency is reinforced in societies (like Kenya) by the fact that women, after nurturing and caring for their children, do not have exclusive rights over them. Frank and McNicoll (1987) point out that, in Kenya, the bride wealth exchanged for a woman entitles the husband (and his lineage) to exclusive rights over the children. Furthermore, women are frequently denied access to land and are allowed only usufructuary rights to the land they labor on with their children.

Child fosterage, where children are raised and cared for by their grandparents or foster parents, is common practice in traditional African society. In west Africa, one-third of children are found living with others than their parents at any given time. According to Isuigo-Abanihe (1985), children are fostered to strengthen family ties, to offer companionship to widows, to enhance their education, and to assist with domestic chores. Fostering can enhance fertility over all, especially because parents do not have to bear the economic burden of caring for all the children they bear. In Caldwell and Caldwell's (1987) estimation, the institution of fosterage so weakens the link between biological parentage and the number of children that any discussion about the net economic costs of high fertility is rendered meaningless.

For the most part, cultural practices enhance fertility, but there are other instances where they act as constraints to fertility. Examples include the prohibiting of premarital fertility, the practicing of postpartum female sexual abstinence, and resort to grandmaternal terminal sexual abstinence (Page and Lesthaeghe, 1981; Caldwell and Caldwell, 1987). In Rwanda, where fertility rates are among the highest in Sub-Saharan Africa and where family-planning efforts have been ineffective, virginity is strictly enforced until marriage. If an unmarried woman carries a child, the child is believed to cause the death of all his or her maternal family (May, 1990). In such instances, people resort to such practices as hurriedly arranging a marriage or reducing the amount of food given to the expectant mother so she delivers a small, "premature" baby; furthermore, there is an extreme alternative, the inducing of abortion by means of plant extracts.

Anthropological and demographic evidence reveals that many women have definite ideas about desirable child spacing. Evidence from Nigeria and Togo suggests that the mean length of child spacing considered "healthy" was more than two-and-a-half years (Acsadi and Johnson-Acsadi, 1990). In some instances, traditional methods of abstinence and breast-feeding are reinforced by the existence of postpartum taboos. Chaga women of northern Tanzania, for instance, consider it a misfortune and are ashamed when they bear children in quick succession. Among the Hausa of northern Ghana and Nigeria, if a woman goes to her husband while she is still breast-feeding, it is believed that the child will become thinner, weaker, and less healthy. The Kung bushmen, on the other hand, reportedly resort to infanticide of the last-born child if a spacing failure occurs (Schoenmaekers, et al., 1981). Grandmaternal terminal abstinence is practiced to avoid conflict in the performance of motherly and grandmotherly duties. Indeed, abstinence, child spacing, and premarital infertility are encouraged, yet they are overwhelmingly overshadowed by other cultural forces: early marriage, polygyny, child fosterage, and the reproduction of the family lineage.

Another fertility-inducing factor related to culture is the ethnic rivalry that exists among Sub-Saharan societies. In countries like Kenya, Nigeria, Ethiopia, and Uganda, where ethnic tensions run high, there is intense competition for economic and political resources. According to Mazrui and Mugambwa (1986), communities are reluctant to commit ethnic suicide by reducing their pace of fertility, because numbers are relevant in the sharing out of resources and power. For example, the Baganda of Uganda, who constitute less than 20% of the population, are the most enterprising and productive and usually trigger interethnic competition. Likewise, the Kikuyu of Kenya and the Amharics of Ethiopia have assumed hegemonic roles and created resentment among other ethnic groups.

Cultural factors often are viewed as dominant in the explaining of fertility, but historical and colonial factors, although often ignored, have also had some impact. Mbacké (1994) points out that, in Francophone West Africa, the French influenced fertility through the Law of 1920, which forbade publicity about, and use of, modern contraceptives. Also, the emphasis on cash-crop agriculture and the imposition of forced labor placed undue burdens on Africans to fulfill the labor demands of external markets.

Mortality Levels

Mortality levels in Sub-Saharan Africa have declined substantially over the years, converging toward the levels associated with more developed countries. Improvements in health, sanitation, and nutrition standards; massive vaccination campaigns against measles, smallpox, and other diseases; and increased efforts on the part of World Health Organization and the International Red Cross have all contributed to this downward trend.

In spite of the falling death rates, there are still slight regional variations in mortality levels that reflect environmental, epidemiological, economic, and sociocultural factors. Death rates in the drought-prone central Sahel states, Niger, Mali, and Chad, are above the

Sub-Saharan average, 14 deaths per 1000. Countries with a high incidence of AIDS-related cases (Malawi, Zambia, South Africa, Botswana, Central African Republic, and Uganda) and those that have experienced social unrest, civil war, and political upheaval (Angola, Sierra Leone, and Mozambique) also have relatively high mortality rates.

A major portion of deaths occurs among infants. *Infant mortality* (infant deaths between 0 and 1 year) and *child mortality* (deaths between 0 and 5 years) rates have declined in Sub-Saharan Africa since the 1950s. In Cameroon, infant deaths declined from 172 per 1000 live births in 1960 to 77 per 1000 in 2000. In Namibia and Tanzania, infant deaths were as high as 150 per 1000 in the 1950s, but have since declined to 68 and 99 per 1000, respectively. In spite of these declines, the average rate, 94 infant deaths per 1000, remains relatively high compared with the average rate in more developed countries, 10 per 1000. The southern African states of Botswana, Zimbabwe, and South Africa, countries with better health care and nutritional programs, have the lowest infant mortality rates in Sub-Saharan Africa; Angola, Sierra Leone, Liberia, and Mozambique have among the highest. In Botswana, for example, the probability of dying by the age of 5 is less than 0.05; in wartorn Mozambique, it is four times higher (0.27) (Hill, 1993). Angola's high infant-death rate is attributable largely to shortage and poor quality of water, poor hygiene, and the underutilization of existing health-care facilities.

Malnutrition, disease, and poverty are the more common causes of infant and child mortality; one cause that is becoming problematic is the growing trend of pregnancies associated with those who have escaped the traditional social controls of elders. LeGrand and Mbacké (1993), who studied these trends in the urban Sahel, conclude that infants born to teenage mothers suffer greater health problems and mortality risks, because teenage pregnancy is associated with worse health-care and vaccination behavior and lower birth weights. Hobcraft (1983) states that the average relative risk of death before age 5 is about 46% higher for children born to mothers under 18 than that for children with mothers between 20 and 34 years of age.

Africa and the Demographic Transition

The demographic transition model was utilized to describe the population history of industrializing Europe. It hypothesizes that European countries went through four demographic stages associated with their level of economic development (Figure 7.2). The first, the *high stationary stage*, is characterized by high birth and death rates; it was characteristic of preindustrial Europe, which was vulnerable to diseases and plagues. Currently, there are no whole countries in this stage, but there are some primitive and isolated societies, such as the Hottentots of the Kalahari and the Pygmies of the rain-forest region. Most African countries are in the second stage, *a high expanding stage*, characterized by declining death rates, but consistently high birth rates. At the third stage, the *late expanding stage*, birth rates begin to decline, while death rates decline further; China, Taiwan, Singapore, and South Korea best typify this stage. The declining birth rates are a latent response to economic development, as people adjust to economic growth, urbanization, and industrial development. In Africa, South Africa, Zimbabwe, Ghana, and Botswana seem to be approaching the third stage. At

The Dynamics of Population Change

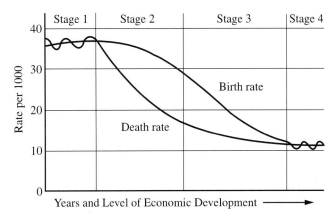

Figure 7.2 Demographic Transition Model.

the final stage, the *low stationary stage*, birth and death rates are low, indicating some degree of economic stability associated with industrial development.

Age Composition and Dependency Burdens

Aggregate population estimates conceal the behavioral attributes of various demographic subgroups. Geographers utilize graphics in the form of pyramids to disaggregate populations into *age* and *sex cohorts*; this allows a more accurate (policy) assessment of short- and long-term trends in fertility and mortality. Along the vertical axis of population pyramids, female and male populations are divided into 5- or 10-year age groups (Figure 7.3). The base of the pyramid represents the youngest age group, the apex the oldest. The horizontal dimension of the pyramid reflects the size of female and male age groups as a percentage of total population.

Sub-Saharan Africa's population is very young. (See Table 7.1.) The majority of countries have more than 43% of their populations under 15 years of age; the Sub-Saharan region averages 44%. A comparison of Figures 7.3a and 7.3b shows that Sub-Saharan Africa has a markedly broader-based pyramid than the United States, a reflection of the higher fertility rates and dependency ratios in the former. The US pyramid is also "top heavy," reflecting higher life expectancy rates relative to Africa. Figures 7.3c and 7.3d show the effects of family-planning programs on the shape of pyramids. In Angola, where family-planning programs have been difficult to implement, the base of the pyramid is wider than in Kenya, where programs have been implemented with relative success.

These broad-based pyramids imply that the Sub-Saharan African work force faces the economic burden of supporting a large segment of its youth. It also implies the need for large subsidies for health care, education, and job-training programs to accommodate future employment. This momentum of younger populations will carry over into the future and will produce demands for future housing, employment, and job benefits. It will also translate into increased fertility rates when these children reach their reproductive years. Even in

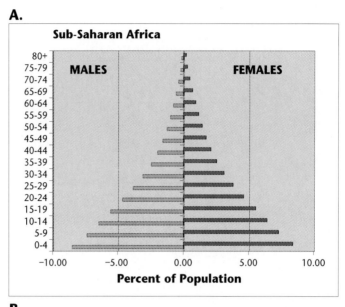

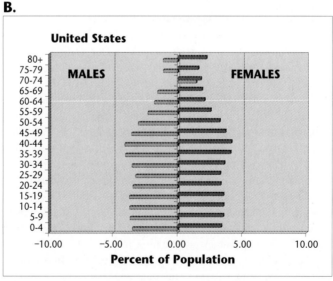

Figure 7.3 Age–Sex Structure, 2000.

those situations where African countries begin to reduce their age-adjusted fertility rates, broad-based pyramid populations will still increase substantially before leveling off.

Most population analysts recommend a *replacement fertility rate* of 2.1 as a standard for populations in developing countries to level off. The replacement fertility rate is the level of fertility at which childbearing women have just enough children to replace themselves in the population. This keeps the existing population size constant through an infinite

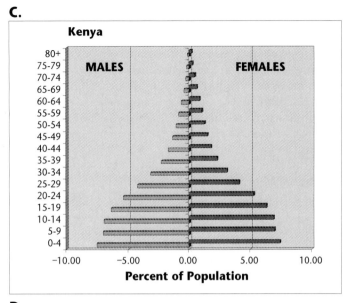

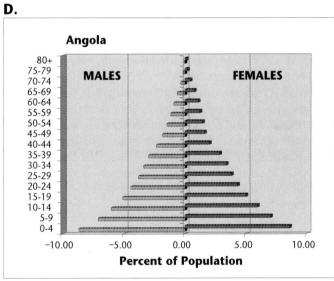

Figure 7.3 (*Continued*)

number of succeeding generations. The World Bank's fertility projections indicate that the only Sub-Saharan countries that will reach these levels by 2030 are Botswana, South Africa, Kenya, Namibia, Lesotho, Zimbabwe, and the small island countries: Cape Verde, Réunion, Seychelles, and São Tomé and Príncipe (Keyfitz and Flieger, 1994). By 2030, the rest of Sub-Saharan Africa will have reached a fertility level of 3.0 and will have added at least 900 million more people to its current population base.

INTERNATIONAL MIGRATION AND REFUGEE ISSUES

Migration involves the movement from one administrative unit to another, resulting in a change in permanent residence. Geographers study migration because it is a dynamic process involving mobility and interaction between places. Migration can be internal, involving movement from residence to residence (intraurban mobility), from rural to urban places, from rural to rural places, from urban to rural places, or from rural to urban to rural places (circular migration). Rural–urban migration is addressed in Chapter 10, which deals with urban geography. This section touches on international and refugee migrations.

International Migration Trends

Official statistics on international migration are difficult to come by and are often plagued with deficiencies; however, recent estimates from the International Organization of Migration indicate that of the world's 150 million migrants, more than 50 million are Africans. Of these, 3 to 5 million are refugees and 20 million are displaced persons. Most of the migration is intracontinental, and the majority of countries have more emigrants than immigrants (Figure 7.4). Migratory flows are dictated by regional variations in economic development. In west Africa, the cocoa farms of Côte d'Ivoire attract migrants from neighboring Burkina Faso, Liberia, and Mali. About 40% of Côte d'Ivoire's 16 million people are foreigners, including 4 million from Burkina Faso (Migration News, 2001). Oil-rich Nigeria has been a magnet for migrants in the past, but problems with the Nigerian economy have forced the government to repatriate out foreign workers. In central Africa, the oil and forest resources of Gabon have attracted migrants from as far away as Nigeria and Mali and from as near as Cameroon and Equatorial Guinea. In east Africa, Kenya has been the benefactor of educated Ugandans, in spite of its current negative net migration rate. South Africa has traditionally been a magnet for migrants in southern Africa; however, the country has had to clamp down on the from 2 to 4 million illegal migrants. In 1996, the Southern African Migration Project, a multifaceted research, policy, and training program, was established to facilitate the formulation and implementation of new initiatives on cross-border population migration in the region (McDonald, 2000).

Labor mobility, a prominent aspect of international migration in Africa, has disadvantages: labor-exporting countries lose able-bodied and skilled people, and the influx of labor into the host country overburdens its job market, making it spend more on welfare services for the unemployed. On the positive side, migrants can provide needed skills and contribute to the host country's economic growth. As the workers leave, they relieve the source country of pressure on its labor market, and they send home remittances of the income they earn. Labor-mobility immigration was considered an important part of Côte d'Ivoire's economic growth, contributing to its plantation agriculture, industry, and commerce (World Bank, 1989). Also, Gabon's oil and mineral wealth requires a supply of manpower roughly double the internal supply; therefore, it attracts labor from Cameroon and the Congo (Russell, *et al.*, 1990).

International Migration and Refugee Issues

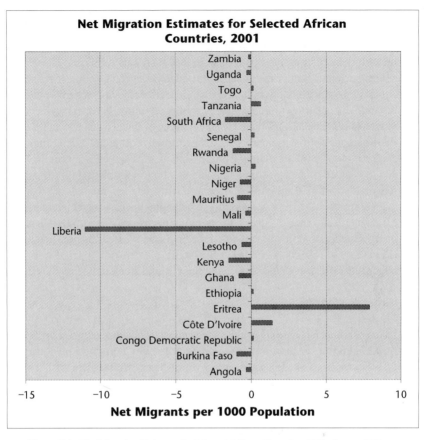

Figure 7.4 Net Migration Estimates in Selected African Countries, 2001. *Source:* U.S. Central Intelligence Agency, (2001). *The World Factbook 2001*, Washington, DC: Brassey's.

Remittances are of particular relevance to source countries. Between 1997 and 1999, remittances to the Sudan averaged about $665 million[1] a year (IMF, 2000). Average remittances into Ethiopia and Ghana exceeded $30 million; those for Benin, Mali, and Côte d'Ivoire exceeded $90 million. Nigeria's remittance income in 1999 was $1.5 billion.

Another concern about international migration has been the "brain drain" of African intellectuals and students. In 1987, it was estimated that 70,000 high-level Sub-Saharan Africans were officially resident in European Community countries; in 1984, the number of highly skilled Nigerians in the United States was about 10,000 (Ricca, 1989). Somalia and Ethiopia, for instance, have lost skilled workers to the Gulf states. The International Labor Organization estimated that, by 1985, two-thirds of Sudan's professional and technical workers had left the country.

---

[1] In US dollars.

Ricca (1989) offers a number of reasons why the extent of the brain drain has perhaps been minimized in Africa. First of all, an increase in highly skilled personnel and increased intra-African migration mean that some of the exodus of highly skilled personnel is now being offset by equally qualified people from other African countries. Therefore, the net loss of skilled personnel is not as dramatic as it used to be. Second, people leave because the jobs they are trained for are already saturated; therefore, their leaving actually relieves pressure on the job market, while the source country benefits additionally from income remittances, as previously mentioned. Third, there have been international efforts to facilitate the return of qualified personnel to Africa. In 1983, the Intergovernmental Committee for Migration, in conjunction with the Economic Commission of Africa, the European Economic Community, and the United States government, launched the Return of Talent Program. The program offers highly qualified Africans who want to return home a number of incentives, such as payment of travel costs; shipping of personal belongings; salary matching for at least eight months; a transition allowance to cover costs of living for two months; a $10,000 allowance for equipment, supplies, and instruments; and medical and accident insurance coverage for one year (Russell, et al., 1990). As of 1987, Botswana, Cameroon, Ghana, Kenya, Lesotho, Malawi, Nigeria, Sudan, Swaziland, Tanzania, Zambia, and Zimbabwe had signed on.

Refugee Migration

Another concern regarding international migration is the refugee crisis, which has taken on further enormity recently. Estimates from the US Committee for Refugees (2001) reveal that there were 3,346,000 refugees (23% of the world total) in Africa south of the Sahara by the end of December 2000.

The principal sources of refugees in 2000 were eastern and central Africa (Figure 7.5). Religious and ethnic strife in Sudan has resulted in the export of 460,000 refugees. Ethnic genocide of Hutus and Tutsis in Burundi and civil war between UNITA and MPLA forces in Angola (see Chapter 5) has produced many refugees in neighboring countries. In Sierra Leone, conflicts over the control of diamonds and several occasions of human-rights violence incited 400,000 people to seek refuge in Guinea. Clan warfare, droughts, and famine drove an estimated 370,000 Somalian refugees into Kenya, Ethiopia, Djibouti, and Egypt. Genocide and political persecution had at one point caused several thousand Rwandans, many of whom have since returned, to seek refuge in Uganda, the Democratic Republic of Congo, and Burundi (Figure 7.6). The civil crisis in Liberia continues to destabilize the country. The rebel forces of the National Patriotic Front of Liberia ravaged the countryside and engaged in battles with the United Liberation Movement of Liberia for Democracy and west African peace-keeping troops. Civil unrest continues as the government in Monrovia struggles to restore civic order.

Measuring actual numbers of refugees can be problematic in view of the varied definitions offered by various international and regional entities. The most widely used definition comes from the 1951 Geneva Convention, which characterizes refugees as anyone who, "owing to well-founded fear of being persecuted for reasons of race, religion, nationality, membership of a particular social group or political opinion, is outside the country of his/her nationality and is unable or unwilling to avail him/herself of the protection of that country."

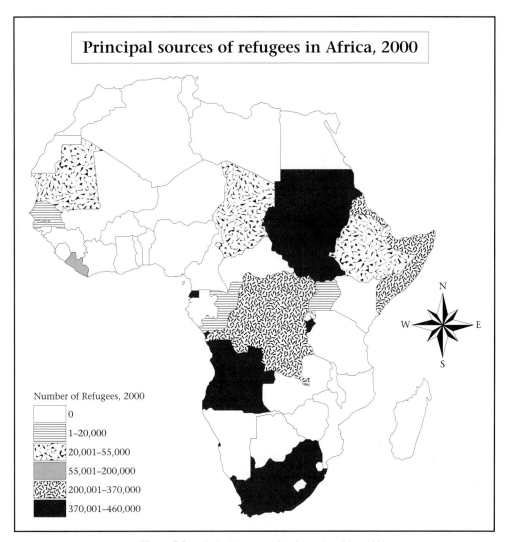

Figure 7.5 Principal Sources of Refugees in Africa, 2000.

This definition excludes *economic refugees*, who flee from economic deprivation and famine. In 1967, the Organization of African Unity's (OAU) Convention on Refugees modified the definition to include those who are forced by external aggression, occupation, or foreign domination to leave their country (Black, 1993). Then again, there are *internal refugees* who, in response to religious, ethnic, or political persecution, flee to havens within their country. Because they have not left the confines of their country, they are not provided with the guarantees and protections provided by the Geneva Convention, and they are ineligible for any benefits from the UN High Commissioner for Refugees (UNHCR). Bascom (1993) estimates that there are at least 4 million such internal refugees in Sudan.

Figure 7.6 Refugee Children in Burundi.

Granting refugee status is a humanitarian and peaceful act that cannot be regarded as hostile by the refugee's country of origin. Refugee status is also not an inherent right of an individual, but a sovereign right reserved for the state to grant or refuse (Ricca, 1989). The human rights of refugees in Sub-Saharan Africa are protected by legislative instruments that ensure legal recognition by the issuing of identity and travel documents.

The scale of refugee crises in Sub-Saharan Africa is cause for concern. Countries that host refugees inherit the responsibility of ensuring their social, economic, and political well-being. There are the logistical difficulties of allocating land for settlement, extending relief efforts to those settlements, and coordinating economic programs with private voluntary organizations and nongovernmental organizations (NGOs). Also, while governments try, they cannot completely protect refugees from acts of discrimination and violence.

The complexities of the refugee problem render programs and solutions difficult to implement. However, major sponsors of refugees, such as the UNHCR, NGOs, and the OAU, continue to provide assistance through emergency relief efforts, job and educational placement, investigation of individual cases, organizing of settlement camps, and organizing of repatriation. These organizations continue to explore various solutions to the crisis; three are (1) voluntary repatriation, (2) local integration, and (3) third-country settlement (Russell, *et al.*, 1990).

Voluntary repatriation is a preferred route, but is contingent upon the social, political, and economic stability of the source country. Such repatriations have taken place in Zimbabwe and, quite recently, in Namibia. In 1989, the UNHCR budgeted $38.5 million toward the Namibian repatriation operation. Of this, $9.5 million was for international transport, $7.8 million for predeparture and postarrival supplies and services, $6.0 million for program support and administration, and $5.2 million for the purchase and prepositioning of foodstuffs (Simon and Preston, 1993). The UNHCR was primarily responsible for registering Namibians abroad, administering vaccinations and medical treatment, and negotiating amnesty agreements for political offenses.

Local integration into host countries can be problematic, especially in cases where land has to be set aside for settlement. Also, the granting both of economic subsidies and of various forms of education and training assistance to refugees sometimes incurs the wrath of local residents who compete for similar forms of government assistance. Local integration can be appealing if development assistance from international sources is forthcoming. Third-country settlement is a last resort and is rarely employed.

GOVERNMENT POLICY AND FAMILY PLANNING

Sub-Saharan Africa has for some time taken a lukewarm attitude toward family planning. In view of their limited financial resources, governments understandably focused their attention on economic-development ventures. For a time, in fact, countries like Zambia pursued a pronatalist agenda, arguing that population growth was needed to sustain economic growth. Countries like Tanzania, Kenya, and Botswana at one time argued that they were underpopulated; Botswana's having 1.4 million people living in a country that is slightly larger than France (224,606 vs. 218,000 square miles) certainly makes this argument at least plausible.

There is also the argument that family planning per se cannot solve Africa's problems, which are deeply rooted in culture and tradition. Therefore, a better approach would be to devote scarce resources to economic development that, in turn, will correct the attendant problems associated with population. Population and economic development are not mutually exclusive issues; a strategy that tackles each problem in isolation is certainly not viable. An integrative policy that deals with the demographic and socioeconomic aspects of family planning is the logical route to take.

Sub-Saharan countries are now beginning to change their attitude toward family planning. They have taken their cue from the three World Population Conferences held in Bucharest (1974), Mexico (1984), and Cairo, Egypt (1994). The first conference at Bucharest laid the foundation for the adoption of the World Population Plan of Action, which is the main instrument that provides guidance and serves as the standard reference on population issues. The document covers a range of topics: the sovereign rights of nations; human rights; the status of women; international equity; the interrelations between population, the environment, and development; population redistribution; and the role of nongovernmental and intergovernmental organizations (United Nations Secretariat, 1994).

At the Third African Population Conference, held in Dakar, Senegal in December of 1992, African leaders pledged to improve the quality of lives of Africans and addressed most of the issues covered in the Plan of Action document. A major goal set at the conference was to reduce the regional natural growth rate from 3.0% to 2.5% by the year 2000 and to 2% by the year 2010. At this juncture, these seem to be pretty ambitious goals. Current evidence suggests that fertility decline is slow, although countries such as Zimbabwe, Botswana, and Kenya have made some headway. Other goals set at the Dakar conference include (1) doubling the regional contraceptive prevalence rate from 10% to 20% by the year 2000 and again to 40% by 2010 and (2) achieving a life expectancy of 55 years and an infant mortality rate of 50 per 1000 live births by 2000. To fulfill these goals, African governments must have the political will to commit themselves to comprehensive family-planning strategies that are consistent with broader environmental and socioeconomic development policies.

The following areas will also need to be addressed:

1. Balancing national sovereign rights with individual human rights.
2. Creating effective delivery of modern birth control methods and information systems by integrating family planning with maternal and other health-care programs, through community-based, market-based, and social-marketing strategies.
3. Improving the status of women.
4. Targeting at-risk populations effectively.
5. Forging effective partnerships with nongovernmental organizations and the private sector.
6. Developing appropriate strategies to deal with the refugee crisis.

Sovereign Rights of Nations and Human Rights

The World Population Plan of Action and the draft document for the 1994 Cairo conference emphasize the sovereign right of each nation around the world to formulate and implement population policies that are consistent with its moral, cultural, and life experiences. All countries are also advised to respect and ensure the rights of couples and individuals to freely choose family size and spacing of children. Any form of forced or imposed family-planning programs is considered a breach of human rights. This presents a dilemma for countries like China, which has enforced a one-child policy. In the context of Sub-Saharan Africa, it implies enabling families to make informed choices by offering them a set of comprehensive and integrative strategies.

Integrative Family Planning Policies

One way of offering men and women the information and means to achieve their family goals is to integrate family-planning programs with maternal and child-care programs. Such programs are cost effective, offering clients both services simultaneously and avoiding long delays between referrals from maternal health-care clinics to family-planning clinics. One of the constraints on Rwanda's family-planning policy, which was adopted in 1981, was the slow implementation of an integrated family-planning service-delivery system (May, *et al.*, 1990). Kenya, which experienced a 46% decline in fertility between the early 1970s and 2001, had implemented integrated family planning with maternal and child health services as early as the mid-1960s.

Community and Market-Based Strategies and Social Marketing

Family-planning service delivery can be further enhanced through community-based, market-based, and social-marketing strategies. Community-based approaches are geared more toward localities and rural districts; they frequently incorporate nonphysicians (midwives and other volunteer workers) in the delivery of family-planning and health-care services. A key objective is to improve access to family-planning services by removing geographic, financial, bureaucratic, and communication barriers (Kols and Wawer, 1982). Zimbabwe, which has one of the most effective family-planning programs in Sub-Saharan Africa, operates a community-based distribution system with a cadre of mobile clinics and field educators who make home visits to talk with women about family planning (Piotrow, *et al.*, 1992). This partly explains the drop in Zimbabwe's fertility rate from 7.4 in the early 1970s to 4.0 in 2001 (Table 7.1). Rwanda's National Office of Population trains personnel to conduct information, education, and counseling activities at the grass-roots level. In Ghana, a midwives' project trained more than 200 midwives countrywide in family-planning delivery. In Nigeria, a market-based distribution project utilized traders in traditional markets to sell modern contraceptive devices (Ladipo, *et al.*, 1990).

Another way of disseminating information is through social marketing, which employs commercial-marketing techniques, such as popular radio and TV shows, to achieve family-planning goals. In Ghana and Nigeria, information on family planning was incorporated into story lines of popular TV programs (Olaleye and Bankole, 1994), which was complemented with information on the location of clinical facilities.

Female Empowerment

As indicated in the previous section, family planning must take into consideration the respect and dignity of all persons involved; this includes the prohibition of all forms of discrimination against women and the changing of laws and instruments that are detrimental to the advancement of women. The 1992 African Population Conference called for the passing of legislation to enhance the legal status of women within the family and the community, in order to enable women to contribute more directly in decision making and in the formulation of strategies aimed at upholding family values and at contributing to the stability of African societies. The conference also recommended the participation of women in all production sectors and the provision for them of appropriate technologies, better access to credit facilities, and greater economic independence. These are laudable recommendations; however, for African countries caught in the legal tangles of civil versus customary law, they are difficult to implement in traditional societies. In Kenya, for example, customary law regarding marriage, property rights, and inheritance dictates the division of responsibilities between husband and wife (Frank and McNicoll, 1987). The Kenyan government has tried to establish a uniform set of national laws, but with little effect. The government is reluctant to offend the customary principles and traditions of the ethnic groups that reside within its legal bounds.

Effective Targeting of At-Risk Populations

While most family programs in Sub-Saharan Africa target women, it should also be emphasized that men must not be left out of the family-planning equation, because they make important decisions about family size and initiate traditional methods, such as rhythm. Zimbabwe, in 1988, modified its female-biased family-planning program and developed a three-year male-motivation project, the first of its kind in Sub-Saharan Africa (Piotrow, 1992). The project was designed to encourage more favorable attitudes about family planning among men and to promote joint decision making about fertility control and family size. Social-marketing techniques were employed to produce a radio drama series on family planning and to create a soap opera conveying motivational messages about socioeconomic hardships. Other at-risk groups that warrant attention are teenage adolescents and high-parity women. Older women who tend to get pregnant are often ill served by the health-care system.

Partnerships with NGOs and the Private Sector

Governments in Sub-Saharan Africa must cooperate with NGOs and agencies in the private sector to coordinate the delivery of services just mentioned. The International Planned Parenthood Federation, the Population Council, Family Planning International Assistance, Pathfinder International, Population Information Network, UN Population Fund, Ford Foundation, and Oxfam are among several NGOs devoting financial, technical, and human resources to population activities. The World Fertility Surveys and Demographic and Health Surveys provide evidence of increased NGO-sponsored research in biomedical and social activity.

Nongovernmental organizational efforts should be complemented by private-sector activity, which offers flexibility and innovation in health-care delivery. Enhanced private-sector and NGO activity facilitates information, education, and communication programming by means of social marketing and other devices. Given the fact that per capita funding for family planning in Africa is at less than half the level in Asia and Latin America, African governments must have the political will to forge effective partnerships with NGOs and the private sector to embark upon more effective and efficient family-planning strategies.

Programs for Refugees

Finally, while the solutions presented address problems associated with fertility and mortality, we must not discount the problems emanating from refugee flows. African governments must address the root causes of the refugee crisis by providing the appropriate mechanisms for conflict resolution, by respecting the human rights of refugees, by supporting voluntary repatriation efforts, and by assisting host countries with development aid packages specifically targeted for refugees.

CONCLUSION

Populations in Sub-Saharan Africa are still growing rapidly. This growth is being spurred by high birth and fertility rates, which, in turn, are driven by strong cultural and traditional forces. This growth is reinforced by the increasing number of young dependents and the failure of some governments to respond in a timely and appropriate manner to the dilemma. The resultant effect is the continued pressure on arable land and an increase in the demand for scarce governmental resources. These problems are further compounded by the crises associated with migration and especially with refugee migration.

Some governments, such as those in Kenya, Zimbabwe, Ghana, Botswana, and South Africa, have responded quite well by adopting comprehensive family-planning programs that have led to some declines in fertility rates. Community-based and integrated health-care approaches have had a positive impact. Most policies tend to target women, but men should also be included in family-planning counseling and advisory sessions. It is important, however, to realize that family-planning programs cannot occur in isolation. They need to be pursued in conjunction with a broad set of social and economic development strategies. Creating a slew of family-planning programs is ineffective as long as families are deprived of socioeconomic benefits and opportunities. Therefore, governments need to adopt an integrative and holistic approach that reflects the social, cultural, economic, and institutional factors that account for the population dilemma in Sub-Saharan Africa.

KEY TERMS

Arithmetic population density
Physiological population density
Age and sex cohorts
Birth rates
Child fosterage
Death rates
Demographic transition model
Dependency ratios
Doubling time
Infant and child mortality

Impaired fertility
Sterility
Polygyny
Postpartum abstinence
Refugee
Internal refugee
Economic refugee
Rate of natural increase
Fertility rates
Replacement fertility rate

DISCUSSION QUESTIONS

1. What are the major indicators that population geographers employ in analyzing demographic trends?
2. Describe the patterns of population distribution in Sub-Saharan Africa.
3. What are the major causes of high fertility in Sub-Saharan Africa?

4. Discuss the effect that sociocultural factors have on fertility rates in Sub-Saharan Africa.
5. Describe the age and sex characteristics of African populations. What are the policy implications of these characteristics?
6. Discuss and evaluate the refugee crisis in Sub-Saharan Africa. What, in your opinion, are the root causes of this crisis?
7. Outline and assess the major population problems confronting African governments. To what extent can these problems be solved?
8. Design a feasible family-planning strategy for African countries, one that takes into consideration human-rights issues.

REFERENCES

ACSADI, G., et al. (1990). *Population Growth and Reproduction in Sub-Saharan Africa*, Washington, DC: World Bank.

ACSADI, G. & JOHNSON-ASCADI, G. (1990). "Demand for Children and Child-Spacing," in *Population Growth and Reproduction in Sub-Saharan Africa*, Acscadi, G., Johnson-Ascadi, G. & Bulatao, R., eds., Washington, DC: World Bank.

BASCOM, J. (1993). "Internal Refugees: The Case of the Displaced in Khartoum," in *Geography and Refugees*, Black, R. & Robinson, V., eds., London: Belhaven Press.

BONGAARTS, J., FRANK, O., & LESTHAEGHE, R. (1990). "The Proximate Determinants of Fertility," in *Population Growth and Reproduction in Sub-Saharan Africa*, Acsadi, G., Johnson-Ascadi, G., & Bulatao, R., eds., Washington, DC: World Bank.

BLACK, R. (1993). "Geography and Refugees: Current Issues," in *Geography and Refugees*, Black, R. & Robinson, V., eds., London: Belhaven Press.

CALDWELL, J. (1994). "Fertility in Sub-Saharan Africa: Status and Prospects," *Population and Development Review*, 20(1):179–187.

CALDWELL, J. & CALDWELL, P. (1987). "The Cultural Context of High Fertility in Sub-Saharan Africa," *Population and Development Review*, 13(3):409–437.

DOENGES, C. & NEWMAN, J. (1989). "Impaired Fertility in Tropical Africa," *Geographical Review*, 79(1):99–111.

FRANK, O. & MCNICOLL, G. (1987). "An Interpretation of Fertility and Population in Kenya," *Population and Development Review*, 13(2):209–243.

HILL, A. (1993). "Trends in Child Mortality," in *Demographic Change in Sub-Saharan Africa*, Foote, K., Hill, K., & Martin, L., eds., Washington, DC: National Academy Press.

HOBCRAFT, J. (1983). "Child-Spacing Effects on Infant and Early Child Mortality," *Population Index*, 49(4):585–618.

INTERNATIONAL MONETARY FUND (IMF) (2000). *IMF Balance of Payments Statistics Yearbook*, Annual Report, Washington, DC: IMF.

ISUIGO-ABANIHE, U. (1985). "Child Fosterage in West Africa," *Population and Development Review*, 11(1):53–73.

JOLLY, C. & GRIBBLE, F. (1993). "The Proximate Determinants of Fertility," in *Demographic Change in Sub-Saharan Africa*, Foote K., Hill, K. & Martin, L., eds., Washington, DC: National Academy Press.

KAMUZORA, L. (1987). "Survival Strategy: The Historical and Economic Roots of an African High Fertility Culture," in *The Cultural Roots of African Fertility Regimes: Proceedings of the Ife Conference*, Ebigbola, J. & van de Walle, E., eds., Ife, Nigeria: Obafemi Awolowo University.

KEYFITZ, N. & FLIEGER, W. (1994). *World Population Growth and Aging: Demographic Trends in the Late 20th Century*, Chicago: University of Chicago Press.

KOLS, A. & WAWER, M. (1982). "Community-Based Health and Family Planning," *Population Reports*, Series L, 3.

LAPIDO, O., MCNAMARA, G., DELANO, G., WEISS, E., & OTOLORIN, E. (1990). "Family Planning in Traditional Markets in Nigeria," *Studies in Family Planning*, 21(6):311–321.

LARSEN, U. (1994). "Sterility in Sub-Saharan Africa," *Population Studies*, 48:459–474.

LEGRAND, T. & MBACKÉ, C. (1993). "Teenage Pregnancy and Child Health in Urban Sahel," *Studies in Family Planning*, 24(3):137–149.

MAY, J. (1990). "Family Planning in Rwanda: Status and Prospects," *Studies in Family Planning*, 21(1):20–32.

MAZRUI, A. & MUGAMBWA, J. (1986). "Population Control and Ethnic Rivalry: Local and Global Perspectives," *Ethnic and Racial Studies*, 9(3):334–358.

MBACKÉ, C. (1994). "Family Planning Programs and Fertility Transition in Sub-Saharan Africa," *Population and Development Review*, 20(1):188–193.

MCDONALD, D., ed. (2000). *On Borders: Perspectives on Cross-Border Migration in Southern Africa*. Cape Town, South Africa: SAMP/Idasa.

MIGRATION NEWS (2001). Africa: Ivory Coast and South Africa, vol. 8, #3. March 2001. http://migration.ucdavis-edu/mn/cir_mn.html

MWAMBIA, S. (1973). "The Meru of Central Kenya," in *Cultural Source Material for Population Planning in East Africa, vol.3, Beliefs and Practices*, Molnos, A., ed., Nairobi, Kenya: East Africa Publishing House.

OLALEYE, D. & BANKOLE, A. (1994). "The Impact of Mass Media Family Planning Promotion on Contraceptive Behavior of Women in Ghana," *Population Research and Policy Review*, 13:161–177.

PAGE, H. & LESTHAEGHE, R. (1981). *Child-Spacing in Tropical Africa: Traditions and Change*, London: Academic Press.

PEBLEY, A. & MBUGUA, W. (1989). "Polygyny and Fertility in Sub-Saharan Africa," in *Reproduction and Social Organization in Sub-Saharan Africa*, Lesthaeghe, R., ed., Berkeley: University of California Press.

PIOTROW P., et al. (1992). "Changing Men's Attitude and Behavior: Zimbabwe Male Motivation Project," *Studies in Family Planning*, 23(6):365–375.

POPULATION REFERENCE BUREAU (2001). Human Development Report, 2001, Washington, DC: Population Reference Bureau, Inc.

RICCA, S. (1989). *International Migration in Africa: Legal and Administrative Aspects*, Geneva, Switzerland: International Labor Office.

RUSSELL, S., JACOBSEN, K., & STANLEY, W. (1990). *International Migration and Development in Sub-Saharan Africa*, vol. I, Washington, DC: World Bank.

SCHOENMAECKERS, R., SHAH, R., LESTHAEGHE, R., & TAMBASHE, O. (1981). "The Child-Spacing Tradition and the Post-Partum Taboo in Tropical Africa: Anthropological Evidence," in *Child-Spacing in Tropical Africa: Traditions and Change*, Page, H. & Lesthaeghe, R., eds., London: Academic Press.

SIMON, D. & PRESTON, R. (1993). "Return to the Promised Land: Repatriation and Resettlement of Namibian Refugees, 1989–1990," in *Geography and Refugees*, Black, R. & Robinson, V., eds., London: Belhaven Press.

SONKO, S. (1994). "Fertility and Culture in Sub-Saharan Africa: A Review," *International Social Science Journal*, 141:397–411.

UNITED NATIONS HIGH COMMISSION FOR REFUGEES (1993). *The State of the World's Refugees: In Search of Solutions*, New York: Oxford University Press.

UNITED NATIONS SECRETARIAT (1994). "Third African Conference," *Population Bulletin of the United Nations*, 37–38:37–46.

UN COMMITTEE FOR REFUGEES (2001). *World Refugee Survey 2001*, Washington, DC: USCR.

VAN DE WALLE, E. (1993). "Recent Trends in Marriage Ages," in *Demographic Change in Sub-Saharan Africa*, Foote, K., Hill K., & Martin, L., eds., Washington, DC: National Academy Press.

WAWER, M., MCNAMARA, T., & LAURO, D. (1991). "Family Planning Operations Research in Africa: Reviewing a Decade of Experience," *Studies in Family Planning*, 22(1):279–294.

WORLD BANK (1989). *Sub-Saharan Africa: From Crisis to Sustainable Growth*, Washington, DC: World Bank.

PART III

Development Context

8

Geography and Development in Sub-Saharan Africa

Samuel Aryeetey-Attoh

INTRODUCTION

Development is a multidimensional phenomenon involving a broad set of economic, social, environmental, institutional, and political factors. It is not surprising, therefore, to find different interpretations of development from respective disciplines. Economists employ income and productivity variables to measure development; sociologists focus on social pathologies and cultural and behavioral attributes. Political scientists emphasize variables that determine the degree of political freedom, the effectiveness of political institutions, and the level of political participation; ecologists study the incidence of pollution and the effects of forest and land degradation.

Geographers, by contrast, examine development from a variety of perspectives. They approach development from a more holistic perspective, integrating as many dimensions as possible to provide a much broader perspective. Geographers go beyond aggregate measures of development to examine its *distributional* impacts on various segments of society. Also, geographers study the *spatial patterns* of uneven development between areas of modern (core regions) and marginal development (peripheral regions) and the underlying processes that explain these inequalities. Closely related to this is an analysis of the degrees and levels of *spatial interaction* between regions, such as rural–urban, urban–urban, and rural–rural interactions. These levels of interaction can be determined, for example, by examining the *spatial diffusion* of ideas and innovations from areas of modern development to areas of marginal development—a process referred to as *cascade diffusion*.

Cross-cultural analyses allow geographers to appreciate the value-laden aspects of development. Development priorities in Sub-Saharan Africa, which tend to focus on basic needs and survival strategies for the poor, are quite different from those in the United States. Development is also a normative concept, specifying what "ought to be done" under the present conditions and circumstances. Such normative questions are driven by the value judgements and behavioral norms of different societies.

Geographers further appreciate the importance of *scale* in their analyses of development, recognizing that communities at the local level are affected by regional, national, and global forces. Sub-Saharan Africa, like any other region in the world, is susceptible to global forces, and the critical question is whether the continent is converging toward or diverging away from our global economy.

Aspects of these geographic attributes of development are highlighted in this chapter. First, the definitional dimensions of development are addressed; then, the geographic patterns of inequality in Sub-Saharan Africa are examined. Historical, economic, institutional, and sociological explanations for these inequalities are then presented. The chapter concludes with a review of development strategies that have been employed to address development inequities in Sub-Saharan Africa.

DEFINING AND MEASURING DEVELOPMENT IN SUB-SAHARAN AFRICA

It is often difficult to arrive at a consensus on the definition of development, given its multidimensional attributes; however, Seers (1969) came up with a definition that captures the essence of development. He states that development involves "creating opportunities for the realization of human personality." In the context of Sub-Saharan Africa, this translates to providing adequate educational, health, and nutritional benefits to enhance human capabilities; creating avenues for women to ponder alternative life experiences; upgrading the physical infrastructure to facilitate the exchange of goods and services; conserving vital nonrenewable resources and sustaining stable environments; advocating human rights and creating avenues for the free expression of ideas; and marshaling all human and natural resources to build a better Sub-Saharan Africa and one that is more competitive in the global economy. For the African woman who labors several hours a day in the agricultural fields, the definition raises questions about whether she has enough time in her day to ponder opportunities for the realization of her human potential: And what about the children, urban poor, and peasant farmers—what kinds of opportunities are available to them?

To address these questions, we need to examine current development trends in the region from as comprehensive a perspective as possible. We begin with more conventional economic measures and then examine broader human and social dimensions.

Economic Dimensions of Development

The *gross national product (GNP)* and *gross domestic product (GDP)* are the more conventional measures of economic development. The former measures the total domestic and foreign value-added claimed by residents; the latter measures the total value of output of goods and services produced in an economy, by residents and nonresidents, regardless of the allocation to domestic and foreign claims. In 1999, Sub-Saharan Africa's 642 million people had a total GNP of $320[1] billion (World Bank, 2001a). This was below South Korea's annual productivity, $397.9 billion from its 47 million people, and it was less than half of the annual productivity level of Brazil ($743 billion) and its 168 million people.

Analysts caution against the use of the GNP for comparative purposes. In Sub-Saharan Africa, productivity in the urban informal and rural subsistence sectors is usually ignored and left unaccounted for in official statistical reports. Moreover, the significant contributions that women make to agricultural production and the household economy are underestimated. These factors account for the low GNP values in government and international reports. Nevertheless, the understatement in productivity levels recorded by official records should be cause for concern. To account for differences in exchange rates, price levels, and cost of living between countries, *Purchasing Power Parity* (PPP) estimates are now published in international development reports for cross-comparative purposes. Poorer countries usually have PPP estimates higher than the exchange rate GNP; richer countries tend to have lower rates. For example, Angola had a GNP per capita of $220 in 1999, yet its PPP GNP per capita was $632. In the same year, Japan's PPP GNP per capita, $24,041 was lower than its GNP per capita, $32,230.

The per capita GNP for Sub-Saharan Africa in 1999 was $500—about one-quarter of the average in Latin America and the Caribbean and in East Asia and the Pacific. The evidence from Table 8.1 does suggest variations among countries, from a high of $3,350 in Gabon to a low of $100 in Ethiopia. In addition to Gabon, countries on the high end of the GNP per capita scale include Botswana, South Africa, and Namibia. These countries are well endowed with strategic mineral resources, including oil, diamonds, and gold. Countries on the low end of the scale include the drought-stricken Sahel countries of Chad, Mali, and Niger and the war-torn economies of Sierra Leone, Burundi, and Rwanda.

Over all, Sub-Saharan Africa's economic performance in the 1990s was a slight improvement over its performance in the 1980s. However, it continued to experience slow economic growth in the 1990s: an average rate of 2.4% (Table 8.1). Sub-Saharan Africa's average growth rate in the 1990s was below the rate achieved in Latin America and the Caribbean and well below the rate in East Asia and the Pacific region. A number of countries, among them Benin, Botswana, Ethiopia, Ghana, Guinea, Lesotho, Mauritania, and Malawi, showed moderately strong growth in the 1990s, while Mozambique and Uganda experienced significant growth.

As Table 8.1 reveals, the strong economic performances in the latter countries are closely associated with increased domestic investment and significant growth in exports of goods and services. All the aforementioned countries are also rated highly on the World Bank's Country Policy and Institutional Assessment Index—an index that rates a country's social, economic, and political performance on the basis of 20 criteria clustered into four major categories: economic management (debts, inflation, fiscal policy); structural policies (trade, foreign exchange, banking, private sector incentives); social inclusion (equality of economic opportunity, building of human capacity, monitoring of poverty); and public sector management and institutions (property rights, efficiency in public expenditures, corruption). Table 8.1 shows that countries with moderate-to-strong economic growth received a high

---

[1] Monetary value is US dollars.

TABLE 8.1 Economic Dimensions for Selected African Countries

Country	GNP per Capita 1999	GNP per Capita (in US dollars)	Average Annual % Growth Gross Domestic Product, 1980–90 & 1990–99	Average Annual % Growth Gross Domestic Investment, 1990–99	Average Annual % Growth Export of Goods & Services, 1990–99	External Debt as a Percentage of GNP, 1998	Gini Index, 1999	CPIA* Rating, 1999
Sub-Saharan Africa	500	1450	1.7	2.4	3.6	4.4		
Latin America	1955	6280	1.7	3.4	4.9	8.0		
East Asia & Pacific	1833	3500	8.0	7.4	7.0	12.6		
Angola	222	632	3.4	0.8	12.9	8.2	279		F
Benin	380	886	2.5	4.7	5.3	1.9	46		B
Botswana	3240	6032	10.3	4.3	−1.3	2.5	10		
Burkina Faso	240	898	3.6	3.8	4.8	0.4	32	48.2	C
Burundi	120	553	4.4	−2.9	−12.4	2.4	72	33.3	F
Cameroon	580	1444	3.4	1.3	0.0	2.7	98		C
Central Africa Rep	290	1131	1.4	1.8	−1.7	6.7	55	61.3	F
Chad	200	816	6.1	2.3	4.4	5.0	38		D
Congo Republic	670	897	3.3	0.9	4.7	4.3	280		F
Côte d'Ivoire	710	1546	0.7	3.9	17.6	4.7	122	36.7	A
Democratic Republic of Congo			1.6	−5.1	−3.5	−5.5	196		F
Eritrea	200	1012	—	5.2		0.5	149		C
Ethiopia	100	599	1.1	4.8	13.4	9.3	135	40.0	B
Gabon	3350	6435							
Ghana	390	1793	3.0	4.3	4.2	10.8	55	32.7	A
Guinea	510	1761		4.2	2.4	4.7	69	40.3	
Kenya	360	975	4.2	2.2	4.9	0.4	45	44.5	D
Lesotho	550	2058	4.6	4.4	2.3	11.3	42	56.0	B
Madagascar	250	766	1.1	1.7	0.9	3.6	89	46.0	C
Malawi	190	581	2.5	4.0	−7.5	4.9	77		B
Mali	190	693	0.8	3.6	−0.8	9.6	84	50.5	C
Mauritania	380	1522	1.8	4.1	6.8	1.6	148	38.9	A
Mozambique	230	797	−0.1	6.3	13.1	13.4	74	39.6	B
Namibia	1890	5369	1.3	3.4	2.5	4.3			
Niger	190	727	−0.1	2.5	5.4	1.7	55	50.5	D
Nigeria	310	744	1.9	2.4	5.8	2.5	74	50.6	D
Rwanda	250		2.2	−1.5	2.1	−6.0	34	28.9	C
Senegal	510	1341	3.1	3.2	3.1	2.6	58	41.3	B
Sierra Leone	130	414	1.2	−4.8	−10.3	−12.2	126	62.9	F
South Africa	3160	8318	1.0	1.9	3.0	5.3	18	59.3	
Sudan	330	1298							F
Tanzania	240	478		3.1	−1.7	9.5	71	38.2	B
Togo	320	1346	1.7	2.5	11.6	1.5	68		F
Uganda	320	1136	2.9	7.2	9.9	16.3	35	39.2	A
Zambia	320	686	1.0	1.0	11.3	1.8	181	49.8	B
Zimbabwe	520	2470	3.6	2.4	−0.7	11.0	69	56.8	D

*CPIA is the Country Policy and Institutional Assessment Rating.

Source: World Bank. (2001) *World Development Report, 2001*, Washington, DC: World Bank.

CPIA rating (As or Bs) from the World Bank. The rating is conducted each year by the World Bank and is used as a benchmark for allocating loans and assistance to countries.

Both Uganda and Mozambique demonstrate that countries perceived to be "at-risk" or "beyond help" can overcome substantial development problems. Uganda has overcome a despotic and volatile political system (under Idi Amin) and a host of social problems, including the AIDS epidemic, by investing in strategic production sectors, social programs, and poverty-eradication programs. For example, the liberalization of crop prices and investments in agriculture have resulted in the diversification and growth of agricultural products. Besides coffee and tea, which are the principal cash crops, the agricultural sector has diversified to include nontraditional exports, such as fruits, vegetables, tobacco, and flowers. Fruit and vegetable exports, increased from $0.4 million (1% of total noncoffee export receipts) in 1994 to $30.6 million (18% of noncoffee receipts) in 1998 (IMF, 1999a). In addition, Uganda has embarked on a poverty-eradication action plan that targets investments in health and education, in micro- and small-enterprise development, in rural infrastructural improvements, and in the empowerment of women. Uganda has also been the beneficiary of an enhanced Heavily Indebted Poor Countries (HIPC) debt-relief program, designed to ease debt burdens and provide additional finance for basic health and education.

Mozambique's recovery from more than 20 years of civil war and a chaotic economy is attributable to a series of World Bank reform packages designed to rehabilitate the country's overall infrastructure and revive its agricultural and industrial sectors. Following an economic crisis in the 1980s, Mozambique witnessed an increase in manufacturing output by almost 66% in 1998 (Fauvet, 2000). The mining industry, comprising marble, bauxite, and graphite, grew by 30% in 1997, in part through an increase in foreign investment that was lured by a lowering of mining taxes.

While aggregate economic indicators on GDP provide some perspective on growth trends in Sub-Saharan Africa, it is important to realize that they mask important *distributional effects* that are central to geographic study. The *Gini Index of income inequality* measures the extent to which the actual distribution of income differs from a hypothetical distribution in which each person receives an identical share. The index shows how close a given income distribution is to absolute equality, ranging from 0 (equality) to 1 (inequality). Higher economic growth rates do not necessarily come with greater income equality. Although Mozambique and Uganda have higher economic growth rates than Côte d'Ivoire and Ghana, they both have high income inequalities. Countries with extreme inequalities include war-torn Sierra Leone, landlocked Central African Republic and Lesotho, and the racially polarized states South Africa and Zimbabwe.

In addition to dealing with growing income inequalities, African countries continue to be mired in debt. A troubling number of African countries have external debts that total more than twice their annual GNP (Table 8.1). Sub-Saharan Africa's $230 million of debt in 1998 soaked up much of its export earnings; more money is spent on debt repayment than on health care and education. Besides having to worry about social spending, countries like Angola and Zambia have borrowed heavily to revive their struggling mining industries. Others, like Sierra Leone and Democratic Republic of Congo, unfortunately, have used borrowed money to finance civil wars. These heavily indebted countries experienced economic stagnation in the 1990s.

To alleviate the debt burden, the International Monetary Fund (IMF) and World Bank launched the HIPC initiative in 1996, and later an enhanced HIPC. To be eligible for support, countries must adopt reform programs sponsored by the IMF and World Bank and must implement them regularly and satisfactorily; must have an "unsustainable" debt (meaning that the external debt ratio is above 150% for the net present value of debt to exports); and must have instituted a poverty-reduction program. In 1996, 33 countries in Sub-Saharan Africa, with a debt stock totaling over $200 billion, were eligible for HIPC status.

Human and Social Dimensions

Economic measures provide us with a unidimensional view of development. Recently, there has been a concerted effort to broaden our perspective on development by emphasizing dimensions of human development and sustainable development.

The UN Development Program, in a series of annual reports, developed a Human Development Index (HDI), an unweighted average of the combined effect of life expectancy, literacy, mean years of schooling, and per capita GDP. The index is based on the assumption that, at all levels of development, the three essential outcomes for people are to lead a long and healthy life, to acquire knowledge to communicate and participate in the life of a community, and to have access to resources needed for a decent standard of living (UNDP, 1999). Central to the theme of human development is the process of enlarging people's choices, which is consistent with Seers' definition of development.

The HDI is computed by taking into account maximum and minimum values for each variable and the relative distance of an actual value within this range. The index for a country with very highest values approaches 1.0. For example, the United States has the index 0.934; most of Sub-Saharan Africa falls in the category of "low human development," with indices below 0.40 (Table 8.2). Over the last 20 years, Ghana, Kenya, Cameroon, and Congo have joined Botswana, Gabon, Lesotho, South Africa, and Zimbabwe in the "medium human development" category, with indices between 0.5 and 0.8. The HDI is not strongly correlated with per capita GDP. The "HDI rank minus GDP rank" column measures how well African countries translate the wealth they have into benefits for their citizens. A positive number indicates that the country makes good use of its resources to help its people. Cameroon's HDI ranking is 29 points higher than its GDP ranking (Table 8.2). Namibia (-39) and South Africa (-49), on the other hand, have HDI rankings significantly lower than their GDP rankings. The inconsistency between GNP rates and HDI levels reinforces the need to supplement economic measures with human and social dimensions.

Another index that measures human capability is the Human Poverty Index (HPI), an index that focuses on people who lack human essentials—longevity, knowledge, and a decent human life. The World Bank (2001a) reports that there are approximately 300 million Africans living in poverty today. Niger, Sierra Leone, Burkina Faso, Ethiopia, and Chad have more than half of their populations deprived in the three dimensions previously mentioned.

Sub-Saharan Africa's performance on the HDI and HPI confirms the limited choices that people have in maximizing their opportunities and realizing their human potential. Their needs and desires at the most basic level are for longer life expectancy, reduced illness, and

TABLE 8.2 Human and Social Dimensions for Selected African Countries

Country	Human Development Index, 1999	GDP per Capita (in US dollars) Rank minus HDI Rank*	Human Poverty Index	Percentage of Population Undernourished
Sub-Saharan Africa	.467			34
Latin America/Caribbean	.760			12
East Asia & Pacific	.719			12
Angola	.422	−44	NA	43
Benin	.420	−4	45.8	14
Botswana	.577	NA	27.0	27
Burkina Faso	.320	−17	58.2	32
Burundi	.309	0	49.5	68
Cameroon	.506	29	31.1	29
Central Africa Rep	.372	−16	46.1	41
Chad	.359	−7	53.1	38
Congo	.502	29	30.7	32
Côte d'Ivoire	.426	−20	42.9	14
D. R. Congo	.429	8	40.0	61
Eritrea	.416	−3	44.0	65
Ethiopia	.321	0	57.2	49
Gabon	.617	−44		8
Ghana	.542	0	29.1	10
Guinea	.397	−32		29
Kenya	.514	18	31.8	43
Lesotho	.541	1	25.8	29
Madagascar	.462	16	38.6	40
Malawi	.397	8	43.4	32
Mali	.378	0	47.8	32
Mauritania	.437	14	47.2	13
Mozambique	.323	−11	48.3	58
Namibia	.601	−39	34.5	31
Niger	.274	−7	63.6	46
Nigeria	.455	11	36.1	8
Rwanda	.395	−8	44.2	39
Senegal	.423	−13	45.9	23
Sierra Leone	.258	0	58.2	43
South Africa	.702	−49	18.7	NA
Sudan	.439	19	34.8	18
Tanzania	.436	−13	32.4	41
Togo	.489	5	38.3	18
Uganda	.435	−4	41.0	30
Zambia	.427	9	40.0	45
Zimbabwe	.554	−13	36.2	37

*A positive figure indicates that the HDI rank is higher than the GDP per capita (in US dollars); a negative indicates the opposite.

Source: UNDP. (2001) *Human Development Report,* Oxford: Oxford University Press.

greater economic opportunity. Improvements in meeting these basic needs must be achieved by better nutrition, improved medical care, better income distribution, and increased levels of education and employment. These are all essential prerequisites for enhancing the region's human-resource base.

Investing in human development requires commitments at the early stages of human life, that is, at the infant stage. The World Health Organization states that the first three years are critical, because the human brain develops to 80% of its adult capacity during these formative years. Unfortunately, Sub-Saharan Africa's record here has been abysmal. The UN Food and Agriculture Organization (FAO) specifies 2360 calories as the minimum necessary daily consumption level. Only a few countries meet this standard, as shown in Table 8.2. In those situations where food is scarce, children are the immediate victims.

Technological Dimensions

Adult literacy has improved considerably in Sub-Saharan Africa— from 28% in 1970 to 59.6% in 1999—although some countries, such as Niger (15%) and Burkina Faso (23%), continue to lag behind. What continues to be of concern in Africa is the limited development in technological capacity and achievement, a factor that contributes to the low levels of productivity. As advanced countries strive toward new innovations in telecommunications, information systems, and biotechnology, the digital divide continues to widen, because Sub-Saharan Africa lags further and further behind. Research and development (R&D) is critical for assimilating, adapting, and improving on new technologies and for improving the productive capabilities of industry and agriculture. The proportion of R&D scientists in the region is not very encouraging. With the exception of South Africa, which recently reported a competitive ratio— 1031 scientists and engineers in R&D per million people—ratios for the rest of Africa range from a high of 234 in Gabon to a low of 3 in Senegal. (Data are compiled from World Bank, 2001b.) Compare these to 2193 per million in South Korea, 660 per million in Argentina.

The 2001 Human Development Report (UNDP, 2001) introduces a Technology Achievement Index (TAI), which is designed to measure the ability of a country to create and diffuse technology and to build a critical mass of human skills. The index attempts to capture the full range of technologies by focusing on only eight indicators: patents granted per capita; receipts of royalty and license fees; Internet hosts per capita; technology exports; telephones and electricity consumption per capita; mean years of schooling; and enrollment in science, math, and engineering. The countries of the world are then classified into four categories, with the advanced economies as "leaders" and emerging economies as "potential leaders." South Africa and Zimbabwe are the only two countries in Sub-Saharan Africa classified as "dynamic adopters"—countries with important high-tech industries and technology hubs. For example, Gauteng province in South Africa is making a bid to become an innovation hub for knowledge-intensive industries, technological entrepreneurship, venture capitalists, and information and communication technologies. Zimbabwe is exploring innovations in agricultural science and technology, including genetic engineering. The majority of the countries in Sub-Saharan Africa are classified as "marginalized" in terms of their technological achievement, although some countries, such as Senegal and Ghana, are fast approaching the threshold of "dynamic adopters." Both countries are laying the foundation for improvements in science and technology.

As Sub-Saharan Africa embarks on strategies to enhance human development, it is important that such strategies proceed along a path of *sustainable development*. The World Resource Institute, the UN Development Program (UNDP), and the World Bank caution the world about the need to protect the productive capacity of the environment, and that is certainly admonitory for those Sub-Saharan countries that are eager to develop as rapidly as possible without considering the social and environmental consequences. Sustainable development, which means managing resources efficiently to meet current and future needs, has several dimensions. From an economic perspective, it implies committing resources more equitably toward continued improvement in African living standards (WRI, 1994). From a human perspective, it means achieving progress toward stable populations (as noted in the chapter on population); ensuring full use of human resources, by improving education, health services, and nutrition; and promoting the full participation of people in the planning and execution of the policies that affect their daily lives. From an environmental perspective, sustainable development stresses the judicious use and protection of natural resources (see Chapter 3); from a technological standpoint, it means the introduction of improved and appropriate technologies.

Other scholars have gone a step further and measure development from a "subjective" standpoint, arguing that the "objective" or more tangible indicators do not offer any perspective on people's perceptions about their levels of satisfaction or happiness. For example, one objective indicator, the number of people employed, does not exactly tell whether they are satisfied with their jobs. Norman (1975), for example, developed a "hedonometer" to compare happiness levels in England and Botswana. She employed six criteria to measure people's perceptions about their immediate environment, their family and community support system, their level of well-being, their aesthetic environment, and their exploratory drives. She concluded that people in Botswana are much happier than people in England. Although this study might seem highly subjective, there are those who believe that modern development should not interfere with the daily life experiences of Africans in traditional societies, because it is difficult to determine whether such development is welcome or not.

GEOGRAPHIC PATTERNS OF DUALISM

The aggregate indicators analyzed so far have provided us with a framework for a country-by-country comparison; however, they conceal the intraregional inequalities that are so characteristic of Sub-Saharan countries. Geographers acknowledge that, while there are spatial variations in development patterns among Sub-Saharan African countries, important intraregional differences exist within countries. All countries are characterized by a dualistic pattern of development, one in which a small, modern core sector coexists with a relatively large peripheral and rural sector. Groups of people in these sectors have different experiences with respect to conditions that have a bearing on their well-being. Generally, geographers identify four levels of dualism:

1. International Dualism
2. The Dual National Economy
3. Urban Dualism
4. Rural Dualism

International Dualism

International dualism reflects the gaps that exist between the northern, high-income countries and the "southern," low-income countries, including those of Sub-Saharan Africa. The north–south distinction was defined by an Independent Commission on International Development Issues created under the auspices of the World Bank. The Commission broadly defined the advanced countries in the northern hemisphere, plus Australia and New Zealand, as the north and the rest of the world as the south. In addition to the widening digital divide, there is a wide gap in GNP per capita, human development, and life expectancy, not only between Sub-Saharan Africa and the northern (high-income) countries, but also between Sub-Saharan Africa and its counterparts in other developing regions (Figure 8.1). The gap in GNP per capita between Sub-Saharan Africa and the north in fact widened between 1985 and 1999, from 3.86% to 1.94%.

World systems and dependency theorists have argued that this gap is perpetuated by a dominance–dependence relationship that results in an unequal exchange of resources between north and south, such as the exchange of raw materials for industrial products (Knox and Agnew, 1998; Wallerstein, 1982; Frank, 1980). Sub-Saharan Africa ends up in

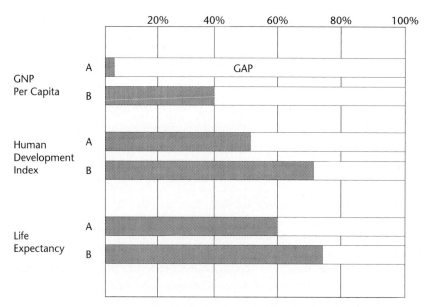

A = Sub-Saharan Africa average as a percent of northern (high-income) countries
B = Sub-Saharan Africa average as a percent of developing countries

Source: Compiled from Human Development Report (2001) and World Development Report (2001)

Figure 8.1 Disparities between countries in Sub-Saharan Africa, developing countries, and high-income countries.

Geographic Patterns of Dualism

a subordinate position and becomes vulnerable to the rules and regulations of an international system controlled by advanced countries. Galtung (1971) takes this argument a step further in his "structural theory of imperialism" and argues that there is a harmony of interest between privileged elites in the north and the south. As a result, ruling elites in Sub-Saharan Africa facilitate the unequal exchange of resources by investing in foreign goods and services, thus enabling rich countries and multinational corporations to appropriate much of the wealth from African economies. Reports about scandal, corruption, and embezzlement in African governments are not uncommon. Every now and then, we hear about African heads of states and government officials funneling hard-earned money generated from rural areas into foreign banks. It is hard to say whether the north–south gap is a result of "imperialist" forces or of poor judgement and bad internal management on the part of African governments and urban elites.

The dual national economy (core–peripheral disparities in African countries) Severe disparities exist between core urban areas and peripheral rural areas in Sub-Saharan Africa. Figure 8.2 shows the disparities in access to improved sanitation facilities. Ethiopia, Niger, Burkina Faso, Burundi, Chad, Namibia, and Mozambique exhibit

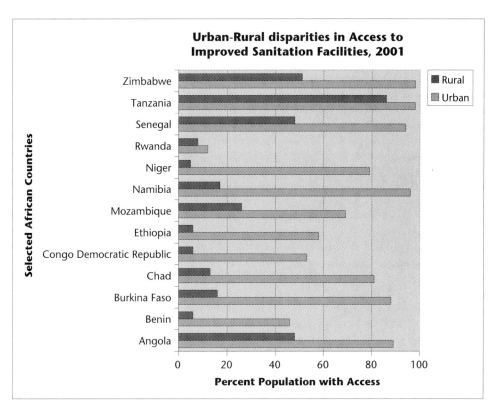

Figure 8.2 Urban–Rural disparities in access to improved sanitation facilities, 2001.

severe disparities in the distribution of sanitation services. There are a few countries—for example, Tanzania—where there is equal access in the distribution of services.

These rural–urban disparities are a function of the distribution of economic activity. Figure 8.3 shows the distribution of major core centers of economic activity, including manufacturing, agricultural export and processing, and mining centers. Several key manufacturing centers are located in major cities to take advantage of *urbanization economies*—production cost savings derived from locating in close proximity to urban amenities and communications technologies.

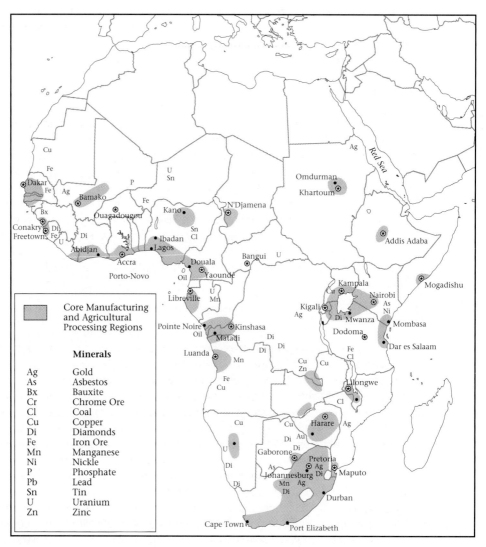

Figure 8.3 Core regions in Sub-Saharan Africa.

The map shows that these centers of modern development are clustered along the west coast of Africa, in the Witwatersrand region of South Africa, in the copper belt of the Democratic Republic of Congo and Zambia, in the Lake Victoria region, and in central Ethiopia. Fair (1991) provides examples of the kinds of activities associated with these core areas. Southwest Cameroon, for example, accounts for only 15% of the country's land area, but it contains 75% of its population; a rich forest zone; a diverse agriculture based on cocoa, coffee, palm oil, timber, and rubber; and a core of manufacturing activities anchored by aluminum smelters and petroleum refineries. Northwest coastal Gabon and the lower Congo Democratic Republic regions account for most of the oil production in their respective countries. The Lake Victoria borderland, which covers part of Kenya, Tanzania, and Uganda, contains rich agricultural land that supports export crops, such as coffee, tea, and cotton. Other examples of core regions include the Blantyre–Lilongwe axis of Malawi, the Luanda–Malanje corridor of Angola, and southern Chad.

Urban Dualism

Another type of dualism manifests itself in the cities of Sub-Saharan Africa in the form of the formal–informal sector dichotomy. The urban informal sector consists of those individuals and enterprises who operate outside the mainstream of government regulation and benefits. Sub-Saharan cities do not have enough jobs to accommodate the throngs of migrants who come in from rural areas with expectations of securing formal-sector jobs. As a result, most of them develop adaptive strategies and set up small- or medium-sized businesses in the informal sector, such as barber shops, basket weaving, pottery making, and car-repair stations.

The irony of the informal sector is that it constitutes a pool of creative and talented people whose resources can be tapped to secure badly needed government revenue. The potential of the informal sector and its contributions to the urban economy are discussed in more detail in Chapter 10.

Rural Dualism

In the rural areas, another form of dualism exists: Relatively wealthy commercial or cash-crop farmers in cocoa, coffee, cotton, groundnuts, rubber, and tea coexist with subsistence farmers who produce just enough food for themselves and their families. Examples of the latter include the shifting cultivators of the tropical rain forest region and the Fulani and Masai pastoralists of central West Africa and east Africa, respectively. (See Chapter 13 for further details.)

This pattern of rural dualism works to the detriment of subsistence farmers. More often, commercial farmers are the recipients of government benefits and incentives and new agricultural technologies. An example of a discriminatory policy against subsistence farmers is the *green revolution* discussed in Chapter 13. The green revolution is a technological innovation in agriculture that has been creating new, high-yielding, nutritious varieties of rice, wheat, and corn, resulting in increased agricultural yields. Scholars have often wondered why the green revolution has not been adopted as widely in Africa as it has been in India and China. The major reason is the dual structure in rural economies. Often, it is the richer commercial farmers who secure credit from financial institutions to purchase those vital inputs, such as water pumps and nitrogen fertilizers, required for the new seeds. This is a classic example of a *hierarchical diffusion* process, whereby a new innovation leapfrogs from one important person to the next or from one major sector to the next before "cascading" down to minor sectors.

These patterns of dualism in Sub-Saharan Africa create a three-way interactive process among the commercial sector in rural areas, the formal institutions in urban areas, and the world economy in the north (Figure 8.4). The commercial farmers benefit from government incentives in urban areas; formal institutions, in turn, benefit from the foreign exchange derived from the sale of cash crops through their marketing boards. In several instances, the foreign exchange earned is internalized in urban areas. This three-way interactive process reinforces the disparities that exist between core and peripheral regions at all scales and polarizes people who reside in the urban-informal and rural-subsistence sectors. In many respects, the constituents from the urban-informal and rural-subsistence sectors have limited opportunities to realize their human potential and to participate in the global exchange economy.

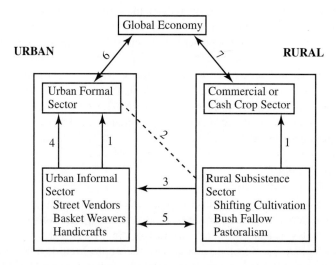

1. Inequality between the two sectors widens, because formal sector and cash-crop sector enjoy more government incentives and benefits.
2. Inadequate investment; lack of extension services, credit, technical assistance.
3. Severe backwash effect as people migrate to urban areas.
4. Low-cost services in informal sector are exploited by formal sector.
5. Rural–urban linkages: remittance of incomes, rural–urban networks formed.
6. Strong interrelationships between urban elite and global economy: foreign investments, demand for imported goods and services.
7. Proceeds for primary products (cocoa, coffee, cotton, tea) and minerals exported.

Figure 8.4 Different scales of dualism in Sub-Saharan Africa. Adapted and modified from Obudho, R. A. & Taylor, D. R. F., eds. (1979). *The Spatial Structure of Development: A Study of Kenya,* Boulder, CO: Westview Press, 20.

THEORETICAL EXPLANATIONS OF CORE–PERIPHERAL DISPARITIES

African geographers and development analysts offer a number of explanations about how and why core–peripheral disparities continue to evolve in Sub-Saharan Africa. These explanations fall under four major theoretical categories: historical–colonial, economic–modernization, structural–institutional, and sociopsychological theories.

Effect of Colonialism

The colonial legacy of foreign occupation and the superimposition of European rules, regulations, and values on Sub-Saharan countries is detailed in Chapter 4. With colonialism came land alienation, the plantation economy, foreign control of the means of production, and the construction of a colonial infrastructure that had an adverse impact on the spatial structure of African economies. Of these factors, it was the superimposition of a colonial infrastructure that laid the foundation for dislocating the spatial structure and exacerbating the core–peripheral disparities in Sub-Saharan Africa. Most of the physical (roads, rail, bridges) and social (schools, hospitals) infrastructure constructed during the formal colonial era was concentrated in key urban centers that acted as outposts and headlinks for European centers. Several of these administrative centers were located along the coast of west Africa and have now evolved into capital cities. (See Chapter 10 on urban geography.) The roads and railways were designed in a "tree-shaped" pattern, with major highways connecting raw-material and mineral sites directly to colonial centers. Only a few secondary roads branched out from these major highways in an attempt to link smaller communities.

Sub-Saharan countries have had difficulties in extending and improving on this infrastructure. As a result, governments have tended to concentrate the provision of basic infrastructural and social services in the former colonial centers, and they have failed to provide their countries with a spatial system that facilitates a more equitable distribution of services and facilities, one based on criteria of social justice and economic efficiency.

Economic Theories

Most of the economic explanations offered about core–peripheral disparities have centered around modernization theories, unilinear growth models, and efficiency–equity debates. *Modernization theory* was attractive to African governments in the 1950s and 1960s because it offered prescriptions on how to catch up with more developed countries as quickly as possible. In the eagerness, anticipation, and high expectations that accompanied the advent of independence in the 1950s and 1960s, these models of economic development provided considerable appeal. Modernization theorists assumed that, with the right combination of capital, know-how, and attitude, economic growth would proceed on a unilinear path toward self-sustenance and prosperity and that countries would make a transition from a traditional to a modern state (Todaro, 1989).

One of the more common models was Rostow's theory of economic growth. Rostow (1963) assumed that developing countries could proceed along a development continuum involving five stages: (1) the traditional or preindustrial stage, (2) the preconditions for takeoff stage, (3) the takeoff into self-sustaining growth stage, (4) the drive to maturity, and (5) the age of mass consumption. A number of conditions were assumed to be necessary for takeoff, a critical one being the mobilization of domestic and foreign savings to generate investment to accelerate economic growth.

The Harrod-Domar model offered some guidelines, hypothesizing that the rate of economic growth in a country is determined by the national savings ratio and the national capital-output ratio. Simply put, the more of its GNP that a country saved and invested (to rehabilitate and maintain its infrastructure, for example), the faster it would grow (Todaro, 1989). Rostow recommended a critical amount of investment, in the range from 15% to 20%, as a necessary prerequisite for takeoff into self-sustaining growth. In 1997, Sub-Saharan Africa averaged as a savings ratio 8.4%, with Gabon registering the highest (31%) and Angola the lowest (-14.4%). Only 9 out of 32 African countries had ratios that fell within Rostow's guidelines (World Bank, 1998). Rostow's contention was that limited domestic savings could be augmented by foreign aid and investment. This increased the vulnerability of African countries and forced them to make several concessions to advanced countries.

Rosentein-Rodan (1964) proposed a more "balanced" pattern of investment in a wide range of industries to help African countries break out of their cycle of sluggishness. This balanced approach to "modernization" would allow people to work more productively, with more capital and improved techniques, and would create opportunities for people to support each other as customers. A single industry could not create a market demand for its own product; in that case, demand would end abruptly. Therefore, rather than proceed "bit by bit" or industry by industry, a "big push" was necessary to create industries that complemented one another and satisfied a range of human needs and demands. Essential to this big push were large investments in social and physical infrastructure—power, transport, communications, and housing.

Development theorists like Hirshman (1958) agreed on a big push of some sort, but advocated a less balanced approach to economic growth. Recognizing the scarcity of the financial resources available to African countries, they concluded that the essential task here would be to select and then concentrate on those strategic sectors of the economy that had the strongest linkages or the best chance to stimulate development in other sectors of the economy. African countries were naturally inclined to go with an "industry-first" strategy, in view of its strong backward and forward linkages. Most countries, in fact, pursued import-substitution strategies as a first step toward achieving industrial self-sufficiency. Côte d'Ivoire's approach was to promote market liberalization, develop the industrial sector, and encourage local entrepreneurship (Discussion Box 8.1).

These economic-development theories have been criticized as being highly unidimensional and ethnocentric. Modernization implied that the underlying social, economic, technological, and political structure of traditional sectors in Africa could be completely transformed to assume the characteristics of those in more developed countries, once certain underlying economic parameters (such as increased investments) were taken care of. Not many attributes of Africa's sociocultural order conformed with these models. Furthermore, these Western models of economic development did not take into account the dualistic structure of African countries. The assumptions of the Rostow and of the balanced and unbalanced

DISCUSSION BOX 8.1:
AFRICAN SOCIALISM

President Julius Nyerere of Tanzania spent the 1960s and 1970s building a society based on socialist principles. His *raison d'etre* of African socialism was articulated in a document called the Arusha Declaration, which was released in 1967. The year 1967 essentially marked the implementation of a socialist policy that was geared toward the elimination of poverty and income inequalities and toward the improvement in the quality of rural lives. Major commercial, industrial, and financial institutions were nationalized, and public officials were restricted in their ownership of property. In 1968, the Ujamaa policy was instituted to promote self-help, self-esteem, and local initiatives along family, communal, and cooperative lines. The Ujamaa, modeled along the same lines as the Soviet collective or the Chinese commune, was a consolidated, self-sufficient settlement that provided educational, health, and social services to meet the basic needs of its resident population. Ujamaa residents pooled their labor, land, and capital resources for agricultural and other ventures. Proceeds from the cooperative effort were employed to finance a range of development projects. To encourage rural participation in the decision-making process, Tanzania embarked on a policy of decentralization in the early 1970s, as regional and local district offices were dispersed into rural villages. Tanzania's socialist strategy had mixed results. There was some success in achieving social equity, but the country became one of the poorest in Sub-Saharan Africa, with slow growth in its productive sectors. Tanzania has since implemented a World Bank–sponsored enhanced structural-adjustment program aimed at restructuring and privatizing public enterprises, reforming its civil service, improving the social infrastructure, and eradicating poverty.

Ethiopia also adopted a socialist policy when, in 1975, the new socialist military council replaced Emperor Haile Selassie. The new government immediately nationalized all banks and insurance companies and several industrial and commercial enterprises. A new land-reform program was instituted to bring all rural and urban land under government control. This had devastating effects on communal and traditional forms of land ownership in villages, as rural farmers were forced into collectives. Ethiopia continues to be in a state of social and economic destitution, with agriculture accounting for more than 90% of its foreign earnings. After the overthrow of Mengistu's military dictatorship in 1991, the transitional government adopted a policy of reconstruction and rehabilitation based on market-driven prescriptions set by the World Bank, the IMF, and the African Development Bank.

growth models were applicable only to the formal urban or modern sectors of African countries, which exhibit most of the necessary prerequisites for a takeoff or a big push. Therefore, existing inequalities between the core and periphery were reinforced.

In spite of these problems, the earlier models of economic development were widely accepted by African governments. Ghana, between the years of 1957 and 1966, pursued a policy of rapid economic growth, one in line with the ideas of the big-push development theory prevailing at the time. Ghana's head of state at the time, Kwame Nkrumah, firmly believed in the proposition that industrial development was a precondition for agricultural progress— it provided nonagricultural avenues of employment, relieved pressure from land, and generated demand for agricultural goods. Agriculture, therefore, played a secondary role, and investment was channeled to physical infrastructure and industry in the core region of southern Ghana. But capital-intensive industrialization was accompanied by slower growth, partly because the food supply had failed to keep pace with demand, and led to inflation and to pressures for quantities of food imports so large as to form 15% to 20% of total imports in 1976.

In the 1950s and 1960s, Nigeria pursued an import-substitution policy of industrial development. Inputs such as machinery, equipment, capital, technical know-how, and semi-processed materials were imported to produce finished goods. Nigeria's First National Development Plan (1962 to 1968) aimed to achieve 4% as an annual GDP growth rate and 15% of the GDP as a savings and investment ratio to achieve self-sustained growth by 1980 (Filani, 1981). The industry-intensive nature of the planning process meant that very few benefits went to the agricultural sector, thereby intensifying the inequalities between urban and rural areas.

For most of the Sub-Saharan African countries emerging from the colonial era, development policy was a question of whether to pursue *efficiency-oriented* growth strategies or *equity-oriented* growth strategies. Greater efficiency implied increased economic output; greater equity implied fairer distribution of the fruits of economic growth. Advocates of the former argue that core areas have all the necessary prerequisites for growth: adequate infrastructure, skilled professionals, and means of communication. Pursuing an equity-now approach would retard economic growth in the core, eventually producing merely an equitable distribution of poverty and impairing a country's competitive position in the global economy (Reitsma & Kleinpenning, 1985).

Therefore, at the initial stages of development, this view found it wise to concentrate scarce resources in core regions, so that, as these regions mature and expand at a later stage, peripheral regions will open up and begin to attract economic activities. With the spread of the manufacturing from the core to the periphery, the latter becomes increasingly integrated into the national economy, and the fruits of economic development become more evenly distributed.

The problem with an efficiency-now approach is that it assumes that growth will eventually trickle down to peripheral regions. Are urban areas in Sub-Saharan Africa centers of innovation and activity to the extent that they can generate the necessary growth impulses in the periphery? In reality, *backwash effects* in Sub-Saharan Africa are much stronger than the trickle-down effects. Backwash effects occur when human and physical resources are drained from peripheral areas into core areas. For example, foreign exchange derived from the fruits of rural labor are internalized in urban areas to the detriment of rural areas. As rural areas become more impoverished, young and able-bodied people leave and thereby deprive rural areas of essential labor required to rehabilitate their environment. Aside from these backwash effects, the process of *hierarchical diffusion* is prevalent, as new technologies and innovations are diffused from one major core area to the next before eventually trickling down, or cascading, to smaller towns and rural communities. As these backwash effects and processes of hierarchical diffusion prevail, the gap between the core and peripheral areas continues to widen.

Therefore, for most African governments, the dilemma is whether to continue an efficiency-oriented strategy, knowing full well the consequences of greater inequalities, or to opt for equity-oriented strategies that might retard growth in core areas and reduce overall growth. It is not surprising that most countries continue to pursue efficiency-oriented strategies, in one form or another, because they have few resources and because there is less political risk involved. Equity-oriented strategies have longer-term rewards but require patience and tolerance from urban political constituents.

Structural–Institutional Theories

Core–peripheral disparities continue to prevail in Sub-Saharan Africa because there are underlying structural and institutional barriers that retard the diffusion of economic benefits and growth impulses from core regions. These barriers have to be removed to facilitate any trickle-down or diffusion process.

One problem relates to the economic, political, and power structure in those urban areas where powerful elites reside. The urban elite forms a strong political coalition that usually dictates policy to suit their own parochial interests. Sometimes, the ruling elite is a powerful and wealthy ethnic group, such as the Kikuyu in Kenya, the Batswana in Botswana, or the Baganda in Uganda. Often, they have created policies that are discriminatory against other ethnic groups, prompting African scholars to coin the term *internal colonialism*.

The low savings and investment levels in the region are also attributable to the consumption habits of the urban elite. Rather than investing in domestic economic ventures, they engage in conspicuous and luxurious consumption that patronizes imported goods, such as expensive cars, food items, and building materials.

The structure of food prices also works to the detriment of African farmers. Governments have deliberately kept the price of food artificially low in urban areas to appease their urban constituents. This policy acts as a disincentive for African farmers to produce more crops for commercial purposes, thus depressing their farm incomes. It is not surprising, therefore, to find that farmers prefer to subsist rather than to waste their time, energy, and effort on marketing surplus crops.

Bank, credit, and land institutions in Sub-Saharan Africa also work to the detriment of residents in the rural periphery. Banks and credit institutions are often unwilling to invest in rural development ventures because of the perceived risk associated with small business and other ventures. Small-scale farmers often lack the financial and technical assistance to purchase the required inputs and technologies to raise their production levels. Also, very few agricultural extension and service centers are located in rural areas to provide the necessary counseling and advice on effective farm management and on obtaining nonagricultural employment opportunities. In the absence of assistance from core regions, several informal networks and cooperative societies have cropped up in rural areas to provide the necessary support mechanism. (Examples are provided later in the chapter.)

Complex land-tenure arrangements also limit opportunities for economic advancement. With increased population pressure on land, traditional authorities are becoming stricter with allocating usufructuary rights on community-owned land. Women are the most frequent victims of the land-tenure system. The complexities of these land-tenure systems are discussed in more detail in Chapter 6.

Sociopsychological Theories

The sociopsychological explanations for marginal development in peripheral regions focus on social behavior, attitudes, and propensities; therefore, they tend to be controversial and, in some cases, narrow minded. Hoselitz (1960), following Parsons (1951), attributed the underdeveloped nature of traditional societies to particularism, ascription, and functional diffuseness as opposed to universalism, achievement orientation, and functional specificity. *Particularism* occurs when the interests of particular groups direct the movement and processes of society. Such practices inhibit social mobility between cultural groups, between different professions, and sometimes between localities. The caste system in India and apartheid in South Africa exhibit particularistic tendencies. It also occurs when the ruling elite or a dominant ethnic group advances its own causes at the expense of others. Particularism can also be found in more developed countries when discriminatory and exclusionary policies work against certain underrepresented groups.

Inadequacy in level of development is also attributed to situations where recruitment and reward are determined by *ascription*, or one's status, rather than by achievement and competence. Political appointments made by military dictators in Africa are designated in terms of ascriptive norms. Also, roles in the traditional sector are *functionally diffuse* rather than specific. In other words, the same person performs several roles—perhaps farmer, trader, and house builder—whereas, in the more formal sector, there are highly specialized and more specific roles.

Other sociological theories focus on social and individual qualities that are perceived to be more conducive to economic development, such as the development of an achievement motivation to assume entrepreneurial qualities, as opposed to that of a need to conform to other expectations or to become more dependent on others (McClelland, 1961). The entrepreneurial abilities of plantation farmers and other occupational groups in Africa have rendered such theories useless, as is noted in Discussion Box 8.2.

Others have gone on to suggest that traditional societies resist innovations and new technologies because the people are conservative and fatalistic. This viewpoint is particularly naive in ignoring the structural and institutional barriers that farmers confront. Peasant farmers are reluctant to adopt new technologies, not because they are fatalistic, but because they lack the financial means and necessary technical advice that should accompany the introduction of a new technology. Theories that reduce the development of societies to

DISCUSSION BOX 8.2: AFRICAN CAPITALISM

Some scholars have suggested that capitalism in Africa has not manifested itself in its truest sense. African capitalism is often viewed as having been small in scale (e.g., small volume of assets) and as lacking dynamism in terms of growth in assets and a willingness to venture upon new enterprises and operations. It is also tempting to view African capitalism as state capitalism, dominated and coopted by state bureaucracies. This view may hold true in some countries, but Côte d'Ivoire is probably the one county that comes closest to exhibiting the characteristics of a capitalist state. Rapley, in his book on Ivoiria capitalism, states that indigenous capitalists in Côte d'Ivoire emerged as early as the 1920s and 1930s. These were wealthy plantation capitalists who utilized the entrepreneurial attributes they acquired from French colonists to develop an effective, well-organized political machinery that gained control of the state during the decolonization process.

Since independence, the government has orchestrated capitalist development in Côte d'Ivoire—a deliberate intent of the late President Felix Houphouet-Boigny. The state, acting essentially as guardian, manager, and facilitator of capitalist development, adopted a two-pronged strategy based on encouraging foreign investment and mobilizing domestic savings. In 1962 and 1963, the government founded the National Investment Fund and the National Finance Society to nurture and support the development of local enterprise. This was followed in 1965 by the institution of the Ivoirian Bank of Industrial Development, which encouraged investments from local and foreign shareholders. In the late 1960s, the government's import-substitution strategy was replaced by an export-oriented approach, to improve the competitive advantage of agroindustries catering to European markets. These earlier strategies were supplemented with a host of aggressive strategies in the 1970s to further stimulate the private sector. These included the creation of a stock market, the provision of incentives to encourage public-sector employees to become private entrepreneurs, and the institution of counseling and financial programs to stimulate the development of local capital. For instance, between 1968 and 1976, the Office for the Promotion of Ivoirian Enterprises created 246 private Ivoirian companies. Furthermore, between 1974 and 1978, private Ivoirian capital assets in the modern sector grew by 43%, and, between 1970 and 1980, the value of private Ivoirian capital's industrial assets grew at an annual rate of 29.6%, which was accompanied by a 579% increase in investment by enterprises and a 713% increase in industrial exports.

Côte d'Ivoire's economic "miracle" took a jolt in the 1980s as international prices in coffee and cocoa slackened. The GDP growth rate in 1983 and 1984 averaged between 2 and 3%, small compared with the 11% and

> 6–7% growth rates in the 1960s and 1970s, respectively. Like several other African countries, Côte d'Ivoire is currently engaged in the IMF–World Bank's structural-adjustment programs of fiscal austerity and privatization. In spite of the temporary setback in the 1980s, the country continues to encourage market liberalization and local entrepreneurship. Recent discoveries in natural gas and oil have presented opportunities for economic diversification.
>
> Source: Adapted from Rapley, J. (1994). *Ivoirien Capitalism: African Entrepreneurs in Côte d'Ivoire*, Boulder, CO: Lynne Rienner.

changes in consciousness or to vague interpretations of a society's propensities and motivations are unsupported—for example, the sociological explanation of underdevelopment. Such theories generalize about certain customs and institutions without taking into consideration the heterogeneous character of behaviors, social roles, and institutions of developing societies.

DEVELOPMENT STRATEGIES

The widening disparities between core and peripheral areas in Sub-Saharan Africa is cause for concern among geographers and development analysts. What kinds of appropriate strategies can be designed to narrow this gap and create more opportunities for the development of human capacity? A number of strategies are considered next.

Growth with Equity

The debate on whether to pursue a growth-first or equity-later policy rages on. Core–peripheral disparities persist in Sub-Saharan Africa, and they will continue to intensify as long as rural areas are ignored. Streeten and Stewart (1981) propose a number of alternative forms of redistribution. Nonincremental redistribution is referred to as a policy of redistributing existing assets. It includes land reforms; wider spread of ownership or nationalization of industrial property; and radical reforms of institutions to give the poor greater access to educational, health, credit, and technological resources. This policy is not likely to be feasible in Sub-Saharan Africa in the wake of recent structural reforms by the World Bank calling for increased privatization. Besides, redistribution of assets through land reform not only challenges cultural institutions in the region, but also requires ancillary institutional reform in areas of credit, water availability, energy, and so on. So, in and of itself, nonincremental redistribution is likely to increase inefficiency without improving the living conditions of the poor, a pattern that has been recorded in several Latin American countries.

Other forms of redistribution, such as incremental redistribution, which involves taxing those who are better off to redistribute wealth to those who are worse off, and redistribution through growth are alternatives available if a deliberate interventionist strategy is pursued by a Sub-Saharan country. Incremental redistribution is the policy most commonly adopted in advanced economies, and it involves progressive taxation with redistribution through social welfare expenditures and subsidies. However, African countries have limited ability to collect taxes; the absolute number of taxable firms and individuals is relatively small compared with the total population, so this strategy is not likely to be effective in changing unequal regional distributions.

Redistribution with growth (growth with equity) offers a more viable policy alternative. This strategy involves redirecting the extra income produced by growth in core regions to weaker peripheral regions. Such a redistribution emphasizes growth in the peripheral areas, and investments would be channeled toward increasing productivity and making better use of existing natural resources. If pursued consistently, such an approach could bring about regional convergence and yet would not be disruptive, because it would conserve existing investments in core regions. Such a strategy could increase average income, average value added, and average output in peripheral regions, as well as strengthen and diversify local economies. Regional policies must also attack the root causes of poverty, such as illiteracy, ill health, and unemployment, to substantially modify total-income inequality. Thus, a basic-needs approach should accompany the adoption of a redistribution-with-growth strategy.

Basic-Needs and Poverty-Reduction Strategies

A basic-needs strategy implies identifying target groups and providing opportunities for the full physical, mental, and social development of their human personality. The strategy spells out in detail human needs in terms of affordable housing; of pertinent education (including vocational and technical training); of feasible energy options; and of preventive health care, as opposed to more costly curative health measures. Preventive health care involves health education and the enforcement of sanitation programs: improving domestic water supply; improving housing, cooking and meat inspections; and identification and elimination of potential insect breeding sites. A basic-needs strategy could also involve nonmaterial needs, such as self-determination, security, participation in decision making, and gaining a sense of purpose in life and work (Streeten, 1981).

Investment channeled toward increased productivity in weaker peripheral regions should be devoted to production of basic goods that can be used to increase the consumption levels of the local population. This could include production of agricultural and nonagricultural commodities—for example, of food and of improved building materials that use local resources.

Both the World Bank and IMF, through their Poverty Reduction and Growth Facility programs, recognize that economic growth and poverty reduction must proceed simultaneously if African countries are to break the cycle of poverty. Mozambique recently developed a poverty-reduction strategy that focuses on universal, informal, and technical–vocational training; on primary health-care improvements targeted towards women, children, and HIV/AIDS and endemic diseases; on agricultural and rural-development efforts to support rural extension programs, crop technologies, food security, financial credit and market access, and upgrades in the rural infrastructure; on promoting good governance to address the needs of the poor; and on improving macroeconomic and financial management to reduce debt, promote trade, and enhance efficiency in the allocation of public expenditure (Republic of Mozambique, 2001). In Chad, efforts to protect vulnerable groups emphasize environmental protection and desertification management, improving of cotton and livestock production, improving of efficiency and effectiveness of education spending, and organizing of a participatory system to regularly assess and monitor poverty programs.

It is often posited that there exists a fundamental conflict between growth objectives, which emphasize efficiency, and the basic-needs and poverty-reduction strategies, which emphasize equity and tend to consider a longer term. The conflict exists when there is an increase in the productivity of commodities consumed by the wealthy at the cost of commodities consumed by the poor. This is particularly aggravated if the commodities consumed by the poor have to be imported, because the increase in the prices of their consumption items usually outstrips the wage increases accompanying growth. Increasing incorporation of peripheral regions into a monetized market economy is harmful when national prices are transmitted to weakly monetized households, especially when this is accompanied by a decrease in the supply of basic goods and services needed by the poor, particularly food, shelter, education, and basic health. The basic-needs and poverty-reduction approaches advocated here emphasize increased productivity and the growth of commodities consumed by the poor.

Intermediate or Appropriate Technology

Intermediate technology uses labor-intensive, affordable technologies rather than capital-intensive, sophisticated, highly energy-dependent technologies. It supports the mobilization of priceless resources possessed by humans: their ingenuity and skillful hands. It economizes on the use of scarce resources and is designed to serve humans instead of making them "the servant of machines." Intermediate technology is vastly superior to primitive technology, and yet it is much simpler and cheaper than the supertechnology of more advanced economies (Schumacher, 1974). It is a technology to which most Africans can gain access. African governments need to adopt appropriate technological policies and encourage more research activity in this area. The UN Educational, Scientific, and Cultural Organization (UNESCO) has sponsored a series of conferences to encourage African states to formulate national science and technology policies (Vitta, 1990). Also, two Technology Policy Studies in east and west Africa were established in 1982 and 1984 to assess the impact of existing technology policy in their respective countries and to gather useful information to design and implement future policy in technology.

Self-Reliance Strategy

A self-reliance strategy involves relying on one's own capabilities, judgment, resources, and skills to enhance social, political, economic, cultural, attitudinal, and moral independence. It involves self-reliance in decision making and planning, popular participation and collective action, internalizing of the benefits of and minimizing of the drainage of resources, and equal sharing in development (Galtung, 1982).

Most self-reliant and self-help efforts in Sub-Saharan Africa have occurred on a more localized scale. They are often formed to fill a vacuum created by "centralist" governments that fail to design interventionist strategists to meet rural needs. In Kenya, *Harambee* (meaning "let's pull together" in Swahili) self-help groups have been instrumental in providing primary and secondary schools, technological institutes, basic health services, water, social welfare, and economic opportunity. These grassroots organizations accounted for 30% of all capital formation in rural areas through the mid-1970s, and 10% of total national development expenditure between 1979 and 1983 (Thomas, 1987). In Zimbabwe, a Savings Development Movement, totaling 5700 clubs nationwide in 1984, mobilized local financial resources to invest in a variety of rural projects. Proceeds from savings were rotated among members, and the more advanced savings clubs were able to diversify into agroprocessing enterprises, consumer stores, community wells, and dams (Bratton, 1989). These clubs were successful in devising effective low-cost methods for promoting capital formation and agricultural investment not only among the poor but also among women, who constituted the majority of their membership.

After severe droughts in 1973 in the village of Wuro-Sogi, Senegal, a community-development association was created to finance a community water system, a millet mill for women, a village pharmacy, and a communal vegetable garden managed by women. Other successful initiatives include efforts by women in the village of Saye, Burkina Faso to organize and build an earth dam; the Kenya Water Health Organization's efforts to organize a cooperative program to train women in Mombasa to build and maintain simple water systems; and the Green Belt Movement in Kenya, which was organized by a women's rights advocate, Wangari Maathai, to encourage people to plant trees and form tree belts (WRI, 1992).

At the state level, self-reliance has been difficult to implement as a development policy. The country that has come closest to pursuing a national self-reliant policy is Tanzania. In an attempt to attack poverty and destitution among the rural poor, Tanzania launched the Arusha Declaration in 1967, the government's policy statement on self-reliant rural development. The main focus of the declaration was to restructure the Tanzanian economy in a socialist direction by encouraging public ownership of the means of production and creating *Ujamaa* (familyhood and brotherhood) villages, which would become the basis for social organization and cooperative activity in Tanzania. (See Discussion Box 8.1.) In response to this program, the government increased its capital expenditures in rural areas. Between the first (1964–1969) and second (1969–1973) Five Year Plans, the development budget for primary schools and adult literacy programs increased from 9% of the total to 29%. Expenditures on rural water programs also increased from 29% of the development budget to 45%, and for rural health centers, they increased from 30% to 39% (Clark, 1975).

The Ujamaa village program was designed to consolidate scattered rural settlements into self-sufficient nucleated sites, ranging from 100 to 700 families. Initially, resettlement into villages was voluntary, but, after a slow response, the government decided to make it mandatory. There were some achievements in social equity, particularly in adult literacy, health, and income distribution. In 1962, the ratio of the highest to the lowest salary was 33:1. By 1980, the ratio had narrowed to 6:1 (Lonsdale 1995). However, the Ujamaa system was plagued by ineffective management and the lack of adequate investment in rural infrastructure. A major problem was that the policy of socialism was instituted from the top by a bureaucracy that was out of touch with the needs of rural people and inexperienced in handling the complexities of rural development. The government's emphasis on social equity took its toll on the productive sectors of the Tanzanian economy. Tanzania not only has one of the lowest per capita GNP rates in Sub-Saharan Africa, but, between 1980 and 1985, it averaged 0.9% in GDP growth rate. With the resignation of Nyerere, there has been a shift in government policy from socialism to a more liberalized economic approach based on agreements with the World Bank and the International Monetary Fund to institute Structural Adjustment and Economic Recovery Programs.

Self-reliance at the state level implies total autonomy and independence in decision making and the allocation of resources. No single Sub-Saharan country is in a position to break away and develop a totally autonomous stance in our globally interdependent world. Sub-Saharan Africa has been firmly entrenched in the global exchange economy since colonial times. Therefore, as Galtung (1982) suggests, self-reliance must be redefined to combine independence with interdependence and autonomy with equity and mutual cooperation.

Interdependent Development

The type of interdependent development advocated here calls for cooperation among governmental and organizational units, at all geographic scales, to facilitate development efforts. This implies developing cooperative and compatible arrangements between local and national governments, local and international organizations, national and regional organizations, and national and international governments and organizations (Figure 8.5). At the local level, independence and autonomy are strongest. The level of interdependence increases as we move from the local to the national, the regional, and the international scales.

It is important for rural constituents in Sub-Saharan Africa to maintain a certain degree of autonomy in their decision-making and planning processes. Self-help efforts, however, must be complemented by committed efforts on the part of national governments to provide the necessary technical and financial assistance. There are a number of examples where this interdependent relationship between village and national governments has occurred.

In western Nigeria, hometown associations function as intermediaries in linking central government institutions to local communities (Barkan, *et al.*, 1991). The associations encourage voluntary participation between village descendants and local residents. Emigrants provide cash contributions to sponsor development projects in their local communities. Those who leave maintain strong ties by forming networks of hometown associations in urban areas. The associations have provided a range of basic education, health, communication, and infrastructural services. Through the success of hometown associations, the Nigerian government is now providing more support to village community-development organizations.

In Zimbabwe, the Organization of Rural Associations for Progress has worked effectively with the poor at the local village level and expanded its efforts to the national and international level (Nyoni, 1987). The organization mobilizes rural groups and associations to explore and articulate their development priorities. The decision-making structure begins at the family level: Families convene to discuss and design strategies. Representatives of the family unit then meet at the group or production-unit level to discuss community projects. Production units are further consolidated into associations at the district level. For each

Development Strategies

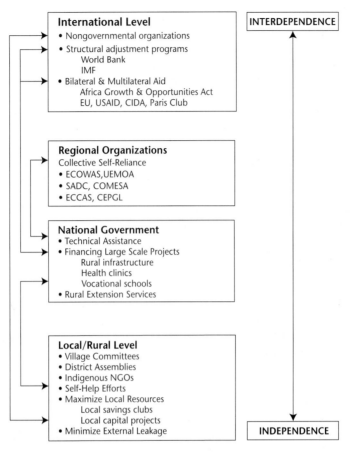

Figure 8.5 Model of interdependence.

association there is designated a development center, which offers technical, counseling, and educational services. Representatives from each association form an advisory board, which promotes the exchange of ideas and acts as an intermediary between the rural associations and the national and international organizations.

Zambia passed a Local Administration Act in 1980, which recommended the creation of an integrated district council to administer political, economic development, scientific, technological, and social affairs at the local level. The district councils were part of the Zambian government's efforts to decentralize administration and empower local communities to formulate short- and long-term development programs (Noppen, 1990). Although similar attempts at decentralization have been attempted in Botswana, Kenya, Tanzania, and Zimbabwe, there are a number of problems. For instance, local district councils and their professional staffs usually lack the expertise to plan and manage policy effectively. Furthermore, central governments usually fail to clarify the role of district governments in the overall scheme of political and administrative management (de Valk and Wekwete, 1990).

There is hardly any devolution of power to local districts that would grant them complete autonomy and authority to design, execute, and finance local projects without central government oversight and approval. Usually, decentralization in Sub-Saharan Africa takes the form of deconcentration, which passes down only administrative discretion to local offices of central government ministries and requires local districts to refer proposals to the center.

In spite of administrative failures in local district councils, they need to be encouraged. Their success is predicated upon effective local organization and leadership to mobilize local resources. These local efforts and initiatives should by no means be divorced from national goals and objectives, which should be positioned to complement local initiatives. (See Figure 8.5.)

At the same time, if national governments in Sub-Saharan Africa are to be effective partners in assisting with development projects in rural areas, they will require assistance from regional and international organizations. At the regional scale, there have been various attempts at Pan-Africanism; at the international level, assistance has come via structural-adjustment programs and nongovernmental organizations.

MULTILATERALISM

Bilateralism and the Africa Growth and Opportunity Act

In May of 2000, the United States sought to strengthen its bilateral ties with a number of African countries: President Clinton signed the Africa Growth and Opportunity Act (AGOA). The act seeks to strengthen American–African trade and investment ties by allowing African states more access to US markets, credit, and technical know-how and by providing social and economic opportunities to vulnerable groups in order to ensure stable and sustained growth and development (US Congress, 1999). In 1999, the United States imported $18 billion worth of goods from Africa—almost half from Nigeria and South Africa (WTO, 2000). Total trade with Africa amounted to about $28 billion—2.5% of overall U.S foreign trade. To be eligible to receive benefits from the AGOA, African countries must have demonstrated a commitment to economic and political reform by engaging in political pluralism and the rule of law, by developing policies to reduce poverty and combat corruption, by protecting human rights and worker rights, and by adhering to the IMF and World Bank structural-adjustment programs.

Critiques of AGOA state that it caters to the interests of a few countries, it does not demonstrate a commitment towards reducing the debt burden in Africa, and it does not provide a mechanism to ensure adequate labor and environmental standards. To address these shortcomings, an alternative bill—the Human Rights, Opportunity, Partnership, and Empowerment for Africa Act (the HOPE for Africa Act)—was proposed.

Regional Integration and Pan-Africanism

Regional cooperation and interdependence among Sub-Saharan countries needs to be encouraged to promote a greater sense of collective self-reliance. Calls for African unity have been prevalent since the early 1900s. The first Pan-African conference was convened in London in 1900 to protest colonial rule in Africa; the theme of Pan-Africanism has persisted to this day. A number of prominent African leaders, such as Kwame Nkrumah of Ghana and Sekou Toure of Guinea, took up the challenge of Pan-Africanism right through to the time of independence; they envisioned an ultimate form of Pan-Africanism, involving a United States of Africa with one government. Other leaders, such as Jomo Kenyatta of Kenya and Leopold Senghor of Senegal, preferred a more moderate and gradual approach that preserved the sovereignty of individual governments.

The Organization of African Unity (OAU) was formed in 1963 to champion the cause of African unity and defend the sovereign rights of African states. It continues to endorse efforts to promote greater economic integration. At the 21st Assembly, held in Addis Ababa in 1985, African countries reaffirmed their commitment to the 1980 Lagos Plan of Action on the establishment of an African Common Market by the year 2000—a pretty lofty goal at the time. This was followed in 1991 by the signing of a treaty in Abuja, Nigeria to establish a Pan-African Economic Community (AEC) by the year 2025. The AEC was set up to enhance collective self-reliance and self-sustaining development in Africa; to promote greater social, cultural, and economic integration among African states; to coordinate and harmonize the activities (such as tariff systems) of existing regional organizations in Africa (as in Table 8.3); and to eventually establish an African Central Bank, an African Economic and Monetary Union, and a Pan-African Parliament.

Part of the Abuja Treaty's agenda is to encourage three subregions—west Africa, central Africa, and east and southern Africa—to proceed through various stages of economic integration, beginning with a free-trade area and followed by a customs union, common market, and economic community. A *free trade area* enables member countries to trade freely among themselves, yet levy different external tariffs. Free internal trade is encouraged in a *customs union*, but members must agree on a common external tariff. A *common market* has all the attributes of a customs union plus the free movement of labor and capital; and an *economic union* is a more elaborate form of a common market, usually including a central bank, a common currency, and a common economic policy (Balassa, 1961). Given the current status of Sub-Saharan African countries, it will take a while before they reach the ultimate form of economic integration that intended to be characteristic of the European Community.

Collective self-reliance provides Sub-Saharan Africa with a framework to pursue common goals and common interests that ensure economic and political sustainability. Promoting collective self-reliance through regional economic integration is relevant in today's global economy, considering the numerous trading blocs cropping up in North America (the North American Free Trade Agreement, or NAFTA), Europe (the European Union or EU), the Asian–Pacific realm, Latin America (the Latin America Free Trade Area, or LAFTA), the Caribbean (the Caribbean Comunity, or CARICOM), and Asia (the Association of Southeast Asian Nations, or ASEAN). Africa needs to strengthen its competitive position in the global economy by forging viable regional economic alliances.

Regional economic integration is also warranted given the poor record of intraregional trade in Africa. In 1999, only 9.9% of Africa's merchandise exports went to other African countries, while 51% went to Europe (WTO, 2000). Even amongst the regional groups listed in Table 8.3, there is minimal trade. For example, among the countries that make up the Economic Community of West African States (ECOWAS) regional trade accounts for just 6%; the figures are almost similar (8%) for the Common market of Eastern

TABLE 8.3 Examples of Regional Integration Efforts in Africa

Regional Organizations and Headquarters	Member Countries	Goals and Objectives	Share of Intra-African Trade* 1998
1964: Customs and Economic Union of Central Africa (UDEAC). Headquartered in Bangui, Congo.	Cameroon, Central Africa Republic, Chad, Congo, Equatorial Guinea, and Gabon.	To achieve economic integration and the harmonization of development planning, fiscal legislation, and industrial investment and transport policies.	2.5%
1973: Mano River Union States (MNU). Headquartered in Freetown, Sierra Leone.	Sierra Leone, Liberia, and Guinea.	To establish a customs and economic union between the member states in order to improve living standards.	Less than 0.5%
1973: West African Economic Community (CEAO). Headquartered in Ouagadougou, Burkina Faso. Abolished in 1994 to form the West African Economic and Monetary Union (UEMOA).	Franco-phone speaking countries of Benin, Burkina Faso, Côte D'Ivoire, Mali, Mauritania, Niger, and Senegal. Togo had observer status.	To gradually eliminate customs tariffs on goods imported from member countries and encourage free movement of individuals amongst countries. Has a free trade area for agricultural products and raw materials and as a preferential trading area for approved industrial products, with a regional cooperation tax.	7–8%
1975: Economic Community of West African States (ECOWAS). Headquartered in Lagos, Nigeria.	Benin, Burkina Faso, Cape Verde, Côte D'Ivoire, Gambia, Ghana, Guinea, Guinea-Bissau, Liberia, Mali, Mauritania, Nigeria, Senegal, Sierra Leone, and Togo.	To establish a common customs tariff. Free movement of capital and people, a common trade policy, and harmonization of agricultural, communications, energy, and infrastructural policy. Members have also signed a pact of nonaggression and mutual defense.	5–6%
1976: Economic Community of Great Lakes Countries (CEPGL). Headquartered in Gisenyi, Rwanda.	Burundi, Rwanda, and the Democratic Republic of Congo.	To pursue common economic, scientific, cultural, political, military, and technical goals. Remove trade barriers and cooperate to develop and implement joint projects.	0.6%
1980: Southern Africa Development Community (SADC). Headquartered in Gaborone, Botswana.	Angola, Botswana, Democratic Republic of Congo, Lesotho, Malawi, Mauritius, Mozambique, Namibia, Seychelles, South Africa, Swaziland, Tanzania, Zambia, and Zimbabwe.	To achieve development and economic growth, alleviate poverty, enhance the standard and quality of life of the peoples of southern Africa, and support the socially disadvantaged through regional integration. To promote and defend peace and security by consolidating the long-standing historical, social, and cultural affinities and links among the peoples of the region.	10.2%–22%

TABLE 8.3 *(Continued)*

Regional Organizations and Headquarters	Member Countries	Goals and Objectives	Share of Intra-African Trade* 1998
		To promote self-sustaining development through collective self-reliance and interdependence of member states. To achieve complementarity between national and regional strategies and programs.	
1983: Economic Community of Central African States (ECCAS). Headquartered in Libreville, Gabon.	UDEAC and CEPGL members, as well as São Tomé and Principe.	To promote regional economic cooperation and establish a central African common market.	2.0
1994: Common Market for Eastern and Southern Africa (COMESA). Formerly the Preferential Trade Area (PTA). Headquartered in Lusaka, Zambia.	Angola, Burundi, Comoros, Democratic Republic of Congo, Djibouti, Egypt, Eritrea, Ethiopia, Kenya, Lesotho, Madagascar, Malawi, Mauritius, Mozambique, Namibia, Rwanda, Seychelles, Sudan, Swaziland, Tanzania Uganda, Zamibia, and Zimbabwe.	To cooperate in developing the natural and human resources for the good of all member countries. To promote joint economic development activity. To promote peace, security, and stability among member states. To create an enabling environment for foreign, cross-border and domestic investment, including the joint promotion of research and adaptation of science and technology for development.	7.7%
1994: West African Economic and Monetary Unit (UEMOA).	Benin, Burkina Faso, Côte d'Ivoire, Guinea–Bissau, Mali, Niger, Senegal, and Togo.	To create a common market to promote economic integration. Trade liberalization—common external tariff rates. Single financial market and regional stock exchange.	11.1%

*Share of African trade estimates derived from IMF. (1999b). *Direction of Trade Statistics,* Washington, DC: World Bank and the World Bank. (2000). *World Development Indicators.*

and Southern African (COMESA) (Hardy, 1992). Intraregional linkages need to be strengthened to enhance economies of scale in production, to support production technologies, and to promote greater functional specialization and technological efficiency. This is especially relevant for a continent that has approximately 50% of the world's landlocked countries and a relatively high proportion of small-sized countries with limited domestic markets.

Efforts at economic integration have largely failed, hampered by uneven levels of development between countries and unequal distribution of benefits and costs. ECOWAS created the Fund for Cooperation, Compensation, and Development (FCCD) to ensure a more equitable distribution of costs and benefits among member states and to mobilize resources to support regionwide development projects (Ojo, 1999). In spite of the shortcomings in intraregional trade, ECOWAS has made some progress in the areas of transportation and telecommunication development, the free movement of persons, and regional security. For example, much of the transcoastal west African highway network from Lagos to Nouakchott and the trans-Sahelian highway from Dakar to N'Djamena have been completed. Also, efforts to enhance the region's telecommunications with multimedia and broadband technologies are being realized. In addition, ECOWAS travel certificates have been introduced to facilitate intraregional travel, and an ECOWAS Peace Monitoring Group (ECOMOG) is in place to mitigate conflicts and preserve peace and stability in the region.

In spite of some success stories, regional economic organizations in Africa must still deal with the obstacles that are impeding progress towards enhanced trade and full economic integration. Among the lingering problems are limited manufacturing capacity, weak financial sectors, poor transportation and communications infrastructure, low per capita income, weak governance, and a lack of technical, financial, and management capacity to deal with the full slate of mechanisms and responsibilities required for successful integration (World Bank, 2001a). Also problematic is the existence of overlapping and competing groups, as exemplified by Common Market for Eastern and Southern Africa (COMESA) and SADC and by ECOWAS and West African Monetary and Economic Union (UEMOA). With the exception of Botswana and South Africa, all the 14 members of SADC also belong to COMESA. This overlap has prompted some countries to consider withdrawing from COMESA, primarily Lesotho, Mozambique, and Zambia, possibly because Kenya and Zimbabwe dominate trade activity within COMESA by accounting for a disproportionate share of exports (65%) (Muuka, *et al.*, 1998).

Given these problems, prospects for any form of free trade, let alone a common market or economic union in the near future, are not very encouraging. In spite of these difficulties, Sub-Saharan Africa should persist with the aims and objectives of the Abuja Treaty. Further collaborative opportunities in transportation and communication, research and technology, environmental management, resource conservation, and food security should be encouraged. One of the dilemmas of regional integration schemes in Africa is that countries seem to prefer formal trade and factor market integration, rather than basic policy coordination and collaboration in regional projects (World Bank, 2001a). African countries can benefit from SADC's model of economic cooperation, where each member country is responsible for coordinating key economic sectors: Angola is responsible for energy; Botswana coordinates agricultural research, livestock production, and animal disease control; Malawi coordinates inland fisheries, forestry, and wildlife; Namibia takes care of marine fisheries and resources; South Africa manages finance and investment; Tanzania handles industry and trade; and Zambia coordinates mining, employment, and labor. SADC has been more successful than any of the other regional organizations in Africa in securing assistance from international agencies. In 1999, SADC signed a trade, development, and cooperative agreement with the European Union to reduce tariffs and to develop small enterprises, improve telecommunications and information technology, and enhance the tourism, mining, and energy sectors. It is no surprise that intra-SADC trade increased from 1.4% in 1985 to 10.2% in 1998. Simon (2001) indicates that intra-SADC trade is currently at 22% and projects an increase to as much as 35% once the SADC Free Trade Area becomes fully operational. The top products exported from SADC member countries are diamonds, tobacco, cotton, bovine meat, tuna, and maize.

Several opportunities exist for regional entities in Africa to collaborate on a variety of initiatives. The African Economic Research Consortium is an example of a successful venture that endeavors to build regional capacity (World Bank, 1989). It provides a forum for African researchers to discuss and evaluate research on a variety of economic topics ranging from debt management and taxation policy to structural-adjustment problems. Other forms of cooperation include the Association for the Advancement of Agricultural Science in Africa, the African Association for Literacy and Adult Education, the Intergovernmental Authority on Drought and Development, and the African Regional Center for Technology.

Multilateralism: Structural-Adjustment Programs and Nongovernmental Organizations

Structural-adjustment programs At the international scale, aid to Sub-Saharan Africa has taken several forms; financial and technical assistance has been forthcoming from bilateral and multilateral sources. However, since the 1980s, financial and technical assistance has been conditioned by the World Bank's structural-adjustment programs (SAPs). The SAPs were initiated to adjust malfunctioning economies and to promote greater economic efficiency and economic growth, as a way to make Sub-Saharan economies more competitive in today's global economy. By 1990, 32 African countries had launched structural-adjustment programs or borrowed from the International Monetary Fund (IMF) to support policy reforms.

Candidates for structural adjustment are countries with budget deficits, balance-of-payment problems, high inflationary rates, ineffective state bureaucracies, inefficient agricultural and industrial production sectors, overvalued currencies, and inefficient credit institutions. As a result, countries that buy into the program are required to do the following:

1. Adjust domestic demand to reduce expenditure on imports and release resources for exports, which can be done by devaluing the currency and limiting money and credit growth (Stewart, *et al.*, 1992). Currency devaluation involves lowering the official exchange rate of an African country relative to the rest of the world. This has the effect of raising the local price of imported goods, thereby lowering domestic demand as well as lowering the external price of African exports with the expectation of increasing foreign demand for African products.
2. Restructure the public sector and state-owned enterprises by trimming overstaffed bureaucracies and padded payrolls, improving institutional management, and encouraging privatization.
3. Eliminate price controls and subsidies, such as artificially reducing the price of food in urban areas, that lower farm incomes.
4. Restructure the productive sectors of the economy by liberalizing trade and removing import quotas and high tariffs that protect uncompetitive firms and by providing export incentives to promote export growth, particularly in agriculture.

These are the main policy conditions set by the World Bank as a basis for lending. Bilateral aid agencies and multilateral lending institutions, like the IMF, now coordinate their policies with the World Bank and make assistance conditional upon the host governments' acceptance of structural-adjustment programs. Structural adjustment has now become the premier internationally supported, long-range development policy for Sub-Saharan Africa (Weismann, 1990).

However, studies have shown that the social, human, environmental, and equity dimensions of SAPs have been largely ignored in Africa. Weismann's (1990) analyses of Ghana and Senegal show that, in spite of unprecedented growth in their economies in the late 1980s, rural farmers, the urban poor, and women in general were negatively affected by SAPs. In Ghana, a series of currency devaluations and strict credit restrictions had the effect of raising the prices of consumer goods and agricultural inputs, thus affecting women and children who rely on food trading and production. In urban areas, policies that were geared toward trimming government bureaucracies were not complemented with adequate retraining or job placement programs for the redeployed.

In Senegal, reduction in government support for agricultural credit and subsidies had an adverse impact on farmers who could no longer acquire fertilizers and seeds to boost their production levels. Among the urban dwellers, wage-restraint policies and salary freezes reduced the average worker's purchasing power by 30% in 1985. Restrictions on public-sector hiring in Tanzania resulted in only 7,300 additional jobs created for about 200,000 individuals between 1986 and 1990 (Wagao, 1992). Declines in government expenditure induced by adjustment also led to a reduction in educational resources and associated declines in enrollments and the quality of education in Tanzania. A recent report by the World Bank (2001) acknowledges increasing levels of poverty, vast inequality in incomes, and higher debt burdens in African countries.

These adverse social and economic impacts on disadvantaged groups have prompted scholars to call for a reexamination of structural adjustment policies. There also needs to be more effective participation and representation of disadvantaged groups in the policy-formulation and planning process of adjustment programs. Ghana has responded with the institution of PAMSCAD, or Program of Action to Mitigate the Social Costs of Adjustment, which is designed as a supplemental basic-needs strategy to improve primary health care, reduce childhood diseases, enhance child nutrition, and compensate for adjustment-related job losses. The World Bank has tried to cushion the effects of SAPs by introducing two major initiatives in the 1990s: the Heavily Indebted Poor Countries (HIPC) initiative, and the Poverty Reduction and Growth Facility program. Program beneficiaries must prepare plans that address the needs of vulnerable groups and devise short- and long-term strategies to mitigate the problems associated with poverty.

Critics of the SAPs (Adedeji, *et al.*, 1990; United Nations Economic Commission for Africa, 1989) have come out with an *African Alternative to the Structural Adjustment Programs for Socio-Economic Recovery and Transformation* (AAF-SAP). The policy document calls for a more human-centered and holistic approach that reflects the political and economic realities of Africa and is flexible enough to acknowledge the uniqueness of each country. Wai (1992) is also quick to point out that 70% of the World Bank and International Development Association's resources goes toward investment projects in agriculture, industry, water supply, education, and health. Furthermore, aside from structural-adjustment programs, the World Bank sponsors social adjustment, education, food security, agricultural extension, and African capacity-building initiatives.

The role of nongovernmental organizations Nongovernmental organizations (NGOs) have been instrumental in providing a wide variety of aid packages to Sub-Saharan countries. NGOs operate neither as government nor as for-profit organizations; they are a diverse group of largely voluntary organizations that work with people organizations to provide technical advice and economic, social, and humanitarian assistance. They can be professional associations, religious institutions, research institutions, private foundations, or international and indigenous funding and development agencies. The World Bank is collaborating with an increasing number of NGOs, particularly grassroots organizations, to address rural-development, population, health, and infrastructural issues that are pertinent to the human costs of its structural-adjustment programs. Several NGOs also have consultative status with the United Nations Economic and Social Council.

Some well-known NGOs are Catholic Relief Services, the Salvation Army, CARE, World Vision, Save the Children, Ford Foundation, and Oxfam. In Sub-Saharan Africa, Kenya hosts the highest number of NGOs, which is estimated to be over 400. There are estimated to be 94 in Uganda, 80 in Zimbabwe, and 46 in Ethiopia (Bratton, 1989). Most of the "indigenous" NGOs in Africa are community-based grassroots and service-based organizations. They usually fill a void where governments have been largely ineffective and out of touch with local needs. Some of their specific developmental objectives in Sub-Saharan Africa include tackling poverty, providing financial credit and technical advice to the poor, empowering marginal groups, challenging gender discrimination, and delivering emergency relief.

There is, however, the potential for conflict between African governments and NGOs. By filling a void in terms of additional investments, capital, and services to rural areas, NGOs undermine the ability of governments to perform as effective leaders and policy makers. In response to this potential threat, some African governments have instituted coordinating bodies to supervise NGO activity. In Burkina Faso, the government relies on the Permanent Secretariat of NGOs to monitor NGO activity; in Togo, the Council of NGO Activity (CONGAT) coordinates NGOs through government policy. Other examples of coordinating bodies are the Voluntary Organizations in Community Enterprise (VOICE) in Zimbabwe and the Zambia Council for Social Development, both of which coordinate social-service activities.

Nongovernmental organizations have a vital role to play in the social and economic development of Sub-Saharan Africa. More partnerships between international, national, and indigenous NGOs should be encouraged. There is a growing awareness of the importance of voluntary activity. In a conference held in Dakar, Senegal in 1987, delegates from African countries agreed to form a Pan-African umbrella organization—the Forum of African Voluntary Development Organizations (FAVDO)—to provide a forum for groups to exchange ideas, share their expertise and resources, support local initiatives, and establish effective channels of communication and partnerships with governments and with intergovernmental organizations.

CONCLUSION

Sub-Saharan Africa faces important challenges in development policy and planning as it heads toward the 21st century. The socioeconomic problems perhaps seem insurmountable, but opportunities still exist for Africa to tap into its vast reservoir of human and natural resources and carve out policies that are not solely growth inducing, but pay more attention to equity and humanistic dimensions. Self-help efforts at the grassroots level are on the rise; people are beginning to control their own destinies by channeling their energies and creativity toward more positive ventures. These local initiatives must be supported with complementary efforts from national and international levels.

Sub-Saharan Africa needs to seek more avenues for economic, cultural, and political cooperation at the regional level. As regions around the world continue to form alliances and trading blocs, Sub-Saharan Africa needs to carve out its own niche by promoting the legacy of collective self-reliance bequeathed by the architects of African nationalism.

While international efforts at revitalizing Sub-Saharan Africa are welcome, they must be well intended. What is more important, they must address the extreme inequalities that exist between the core and the periphery, and they must design strategies that provide the poor with opportunities to maximize their human potential.

KEY TERMS

Human development
Sustained development
Gross National Product (GNP)
Gross Domestic Product (GDP)
Purchasing Power Parity (PPP)
Hierarchical diffusion
Core–periphery disparities
International dualism
Urban dualism
Rural dualism
Urbanization economies
Colonial infrastructure
Physical quality of life
Poverty-reduction strategy

Modernization theory
Big push
Efficiency–equity debate
Growth with equity
Heavily Indebted Poor Countries (HIPC)
Basic-needs strategy
Appropriate technology
Self-reliance
Interdependent development
Regional integration
Structural-adjustment program
Nongovernmental organizations
Self-help

DISCUSSION QUESTIONS

1. What, in your opinion, are the most appropriate ways of measuring development in Sub-Saharan Africa?
2. Describe the core–peripheral disparities that exist within Sub-Saharan African economies.
3. What are the underlying economic, structural, and sociological explanations for core–peripheral disparities in Sub-Saharan Africa?
4. Is a self-reliant strategy the most appropriate strategy for Sub-Saharan Africa?
5. Discuss and evaluate the equity–efficiency debate in the African development literature.
6. Provide examples of grassroots responses to development problems in Africa's rural areas.
7. Discuss the problems associated with regional economic integration in Sub-Saharan Africa.
8. What, in your opinion, should be the appropriate international response to development problems in Sub-Saharan Africa?

REFERENCES

ADEDEJI, A., et al. (1990). *The Human Dimensions of Africa's Persistent Economic Crisis: Selected Papers,* London; New York: United Nations Economic Commission for Africa.

BALASSA, B. (1961). *The Theory of Economic Integration,* Homewood, IL: Richard D. Irwin.

BARKAN, J., MCNULTY, M., & AYENI, M. (1991). "HomeTown" Voluntary Associations, Local Development, and the Emergencies of Civil Society in Western Nigeria," *Journal of Modern African Studies,* 29(3):457–480.

BRATTON, M. (1989). "The Politics of Government–NGO Relations in Africa," *World Development,* 17(4):569–587.

CLARK, E. (1975). "Socialist Development in an Underdeveloped Country: The Case of Tanzania," *World Development,* Vol. 3, 4:223–228.

DE VALK, P. & WEKWETE, K. (1990). *Decentralization for Participatory Planning?* Aldershot, UK: Gower Publishing Company.

FAIR, D. (1991). "Middle Africa's Economic Islands of Development—Thirty Years on," *Africa Insight,* 21(1):41–47.

FAUVET, P. (2000). "Mozambique: Growth and Poverty," *Africa Recovery,* 14(3):12.

FILANI, M. (1981). "Nigeria: The Need to Modify Center-Down Development Planning," in *Development From Above or Below,* Stohr W., & D. Taylor, D., eds., New York: Wiley & Sons.

FRANK, A. (1980). *Crisis in the World Economy,* London: Heinemann.

GALTUNG, J. (1971). "A Structural Theory of Imperialism," *Journal of Peace and Research,* 8(2):81–118.

GALTUNG, J. (1982). "The Politics of Self Reliance," in *From Dependency to Development: Strategies to Overcome Underdevelopment and Inequality,* Munoz H. ed., Boulder, CO: Westview Press.

HARDY, C. (1992). "The Prospects for Intra-Regional Trade Growth in Africa," in *Alternative Development Strategies in Sub-Saharan Africa*, Stewart, F., Lall, S., & S. Wangwe, S., eds., New York: St. Martin's Press.

HIRSCHMAN, A. (1958). *The Strategy of Economic Development*, New Haven, CT: Yale University Press.

HOSELITZ, B. (1960). *Sociological Aspects of Economic Growth*, New York: Free Press.

INTERNATIONAL MONETARY FUND (IMF) (1999a). *Uganda: Selected Issues and Statistical Appendix*, Washington, DC: IMF Staff Country Report No. 99/116.

INTERNATIONAL MONETARY FUND (IMF) (1999b). *Direction of Trade Statistics,* Washington, DC: World Bank.

KNOX, P. & AGNEW, J. (1998). *The Geography of the World Economy*, New York: Wiley.

LONDSDALE, J. (1995). "Tanzania" in *Africa South of the Sahara,* London: Europa Publications.

MCCLELLAND, D. (1961). *The Achieving Society*, Princeton, NJ: Van Nostrand.

MUUKA, G., HARRISON, D., & MCCOY, J. (1998). "Impediments to Economic Integration in Africa: The Case of COMESA," *Journal of Business in Developing Nations*, 2(3).

NOPPEN, D. (1990). "Decentralization and the Role of District and Provincial Units in Zambia," in P. de Valk & K. Wekwete, *Decentralization for Participatory Planning?* Aldershot: Gower Publishing Company.

NORMAN, G. (1975). "Introducing the Hedonometer," *The Times*.

NYONI, S. (1987). "Indigenous NGOs: Liberation, Self-Reliance, and Development," *World Development,* 15:51–56.

OBUDHO, R.A & TAYLOR, D.R.F., eds. (1979). *The Spatial Structure of Development: A Study of Kenya*, Boulder, CO: Westview Press.

OJO, O. (1999). "Integration in ECOWAS: Successes and Difficulties," *Regionalisation in Africa*, Bach, D., ed., Bloomington, IN: Indiana University Press.

PARSONS, T. (1951). *The Social System*, Glencoe, IL: Free Press of Glencoe.

RAPLEY, J. (1994). *Ivorien Capitalism: African Entrepeneurs in Côte d'Ivoire,* Boulder: CO, Hynne Rienner.

REITSMA, H. & KLEINPENNING, J. (1985). *The Third World in Perspective,* Totowa, NJ: Rowan and Allanheld.

REPUBLIC OF CHAD (2000). *Interim Poverty Reduction Strategy Paper for 2000–2001*, Washington, D.C. IMF.

REPUBLIC OF MOZAMBIQUE (2001). *Action Plan for Poverty Reduction (2001–2005)*, Maputo, Mozambique: Council of Ministers.

ROSENSTEIN-RODAN, P. (1964). *Capital Formation and Economic Development*, Cambridge, MA: MIT Press.

ROSTOW, W. (1963). "The Take-Off into Self-Sustained Growth," in *The Economics of Underdevelopment,* Agarwala A. & Singh, S. eds., London: Oxford University Press.

SCHUMACHER, E.F. (1974). *Small Is Beautiful: Economics as if People Mattered,* New York: Harper & Row.

SEERS, D. (1969). "The Meaning of Development," *International Development Review,* 11.

SIMON, D. (2001). "Trading Places: Imagining and Positioning the New' South Africa within the Regional and Global Economies", *International Affairs,* 77:2:377–405.

STEWART, F., LALL, S., & WANGWE, S. (1992). "Alternative Development Strategies: An Overview," in *Alternative Development Strategies in Sub-Saharan Africa,* Stewart F., Lall, S., & Wangwe, S., eds., New York: St. Martin's Press.

STREETEN, P. (1981) ed., *Development Perspectives,* London: Macmillan.

STREETEN, P. & STEWART, F. (1981). "New Strategies for Development: Poverty, Income Distribution and Growth", in P. Streeten, ed.: *Development Perspectives,* Oxford: Oxford University Press, pp. 148–174.

THOMAS, B. (1987). "Development through Harambee: Who Wins and Who loses? Rural Self-Help Projects in Kenya," *World Development,* 15(4):463–481.

TODARO, M. (1989). *Economic Development in the Third World,* New York: Longman Press.

UNITED NATIONS ECONOMIC COMMISSION FOR AFRICA (1989). *African Alternative Framework to Structural Adjustment Programs for Socio-Economic Recovery and Transformation (AAF-SAP),* Addis Ababa, Ethiopia: United Nations Economic Commission for Africa.

UNITED NATIONS DEVELOPMENT PROGRAM (UNDP) (1999). *Human Development Report,* Oxford: Oxford University Press.

UNITED NATIONS DEVELOPMENT PROGRAM (UNDP) (2001). *Human Development Report,* Oxford: Oxford University Press.

UNITED STATES CONGRESS (1999). *Providing for the Consideration of H.R. 434, Africa Growth and Opportunity Act: Report,* Washington, DC: US Government Printing Office.

VITTA, P. (1990). "Technology Policy in Sub-Saharan Africa: Why the Dream Remains Unfulfilled," *World Development,* 18(11):1471–1480.

WAGAO, J. (1992). "Adjustment Policies in Tanzania, 1981–89: The Impact on Growth, Structure, and Human Welfare," in *Africa's Recovery in the 1990s,* Cornia, G. van der Hoeven, R. & Mkandawire, T., eds., New York: St. Martin's Press.

WAI, D. (1992). "The View from the World Bank," in *Structural Adjustment and the Crisis in Africa,* Kennet D. & Lumumba-Kasongo, Lewiston T., NY: Edwin Mellen Press.

WALLERSTEIN, I. (1982). "Dependence in an Interdependent World: The Limited Possibilities of Transformation within the Capitalist World Economy," in *Dependency and Development,* Munoz, H. Boulder, CO: Westview Press.

WEISMANN, S. (1990). "Structural Adjustment in Africa: Insights from the Experiences of Ghana and Senegal," *World Development,* 18(12):1621–1634.

WORLD BANK (1989). *Sub-Saharan Africa: From Crisis to Sustainable Growth,* Washington, DC: World Bank.

WORLD BANK (1998). *African Development Indicators: 1998–99,* Washington, DC: World Bank.

WORLD BANK (2000). *World Development Indicators, 2000,* Washington, DC: World Bank.

WORLD BANK (2001a). *Can Africa Claim the 21st Century?* Washington, DC: World Bank.

WORLD BANK (2001b). *World Development Report,* 2001, Washington, DC: World Bank.

WORLD RESOURCES INSTITUTE (WRI) (1992). *World Resources 1992–1993,* New York: Oxford University Press.

WORLD RESOURCES INSTITUTE (WRI) (1994). *World Resources 1994–1995,* New York: Oxford University Press.

WORLD TRADE ORGANIZATION (WTO) (2000). International Trade Statistics, Geneva: World Trade Organization.

9

Transport and Communication in Sub-Saharan Africa: Digital Bridges Over Spatial Divides

Joseph R. Oppong

TRANSPORTATION SYSTEMS IN AFRICA

Transport systems are composed of various transport modes, such as road, rail, water, and air, that have developed and been implemented over time and under different human geographic conditions. These modes vary geographically in complexity, in response to technological advances and to the type of social, political, and economic context. For example, modern road transportation includes anything from footpaths to multilane highways, each of which can have different characteristics and all of which could combine to create a coherent, coordinated system. Transport systems facilitate movement of goods and services and enhance the degree of spatial interaction to link centers of supply and demand.

    Unlike the excellent interstate road and rail transportation systems and the numerous airports and urban transit facilities of the United States, transportation systems in Africa are problematic. Intercountry networks are poorly developed, and the quality of transportation systems—state of repair, availability, and efficiency—varies significantly between and within countries. Even when reliable transportation infrastructure exists, political conflicts, such as civil wars, sometimes disrupt usage. The difficulties associated with intermodal transportation could make the use of electronic communication systems an ideal alternative that provides digital bridges over spatial divides. Unfortunately, while growing rapidly, the Internet and other electronic communications continue to be confined to the major cities. This chapter examines the geography of transport, communication, and Internet access in Africa with respect to the aforementioned issues.

The transportation of goods and people has never been easy in Africa. The terrain, rivers, animals, and diseases frequently present major obstacles. Prior to European contact, transportation was primarily on foot, and goods were carried on the head or back because insect pests such as the tsetse fly prevented the use of animals for lading and hauling. River transport was similarly restricted and problematic. Cataracts, rapids, falls, seasonal flows, and shifting river channels were a major challenge for dugout canoes. For example, Victoria Falls, on the Zambezi River, is 1600 m (1 mi) wide and 90 m (300 ft) in elevation. Lacking natural shelter and exposed to heavy surf, the smooth and even coasts of Africa required expensive breakwaters and harbors. Torrential rains produced washed-out roads and broken bridges; falling trees blocked roads and footpaths in the rain forest. Nevertheless, well-established transport networks, including trade routes, emerged in many parts of Africa during precolonial times, based primarily on paths beaten out of the bush by human feet. The major thoroughfares today are still based on these precolonial paths.

During the colonial era, these networks were restructured to facilitate commerce and administration, as defined by the colonial powers. Seaports, roads, and railways were constructed to penetrate into the interior and permit the extraction of such natural resources as minerals and timber. Needing to accommodate only the low traffic densities of the time, these early roads were built cheaply and very simply. Consequently, running speeds were low, vehicle life was relatively short, and large sections of the network were unusable for varying periods during the rainy season. The seasonal nature of the road network remains an important feature of the system today—the dry season network is several times more extensive than the all-weather network.

Since independence, the most intensively used links in the road system have been upgraded. Gravel or laterite surfaces have been replaced, sometimes by two-lane bitumen surfaces. Few motorways exist in the region; the Lagos–Ibadan expressway in Nigeria and the Accra–Tema motorway in Ghana are good examples. Bridges have replaced ferries, and roads have been realigned to take out sharp bends and steep gradients. Nevertheless, road densities vary significantly between locations, particularly because of population density, topography, proximity to urban centers, economic activity, and competing forms of transport.

Road Transportation

Roads in African countries are generally of 3 classes: primary, secondary, and tertiary (or rural) roads. Primary and secondary roads are main arterial highways of relatively high standard, connecting major population centers and provincial capitals.

Rural transportation   Rural roads, which are usually in a poor state of repair, include

- penetration roads that provide access to potential development areas;
- provincial roads that connect small districts or communities; and
- feeder roads that link agricultural areas to market centers directly or via main arterial roads.

Rural travel is filled with difficulties. Roads and seasonal tracks are rarely maintained, and people walk along treacherous paths and footbridges to obtain water and firewood and to reach markets, schools, and clinics. These same tracks, paths, and footbridges are used to transport export crops and food destined for urban populations. In rural Africa, most people walk, carrying their burdens; frequently, women carry the bulk of the burden. Recent studies in Burkina Faso, Uganda, and Zambia indicate that walking is the principal means of transport for 87% of rural households (Barwell, 1996). Besides walking, nonmotorized vehicles—bicycles, wheelbarrows, donkeys, and carts, both hand pulled and animal drawn—are the primary means of transport.

Seasonal inaccessibility is a perpetual problem that plagues rural transportation. During the dry season, when roads are generally motorable with little difficulty, availability of transportation is difficult. In some villages, cars rarely pass by for days; in others, a taxi might arrive once a day. The rainy season compounds the problem of unavailable transport (Figure 9.1). Road conditions deteriorate so much that, in some places, access is frequently only by walking along muddy paths and sometimes by wading through water. Other problems include broken bridges, flooded roads, washed-out paths, and overgrown tracks providing good hiding places for poisonous snakes.

One reason for the poor maintenance of rural roads is that ownership and responsibility for managing and maintaining them is often not clearly defined (Figure 9.2). The cost of

Figure 9.1 In many parts of Africa, road transportation is problematic during the rainy season.

Transportation Systems in Africa

Figure 9.2 Torrential rain makes road maintenance a constant necessity.

construction, major rehabilitation, or periodic maintenance is frequently financed through donor resources by the central government, but routine maintenance falls to cash-strapped local government authorities, which rarely have the expertise or financial resources to independently maintain their road networks. Furthermore, under cost-recovery considerations, the application of economic criteria in the selection of roads for investment leads to the improvement of individual high-traffic roads and the neglect of the network—a highly inefficient use of public funds (Fishbein, 2001).

Urban transportation Urban transport consists of three main types of transit: government-operated bus systems; privately owned and operated minibuses known as poda-podas (Sierra Leone), tro-tros (Ghana), or matatus (Kenya); and taxis (Figure 9.3). Government-operated buses are generally unreliable, require users to wait at unsheltered terminals, and overload passengers in a disorderly and unorganized manner, particularly during peak periods. Moreover, most trips require multiple transfers, and the buses move very slowly, frequently going only about 8–12 km (5–8 mi) per hour through the ever-present congestion on the few available roads (Figure 9.4). Street trading is common along the main routes, and there is often a failure to enforce regulations against on-street parking, thereby exacerbating the congestion problem.

For most poor people in urban areas, transport is time consuming, costly, and unsafe. On average, 50% of transport in African cities is by walking or bicycling, but people who

Figure 9.3 A motor taxi in Lome, Togo. The client sits on the back seat, sometimes with a piece of luggage on the head.

Figure 9.4 A familiar scene of traffic congestion on urban roads in Africa.

Transportation Systems in Africa

Figure 9.5 Traffic accidents are a frequent sight, as a result of poor road conditions and inadequate automobile servicing.

take some form of motorized transport spend a high proportion of their household income (up to 30% in Dar-es-Salaam, for example) on mobility. Moreover, about two-thirds of urban traffic fatalities in Sub-Saharan Africa (a much higher figure per capita than in developed countries) occur among pedestrians, with half of those fatalities being children (Figure 9.5).

The Pan-African highway Some progress was made in highway building in Africa during the 1990s. World Bank loans and national budgets financed the building and improvement of road networks in many African countries. Some countries have been spending substantial amounts of money on transport development—such as Gabon, which invested a massive portion of its oil revenues in constructing all-weather roads and a railway that connects to the interior. Nevertheless, the road network in Africa in 2000 was still inadequate, and only about one-third of the African road system was up to the standard of all-weather roads (Figure 9.6).

The Economic Commission of Africa (ECA) initiated an action plan in 1962 to develop two major highway networks in North Africa. A trans-Saharan route would link north Africa with west Africa, and an east–west route—from Cairo and passing through Tripoli, Tunis, Algiers, and Rabat—would link the countries of the Mediterranean coast. The

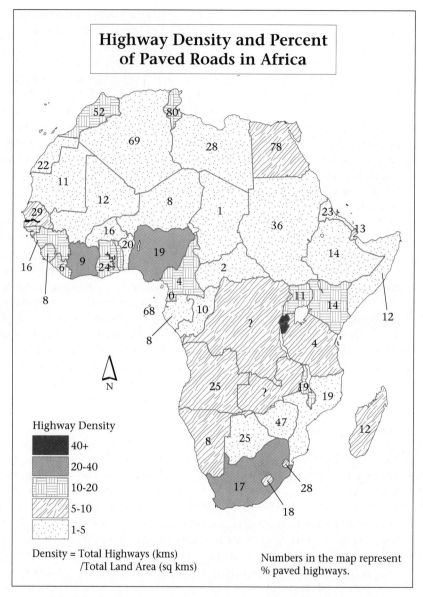

Figure 9.6 Highway density and percent of paved roads in Africa: 1996–1999.

east–west route essentially exists now, and the north–west highway is currently being built with international assistance.

In 1967, ECOWAS established highway priority projects to improve and expand on the west African road network. They include one coastal 5400 km (3356 mi) road that runs through Lagos, Nigeria to Dakar, Senegal, and one in the Sahel hinterland about 6800 km

(4226 mi) long that roughly follows the 12th parallel, linking Nouakchott in Mauritania and Fort Lamy in Chad. International assistance to support these projects is well under way. The French-speaking countries Côte d'Ivoire, Togo, Benin, Burkina Faso, and Niger, with help from bilateral donor agencies, are filling in other details of a west African network.

In east Africa, links already exist between Kenya, Uganda, and Tanzania. Others are being developed between Zambia and Botswana, between Ethiopia and Kenya, and between Ethiopia and the Sudan, with a possible extension to Egypt. These are the arteries of the Trans-East African Highway, the most important link of all. When completed, it will measure about 6400 km (3977 mi) in length and will link Mombasa (Kenya) with Lagos (Nigeria), passing through Uganda, Democratic Republic of Congo, the Central African Republic, and Cameroon. In addition to these 6 countries through which it will run, the projected highway is also of interest to 11 neighboring ones: Burundi, Chad, Congo, Ethiopia, Gabon, Niger, Rwanda, Somalia, Sudan, Tanzania, and Zambia.

Serving as a backbone to the Pan-African road system, these four highway projects will be linked by feeder roads to complete the arterial highway system. When the improvements of these four major highways are completed, it will be possible to travel from Algiers on the Mediterranean Sea to Mombasa on the Indian Ocean and thence to Dakar on the Atlantic Ocean. It will also be possible to travel from Cairo to Gaborone, Botswana.

Once these roads are constructed, regular maintenance is critical. Delay or neglect in maintenance inevitably exacts costly repairs. The 350-km (220-mi) road from Kinshasa, Democratic Republic of Congo to the port of Matadi, which once had made possible a travel time of 5 hours, became so bad that, in the mid-1990s, trucks needed five days to complete the trip. Amazingly, there is no paved road linking Brazzaville, the national capital of the Republic of the Congo, and Pointe Noire, its major seaport. Unfortunately, such problems— poor maintenance and missing transport links—are quite widespread.

Railways

Much as in Europe or North America, the major period of railway development in Africa extended from the end of the 19th century to the end of World War I. This expansion, however, was not coordinated. Railways were built as common carriers, usually under government ownership between 1895 and 1914, primarily to stimulate cash cropping of export crops, to facilitate mineral development, and to demonstrate colonial occupancy. Most railways had single lines, commonly used light track and simple signaling systems, and avoided embankments and the cutting of tunnels wherever possible. The result was a preponderance of sharp curves, steep gradients, circuitous routes, and, consequently, low running speeds. Moreover, track gauges varied between countries; so did the braking and coupling systems. For example, the Tanzania–Zambia Railway Authority (TAZARA) operates 1860 km (1156 miles) of narrow-gauge track between Dar es Salaam and Kapiri Mposhi in Zambia [with 969 km (602 miles) in Tanzania and 891 km (554 miles) in Zambia], but, because of the difference in gauge width, this system does not connect to the Tanzanian Railway system. Thus, the colonizing powers left a difficult and costly legacy for independent African countries wishing to link their rail services. As with roads, rail network improvement since the 1960s has led to lower transport costs. Recent improvements include the realigning and strengthening of the track, better signaling systems, and extensions of the original network to provide missing links and to connect previously unserved locations.

The early railways were constructed partly to facilitate the administration of interior regions and bring supplies from ports to central consumption or distribution points and partly—especially in the south—to enable valuable minerals or commodities to reach the coast for export. Railways are much more important than rivers in the transport structure of west Africa, but are few and widely spaced. Instead of forming a network, the rail pattern consists of a series of individual fingers that extend inland from various ports, but which usually do not connect with other railway lines. Many rail lines are not properly connected, and this limits their utility. For example, Cameroon's seaport of Douala is not connected to the inland capital, Yaounde, by rail. Apart from a few rail fingers reaching inland from seaports, the region is devoid of railways, and roads provide inadequate links to underserved regions. Isolation and poor transport are major obstacles to development in this region. Similarly, mineral production in the Shaba province of Democratic Republic of Congo has no rail outlet northward. Mineral exports routed through Angola to Lobito or Beira in Mozambique have been disrupted by local wars. Consequently, most exports now take the long, expensive route through Zambia and Zimbabwe to South African ports.

The most productive parts of east Africa are linked by a connected system of railways leading inland from the seaports of Mombasa in Kenya and Dar es Salaam and Tanga in Tanzania. Mombasa, the most important seaport in east Africa, has two ports—a picturesque old harbor used primarily by small vessels, and a modern deep-water harbor with facilities for handling large vessels. From Mombasa, the main line of the Kenya–Uganda railway leads into Nairobi, Kenya, and continues westward to Kampala, Uganda. Dar es Salaam has rail connections to Lake Tanganyika, Lake Victoria, and Zambia. The major railways of Africa can be grouped regionally, as follows: west Africa, Congo Basin and margins, east Africa, and southern Africa.

West Africa In French-speaking west Africa, a major rail line links the Senegal River port of St. Louis with Dakar; it extends to Bamako, Mali. The railway line linking Abidjan through Bobo Dioulasso to Ouagadougou in Burkina Faso was completed in 1955. Guinea's railway line links the national capital, Conakry, with the interior town of Kankan. In Togo, the rail line connects the capital city of Lome with Blitta in the interior. In neighboring Benin, the main railway line links Cotonou with Parakou; a short coastal line linking in Porto Novo, the capital, was completed in 1930.

In Anglophone west Africa, Ghana's railway lines are centered on the southern part of the country and link coastal Accra, the capital, to coastal Sekondi and Takoradi in the western region and to interior Kumasi in the Ashanti Region, producing the Golden Triangle—reflecting the relative economic development of the region. Nigeria is served by nearly 3218 km (2000 mi) of rail, which covers the economically important regions of the country. The major links include Lagos–Ibadan–Kano via Ilorin and Zaria. The Port Harcourt–Enugu line now connects to the Lagos–Ibadan–Kaduna line. Liberia's only rail line links the capital and leading port, Monrovia, with the vast iron ore deposits of the nearby Bomi Hills.

Congo basin and margins The CFL chain of railways links the Atlantic Ocean with Lake Tanganyika and avoids the series of rapids and falls that peak in the Stanley Falls on the Congo River. The Congo–Ocean rail links the Congo River with the Atlantic Ocean at Pointe Noire (Republic of the Congo), providing efficient passage for passengers and freight, including minerals. The Benguela rail runs almost horizontally along the width of Angola, connecting to the Bulawayo (Zimbabwe)–Port Francqui line at Tenke (Democratic Republic of Congo). The railway line from Kigoma, Tanzania connects to the Indian Ocean at Dar es Salaam, with branch lines from Tabora to Mwanza on the southern shore of Lake Victoria in Tanzania. The Tanzam railway, built with Chinese assistance, links Dar es Salaam to Zambia.

East Africa The Kenya–Uganda line, almost 970 km (600 mi) long, links Mombasa with Kisumu on the shores of Lake Victoria. This line has been extended to link in Kasese, which has now become a major shipping point for copper from the Kilembe mines in the foothills of the Ruwenzori Mountains. The Sudan Railways, more than 3200 km (2000 mi) long, provide bypasses along the cataract-ridden section of the Nile between Khartoum and the Egyptian border, giving Khartoum direct access to Port Sudan on the Red Sea and several large towns in the country. The Somalia–Ethiopia line weaves through wadis and ravines to link Djibouti in the Gulf of Aden with Addis Ababa. The Eritrean railway is of narrow gauge, linking the Red Sea port of Massawa with Asmara, the capital of Eritrea.

Southern Africa Southern Africa is probably the region best served by railways. One line links Port Elizabeth with Windhoek, the capital of Namibia, and Walvis Bay on the Atlantic coast. Another links Cape Town with Pretoria through Bloemfontein and connects to the Bulawayo line. Outside of South Africa, a single system links Bulawayo (Zimbabwe) with Livingstone to the northwest, Harare to the northeast, and the Mozambican port of Beira on the Indian Ocean coast. This same system also runs south to Gaborone, the capital of Botswana, where it connects to the South African system.

Africa's rail networks constantly fall victim to numerous regional political upheavals and wars (Discussion Box 9.1). For example, by 1988, many sections of Uganda's railroad system needed relaying, regrading, and realigning, primarily because war and ongoing rebel activity in eastern and northern Uganda had undermined or disrupted rail service in many areas. The Benguela Railway (1394 km or 866 miles long), a narrow-gauge rail system that links the port of Lobito with the central African rail system serving the mining regions of Shaba (Democratic Republic of Congo) and the Zambian Copperbelt, was severely damaged and required millions of dollars for rehabilitation. Similarly, in 1986, the Luanda Railway linking Luanda to Malanje in Angola carried only one-fifth of 1973's level of freight.

Air Transport

Air transport is well suited to Africa's geographic vastness and has become the primary means of international and, sometimes, national travel in Africa. During the late 1940s and the 1950s, as advancements were made in the extension and improvement of rail and road services, a new transport factor emerged with the introduction of internal and international scheduled air services. The rapid development of air transport increased the movement of goods and people and began to open up the hitherto largely closed interior of the continent.

DISCUSSION BOX 9.1: POLITICAL INSTABILITY AND TRANSPORTATION SYSTEMS IN AFRICA

Besides diverting expenditure from maintenance and from much-needed new construction, political instability, whether in the form of civil war or of ethnic conflicts, disrupts transportation systems. Even after the conflicts are over, roads and railways may remain unused for a long time, primarily from fear of land mines (*http://www.africapolicy.org/bp/lmineall.htm*).

By the end of 2000, most of Angola's roads were known or suspected to be mined, severely restricting access to the countryside and to major areas outside the provincial capital. Supplies, equipment, and personnel had to be flown to these places, but some airports were also mined, so this was a very risky operation, which profoundly increased the cost of transporting humanitarian food aid. Similarly, in Mozambique, large tracts of land are mined extensively. During the war between RENAMO (Mozambican National Resistance) and the government, airstrips, railway tracks, and river crossings were mined extensively. Bridges were frequently destroyed, or their pylons were mined. Different regions of the country have a comparative advantage in growing certain foods, but land mines on many of the major roads severely disrupt trade. Large portions of Mozambique's road system were in such disrepair in the mid-1990s that traffic movement was virtually impossible. By mid-1994, approximately one-quarter of Mozambique's road system, including roads in the provinces of Manica, Sofala, Zambesia, Tete, and Niassa, were listed as, or suspected of being, mined. The mining of the transportation system and the insecurity due to fighting forced Zimbabwe, Zambia, and other landlocked states to reroute much of their freight through South Africa (*http://www.mg.co.za/mg/news/99mar1/4mar-landmines_mozambique.html*).

Most of northern Somalia is extensively mined. In fact, a total of 32 important roads were known to be mined, including the Boroma–Zeyla road, the main link with Djibouti, and 18 major and minor roads around Hargesia alone. Extensive mining of roads and other transport infrastructure has significantly disrupted the movement of goods and services. Vehicles must make detours around mined routes, with increasing cost in time and fuel. For example, mining of the Boroma–Zeyla road increased travel distance from 240 to 370 km (150 to 230 mi) and increased the price of a sack of grain by 25% (*http://www.undp.org/erd/devinitiatives/mineaction/somalia*).

Transport became much quicker and usually cheaper. Since then, internal air services have steadily increased, and intercontinental air transport, especially of passengers, has developed greatly. The largest international airports are at Dakar, Senegal; Abidjan, Côte d'Ivoire; Lagos, Nigeria; Douala, Cameroon; Addis Ababa, Ethiopia; Nairobi, Kenya; and Johannesburg, South Africa.

The dramatic increase in Africa's air traffic has brought the fragile aviation infrastructure under tremendous and unbearable strain. For example, only 20 airlines operated out of South Africa before the end of apartheid in 1994, yet more than 80 were operating by 2000 (Schuler, 2000). Correspondingly, the number of air-traffic accidents and midair near-collisions increased (Ott, 1997). In fact, more airline accidents per takeoff occur in Africa than anywhere else in the world. Between 1990 and 2000, there were 12.6 crashes for every 1 million transport planes departing from airports in Africa, compared with 0.5 in the United States and Canada (Schuler, 2000). In 1996, the International Federation of Airline Pilots Associations warned of poor safety and air-traffic control over large portions of the continent and declared that 90% of the airspace was critically deficient. In 2000, only five countries met the safety and security standards of the International Civil Aviation Organization (ICAO)—South Africa, Ghana, Ethiopia, Egypt, and Morocco. Air-traffic controller training was inadequate in some countries, some airports lacked functioning runway lights, and some navigation beacons were permanently out of service. At times, a total lack of communication between pilots and ground-control centers prevailed, leaving pilots unable to give air-traffic authorities their flight-plan details and so making it difficult to coordinate air traffic (Schuller, 2000).

The January 1, 2000, implementation of the mandatory Traffic Alert and Collision Avoidance Systems (TACAS), which alerts a pilot when another plane is within 40 miles, will help to reduce the frequency of air traffic accidents (Ott, 1997). Nevertheless, better air-traffic infrastructure is required. The ICAO conducts safety audits and requires countries to fix identified safety problems. Unfortunately, the ICAO is a UN body that has no power to enforce its regulations. Establishment and strengthening of systems of aviation law and regulations, development and improvement of levels of safety and security, and establishment of a common aviation authority are all critical components of the changes needed to make Africa's skies safe (Ott, 2001).

In addition to the air-traffic infrastructure problems, Africa's predominantly state-owned airlines have been struggling financially. Handicapped by small fleets of old aircraft that are expensive to maintain, yet lacking resources to buy new aircraft and struggling with debts, Africa's mostly unprofitable airlines see the future as dismal. Air Afrique, with an estimated $444 million debt in 2001, and declared technically insolvent by the World Bank, is typical (Michaud, 2001). Reorganization negotiations between the heads of state of 11 African countries that own shares in the company and Air France continue to hit roadblocks. For example, Air France wants to reduce the "bloated personnel" from the current number, 4200 per aircraft (arguing that it is ridiculous to have that many employees for only six aircraft) to 700 by 2000 (Anonymous, 2001). However, the powerful unions and the heads of state who control the airline will have none of this. Regardless of what happens with Air Afrique, privatization of Africa's skies is clearly inevitable. This is particularly poignant in light of the events of September 11, 2001 and the failure of major airlines such as Swissair. Privatization should also bring much needed vitality to Africa's lethargic air-cargo market.

Navigation

Historically, throughout the vast interior between the Sahara and the Zambezi River, people and goods were transported by canoe or boat (Figure 9.7) on the great river systems of the Nile, Sénégal, Niger, Congo, Ubangi, and Zambezi and on the few, but large lakes, Victoria and Malawi. Where conditions were suitable, engine-powered craft supplemented or displaced canoes. However, water transport remains underdeveloped. Also notable were the construction of lake ports and the installation of rail ferries across Lake Victoria. Meanwhile, on the coasts, artificial harbors have been developed. New berths have been added to established port facilities, and a number of ports have been constructed. In the planning of new ports, the choice of site, probable costs, and the possibilities for using containers or other unitized loads have been taken into consideration.

The Congo River is the largest watershed in terms of drainage and discharge in Sub-Saharan Africa. It flows across a relatively flat basin that lies more than 300 m (1000 ft) above sea level, meandering extensively through the rain forest. Entry from the Atlantic Ocean is rendered impossible by a series of falls and rapids, which make it only partially navigable.

Figure 9.7 Canoes have traditionally been important in Africa, in transporting people and goods and for fishing.

Nevertheless, the Congo River has been the major corridor of travel within the Republic of the Congo and the Democratic Republic of Congo. Heavily laden barges ferry people and cargo between Kinshasa and Kinsangani and to Brazzaville on the opposite bank of the river. Navigation on the Zambezi is limited by the strong rapids that are the basis for the Kariba Dam on the border of Zambia and Zimbabwe and the Cabora Bassa in Mozambique.

River transport is more important in the Congo River basin than elsewhere in Sub-Saharan Africa. In the Democratic Republic of Congo, river transport is an important part of the integrated transport system, which consists of the Matadi-to-Kinshasa rail, the Kinshasa-to-Kinsangani river corridor, the Kinsangani-to-Ukundi rail link, and the river services onwards from Lualaba to Kindu. The other national axis consists of river services from Kinshasa to Ibbo on the Kasai River and onward to Shaba by rail. Transshipments on both routes are numerous, and transit times are long. This increases costs and makes freight vulnerable to damage and pilfering. The Congo and Oubangoui Rivers provide a vital link both for the Republic of the Congo, particularly its northern parts, and for the Central African Republic, where Bangui, the capital, is an important river port that serves neighboring Chad.

On the Niger River, seasonal navigation is possible from the railhead at Kouroussa to Bamako, Mali, and from Niamey, Niger to Yelwa. In the lower reaches, the river is navigable from the Niger Delta to Lokoja all year, and seasonally to Baro and Jebba. On the Benue, the open season to Makurdi is June to November, but only 6–8 weeks in August and September at Garoua. The Senegal and Gambia rivers, the White Nile from Juba to Khartoum, and the east African lakes, particularly Victoria and Tanganyika, are also used as links in the transport network.

Good natural harbors are made scarce in west Africa by the abundance of offshore sandbars and silt-choked river mouths. This necessitates the transfer of goods from ships to shore by small boats. Inland, frequent rapids and annual seasonal fluctuations in the water level limit the utility of rivers for transport. The rivers that carry major traffic include the Niger (and its tributary, the Benue), the Senegal River, the Gambia, and, to a smaller extent, the Volta in Ghana.

Navigation is not well coordinated with other transport modes in the region. For example, the extensive navigable waterways of the Congo River system are used primarily for inland transport, yet lack a direct outlet to the sea. The railways on either bank linking Kinshasa and Brazzaville have narrow gauges, are limited in capacity, and are not connected to each other. Other problems with inland waterways for modern transport abound. For example, most rivers are characterized by alternating sections of low gradients and rapids. Few rivers are navigable for any distance from their mouths, and unreliable rainfall patterns mean that they might be closed to navigation for part of the year. In addition, many rivers flow through areas of low transport demand. Moreover, the location of political boundaries frequently reduces river traffic.

Telephone Communications in Africa

Given Africa's difficulties in transportation, developments in telecommunications were expected to play a major part in overcoming the physical divide. Unfortunately, although expectations remain high for the rapidly changing technology, significant obstacles remain.

In general, costly and inconvenient fixed-line telephone systems dominate throughout Africa. Getting a line installed is extremely difficult; once installed, its quality of service is usually poor. Besides, such lines are almost useless for Internet activity. Phone calls are extremely expensive, particularly international calls, because many governments see telephone systems as an income-generating business. In regions of political instability, phone lines are frequent targets during war.

By the end of 2000, an obsolete and dilapidated telecommunication infrastructure was the norm in many countries. Frequently, there are no telephones or other communication facilities available for several miles. Telephone lines primarily serve urban areas; very few villages have even a single telephone (Figure 9.8). In fact, the average teledensity (number of telephone lines per person) in urban areas is estimated to be 25 times that in rural areas (Fishbein, 2001). Nonetheless, the demand for telecommunication services among the rural poor in Africa is high. Historically, telecommunication services have been provided by government-owned monopolies. Urban areas have dominated the service, with rural extensions consisting primarily of pay phones and public call offices at national post and telecommunications offices.

Nevertheless, telecommunication technology is changing rapidly in Africa. A massive explosion of wireless systems based on cellular and personal communications is underway. In fact, wireless systems are growing more rapidly in Africa than in any other part of the world—by 150%, compared with 30–50%, per year. With the advent of mobile and

Figure 9.8 A few African villages are now gaining access to telephone systems.

wireless technologies, new forms of service provision have emerged, including telephone shops, multipurpose telecenters (communication centers), prepay cards, messaging services, and Internet access. Nevertheless, these services are still more widely available in urban than in rural areas.

THE DIGITAL DIVIDE AND AFRICA

The Internet could provide a way around one of Africa's greatest weaknesses, its feeble infrastructure. Poor roads, uncertain power supplies, an unreliable postal system, dilapidated telephone lines, and a dearth of telephones (Tokyo has as many telephones as all of Africa) exacerbate the economic and political woes. The Internet promises to make things better and cheaper—e-mailing a 40-page document from Madagascar to Côte d'Ivoire costs 20 cents, faxing it about $45, sending it by courier about $75—but African countries have a long way to go. At the end of 2000, there were more Internet hosts in New York City than on the entire continent of Africa. An obsolete and dilapidating telecommunication infrastructure, the high cost of links to Internet backbones, and a shortage of technical staff are some of the problems. South Africa has emerged as Africa's premier Internet market and is one of the top 20 countries worldwide in Internet connectivity (Ngwainmbi, 2000).

Measuring the total numbers of Internet users is difficult, but figures for the number of dial-up subscriber accounts of Internet Service Providers (ISPs) are readily available; they indicate that there were over 1,300,000 subscribers in Africa by May 2001, mostly concentrated in South Africa, Egypt, and Algeria (Jensen, 2001) (Figure 9.9). Each computer with an Internet or e-mail connection usually supports from three to five users, so recent surveys suggest about 5 million Web surfers, half of whom live in South Africa, and more than half of whom are predominantly white. This works out to about one Internet user for every 200 people, compared with a world average of one user for every 30 people and to a North American and European average of one in every 3 people. Clearly, urban users outnumber rural users. Comparable figures for other developing regions in 1999 were 1 in 125 for Latin America and the Caribbean, 1 in 200 for South East Asia & the Pacific, 1 in 250 for East Asia, 1 in 500 for the Arab States, and 1 in 2500 for South Asia (UNDP, 1999).

About 38 African countries have 1000 or more dial-up subscribers, and half that number have more than 5000. Only 11 countries had more than 20,000 subscribers in 2001—Algeria, Botswana, Egypt, Kenya, Mauritius, Morocco, Nigeria, South Africa, Tunisia, Tanzania, and Zimbabwe. As a result of the extensive use of wireless links and university networks, the following countries also could have up to 20,000 effective users: Côte d'Ivoire, Ghana, Madagascar, Mozambique, Namibia, Senegal, Uganda, and Zambia. Generally the countries with better infrastructure, in north Africa and south Africa, have larger populations of Internet users. Furthermore, most of these countries were also among the first on the continent to obtain Internet access, so their markets are better developed. Local Internet society chapters exist in virtually every African country, particularly those with large Internet user populations.

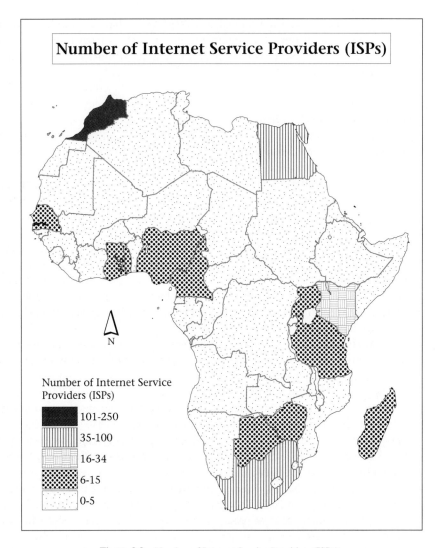

Figure 9.9 Number of Internet Service Providers (ISPs).

Dial-up costs are a major factor restraining Internet access in Africa. In Ethiopia, the cost of the telephone line to the Internet for 20 hours a month amounts to 8.4 times per capita GDP (Hafkin, 2001). In 2001, the average total cost of using a local dial-up Internet account for 20 hours a month in Africa was about $68 a month (usage fees and local-call telephone time included, but not telephone-line rental) (Jensen, 2001). ISP subscription charges vary greatly—between $10 and $100 a month, compared with $29 in the United States—largely reflecting the different levels of maturity in the markets, the varying tariff

The Digital Divide and Africa

policies of telecom operators, the different regulations on private wireless data services, and differential access to international telecommunications bandwidth.

Most African capitals now have more than 1 ISP; in early 2001, there were about 575 public ISPs across the region (Figure 9.9). Fourteen countries had 5 or more ISPs, 7 countries had 10 or more active ISPs (Egypt, Kenya, Morocco, Nigeria, South Africa, Tanzania, and Togo), and 20 countries had only 1 ISP. Although Ethiopia and Mauritius are the only countries where a monopoly ISP is still national policy (i.e., where private companies are barred from reselling Internet services), there are other countries in which this practice still continues, predominantly in the Sahel subregion, where markets are small.

In response to the high cost of Internet services and the slow speed of the Web, and because of the overriding importance of electronic mail, lower cost e-mail-only services have been launched by many ISPs and are continuing to attract subscribers. Similarly, the relatively high cost of local electronic-mailbox services from African ISPs pushes a large proportion of African e-mail users to make use of such free Web-based services as Hotmail, Yahoo, and Excite, most of which are based in the United States. These services can be more costly and cumbersome than standard e-mail software, because extra on-line time is needed to maintain the connection to the remote site, but they do provide the added advantages of anonymity and perhaps of greater perceived stability than a local ISP that might not be in business next year.

There is also a rapidly growing interest in kiosks, cybercafes, and other forms of public Internet access (Figure 9.10). Many existing phone shops are now adding Internet access to their services, even in remote towns where it is a long-distance call to the nearest dial-up access point. The term "telecentre" has been used to describe most of these facilities. In some circles, they are called virtual village halls, telelearning centers, telecottages, electronic cottages, community technology centers, networked learning centers, or digital clubhouses (Share, 1997). Such centers provide a range of community-based activities and services, including access to information and communications technology for individual, social, and economic development (Cisler, 1998).

In addition, a growing number of hotels and business centers provide a PC with Internet access. Regional ISP Africa Online has rolled out hundreds of public-access kiosks as part of its e-touch franchise program, in which local stores are supplied with a PC to provide e-mail and Internet access. Africa Online had approximately 100,000 users spread across 740 outlets in Côte d'Ivoire, Kenya, Uganda, Tanzania, and Zimbabwe before it began its new program of fewer, large, branded I-cafes.

Despite all these significant developments, Africa still lags far behind the rest of the world in information technology. The 29 member states of the Organization for Economic Co-Operation and Development (OECD) represent postindustrial economies and developed democracies and contain 97% of all Internet hosts, 92% of the market in production and consumption of computer hardware, software, and services, and 86% of all Internet users. In contrast, the whole of Sub-Saharan Africa has only 2.5 million Internet users, less than 1% of the world's on-line community. Indeed there are more users within affluent Sweden than in the entire continent of Africa.

(a)

(b)

Figure 9.10 (a) People in Gaborone, Botswana use telecentres downtown to check Hotmail accounts and more. (Courtesy: Dr. Abdul Alkalimat, African Studies Center, The University of Toledo). (b) Internet cafes are now a part of the African urban landscape.

Furthermore, a clear gender imbalance pervades Internet access and use in Africa. In South Africa, where 3% of the population has Internet access, only 19% of the users are female. Similarly, female users make up only 37.5% in Zambia, 31.5% in Uganda, 13.9% in Ethiopia, and 12% in Senegal (Hafkin, 2001).

Distance Learning

Internet access for education is more essential for Africa than for any other part of the world because of the dearth of current journals, periodicals, and books in African libraries, particularly university libraries. With the appropriate online connection and current technology, the massive library collection of the US Library of Congress can be simply a few keystrokes away from a user in any part of Africa. Thus, scholars and students can keep abreast of current developments in their fields relatively easily and inexpensively.

Distance-learning techniques are increasingly being employed by a growing number of higher institutions in Africa. Most of the ongoing distance education initiatives on the continent have been used to upgrade the quality of basic education, and some countries are taking bold initiatives in piloting Internet-based and satellite-linked distance-educational programs in selected courses (ADEA, 1999). The University of Abidjan and the African Virtual University are good examples (Baranshamaje, 1996). National organizations have emerged in some African countries to promote distance education, such the National Association of Distance Education Organizations of South Africa (NADEOSA). Multicountry cooperation in distance education initiatives is also growing rapidly. A World Bank project called World Links for Development links secondary-school students and teachers in Africa to their colleagues around the world for collaborative distance learning (http://www.worldbank.org/worldlinks/).

Multimedia distance learning via the Internet and video conferencing is emerging slowly. The TELESUN program in Cameroon, which uses Internet-based courses in its science program; the FORST program, which links Benin and three other countries with McGill University in Canada; and the RESAFAD program in Djibouti, which provides teacher training from French universities are examples (ADEA, 1999).

The potential of distance learning is promising, but a number of problems remain. Connectivity beyond major cities remains inadequate. Telephone density is less than 2 lines per 1000 people, compared with 48 per 1000 in Asia, 280 per 1000 in America, 314 per 1000 in Europe and 520 per 1000 in high-income countries (Darkwa and Mazibuko, 2000). The absence of clearly defined national distance-education policies in most countries poses another challenge. In fact, the lack of high-level political support for distance education, the lack of recognition of distance learning in assessment of public-service employee qualifications, and the lack of follow-up programs are important constraints.

Computer access is also a major challenge, as is dependable basic infrastructure. For example, the expansion of the electrical power grid in the 1990s facilitated the expansion of the communication infrastructure (e.g., the telephone system) in Ghana. This development has enabled information-technology-based businesses to expand to nonmetropolitan areas of the country, where they provide such basic communication services as telephone, fax, typing, photocopying, printing, training in the use of computers, electronic mail, and electronic networking.

CONCLUSION

Transport and communication networks in Africa are plagued by a difficult historical legacy, a poor and dilapidating infrastructure, harsh economic times, and civil conflict in some areas. Difficulties imposed by the physical terrain, such as the absence of natural harbors, silt-choked river mouths and sandbars, and seasonal fluctuations in water levels due to unreliable rainfall patterns, compound the problems. Poor infrastructure and obsolete regulatory mechanisms hamper the predominantly government-owned air travel industry. Telecommunications suffer from similar problems—obsolete and poorly maintained infrastructure that primarily links urban centers to the rest of the world and neglects the rural areas. Internet access, particularly wireless-based services, promises a digital bridge over Africa's physical divides, but, although they are spreading very rapidly, their impact is felt mainly in the urban centers.

Sweeping policy changes are required to provide a congenial atmosphere for the improvement and sustainable development of Africa's transport and communications infrastructure. For instance, the current practice that governments will profit from telephone systems is no longer sustainable and must give way to a more probusiness, competitive environment. Special efforts to link rural areas to the global network through wireless systems are warranted. Not only are these less expensive, they require less maintenance than the fixed-line systems of antiquity. Increased use of telephones and Internet systems across the continent can release the stress from the clogged arteries of Africa's highways and railway systems and ensure a smooth passage for much-needed food and raw materials. Distance education could be a major solution to Africa's educational needs, but it surely needs some major political help—clearly defined national distance-education policies are critical.

In fact, the key issue in overcoming the physical divides produced by poor transportation infrastructures in Africa may be not technical, cultural, or even financial, but purely political. Obsolete political frameworks, artificially high charges on imported telecommunications and computing equipment, and overregulation surely stultify development. Many governments, aware of the potential of the Internet for disseminating sensitive or embarrassing information, actively oppose a technology that takes press and information control out of their hands. Nevertheless, a nation's benefit from improved Internet access surely outweighs the costs. The global sweep of the World Wide Web is unstoppable, and politicians ignore or oppose it at their own peril and eventual demise. Policies and regulations need to be open and flexible to accommodate new environments and technologies.

REFERENCES

Anonymous (2001). "A New Air Afrique?" *The Economist*, 360, 8236: pp. 55.

Association for the Development of Education in Africa (ADEA) (1999). "Tertiary Distance Learning in Sub-Saharan Africa," *ADEA Newsletter*, 11(1):1–4.

Baranshamaje, E. (1996). *African Virtual University*.

Barwell, I. (1996). "Transport and the Village: Findings from African-Level Travel and Transport Surveys and Related Studies," Washington, D.C.: *World Bank Discussion Paper No. 344, African Region Series*.

References

CISLER, S. (1998). *Telecenters and Libraries: New Technologies and New Partnerships*, http://home.inreach.com/cisler/telecenters.htm.

COEUR DE ROY, O. (1997). "The African Challenge: Internet, Networking and Connectivity Activities in a Developing Environment," *Third World Quarterly,*18(5):883–898.

DARKWA, O. & MAZIBUKO, F. (2000). "Creating Virtual Learning Communities in Africa: Challenges and Prospects," *First Monday*, 5(5), http://firstmonday.org/issues/issue5_5/darkwa/index.html.

FISHBEIN, R. (2001). "Rural Infrastructure in Africa: Policy Directions," *World Bank Africa Region Working Paper Series* No. 18, http://www.worldbank.org/afr/wps/index.htm.

HAFKIN, N.J. (2001). *Gender, Information Technology and the Digital Divide in Africa*, http://www.worldbank.org/gender/info/digitaldivide6.htm.

HODDER, B.W. & GLEAVE, M.B. (1992). "Transport, Trade and Development in Tropical Africa," in *Tropical African Development,* Gleave, M.B., ed., New York: John Wiley.

JENSEN, M. (2001). *The African Internet—A Status Report*, http://www3.sn.apc.org/africa/afstat.htm.

KIMBLE, G.H.T. (1960). *Tropical Africa,* New York: The Twentieth Century Fund.

MICHAUD, P. (2001). Saving Air Afrique. *New African*, Sep 2001, Issue 399:46.

MOYO, L.M. (1996). "Information Technology Strategies for Africa's Survival," *Information Technology for Development*, 7(1):17–27.

NGWAINMBI, E.K. (2000). "Africa in the Global Infosupermarket: Perspectives and Prospects," *Journal of Black Studies*, 30(4):534–552.

OTT, J. (1997). Safety Groups Respond to African ATC Crisis, *Aviation Week and Space Technology*, 146(20):65.

OTT, J. (2001). "Rising African Safety Culture Paves the Way for New Projects," *Aviation Week and Space Technology*, 154(12):106.

OWEN, W., JR. & OSEI, D. (2000). "Role of Multipurpose Community Telecenters in Accelerating National Development in Ghana," *First Monday*, 5(1), http://www.firstmonday.dk/issues/issue5_1/owen/index.html.

SCHULER, C. (2000). "Africa's Skies Not So Friendly," *Christian Science Monitor*, 92(51):1.

SHARE, P. (1997). *Telecentres, IT and Rural Development: Possibilities in the Information Age*, http://www.csu.edu.au/research/crsr/sai/saipaper.htm.

(UNDP) UNITED NATIONS DEVELOPMENT PROGRAM (1999). *Human Development Report, 1999*, New York; Oxford: Oxford University Press.

UNIVERSITY OF WISCONSIN-MADISON LIBRARIES (2000). *Africa Focus*, http://africafocus.library.wisc.edu/.

WORLD BANK (2001). *Gender, Information Technology and the Digital Divide in Africa*, http://www.worldbank.org/gender/info/digitaldivide6.htm.

10

Urban Geography of Sub-Saharan Africa

Samuel Aryeetey-Attoh

INTRODUCTION

Urban geography is the study of how urban phenomena are organized in space. The types of urban phenomena that geographers analyze are (1) the evolution of urban societies, (2) the spatial pattern or internal structure of cities, and the processes that mold and shape the arrangement of land uses within cities, (3) the cause and effect relationships associated with city growth and decline, and (4) the dynamic interaction and functional relationships between cities and their environments in an urban system. These topics form the framework of this chapter.

    The chapter begins with an overview of Sub-Saharan cities in a precolonial context. It then provides a descriptive account of the internal structure of a select group of cities. This is followed by an analysis of the causes, consequences, and policy implications of the urbanization process. Microlevel strategies that deal with improving the internal management of cities are proposed, along with macrolevel strategies designed to create a more balanced system of cities and to enhance the interdependent relationship between cities and their environments.

HISTORICAL EVOLUTION OF AFRICAN CITIES: PRECOLONIAL CITIES

Precolonial Africa experienced a long and rich history. Unfortunately, much of its history, traditions, customs, and artifacts were destroyed during colonialism, and many misconceptions

Historical Evolution of African Cities: Precolonial Cities

about African cultures and institutions arose and became entrenched. It was habitual for colonialists to deny any social and political achievements to Africans. Some scholars at the time even questioned whether technologies were indigenous to Africa or instead imported by external agents. Scholars went as far as denying civilization status to earlier cities on the basis that they lacked a system of writing (Childe, 1950), a cash economy (Braudel, 1976), or organized social and political systems.

Thanks to the efforts of scholars like Connah (1987), Hull (1976), Coquery-Vidrovitch (1991), and Diop (1974), Africa's past is now being reconstructed from folklore, poetry, archeological sites, carbon-dating techniques, linguistics, art objects, and buildings. The evidence clearly shows that there were rules of social behavior, codes of law, and organized economies, and it further suggests that

- towns and cities with diverse sociopolitical organizations have existed for several millennia in Sub-Saharan Africa;
- great cities and towns, such as Napata, Meroe, Axum, Jenne, Timbuktu, Gao, and Great Zimbabwe were major centers of cultural and commercial exchange, religion, and learning;
- precolonial cities had clearly defined divisions of labor, class structures, communication networks, and spheres of influence; and
- the diffusion of technological innovations was widespread—there was a diffusion of iron technology, stone masonry, and other crafts and skills from the early eastern African civilizations of Kush and Axum to the west and south.

There is general consensus that the earliest known cities of Africa emerged around the *central part of the Nile*. One of the most notable was Meroe (4th century to 14th century B.C.), capital of the Black kingdom of Kush. Archeological evidence reveals a civilization that throve on stone and iron technology. Elaborate stone walls, palace buildings, swimming baths, temples, and shrines indicate a culture with an organized social, religious, and political order. It also had an organized agricultural economy based on pastoralism and cultivation, complemented by advancements in irrigation agriculture. Meroe and other Kushite cities are credited with reaching a level of technological sophistication unmatched in precolonial Africa. Building and construction technologies, along with elaborate sculptures, iron-working industries, pottery works, textiles, leatherworks, woodwork, basket weaving, and irrigation technologies are all testimonies to the sophisticated technology that characterized Meroe (Connah, 1987).

Other prominent cities in this region of eastern Africa were Axum, a metropolis of the ancient kingdom of Ethiopia, and the Red Sea port of Adulis (Figure 10.1). Classical writers and ancient Axumite coins provide evidence of a powerful Axumite kingdom that lasted from the 1st century A.D. to about the 10th century A.D., with its influence extending beyond the Ethiopian region into southern Arabia. The kingdom is associated with a remarkable culture of engineering and architectural skills applied to quarrying, stone carving, terracing, building construction, and irrigation. Artifacts recovered from sites also indicate a high level of achievement in metallurgical technology and manufacturing. Axum also had

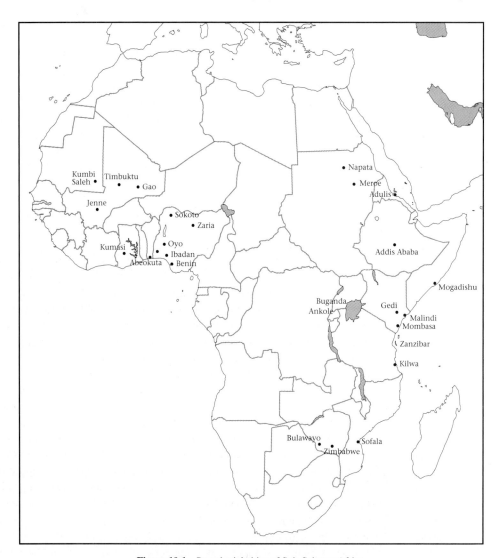

Figure 10.1 Precolonial cities of Sub-Saharan Africa.

an extensive trade network in ivory, precious metals, clothing, and spices with the Roman provinces of the eastern Mediterranean, south Arabia, and India.

Some of the more prominent cities that emerged between 700 to 1600 A.D. in the west African savanna were Kumbi Saleh, Timbuktu, Jenne, and Gao. These cities throve as major centers of the trans-Saharan trade. Arabic scholars (such as Ibn Khaldun) who studied in this region were instrumental in providing historical accounts of the wealth and prosperity of the savanna cities. Gold mining, iron technologies, pottery making, and the productions of textiles are indicative of a high level of technological development in the region. Kumbi Şaleh, the capital of the first powerful state in this region, Ghana (700 to 1200 A.D.), was a major commercial center known for gold, stones, copper, and iron. An elaborate economic system was developed along with a system of taxation. Timbuktu, Jenne, and Gao emerged as great centers of learning and trade in the Mali and, later, Songhai empires. They flourished as intermediate trade centers among the forest zone, north Africa, and Egypt.

Cities in the west African forest region, which extended from Sierra Leone in the east to Cameroon in the west, had well-developed kingdoms prior to the advent of colonialism. Excellent accounts by Mabogunje (1962), Bascom (1955), and Lloyd (1973) attest to the significance of the urban civilizations that emerged in Yorubaland and Hausaland in Nigeria. Other kingdoms in Ashanti and Benin also exhibited multicentered urban networks and organized social and political systems.

The cities in this region functioned not only as political and commercial centers, but also as spiritual centers. Ife, the cradle of the Yoruba civilization, was the spiritual capital of the Oyo empire. Other spiritual centers were Sokoto for Hausaland and Kumasi for the Ashanti. The architectural design of Yoruba cities, like Ife, consisted of a series of concentric city walls with the royal palace at the center and elaborate walls that provided a prestigious element, intimidated potential intruders, and determined a ruler's ability to command his subjects. Inner walls protected the privacy of rulers, and outer walls provided refuge for the masses. Other examples of urban symbolism were passageways and alleys that intersected market plazas and the creation of intimate urban spaces to encourage social interaction and cohesiveness. Cities of the forest region had diverse technologies in iron and smelting, stone building, coarse-mud architecture, brass working, gold mining, and glass melting.

In the central African equatorial region, cities spanned the modern states of Congo, Democratic Republic of Congo, Angola, Zambia, Rwanda, and Burundi. These cities were considered to be the least urbanized in Africa because they lacked the association with large centralized empires and intricate urban networks that characterized the savanna and forest regions of west Africa. However, a number of significant cities did evolve, such as Musumba, the capital of the Lunda empire; Mbanza-Congo, the capital of the Kongo empire; Ryamurari, the old capital of the Ndorwa kingdom of Rwanda; and Kibuga, the capital of the Ganda kingdom. Other prominent states were the "interlacustrine" kingdoms of Buganda and Ankole, situated between Lake Victoria on the east and Lakes Albert, Edward, and Tanganyika on the west. Archeological evidence from this region suggests high levels of craftsmanship and professional artisanship in iron, copper, ivory, pottery, metalwork, and mining.

In coastal east Africa, some of the well-known historic centers were Mogadishu in Somalia; Malindi, Gedi, and Mombasa in Kenya; and Zanzibar and Kilwa in Tanzania. Gedi was one of the first sites to be excavated in Kenya, and the evidence suggests it was founded in the 13th century. Excavations point to elaborate walls, palaces, mosques, and well-designed homes. Kilwa has also been well researched, and the evidence shows it was a well-constructed city with a palace and commercial center, domed structures, open-sided pavilions, and vaulted roofs. A Swahili culture that was urban, mercantile, literate, and Islamic emerged in this region. The technological developments included coin minting, copper works, building craftsmanship, boat building, and the spinning and weaving of cotton. External trade was very active; ivory, gold, copper, frankincense, ebony, and iron were among the goods traded for Chinese porcelain and glazed wares from the Persian Gulf.

In southern Africa, much has been written about the stone-built ruins of Great Zimbabwe. It has been suggested that the city originated as a vital spiritual focal point, in view of the monoliths, altars, and stone tower discovered at its site. Thirty-two-foot elliptical walls that date back to at least the 14th century have been discovered along with the "Great Enclosure," which contains about 182,000 cubic feet of stonework—a structure that is considered to be the largest single prehistoric structure in Africa (Hall, 1976). The technological achievements of Great Zimbabwe go beyond building and stone construction to include mining and metallurgy, manufacturing of pottery, woodcarving, and cotton spinning.

This summary of historic centers suggests high levels of social and technological achievements among precolonial African cities. Archeological evidence suggests the existence of several other historic sites that have not been mentioned here. The advent of colonialism brought about the destruction of much of this rich history. Fortunately, there is a renewed effort to reconstruct the richness and elegance of African civilizations.

Most cities that exist today in Africa are colonial creations. Not surprisingly, the majority of African capitals have a coastal orientation, since they were created as "headlinks" for the exploitation of resources in the interior (Figure 10.2). Most precolonial cities are now virtually extinct or are just relics of the past, such as Timbuktu and Gao. Fortunately, a few cities have survived the ravages of colonialism and still thrive today as historic centers, such as Kumasi, Ghana; Addis Ababa, Ethiopia; Mogadishu, Somalia; and Ibadan and Ife in Nigeria. These cities will continue to be reminders of Africa's proud and rich heritage.

THE INTERNAL STRUCTURE OF AFRICAN CITIES

Compared with North American cities, which display a combination of concentric, sectoral, and multiple-nucleus patterns (Discussion Box 10.1), the physical structure of contemporary African cities exhibits a blend of indigenous and colonial imprints. The shape, structure, and built environment of African cities are a function of their historical legacy. O'Connor (1983) has proposed six types of African cities, based on their history: the indigenous (native) city; the Islamic city; the colonial (administrative) city; the European city; the dual city; and the hybrid city.

The best documented indigenous cities are the Yoruba cities of southwest Nigeria (Mabogunje, 1962; Bascom, 1955) and Addis Ababa, Ethiopia (Pankhurst, 1961; Mehretu, 1993). Oral evidence suggests the existence of sizable Yoruba towns in the 12th and 13th centuries. These towns were enclosed by high walls with gates that linked main routes to central marketplaces. At the center was a palace that was encircled by large compounds of rectangular courtyards. Yoruba cities like Ibadan, Ife, Benin, Oyo, and Abeokuta exhibit some remnants of their precolonial heritage (Figure 10.3). Today, several Yoruba town associations exist to maintain ties with hometowns and to provide a support system for Yoruba migrants. Addis Ababa, Ethiopia, which was founded in 1886 by Emperor Menelik II, is a city that blends modern architecture with a strong indigenous flavor. Italian imprints and

Figure 10.2 Major cities in Sub-Saharan Africa.

ultramodern architecture, such as the modern City Hall, the "Piazza," and shopping and entertainment districts, are superimposed on the city landscape alongside fenced-in, rural-styled homes with cows, sheep, and poultry (Mehretu, 1993).

Examples of Islamic cities in Africa are the Sahelian cities of Kano, Zaria, and Sokoto in the predominantly Muslim northern Nigeria; Niamey in Niger; N'Djamena in Chad;

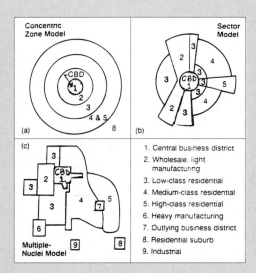

DISCUSSION BOX 10.1: COMPARISON OF URBAN MODELS

Three classical urban land-use models have been utilized to describe the internal structure of North American cities. The concentric zone model was developed by Burgess in 1923 to explain the pattern of social areas in Chicago. The model states that a city grows and expands outward from its central area to form a series of concentric rings. At the center (first zone) is the central business district (CBD), the focus of commercial and social activity. The CBD is surrounded by a transitional zone, which contains abandoned factories and warehouses and high-density low-income housing. This second zone is usually targeted for urban-renewal programs. The third zone is occupied predominantly by working-class families living in modest homes close to their place of work. The fourth zone is predominantly single-family residences interspersed with high-rent apartment complexes, and the fifth zone is a commuter region of low-density residential suburbs and greenbelts. An alternative model, the sector model, was proposed by Homer Hoyt in 1939. Hoyt stated that patterns of land use are conditioned by major routes that radiate from the central city, thereby creating a sectoral, or pie-shaped, pattern rather than a concentric pattern. In 1945, Ullman and Harris proposed a multiple-nuclei model, which suggested that land uses develop around discrete nuclei or centers rather than around a single CBD, as hypothesized by the concentric and sectoral models. These multiple centers evolve because of agglomerative tendencies and because certain activities repel one another. Realistically, most large cities in the United States have a combination of all three elements.

Question: To what extent do Sub-Saharan African cities conform to the models presented in the preceding paragraph?

The Internal Structure of African Cities

Figure 10.3 The ruins of the Maruhubi Palace, built between 1870 and 1888, are an important part of the indigenous architectural landscape and stone culture of the historical town of Zanzibar, Tanzania. Notice the high walls that encircled domed bathhouses and deep stone baths.

Khartoum and Omdurman in Sudan (Figure 10.4), and the east African town of Merca in Somalia. The Islamic imprints consist of a permanent central market (suq), mosques, shrines of saints, a citadel, and public baths. The residential land use is compact, with open courtyards and cellular urban structures. Buildings and alleyways are designed in such a way that they protect privacy. For example, entrances are L shaped, walls are windowless, and alleyways are narrow and irregular to create visual blind spots.

Many present-day African cities were influenced by the colonial expansionist policies of various European powers. (Refer to urban scenes in Figure 10.5.) They were developed primarily as administrative outposts and trade centers and, therefore, several perform significant port functions. These cities essentially formed the apex of a colonial

Figure 10.4 The Mhadi Mosque symbolizes the role of Islam in the urban landscape of Khartoum, the capital of Sudan.

The Internal Structure of African Cities

(a)

(b)

Figure 10.5 Urban scenes in Sub-Saharan Africa. (a) Abidjan is a major economic and cultural center in French-speaking West Africa. (b) International Conference Center in Accra, Ghana. (c) Modern buildings are juxtaposed against colonial German architecture in Windhoek, Namibia, an example of a Dual City. (d) Nairobi, Kenya Skyline. Nairobi's high-rise city center serves as a commercial and communications hub in East Africa.

(c)

(d)

Figure 10.5 *(Continued)*

infrastructure that siphoned resources from their hinterlands to European markets. Capital cities along Africa's coast, such as Dakar, Senegal; Conakry, Guinea; and Freetown, Sierra Leone, are prime examples. Colonial cities were characterized by the social, spatial, and functional segregation of people and land uses. Characteristics of these colonial influences also diffused into provincial towns, such as Bouaké, Côte d'Ivoire; Enugu, Nigeria; Tamale, Ghana; and Lubumbashi, Democratic Republic of Congo.

European cities, like Harare, Zimbabwe; Lusaka, Zambia; Windhoek, Namibia; and Nairobi, Kenya, were developed primarily for European settlement and, consequently, became replicas of cities in Europe. The equestrian statue (erected in honor of German soldiers) and the statue of Curt Von Francois (founder of Windhoek) are remnants of the German colonial influence in Windhoek. In the early stages of its colonial development, Nairobi catered specifically to Europeans. Africans, Europeans, and Asians were segregated on the basis of occupation and residence. The Europeans controlled most of the economic and administrative resources, while the Asians worked as merchants and artisans. Africans, on the other hand, were marginalized and took up menial jobs. As a result, Nairobi was never considered a permanent home for many Africans (Mehretu, 1993).

Dual cities occur as a result of the juxtaposition of two or more of the characteristics already defined. O'Connor (1983) identifies greater Khartoum, Sudan as a city where the indigenous development of the city of Omdurman, west of the Nile, occurred simultaneously with the colonial city of Khartoum on the east side. Kano, Nigeria is another city that has an ancient Muslim city with old walls, surrounded by a newly built-up area that takes on the characteristics of a colonial city. While indigenous and foreign elements are juxtaposed in a dual city, they are integrated in a hybrid city. Most cities in Africa today are hybrids of indigenous, colonial, foreign, and modern development. Accra, Ghana provides a good illustration.

The Internal Structure of Accra, Ghana

Ghana's capital city, Accra, has its origins prior to the 16th century. The city evolved from an original core area centered around Ussher Fort, James Town, and Korle Gonno (Figure 10.6). It was here that the indigenous Ga people settled, and their traditional lifestyles and customs are still prevalent, not only in the core area, but also in the Korle Gonno, Mamprobi, and Labadi districts. The residential land use in these coastal Ga settlements consists of a mix of single- and multistoried structures, with extended families sharing facilities. The single-story structures typically consist of a series of single-banked rooms surrounding a square, unroofed courtyard that forms a spatial arena for social interaction and interhousehold cooperation (Korboe, 1991). These "family houses" have high room density (as high as 4.0 persons per room) and dwelling density (15.7 people per dwelling) and very low vacancy rates (2.7%, including rooms reserved for guests). It is estimated that 25% of the occupants in these family homes are "rent-free consumers." Family homes provide a safe and caring haven for the elderly and for the poor; therefore, homelessness is not as rampant as in large North American cities.

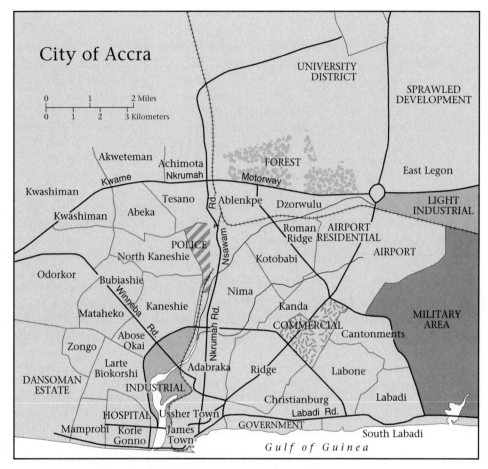

Figure 10.6 Internal structure of Accra, Ghana.

With the advent of colonialism in the late 1800s and early 1900s, Accra assumed a status as a colonial administrative center. Its development was accompanied by the expansion of land uses beyond the coastal core region. A government center quickly emerged immediately to the east of this core area of Ussher Town and James Town. The British imprints on the physical landscape in this area are still very evident, as suggested by current place names and urban symbols. Christiansborg castle was built as the seat of government. Victoriaborg, an exclusive European residential area, was built west of the government center in the mid-1890s. Also in the government center are elaborate structures, some with Victorian and Gothic styles, such as the Parliament House and the Supreme Court building (Figures 10.7a and 10.7b). Other structures include Independence Square, major banks, an arts center, government ministries, and a sports complex.

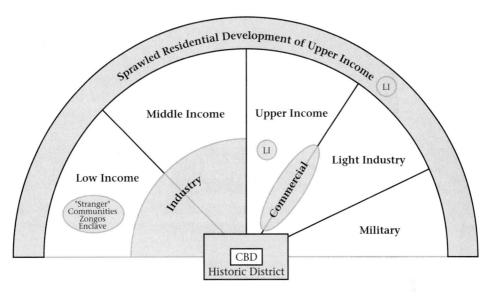

Figure 10.6 *(Continued)*

The residential land use in Accra, particularly south of the Kwame Nkrumah highway, exhibits a sectoral pattern. The region north of the government center and east of the Nkrumah Avenue–Nsawam Road in Figure 10.6 is the upper income region. Low-to-medium density single-family districts, such as Ridge, Cantonments, and Labone were formerly occupied by expatriate civil servants or leased to European businesses for their employees. They are currently occupied by wealthy Ghanaians. This western axis stretches north to the wealthier districts of Airport Residential, Dzorwolu, and Roman Ridge. The anomalies in this region are the slum districts, Nima and Alajo.

The wedge immediately west of the wealthy district (bordered by the Nkrumah Avenue–Nsawam Road and the Winneba Road) is a middle-income district. Kaneshie was originally planned and built by the government for the wealthy, but access has since filtered down to middle-income groups. Tesano, to the north, is a mixture of middle- and upper income groups. West of Winneba Road is a mix of low- to moderate-income households. Some districts, like Korle Gonno, Mamprobi, Dansoman, and Abose Okai, contain government-run housing estates. Mamprobi, for instance, was built after the earthquake of 1939 to house victims and also veterans of World War II. Dansoman Estates was later built to solve the growing housing crisis; however, several homes still lack the necessary utilities and amenities.

Another feature of the Accra urban landscape is that it is interspersed with ethnic and religious enclaves. They are inhabited primarily by the Hausa people of northern Ghana, who began settling in Accra as early as the late 1800s. These ethnic enclaves provide a safe haven for "out of towners." They have become magnets for Hausa traders

(a)

(b)

Figure 10.7 Colonial influence in architecture. (a) Parliament house in Accra, Ghana. (b) The Supreme Court building in Accra, Ghana.

and have spurred the development of other ethnic enclaves in Accra New Town, Nima, Abeka, Adabraka, and Darkoman. Islam is the centripetal force for community identification and continuity in Sabon Zongo (Pellow, 1991). There are several mosques, mutual aid societies, and Koranic schools in the community. Some authors have described the African city as having an inverse concentric pattern, where the wealthy and middle-income groups congregate close to the government center to avoid the hazards of a time-consuming commute and where the poor occupy peripheral locations (Brunn and Williams, 1993; DeSouza, 1990; Hartshorn, 1992). In Accra, a new phenomenon is occurring: In recent years, there has been sporadic and uncontrolled land and residential development in suburban and exurban districts north of the Kwame Nkrumah highway and around the University of Ghana, Legon.

The sprawled development is characterized by unauthorized land development. Most buildings are developed without prior approval from planning agencies. The key agencies responsible for initiating, approving, and executing plans—the Department of Town and Country Planning, the Accra Planning Committee, and the Accra City District Council—have not yet instituted a variety of growth-management ordinances to control this sprawl. The high-income suburb of East Legon, for example, is located in a low-lying area prone to floods and is directly underneath the flight path to the airport. Some homes in East Legon are also located dangerously close to high-tension electrical cables that run through the area. The extension of utilities to these outlying areas is an expensive process and continues to encourage suburbanization. The excess of demand on the existing water system in Accra makes the water pressure in these outlying regions so low that some residents have resorted to well water.

The evidence from Accra, Ghana does not suggest a simple inverse concentric model. It reflects a hybrid of indigenous, colonial, and modern development. A sporadic distribution of ethnic and religious enclaves is superimposed on a sectoral arrangement of residential areas. As the city continues to expand, unmanaged, sprawled developments of high-income areas continue to develop at a rapid pace in the periphery.

CURRENT URBANIZATION TRENDS IN SUB-SAHARAN AFRICA

Urbanization is used in this chapter to reflect the movement from rural areas to urban areas. People who migrate to urban areas usually engage in nonagricultural activities, although studies have shown that the practice of urban agriculture and market gardening is on the rise as an informal activity in African cities. The term *urban* implies a nucleated, nonagricultural settlement. In terms of a minimum population size, its definition varies from country to country. In Denmark and Sweden, the minimum population size is 200; in Canada and Australia, it is 1000; in the United States and Mexico, it is 2500; in Ghana and Nigeria, it is 5000; in Japan, it is as high as 30,000 (Brunn and Williams, 1993).

Most regions of the world, with the exception of Asia and Sub-Saharan Africa, are predominantly urban. Sub-Saharan Africa has been described in several academic circles as the least urbanized region, although evidence from the United Nations Center for

TABLE 10.1 Urban Population by World Regions: 2000 and 2030

Regions	2000 Urban Population		2030 Urban Population	
	Millions	Percent	Millions	Percent
World	2845.0	47.0	4889.4	60.3
Sub-Saharan Africa	219.8	34.3	620.3	52.0
United States and Canada	238.9	77.2	313.5	84.1
North Africa	77.9	53.8	136.4	63.9
China	409.9	32.1	752.1	50.3
Southeast Asia	192.7	37.1	397.4	55.9
South Asia	390.3	28.6	913.1	46.4
Latin America & Caribbean	390.9	75.3	604.0	83.2

Source: Compiled from UNCHS. (2001). *Cities in a Globalizing World*, London: Earth Scan Publishers.

Human Settlements (2001a) suggests that, in 2000, the southern regions of Asia had lower proportions of people living in urban places (Table 10.1).

Over all, about 220 million people, 34% of Sub-Saharan Africa's total population, reside in urban areas. The region exhibits spatial variations in its urbanization levels, as shown in Figure 10.8. Gabon has the highest level of urbanization (81.4%), followed by Congo (62.5%), Mauritania (57.7%), South Africa (50.4%), and Botswana (50.3%). Rwanda is the least urbanized, with an urban population of only 6.2%. Coastal west Africa and much of southern Africa have countries with urban populations above the sub-Saharan average. Among the countries with relatively higher urbanization rates are those that have recently experienced periods of moderate to rapid economic growth, such as Gabon, Cameroon (48.9%), Côte d'Ivoire (46.4%), and the Congo Republic (41%). West African countries, in particular, have benefited from a long tradition of mostly indigenous urbanization associated with the city-based empires of the savanna and forest regions from 1000 to 1800 A.D. Most countries in east Africa, on the other hand, have fewer residents in urban areas. In marginally urbanized countries, such as Rwanda, Burundi, Malawi, and Uganda, urbanization was largely colonial and administrative with little indigenous urban foundation, except in the Arab-influenced coastal areas (Acsadi, *et al.*, 1990).

Sub-Saharan Africa is considered to be among the least urbanized regions, yet its urban population growth rate is among the highest in the world. According to the UN Center for Human Settlements (UNCHS, 2001b), Sub-Saharan Africa's urban population will approach 440 million (46% of the region's projected total, 952 million) by 2020. The average annual growth rate for Sub-Saharan Africa cities in the 1990s was 4.4%, compared with 3.3% for Asian cities and 2.5% for South American cities (World Resource Institute, 1996).

Urban growth rates have remained at least twice as high as the population growth rates, which average 2.5% to 3.0% per year. Some of the fastest growing cities are in Liberia (9.6%), Rwanda (9.4%), Malawi (8.5%), Tanzania (6.3%), Burkina Faso (5.8%), and Uganda (5.2%). This process of rapid urbanization is often referred to as *overurbanization*, a term derived by Breese (1966) to describe a situation where urbanization is running ahead of industrialization. As a result, the city's industrial and administration capacity cannot accommodate the demands and expectations of in-migrants. As the World Bank (2000) notes, African cities are not usually engines of economic growth and structural transformation.

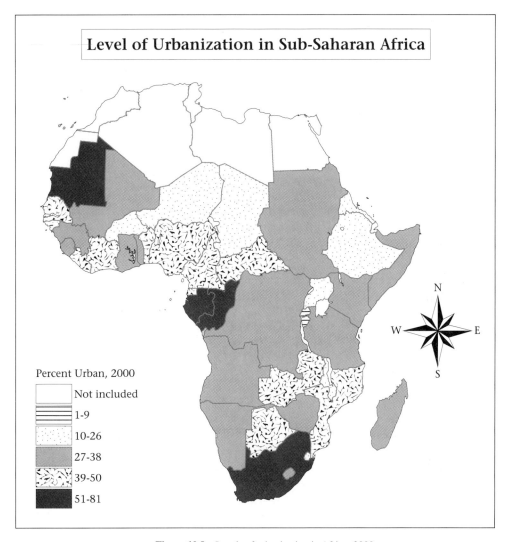

Figure 10.8 Levels of urbanization in Africa, 2000.

They tend to exhibit patterns of "urbanization without growth," which is symptomatic of a broader set of economic and social problems that plague the continent.

The process of urbanization in Sub-Saharan countries can be distinguished from that in more developed countries on the basis that cities in the former are still experiencing an accelerated rate of growth, whereas cities in the latter are at a terminal stage, a stage at which the urban population begins to increase at a declining rate. Suburbanization and exurbanization forces begin to prevail, and the population begins to disperse into smaller communities—a process typical of American cities today.

Another comparison can be made between the historical urban growth rates of more developed countries and current rates in Sub-Saharan Africa. Historically, urban growth rates in the developed countries of Western Europe and North America were more in line with the pace of industrialization, following a smooth progression and matching gradual changes in the industrial and demographic structure. The combined effect of industrial growth, economic growth, and rapid urbanization resulted in declining population growth, enabling cities to expand in step with economic development (Brunn & Williams, 1993). Davis's (1965) estimations for nine European countries show that, during their periods of fastest urbanization, the average annual gain per year was 2.1%, which pales in comparison with the current rates of 5% to 7% in Sub-Saharan cities. Slow economic growth in the region, combined with continued increases in population growth, has resulted in overurbanized cities. (See Discussion Box 10.2 for other comparisons between Sub-Saharan cities and US cities.) This rapid pace of urbanization in Sub-Saharan Africa can be explained by a number of factors outlined next.

DISCUSSION BOX 10.2

In Africa, the migration process is primarily from rural to urban.	In North America, the migration process is primarily from urban to urban.
African cities are experiencing the process of overurbanization.	Central cities are losing populations to the suburbs and exurbs.
African urban systems have a predominantly primary city-size distribution.	Urban systems are well integrated with a rank-size distribution.
African cities have a larger informal economy, including informal housing with a prevalence of renter-occupied units.	Formal sector is prevalent, along with owner-occupied units.
The process of urbanism is less effective. Most migrants end up in the informal sector, where they relive their rural experiences and are therefore not fully integrated into the "urban way of life."	City dwellers adapt quickly to the urban lifestyle and are fully integrated.

Components and Determinants of Urban Growth

The three dynamic components that account for change in Sub-Saharan urban populations are the natural growth of the population residing in the city (the difference between birth and death rates); the rate of in-migration; and, to a lesser extent, the realignment of city boundaries by the annexation or incorporation of adjacent towns. Explosive growth in African cities has been fueled by the combined effect of the first two processes; the accelerating force of migration continues to receive special attention from scholars and is focused on in this section.

A World Bank study conducted by Zachariah & Conde (1981) revealed that rural–urban migration accounted for close to 50% of urban growth in about half the countries surveyed in west Africa. Gambia (65% migration share), Sierra Leone (63%), Liberia (60%), and Côte d'Ivoire (59%) had relatively higher rates of migrants moving into urban areas. The United Nations estimates that natural population increases will be a dominant factor in the growth of urban centers in north and southern Africa, while rural–urban migration will continue to be the most significant urban growth component in west and east Africa.

African scholars have identified a variety of socioeconomic, cultural, and political factors that *push* migrants out of their rural homes and *pull* or attract them toward urban opportunities. A major pull factor in urban areas is the range of economic opportunities offered. Lewis (1954) has shown how labor can be transferred from a traditional, rural subsistence sector with low productivity and "surplus" labor to a highly productive, modern, urban industrial sector. Todaro's (2000) migration model demonstrates how a potential migrant takes into account wage differentials by expecting higher wage levels in urban areas compared with low rural wages in the agricultural sector, prompting migration to occur in spite of urban unemployment.

On average, urban wages in Africa are two to three times higher than rural wages (Eicher, *et al.*, 1970). However, rural–urban wage gaps vary across countries. In 1980, urban wages in Kenya and Zambia were 4.8 and 6.8 times higher than rural wages, respectively. In Ghana (0.5) and Sierra Leone (0.7), on the other hand, the wage gap was much narrower. Recent evidence suggests that some countries are showing signs of narrowing the rural–urban wage gap. Between 1969 and 1983, Tanzania reduced the gap by about 30% (Becker, *et al.*, 1994).

A number of social, cultural, and psychological reasons also factor into the migration decision process. Social prestige and the psychological satisfaction of living in cities were found to be significant pull factors in Nigeria (Adepoju, 1983) and Côte d'Ivoire (Joshi, *et al.*, 1976). Secondary push factors from rural areas include the desire to escape briefly from the social constraints of the extended family, short-term cash needs for bridal wealth, marital instability, birth order, and inheritance laws (Gugler & Flanagan, 1978.) Individuals can be denied access to family-owned farms on the basis of birth order; land passed down to the oldest son limits opportunities for females and younger siblings. It has also been hypothesized that people from the largest ethnic group are more likely to migrate to urban areas where they can depend on social networks already in place.

Government policies with an urban bias also affect rural–urban migration trends. The majority of African cities receive a disproportionate share of funds from governments to invest in the physical and social infrastructure. Also, pricing policies discriminate against rural farmers when the prices of agricultural products are kept artificially low in urban areas to appease the urban political constituency. Depressed agricultural prices in urban areas translate into lower rural incomes and wider rural–urban wage disparities.

Gender migration   A significant variable that is beginning to attract the attention of scholars in migrant selectivity studies is gender. Although the prevailing consensus is that male migrants outnumber females, recent studies show that female migration rates have been rising more rapidly than male migration rates. In Tanzania, for example, 60% to 70% of rural–urban migrants during the colonial era were male. By 1971, 54% were female (Becker, et al., 1994).

A study by Brockerhoff & Eu (1993), which examines the effect of fertility, child mortality, education, ethnicity, age, and marital status on female migrants in Sub-Saharan Africa, revealed that high fertility and child mortality restrains female migration to urban areas. In 75% of the countries surveyed (including Kenya, Togo, Uganda, among others), women with two or more surviving young children were less likely to migrate. Migrating with several children can be burdensome; the alternative, leaving the child home to be fostered by family members, is not always feasible. Having several children also confers certain rights and social privileges (access to land and social status), thus compelling females to remain in villages.

Other studies have found that low fertility rates deter female migration via high premiums placed on children. Females do not leave the village until a desired family size has been achieved (Mholyi, 1987). Child mortality deters female migration to urban areas, because women feel obligated to replace the child. Women with multiple births are less likely to migrate than those with single births. There may be other instances where repeat mortality may ostracize a woman from her husband and kin.

Further evidence on Sub-Saharan Africa reveals that females who migrate to urban areas tend to be more educated, younger in age (20 to 29 years), and unmarried. Also, polygamous women are more likely to circulate than monogamous women, and migrant women are more likely than nonmigrants to foster out their children. The influx of migrants into urban areas is cause for concern, but some scholars have focused on the interdependent relations that exist between rural and urban areas and the positive effects these links have generated for rural households.

Urban–rural linkages   While the rural–urban migration process can be viewed as imposing undue strain on urban environments, rural migrants maintain important and beneficial linkages with their villages or hometowns. The intensity of social links between city and country is one feature shared by all African cities (O'Connor, 1983). A major reason for this is the strength of family and kinship and the wish to retain land rights and diversify rural household incomes. Income and cash remittances are, therefore, a common feature of migrants. It is estimated that goods and money remitted by migrants to rural areas expend 10% to 20% of migrants' household income. Two studies in Kenya revealed that 68% of smallholders in the Central Province remit money to rural relatives and that remittances from relatives formed 9% of smallholder incomes (Hoddinott, 1992).

Although urban–rural remittances cannot capture the total impact of rural development, of particular significance is the contribution return migrants make to the development of farm and nonfarm activities. In Zambia, for example, return migrants serve as innovators and agents of socioeconomic and cultural change by introducing new varieties of crops, by employing new and improved techniques of production, and by encouraging education and innovation (Adepoju, 1983).

Consequences of Urban Growth

Rapid and unmanaged growth in Sub-Saharan African cities has triggered a host of problems that fall under five major categories:

1. urban primacy;
2. the urban informal sector;
3. housing shortages;
4. land management; and
5. infrastructure provision, service delivery, and urban management.

Urban primacy Rapid urbanization in African cities has been accompanied by urban primacy. Primacy can be measured in a variety of ways, including by taking the ratio of the population of the largest city in the country as a proportion of the sum of the remaining cities or as a proportion of the second-largest city. Mark Jefferson (1939), who coined the phrase as early as 1939, stated that a primate city usually had a population that was at least twice as high as the second-largest city. The antithesis to primate is *rank-size regularity*, which states that, in a given national urban hierarchical system, the population of any given town will be inversely proportional to its rank in the hierarchy. In other words, the population of the second-largest city should be one-half the size of the largest city, and the population of the fourth-largest city should be one-quarter the size of the largest city. The assumption is that countries that closely conform to the rank-size regularity usually have a well-balanced and well-integrated system of cities, while those with primate-city-size distributions exhibit a lopsided, loosely integrated urban system.

Figure 10.9 illustrates the prevalence of primacy in Sub-Saharan countries. Primacy is not solely restricted to small countries like Burkina Faso, Guinea, and Burundi, where the capital cities dominate the urban landscape; it is also a characteristic feature of larger countries, such as Angola, Senegal, and Mozambique. Most of the primate cities in these countries evolved as former colonial administrative centers and remain as enclaves of modern development; there has been little effort to extend the polarized colonial infrastructure that was inherited more than 30 years ago. These and other economic, structural, and institutional forces (discussed in Chapter 8) account for the increased disparity between primary cities and their peripheral hinterlands. It is not surprising, therefore, to observe from Figure 10.9 that, between 1985 and 2000, primate cities in Burkina Faso, Chad, Côte D'Ivoire, Somalia, Zambia, and Zimbabwe actually increased their share of population. Rondenelli (1983) has observed that primate cities in Africa grew at a faster rate than secondary cities.

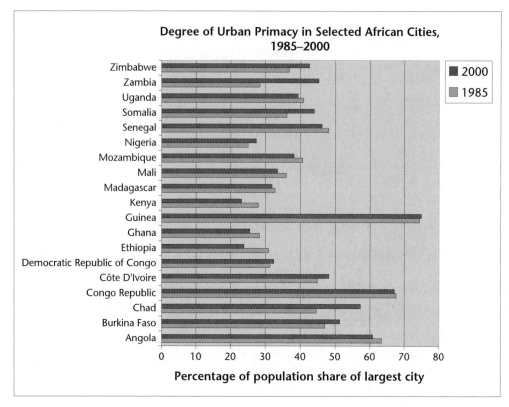

Figure 10.9 Degree of primacy in selected African cities: 1985–2000.

Secondary cities are defined as having populations from 100,000 up to, but not including that of, the largest city in the country (Rondenelli, 1983).

The definition of primacy can be extended beyond population size to reflect the disproportionate amount of social, cultural, economic, and administrative resources concentrated in the primate city (including the best educational institutions, as shown in Figure 10.10). Dar es Salaam, for instance, accounts for a disproportionate share of Tanzania's manufacturing jobs (50.3% in 1978), value-added manufacturing (56.9% in 1974), and social welfare spending (Sawers, 1989). The dominance of primary cities in African urban landscapes is reinforced by severe backwash effects that drain rural areas of vital human and natural resources. The combined effects of primacy and overurbanization account for the existence of a large informal sector in African cities.

The urban informal sector A common phenomenon of African cities is urban dualism—the coexistence of a formal sector with an informal sector. The formal sector includes public- and private-sector enterprises that are officially recognized, nurtured, and regulated by the government (Weeks, 1975). The urban informal economy consists of those

Current Urbanization Trends in Sub-Saharan Africa

(a)

(b)

Figure 10.10 Educational institutions in Accra, Ghana. (a) University of Ghana, Legon Administration building. (b) Achimota College, Administration building, Ghana.

individuals and enterprises that operate outside the mainstream of government activity, regulation, and benefits.

The word "informal" conjures up images of people who engage in social vices and criminal activity; however, there is another side of this sector, one that consists of talented and creative individuals who make worthwhile contributions to the urban economy—artisans, basket weavers, goldsmiths, garment makers. These talented individuals operate without a vendor's license, thereby denying African governments a potential source of badly needed revenue.

The inability of scarce formal-sector jobs to accommodate in-migrants has prompted an increase in informal activities. Although it is difficult to provide exact figures on their extent, with definitional problems and inadequate methods of data collection, it is estimated that the informal sector grew by 6.7% a year between 1980 and 1989 and employed more than 60% of the workforce in Sub-Saharan cities in 1990 (UNDP, 1992). There are variations among African cities in the proportions of informal workers. Charmes (1990) estimates a high of 73% in Burkina Faso; 65% in Kumasi (Ghana) and Niger; 50% in Lagos (Nigeria), Lome (Togo), and Senegal; 44% in Nairobi (Kenya) and Côte d'Ivoire; and a low of 20% in Djibouti.

Informal activities are easy to enter, unregulated, predominantly family owned, small scale, and labor intensive, and they rely on indigenous resources (ILO, 1972). The contribution of informal and small-scale enterprises to the economies of urban Africa is usually underestimated and unappreciated. It provides opportunities for the abundant semiskilled and unskilled labor pools in African cities. The informal construction sector, for instance, contributed 30% to the total value of output from the construction sector in Kenya between 1969 and 1978 and 35% of the construction sector's output in Côte d'Ivoire (UNDP, 1992).

Sethuraman (1981) further demonstrates that the informal sector contributes to the formation of human capital by providing access to training and apprenticeships at a cost much more affordable than that for formal training institutions. The sector also encourages the recycling of local resources, such as tires from automobiles, which are used to make footwear, and caters to the customized needs of residents who cannot afford to purchase items in bulk. Therefore, rather than buy a packet of cigarettes, a person can opt for one or two sticks. Occasionally, the formal sector subcontracts jobs to the informal economy.

Although informal enterprises are frequently criticized for being unproductive, a number of studies have shown that, in addition to generating a surplus, they generate high employment and value added of product per volume of investment when compared with the formal sector. A survey in Nairobi revealed that informal businesses multiplied their initial investments by as much as 6 to 10 times. Furthermore, the average capital required per worker in informal manufacturing was only 11% of that required by the formal sector—yet the informal sector generated an output valued at 26% of that in the formal sector (House, 1981). A similar pattern occurred in Ghana, where the initial capital required per informal worker was 11% of that in the formal sector, and the corresponding figure for value added of product was 19% of that in the formal sector (Steel, 1981).

Traditional views suggest a gap between formal- and informal-sector wages and living standards, but recent evidence suggests that the distinction between the two sectors is becoming blurred (Jamal & Weeks, 1988), which is attributed to a decline in real wages and employment levels in the formal sector. In Tanzania, for example, the real minimum wage fell by 80% between 1974 and 1986; in Uganda, the minimum wage in the 1980s fell to less than 10% of its 1972 value. The combined effect of this has been an increase in the ranks of informal-sector employees and the urban poor.

In spite of the contributions made to urban economies, the informal sector is not without its share of problems. First, the sector is plagued by "benign" neglect on the part of African governments, which continue to concentrate their resources on formal-sector activities. Second, the continued existence of the sector is perpetuated by urban elites who take advantage of the cheap services provided. It is much more affordable to hire a night "watchman" than a licensed security guard, and housemaids and laundry maids from the informal sector offer much cheaper services than those from official agencies. Third, women occupy menial positions in the informal sector. They are usually excluded from manufacturing jobs and are restricted to lower paying service activities, such as petty trading. Fourth, the informal sector accounts for a disproportionate number of the urban poor and is therefore susceptible to social vices and health hazards.

Although studies continue to document the creative potential and contributions of the informal sector, it is unfortunate that very little effort has been made to integrate this sector into the mainstream of government regulation and benefits. After all, the informal sector constitutes a large potential source of badly needed revenue for African governments. With increasing awareness of its potential and contributions, it is hoped more efforts will be directed toward the extension of public services to the informal sector.

Housing A major problem in urban Africa has been the inability of governments to provide adequate and affordable shelter. Housing is a multidimensional phenomenon: as a physical facility, it provides shelter and consumes a fair amount of urban land; as an economic good, it generates equity; as a social good, it enhances one's social status and self-esteem; and as a bundle of services, it offers a range of neighborhood amenities that come with its purchase. Housing is also an important sector of the economy by virtue of the multiplier effects it generates in construction and other related sectors (Bourne, 1981). In African cities, however, the majority of residents are excluded from enjoying these amenities and are relegated to *informal housing settlements* (Figure 10.11).

According to Lim (1987), we can differentiate between three types of informal housing settlements: slum housing, invasion housing, and squatter housing. Slums consist of units built on legally owned or rented land, but their physical structures violate legal minimum standards. In the invasion market, dwellers occupy the land illegally, but their physical structures conform to legal minimum standards. Squatter settlements violate both the legality of land occupancy and the physical standards. Konadu-Agyemang (1991) has argued that squatting is, by and large, absent in parts of west Africa and attributes it to the strong traditional and customary beliefs that people have. Amongst the Ashanti of Ghana, there is a sacred reverence for family-owned and stool-owned land. The belief is that spirits still watch over the properties of their ancestors; therefore, anyone who illegally "squats" on ancestral property is liable to incur possible harm or misfortune.

In Sub-Saharan Africa, informal housing settlements tend to be concentrated in undesirable parts of cities, particularly in the peripheral regions. This is also a typical pattern

Figure 10.11 Informal housing settlements in Ibadan, Nigeria.

in cities in other developing countries, such as Rio de Janeiro (Aryeetey-Attoh, 1992). Estimates of informal settlements in Sub-Saharan cities are as high as 85% in Addis Ababa, 70% in Luanda, and 60% in Dar es Salaam (United Nations, 1984).

In the formal housing sector, major problems occur in state-run housing corporations, most of which consist of large rental and, at times, tenant-purchase units. State housing corporations (SHCs) in Kenya, Tanzania, Senegal, Côte d'Ivoire, and Ghana offer examples of mismanagement, corruption, and inefficiency. For example, a major policy in the 1950s and 1960s was the eradication and replacement of slum and squatter settlements with public-housing estates, most of which centered around key urban and capital districts. Also, SHCs have never been able to keep pace with growing housing deficits. In Tanzania, for example, the government built 5705 low-cost units between 1964 and 1969. Between 1975 and 1980, averages of less than 100 units per year were being built (Stren, 1989a). In Nigeria, only 8616 housing units (or 19%) were constructed out of a proposed total of 46,000 in 1980 (Okoye, 1990).

Most public-housing estates have structural and design defects. Blankson (1988) characterizes public estates in Kumasi, Ghana as having inadequate space and facilities, offering little privacy, and being poorly ventilated and maintained. Due to the inappropriate design of these units, tenants resorted to the illegal extension of their units without consulting local authorities, thereby compounding problems associated with building density, overcrowding, and health and safety hazards.

Demand-side and supply-side factors explain the problems associated with both formal- and informal-sector housing. On the demand side, a critical problem is affordability: the ratio of the housing price to a household's income. As Table 10.2 shows, the ratio of the

TABLE 10.2 Urban Growth Rates and Urban Living Conditions in Selected African Countries

Country	Annual Urban Growth Rate, 1995–2000	House Price/ Income Ratio, 1998	Proportion of Waste Water Treated (%), 1998	Solid Waste Disposal (%)		Transport Mode to Work	
				Open Dump	Sanitary Landfill	Bus/ Minibus	Bike/ Walk
World							
Developing Countries							
Africa							
Angola	5.2						
Benin		3.4	70	75	0	0	10
Botswana	3.0	7.2	95	99	0		
Burundi	5.5	7.5	21.3	33	15	48	39
Burkina Faso	5.8	10.2	18.5	55	0	2	34
Cameroon	4.5	13.4	24.2	31	67	42	28
Côte d'Ivoire	3.2	18.0	45.0	72	0		
Democratic Republic of Congo	3.8			15	16	30	15
Ethiopia	5.2			22	11	13	83
Gabon	4.0		44.0	10	70	25	20
Ghana	3.7	14.0				50	11
Kenya	5.0	15.6	52.0	0	25	70	23
Liberia	9.6	28.0		100	0	80	10
Madagascar	5.0	13.9				60	33
Malawi	8.5	8.3		22	27	27	67
Mali	4.3	3.7		95	2	12	63
Mozambique	5.2	20.0	5.0	100	0	80	14
Nigeria	4.5	10.0				46	—
Rwanda	9.4	11.4	20.0	16	0	32	56
Senegal	4.2	3.5	3.5	100	0	77	13
South Africa	1.9	10.6					
Tanzania	6.3	5.0					
Uganda	5.2	10.4	30.0	75	0	65	
Zimbabwe	3.5	9.8	80	0	65	32	50

Source: UNCHS. (2001). *Global Urban Indicators Data Base*, New York: USA, UNCHS. United Nations. 2000. World Population Prospects: The 1999 Revision, New York: USA, United Nations Division.
UNCHS. (2001). *Cities a Globalizing World*, London: Earth Scan Publications.

median free-market price of a dwelling unit to the median annual household income can be as high as 28 in Monrovia, Liberia; 18 in Abidjan, Côte d'Ivoire; and 16 in Nairobi, Kenya. Affordability denies most Africans the opportunity to own a home. Therefore, unlike in the United States, where approximately 65% of households own homes, in west Africa, about 75% to 85% of urban residents are renters.

Most Africans do not have access to mortgage loans from banks; there are high interest rates and very stringent underwriting standards. Banks provide residential loans only to the very wealthy and prefer to invest in high-yielding industrial and commercial ventures. It is estimated that more than 80% of landlords and owners in west Africa pay for housing out of their personal savings (Konadu-Agyemang, 2001). Interest rates are unusually high, ranging from 20% in Sierra Leone and Senegal to 30% in Ghana.

The most relevant supply-side factors that affect housing in Sub-Saharan cities are the availability of land and the provision of adequate infrastructure and urban amenities. These factors are detailed next.

Land management The availability and management of land are fundamental to any improvement in the quality of urban environments in Sub-Saharan Africa. Access to land facilitates the provision of housing, utilities, transportation, and other relevant urban amenities. It also enables local governments and international agencies to carry out their policies more effectively. Between 1972 and 1982, the World Bank allocated 77% of its loans to site-service and slum-upgrading schemes, which were predicated upon securing land for the urban poor (Mabogunje, 1992).

A major problem with land in African cities has been the inability of municipal governments to efficiently issue, register, and maintain records of land transactions. This has resulted in a proliferation of informal and illegal tenure arrangements. In Cameroon, it is estimated that more than 90% of urban residents do not hold registered title to the land they occupy, though they must pay occupation rights to customary landowners (Mabogunje, 1992). Other World Bank projects indicate further that there have been limited attempts to conduct comprehensive land surveys. Surprisingly, only about 1% of land parcels in the Sub-Saharan region are estimated to have been comprehensively surveyed (Falloux, 1989). Comprehensive land surveys, though, have to be complemented by a broad-based design of regional information systems about all other attributes linked to land management: population, housing, production cost structures, institutional resources, and so on (Chatterjee & Aryeetey-Attoh, 1988).

The complex customary land tenure arrangements (outlined in Chapter 6) have an impact on land development. In the case of communal land or stool land, it is the sacred and customary duty of traditional authorities (chiefs) to represent the interests of their respective communities. They make decisions on the amount and purchase price of land to be leased, decisions that often cause purchase delays that drive up the cost of land. Land-tenure systems have also been affected by former colonial policies that were designed to alienate Africans from land. In French West Africa, for example, private land ownership was restricted to European settlers, missionaries, and private companies, and Africans were allowed only occupancy rights to plots of land. In British colonies like Kenya, Africans were allocated "Crown Land" strictly for occupancy purposes, with no confirmed title. In central and southern African countries like Rwanda, Burundi, Mozambique, Zimbabwe, and South Africa, land was confiscated from Africans, who were thereby transformed into labor reserves, to exploit African labor. These activities by former colonial powers resulted in the social isolation of Africans and created a system of fragmentary and questionable land-tenure arrangements that are still active in African society today.

Some African countries, such as Angola, Ethiopia, Somalia, and Tanzania, have resorted to the nationalization of land to do away with colonial legacies and customary practices of land tenure (Mabogunje, 1992). Such nationalization has, in turn, led to mismanagement practices, corruption, and speculation.

State ownership, customary land practices, and the inability of municipalities to record and maintain registered land titles mean that African local governments continue to lose substantial amounts of property-tax revenues needed to upgrade and provide urban facilities. The result is a continued deterioration of the urban infrastructure.

Infrastructure provision, service delivery, and urban management The rapid and uncontrolled pace of urbanization has incapacitated municipal governments in Africa. Cities continue to be plagued with deficient water-supply and sewage-disposal systems, inadequate solid-waste disposal mechanisms, and inefficient public transportation. Basic infrastructure and services constitute the underlying foundation and support structure of urban environments. It is essential, therefore, for governments to maintain and upgrade the infrastructure to provide a healthy and self-sustaining urban environment. A useful index for determining effectiveness at delivering urban services and infrastructure, and hence the effectiveness of urban governance, is the City Development Index (CDI), developed by the UN Centre for Human Settlement (UNCHS). (See Discussion Box 10.3.) The CDI is a composite index that measures the extent to which cities perform with respect to infrastructure, waste, health, education, and city product (UNCHS, 2001b). The 1998 index for Africa was 42.85, compared with 65.4 in Asia and 66.2 in Latin America and the Caribbean. This calls attention to the pressing problems, to be discussed, that confront African cities in regard to urban management and service delivery.

Water supply and sewage disposal Water is essential to land development and to the production process. Unfortunately, African cities are burdened with numerous shortages in water supply, inequities in the distribution of water, and inadequate treatment and disposal of wastewater.

Reports from the UNDP (2001) and World Bank (2001) demonstrate that the majority of residents in African cities have access to safe drinking water. For example, data from the World Development Report (World Bank, 2001) reveals that 100% of residents in Gaborone, Botswana have access to potable water. The figures are equally impressive for Kampala, Uganda (87%); Lilongwe, Malawi (80%); Mombasa, Kenya (90%); and Harare, Zimbabwe (97%). What is misleading about these figures is that they do not account for such problems as the lack of a continuous supply of water, low water pressure, and dilapidated pipelines in these cities. The UNCHS (2001a) reports that water supply is available only for an average of 2.9 hours per day in Mombasa, Kenya, and some areas in the city have no water in the pipes at all. A number of factors account for these shortages in good-quality water. Faced with insufficient foreign and local revenues, local governments are unable to purchase imported chemicals to treat and purify water and to acquire spare parts to maintain existing water lines. In Kenya and Nigeria, unmaintained, older pipelines burst open when the government tried to install a new, high-pressure water system (Stren, 1989a).

Municipal government interference and low levels of tax collection also explain the limited finances. In Nairobi, Kenya, water finances are controlled by the central government, which tends to transfer proceeds from the water fund into the general fund. Water shortages are compounded by inadequate and unreliable electricity supply, which causes frequent power failures, damaged water pumps, reduced water pressure, and increased maintenance costs.

Another problem is the unequal distribution of water in African cities. In Nigeria, accessibility to reliable water sources improves as one moves from the older core district to

> **DISCUSSION BOX 10.3: THE UNITED NATIONS CENTER FOR HUMAN SETTLEMENTS GLOBAL URBAN INDICATORS DATABASE**
>
> The UN Center for Human Settlements (UNCHS) has developed comprehensive data that provide a framework for a comparative analysis of urban living conditions and best practices across communities in the world. The database includes profiles of cities with respect to historical development, geographical attributes, social and economic development, environmental management, levels of governance and civic engagement, and levels of international cooperation. Also included is a tool kit that comprises 23 key indicators and 9 qualitative measures. The *key indicators* include information on tenure rights; access to land, credit, and basic services; social integration; water consumption, and waste-water treatment; solid-waste disposal; travel time and transportation modes; and government expenditures and revenues. The *qualitative indicators* include housing rights, mechanisms to prepare and implement local environmental plans, disaster prevention and mitigation instruments, urban violence, level of decentralization, extent of citizen participation in local decision making, and transparency and accountability. In addition, there is a "best practices" database, which draws on 1150 initiatives from 125 countries to illustrate innovative and creative approaches to the range of urban issues mentioned.
>
> The UNCHS has also developed a composite index—the *City Development Index (CDI)*—that measures the level of development in cities by five indices: infrastructure (water, sewerage, electricity, and telephones), waste (waste water and solid waste), health (life expectancy and child mortality), education (literacy and combined enrollment), and city product (estimate of the level of economic output in the city, as measured by income or by value added). The CDI can be used to assess the extent to which governments are effective in delivering urban services and enhancing the feeling of well-being in a city. Cities with CDIs above 95% include Stockholm and Melbourne. African cities have the lowest average, an index of 42.84. Cities far below the African average include Lagos, Nigeria (29.3) and Niamey, Niger (21.7).

the suburbs (Onibukon, 1989). Wealthier households and firms purchase imported generators and large water-storage tanks that make them resistant to disruptions. This diminishes any chance of their acting as an effective voice to improve the system. The only other feasible option for low-to-moderate-income households is the communal standpipe, and gaining access to one is time consuming. People who want to avoid long lines end up buying water from private vendors, which is more costly.

Another related problem is the inadequate treatment and disposal of wastewater. In cities like Dakar, Senegal (3.5%); Maputo, Mozambique (5%); and Ouagadougou, Burkina Faso (18.5%), a small proportion of wastewater undergoes some form of treatment. (See Table 10.2.) In African cities, a major distinction can be drawn between the conventional waterborne system, lower cost septic tanks, and pit latrines. Sewage-disposal plants, like water utilities, are susceptible to budgetary, operation, and maintenance-cost constraints, which has led to a shortage of sewage plants and an overload on existing ones. In Dar es Salaam, about 75% of the population uses on-site sanitation, consisting of pit latrines, septic tanks, cesspits, and ventilated pit latrines (UNCHS, 2001a). These are usually poorly maintained and impose a serious hazard of underground water pollution.

Solid-waste disposal The collection and disposal of solid waste is essential for the maintenance of a sanitary and healthy environment and the enhancement of one's socioeconomic well-being. The crowded conditions in African cities have caused many municipal solid-waste agencies to operate below capacity (Figure 10.12). Disposing of solid waste in sanitary landfills is the norm in only a few African cities, such as Libreville, Gabon; Yaounde, Cameroon; and Harare, Zimbabwe. (See Table 10.2.) Dumping and burning are the main disposal methods in several African cities. In Kigali, Rwanda, 84% of solid waste is burned openly.

Problems with solid-waste collection and disposal are largely attributable to inadequate equipment and staff and to the inefficiencies associated with central government agencies.

Figure 10.12 Urban congestions and problems with solid-waste collection in Ibadan, Nigeria.

Various cities have tried a variety of alternatives: Ibadan, Nigeria pursues a policy of limited privatization, where some solid-waste collection is subcontracted to private firms; one Côte d'Ivoire city, Abidjan, contracts with a private firm to clean its streets and collect its solid waste. Despite the improvement in service, private contracts severely drain government revenues and provide more favorable service for upper income districts (Attahi, 1989).

Small-scale scavenging has been suggested by Stren (1989a) as an additional alternative to waste reduction. In Khartoum, 35% of waste is removed by municipal authorities, 21% by domestic animals (goats, cattle, dogs), and 21% by people. Some products, such as bottles, scrap metal, and cartons, are recycled in the informal sector.

Public transportation Public transportation affects the daily operations of urban residents and businesses and is critical in the allocation of urban resources. As Table 10.2 shows, the principal modes of transportation to work in African cities are buses, minibuses, bicycles, and walking. There are a few exceptions: In Libreville, Gabon, 55% of residents take the train to work, and in Lagos, Nigeria, 55% of residents rely on private cars (UNCHS, 2001b). A growing trend in African cities is the declining role of formal public transportation and an increasing prevalence of an informal system of private minibuses and cars. In Conakry, Guinea, there is the virtual nonexistence of public transportation; in Maputo, Mozambique, 60% of commuters rely on informal transportation (Becker, *et al.*, 1994).

Problems in the public-transportation sector have occurred because of frequent breakdowns, inadequate spare parts, and chronic absenteeism by drivers who "moonlight" in the informal sector. The shortfall in public facilities has translated into high transportation costs and long journey-to-work hours for workers. The proliferation of private mini- and taxi buses in African cities has led to severe traffic-congestion and air-pollution problems. Also, the informal fula-fulas, dala dalas, tro-tros, and matatus, as they are called in Kinshasa, Dar es Salaam, Accra, and Nairobi, respectively, are at times operated at dangerous speeds by unlicensed drivers. These buses are not always mechanically sound.

African governments are seeking alternatives to ease congestion. Some are opting for transportation-demand management policies that encourage bike paths, pedestrian walkways, and car- or bus pooling. Linn (1983), for instance, has estimated that walking and cycling account for 60% of trips in Dar es Salaam and Kinshasa. These strategies need to be complemented with an upgrading of the existing road network, which is severely stressed.

The problems outlined are largely a result of a lack of fiscal, technical, and administrative responsibility on the part of African governments. These factors, among others, are essential prerequisites for the successful management of urban problems in Sub-Saharan Africa.

SOLUTIONS TO PROBLEMS IN AFRICAN CITIES

African cities are in a crisis situation. The multitude of problems associated with rapid and uncontrolled urbanization are a function of internal forces within city boundaries and broader forces external to city environments. To devise appropriate solutions, African governments need to address the urban crisis from a two-pronged approach. First, appropriate strategies need to be designed at the microlevel that address the effective management and planning of African cities. Second, appropriate macrolevel and regional strategies need to be devised to maximize the benefits for people outside the dominant city and to retard the flow of migrants to urban areas.

Microlevel Strategies: The Case for Effective Urban Management

As the crises continue to unfold in African cities, there is a need for local governments to respond immediately with more efficient, innovative, and effective ways of managing urban areas. Concerned scholars and institutions have outlined a set of challenges for governments to meet this ever-growing crisis (Mabogunje, 1992; Stren and White, 1989; UNCHS, 2001a; UNCHS, 2001b). The challenges include the following:

1. Enhancing urban governance by decentralizing local government and promoting civic engagement by allowing local communities to participate more effectively in decision-making processes.
2. Enhancing the administrative and technical capacity of local government institutions to coordinate the delivery and operation of services.
3. Improving the financial capability of cities to provide adequate urban services.
4. Seeking appropriate and innovative solutions that not only are cost-effective, but also meet the needs of local constituents.

The UNCHS (2001b) advocates good urban governance that is founded on the principles of decentralization, civic engagement, transparency, accountability, efficiency, and equity. African governments are extremely centralized. A 1982 survey by the International Monetary Fund revealed that local governments in industrialized countries account for 57% of all government jobs, compared with only 6% in African countries (Stren, 1992). In Botswana and Kenya, the share of local-government expenditure is less than 4% of total expenditure, compared with 37% in Brazil and 26% in Mexico. This explains why the loosely structured and poorly functioning local governments in Africa continue to be out of touch with local needs. Quite recently, there have been efforts by African countries to delegate more responsibilities to local governments, but these efforts have been largely ineffective.

Ghana enacted the Local Government Law in 1988 to promote popular participation and ownership of the government machinery by shifting the process of governance from command to consultative processes and by devolving power, competence, and resources to the district level. The decentralization policy was extended in 1994 to include urban-, zonal-, and town-council committees; it embodied community initiatives, training and placement services for workers, basic-needs provisions, and capacity building at the local level (Aryeetey-Attoh, 2001). Unit committees provide community residents with the potential for managing and operating the services that they require. This potential can be realized, however, only if residents receive the necessary education and skill-based training to increase their capacity to meet their own needs.

Côte d'Ivoire and Senegal have enacted significant administrative reforms to transfer power to local elected officials and to improve the coordination of local decision-making efforts. In 1978, communes in Côte d'Ivoire were reorganized and empowered with a wide range of municipal responsibilities. The communes are a decentralized group vested with legal responsibilities and financial autonomy. Residents of communes are "united by a solidarity derived from living together, eager to pursue community interests, and able to find the means for such action within the national community as a whole and within the spirit of national interest" (Ngom, 1989). Stren (1989b) points out that the quality of overall urban services and infrastructure in these French-speaking countries is superior to that in Anglophone countries. In Dakar, however, there is still a lack of clarity regarding the division of power and responsibilities among the various governmental structures.

In addition to creating a more decentralized system of responsible and autonomous local governments, African governments need to forge more effective partnerships between the public and private sectors and community-based organizations. Privatization can range in scope from the contracting or leasing out of tasks and responsibilities to private firms, while municipal governments retain overall supervisory control, to a complete transfer of responsibilities for providing urban services to private-sector firms who set their own price (UNCHS, 1999). Two Francophone countries, Côte d'Ivoire and Senegal, already contract out some urban services to the private sector. Privatization is beneficial if it relieves the financial and administrative burden on the government, satisfies unmet needs, reduces cost of public services to consumers, improves efficiency by promoting competition, enhances innovation and new technology, and improves responsiveness to cost-control measures (McMaster, 1991). Municipal governments in Accra and Kumasi, Ghana have received support from the World Bank to contract with local entrepreneurs to collect household refuse. However, door-to-door refuse collection services are provided to only 5% of households in Accra, Ghana. There are also problems with delays in payments to contractors.

Also important is the potential role that volunteer and community-based organizations can play. In Tanzania, for example, local voluntary associations such as women's rotating savings societies, security teams in informal housing settlements, and welfare associations have been very responsive, particularly to the needs of informal sector residents, when the state and local government response has been minimal (Stren, 1992). The potential exists for community-based organizations and private companies to provide such municipal services as solid-waste management, parking, management of public restrooms, improving of neighborhood roads, and health and education services.

A decentralized local governmental structure cannot function effectively without the appropriate administrative and technical capacity. A qualified and skilled labor force is required to carry out the day-to-day operations and regulations of urban management. Nowhere is this more evident than in the area of urban land management. Most African countries lack the institutional and administrative capacity to record land transactions. The majority of urban residents do not hold registered title to the land, as previously discussed. Furthermore, land-registration procedures in African countries can be lengthy.

Without proper land-registration procedures, African cities cannot administer property taxes, which are an essential revenue source for local governments. There is a need, therefore, to conduct appropriate workshops and technical training programs to upgrade the human resource base required to improve land-management practices. Becker, *et al.* (1994), in reference to a World Bank staff appraisal report in Ghana, call for recognition of a need to restructure salaries and incentives and to develop special postgraduate programs to train both public and private sector managers. Unfortunately, such programs are lacking in Ghana and other African countries. The Economic Development Institute of the World Bank is already involved in organizing a number of workshops and seminars geared toward civil-service improvement, toward administrative-capacity building for policy analysis, coordination, and implementation of national development policies, and toward the developing of training strategies to strengthen institutional management (Adamolekun, 1989).

Improving the administrative and technical capacity of urban governments would also enhance the ability of urban governments to become financially viable and autonomous. In view of the difficulties involved with the administration of property taxes, most local governments continue to be dependent on external sources for finances that are not always reliable. With the exception of a few urban governments in Kaduna and Lagos, local governments rely heavily on state and federal governments for revenue, and this can form as high as from 80% to 90% of their total revenue (Onibukon, 1989).

In addition to improving the capacity to collect property taxes, local governments need to improve their tax- and fee-collecting procedures. This implies taking on the responsibility of cost recovery through user charges and cofinancing or coproduction, where users participate in providing urban services through monetary and labor contribution. In the Accra Metropolitan Area of Ghana, user charges account for just 3% of government revenues, because the central government has not motivated local residents and authorities to assume the responsibility for projects and has thus denied them an opportunity to play a meaningful role in shaping the destinies of their own communities. Successful cost recovery in African cities occurs through self-help and coproduction as residents develop creative ways to pool their resources and talents. In Luanda, Angola, a successful partnership between government, the private sector, and community groups resulted in financing of infrastructure improvements in three districts of the city.

In trying to address these problems of urban management, African governments need not always rely on Western ideals and norms nor be always preoccupied with preserving outmoded colonial structures and procedures. A true test for African urban governments will be their ability to develop appropriate and innovative solutions that reflect the needs of local residents. According to Mabogunje (1992), urban Africa needs a strategy of governance that capitalizes on the institutions and processes with which people are familiar and that entices people to participate actively in the management of cities. Kironde (1992) offers some guidance along these lines in the areas of sanitation, water supply, and urban transportation.

Kironde suggests that urban governments begin to invest in ventilated improved pit latrines as an alternative to waterborne sanitation systems, which are capital intensive and unaffordable for a majority of urban residents. A majority of households in Nigerian cities like Calabar, Kaduna, Ibadan, and Kano (71% to 98%) and, in Tanzania, Dar es Salaam (88.9%) depend on pit latrines. It is unrealistic to expect comprehensive sewer systems to be extended to all urban residents in African cities. Although certain biases exist in regard to the health hazards of pit latrines, the ventilated improved systems, if properly constructed and maintained, may very well be a realistic, cost-effective alternative.

There are further questions about rigid international standards about water consumption and water supply. The World Health Organization's standard, an average household consuming 90 liters of water a day, is quite unrealistic in an African context. Other standards requiring all homes to be connected with a municipal water supply are also unattainable. Urban planners and managers should be more concerned with the central notion of ensuring easy and unimpeded access to safe, clean water. Therefore, as Kironde states, low water consumption and water-supply sharing should not be ignored as viable alternatives.

In the arena of public transportation, the informal modes of public transportation tend to be shunned. True, the dala dalas of Tanzania and the bakassis of Sudan often escape any kind of regulation, yet government-operated transportation cannot meet daily demands, is costly to operate, and drains large subsidies from urban governments. Informal modes of transportation need to be monitored to ensure that they observe traffic regulations, because they offer some flexibility to the passenger by not having fixed routes and because they save governments from incurring huge transportation subsidies.

Several attempts have been made by international agencies and local African governments to deviate from applying rigid standards in the housing sector. Gone are the days when slum eradication, rigid minimum building codes, and forced resettlement were seen as solutions to the housing problem. Donor agencies, such as the World Bank and the US Agency for International Development, have forged partnerships with African governments to encourage local self-help efforts. Site-and-service, core-housing, and squatter-upgrading programs are now regarded as viable alternatives.

Site and service programs are those in which governments purchase and assemble plots of land, install the necessary infrastructure or services (roads, water and sewer lines, electricity), and then sell the plot to low-income households at low interest rates. The purchasers of the plot are then responsible for designing and building a home to conform with their needs. Building, design, and finance costs are kept at a minimum to ensure easy recovery of costs. The program, if carefully implemented, could be a successful partnership between international agencies (who provide low-interest loans and technical advice), national governments (who assemble and develop the site), universities (where engineering and architectural faculty and students offer technical advice), and residents (who have an opportunity to participate in the decision-making process). The program empowers local residents and adds a sense of self-esteem and purpose to their initiatives and efforts. The site-and-services program can be supplemented with a *core house* that consists of the "shell"—walls, floor, and roof.

Site-and-service programs have been adopted in a number of African countries, including Zambia, Zimbabwe, and Kenya. Unfortunately, site-and-service programs do not always reach their intended beneficiaries. The poor sometimes end up reselling their sites to middle- and upper-income households. In Nairobi, for example, several sites were taken up by entrepreneurs who, in turn, sublet rooms at exorbitant rents.

Slum and squatter upgrading schemes have also taken on a new meaning in African cities. Early attempts to eradicate these settlements were rendered futile; the displaced households ended up relocating elsewhere and duplicating their lifestyles. Most governments now accept their existence and are trying to provide the necessary facilities to upgrade them. In addition to these efforts, governments are relaxing rigid building codes and sponsoring more research into alternative building materials that are biotic and fire resistant.

African urban governments need to be more creative and innovative in their approach to urban problems. They also need a change in attitude to incorporate the utilization of local indigenous resources and to encourage more local participation and initiative in the urban revitalization effort. All this can done within a framework that encourages an interdependent relationship between international, national, and local agencies.

Macrolevel Strategies: Regional Policies

The creativity and ingenuity involved in managing the internal problems of African cities can be effective only if in-migration flows are held in check. It is, therefore, necessary to supplement microlevel strategies of internal management with macrolevel policies that retard and reverse the flow of migrants into primate cities.

New towns New towns are comprehensively planned communities designed to be self-sufficient in the provision of urban amenities and services. They act as countermagnets or "interceptor towns" to provide alternative destinations for migrants headed for primate cities. They are also employed as a policy tool to decentralize growth and, therefore, must provide employment and other relevant services that are on a par with those offered in competing cities. New towns can also be satellite cities built close to, but clearly, detached from the commuting zone of an existing city by a greenbelt (park, agricultural land, open space). The purpose of such cities is to provide alternative residential choices for workers and to relieve inner-city congestion.

Governments in Nigeria, Malawi, Côte d'Ivoire, and Botswana have embarked on large-scale capital-relocation programs to serve a variety of economic and political purposes. For example, some new capitals were shifted from a coastal to a more central orientation. The movement of capitals from Lagos to Abuja in Nigeria, Abidjan to Yamoussoukro in Côte d'Ivoire, and Zomba to Lilongwe in Malawi were designed to rid the old capitals of their colonial associations and to promote a new sense of nationalism, pride, and unity. It was felt that a more nearly central location would act as a neutral site in easing regional and ethnic tensions. A central location would also maximize the economic integration of regions outside the older, more developed capital districts and make government more accessible to its constituents. In Malawi, for example, the shift from Zomba to Lilongwe was expected to stimulate agricultural, commercial, and industrial activity in the central and northern regions (Potts, 1985).

Although new capitals in Africa have been promoted as symbols of national unity and economic integration, there have been a number of problems. For example, such projects are very expensive: They require substantial financial commitments to transfer government offices, relocate populations, and fund capital projects. Often, the expenses are compounded by delays in planning and construction. Malawi had to seek assistance from South Africa when its capital expenditures for Lilongwe exceeded its own financial capacity. Initial estimates for building Lilongwe exceeded initial estimates by about 50 to 60 million kwacha. In Abuja, financial mismanagement and questionable allocation of government contracts caused the federal government to curtail its expenditures for projects (Umeh, 1993). New capitals ultimately end up with high living costs to compensate for the substantial financial requirements and delayed costs associated with their construction. As a result, they are not always attractive for low-to-moderate-income households, who end up residing in distant locations or "shadow towns."

Another problem with new capitals is that they do not always stimulate development in peripheral regions. Lilongwe and Abuja ended up as mainly administrative centers, without necessarily providing the requisite multifunctional attributes to stimulate development. As a result, both cities cater largely to an upper-income, white-collar clientele. Furthermore, resettlement schemes associated with these capitals are problematic. In Abuja, an initial plan to resettle and compensate displaced households was altered, because the number of residents affected was underestimated.

In view of these problems, it is not surprising that the Tanzanian government's efforts to transfer its capital from Dar es Salaam to centrally located Dodoma have been stalled. Although some government buildings have been constructed and some bureaus transferred, the country lacks the necessary resources to fully commit to the project.

Growth poles The growth-pole strategy involves the selection of a limited number of urban centers with resource attributes that have the potential to develop. It is assumed that, once these centers receive sufficient levels of investment, they will generate the necessary "trickle-down effect" or spreading of effects, to stimulate development in their surrounding hinterlands.

Most of the earlier development efforts of African countries in the 1950s and 1960s focused on developing large-scale industrial growth centers with the objective of accelerating economic growth. In Uganda, for example, a hydroelectric power plant was developed as an industrial pole to stimulate development of copper-based operations and complement the development of agricultural processing industries to satisfy local demand. The hope was that these critical investments would gradually transform the region (Appalraju and Safier, 1976). Tema, Ghana was also developed as a port and industrial center adjacent to the capital district of Accra.

It was clear that growth-center policies based on large-scale, capital-intensive industrial activities were inappropriate for Sub-Saharan Africa. For example, the backwash effects associated with such growth centers were much stronger than the trickle-down effects they were expected to generate. Growth centers never developed local linkages with their surrounding environments, because the skills required by capital-intensive industries never matched the pools of unskilled and semiskilled laborers in the peripheral hinterlands. Furthermore, growth centers continued to maintain linkages with markets external to their immediate surroundings. Therefore, benefits were never internalized in rural areas, and primate cities continued to drain their rural hinterlands of critical human and natural resources.

Intermediate and small service centers Since the 1970s, there has been a deemphasis on grandiose and capital-intensive industrial poles and a new focus on more feasible and appropriate strategies that promote the development of small and intermediate towns. These strategies are based on central-place principles, which are intended to plan a hierarchy of service centers to ensure the most rational distribution of goods and services to as many people as possible.

Rondinelli (1983) is a major advocate of secondary or intermediate cities. He sees these cities as creating important linkages, because they perform functions found in both urban and rural areas. Most intermediate cities in Sub-Saharan Africa are located inland and require certain physical, environmental, and human resources to become viable.

Both Richardson (1978) and Gaile (1988) have stressed the need to develop small-town service centers to bolster urban hierarchies, expand agricultural productivity, and encourage opportunities for nonfarm employment. Kenya abandoned its earlier emphasis on creating nine growth centers (which was considered too ambitious by analysts) during the early 1970s in favor of Rural Trade and Production Centers (RTPCs), designed to stimulate market and employment expansion in areas of unmet agricultural and livestock potential (Gaile, 1988). Kenya's 1989–1993 plan highlights the RTPC program. Each province in Kenya participated in the program by being assigned one district. Districts were selected on the basis of their agricultural productivity potential. The RTPC received the necessary infrastructural facilities and support services to stimulate development in their immediate hinterlands.

Small towns are not solely focal points for agricultural activity; they can be developed as a framework for the delivery of social services and can act as diffusion nodes for channeling new ideas and innovations to their complementary regions. Gaile and Aspaas (1991) state that small towns are more efficient in the distribution of development benefits throughout their hinterland. They also provide greater per capita opportunities in small-scale enterprise.

The effective integration of small and intermediate centers into the urban systems of African countries requires a genuine commitment from central governments. Genuine commitment translates into providing the necessary infrastructural investments and support services and into enabling small and intermediate towns to function as viable administrative and political entities.

CONCLUSION

The problems in Sub-Saharan cities seem insurmountable, given the rapid influx of rural migrants, the large numbers of informal settlements, the deteriorating infrastructure, the environmental deficiencies, the housing inadequacies, and the dysfunctional administrative structures. Planning, if it ever occurs, is more reactive than proactive; in many instances, efficient urban management is lacking. Under such circumstances, it is difficult to identify a convenient and appropriate starting point. Urban geographers and planners have to begin by developing a comprehensive urban-information system. For example, a better job can be done with conducting comprehensive land surveys to inventory land uses, buildings, and zoning parcels. Such inventories are needed to enhance sorely needed government revenues to administer and enforce urban policy. It is also essential for governments to collaborate with community-based organizations, including informal networks, to devise solutions to the urban crises. Solutions to internal problems must be pursued simultaneously with regional strategies aimed at developing rural and intermediate service centers. This necessitates macroscale planning strategies that incorporate central-place principles to ensure that the maximum number of people have access to basic goods and services. In essence, what it proposed here is a comprehensive approach to the urban crisis—one that addresses the structural, institutional, economic, organizational, and human dimensions of the crisis within its local, regional, and global context.

KEY TERMS

Urban
Urbanization
Urbanism
Precolonial city
Indigenous city
Colonial city
Dual city
Hybrid city
Overurbanization
Suburbanization
Primate city
Rank-size regularity
New towns
Secondary cities

Push and pull migration factors
Urban bias
Urban dualism
Urban formal versus informal sector
Backwash effect
Slum housing
Squatter housing
Invasion housing
Urban management
Site and service program
Squatter upgrading
Growth poles
City Development Index

DISCUSSION QUESTIONS

1. Describe the physical, social, economic, and technological characteristics of precolonial African cities.
2. Describe the internal characteristics of African cities. To what extent are they similar to or different from US cities?
3. Describe the process of urbanization in Africa. How does this differ from the process of urbanization in US cities?
4. Identify the major components of urban growth in African cities.
5. Describe the social and economic characteristics of urban informal sectors in Africa. To what extent is the informal sector a detriment to urban environments in Africa?
6. Identify the principal factors that explain the rapid urbanization of African cities.
7. What are *push* and *pull* factors? Provide appropriate examples from Africa.
8. Discuss and evaluate the major internal problems facing African cities today. How can these problems be addressed?
9. Provide examples of viable solutions to the crises facing African cities today.
10. What is urban management? How can the principle of urban management be implemented effectively in African cities?

REFERENCES

Acsadi, G., et al. (1990). *Population Growth and Reproduction in Sub-Saharan Africa*, Washington, DC: World Bank.

Adamolekun, L. (1989). *Issues in Development Management in Sub-Saharan Africa*, Washington, DC: World Bank.

ADEPOJU, A. (1983). "Issues in the Study of Migration and Urbanization in Africa South of the Sahara," in *Population Movements: Their Forms and Functions in Urbanization and Development*, Morrison P., ed., Liege, Belgium: Ordina Editions.

APPALRAJU, J. & SAFIER, M. (1976). "Growth Center Strategies in Less Developed Countries," in *Development Planning and Spatial Structure*, Gilbert, A., ed., New York: Wiley.

ARYEETEY-ATTOH, S. (2001). "Urban Planning and Management Under Structural Adjustment," in *The Political Economy of Housing and Urban Development in Africa: Ghana's Experience from Colonial Times to 1998,* Westport, Konadu-Agyemang, K. ed., CT: Praegar.

ARYEETEY-ATTOH, S. (1992). "An Analysis of Household Valuations and Preference Structures in Rio de Janeiro, Brazil," *Growth and Change*, 23(2):183–198.

ATTAHI, K. (1989). "Côte d'Ivoire: An Evaluation of Urban Management Reforms," in *African Cities in Crisis*, Stren R.E., & White, R.R., eds., Boulder, CO: Westview.

BASCOM, W. (1955). "Urbanization Among the Yoruba," *American Journal of Sociology*, 60(5):446–454.

BECKER, C., HAMER, A., & MORRISON, A. (1994). *Beyond Urban Bias in Africa*, Portsmouth, NH: Heinemann.

BOURNE, L. (1981). *The Geography of Housing*, New York: Wiley.

BRAUDEL, F. (1976). "Pre-Modern Towns," in *Capitalism and Material Life: 1400–1800*, Weidenfeld & Nicolson, eds., London: Longman.

BREESE, G. (1966). *Urbanization in Newly Developing Countries*, Englewood Cliffs, NJ: Prentice-Hall.

BROCKERHOFF, M. & EU, H. (1993), "Demographic and Socio-Economic Determinants of Rural to Urban Migration in Sub-Saharan Africa," *International Migration Review*, 28(3):557–577.

BRUNN, S. & WILLIAMS, J. (1993). *Cities of the World*, New York: Harper Collins.

CHARMES, J. (1990). "A Critical Review of Concepts, Definitions and Studies in the Informal Sector," in *The Informal Sector Revisited*, Turnham, D., et al., eds., Paris: OECD Development Center.

CHATTERJEE, L. & ARYEETEY-ATTOH, S. (1988). "Information Systems for Housing Planning," in *Informatics and Regional Development*, Giaoutzi M., & Nijkamp, P., eds., Hampshire UK: Gowar Publishing Group.

CHILDE, G. (1950). "The Urban Revolution," *The Town Planning Review*, 21(1):3–17.

CONNAH, G. (1987). *African Civilizations*, Cambridge, UK: Cambridge University Press.

COQUERY-VIDROVITCH, C. (1991). "The Process of Urbanization in Africa: From the Origins to the Beginning of Independence," *African Studies Review*, 34(1):1–98.

DAVIS, K. (1965). "The Urbanization of the Human Population," *Scientific American*, 213(3):41–63.

DESOUZA, A. (1990). *Geography and the World Economy*, Columbus, OH: Merrill.

DIOP, C. (1974). *The African Origins of Civilization: Myth or Reality*, Westport, CT: Lawrence Hill.

EICHER, C., et al. (1970). *Employment Generation in African Agriculture*, Institute of International Agriculture Research Report, Number 9, East Lansing, MI: Michigan State University.

FALLOUX, F. (1989). *Land Information and Remote Sensing for Renewable Resource Management in Sub-Saharan Africa: A Demand-Driven Approach*, Washington, DC: World Bank.

GAILE, G. (1988). "Choosing Locations for Small Town Development to Enable Market and Employment Expansion: The Case of Kenya," *Economic Geography*, 64(3):242–254.

GAILE, G. & ASPAAS, H. (1991). "Kenya's Spatial Dimensions of Development Strategy," *Urban Geography*, 12(4):381–386.

GUGLER, J. & FLANAGAN, W. (1978). *Urbanization and Social Change in West Africa*, Cambridge, UK: Cambridge University Press.

HULL, R.W. (1976). *African Cities and Towns Before the European Conquest*, New York: W.W. Norton.

HARTSHORN, T. (1992). *Interpreting the City*, New York: Wiley.

HODDINOTT, J. (1992). "Modelling Remittance Flows in Kenya," *Journal of African Economies*, 1:2.

HOUSE, W. (1981). "Nairobi's Informal Sector: An Exploratory Study," in *Papers on the Kenyan Economy: Performance, Problems and Policy*, Killick, T., ed., London: Heinemann.

JAMAL, V. & WEEKS, J. (1988). "The Vanishing Rural–Urban Gap in Sub-Saharan Africa," *International Labor Review*, 127(3):271–292.

JEFFERSON, M. (1939). "The Law of the Primate City," *Geographical Review*, 29:226–232.

JOSHI, H., et al. (1976). *Abidjan: Urban Development and Employment in the Ivory Coast*, Geneva, Switzerland: ILO.

KIRONDE, J. (1992). "Received Concepts and Theories in African Urbanization and Management Strategies: The Struggle Continues," *Urban Studies*, 29(8):1277–1292.

KONADU-AGYEMANG, K. (2001). *The Political Economy of Housing and Urban Development in Africa: Ghana's Experience from Colonial Times to 1998*, Westport, CT: Praegar.

KONADU-AGYEMANG, K. (1991). "Reflections on the Absence of Squatter Settlements in West African Cities: the Case of Kumasi, Ghana," *Urban Studies*, 28 (1):139–151.

KORBOE, D. (1991). "Family-Houses in Ghanaian Cities: To Be or Not to Be," *Urban Studies*, 29 (7):1159–1172.

LEWIS, W. (1954). "Economic Development with Unlimited Supplies," *Manchester School*, 22:139–191.

LIM, G. (1987). "Housing Policies for the Urban Poor in Developing Countries," *Journal of the American Planning Asssociation*, 53(2):176–185.

LINN, J. (1983). *Cities in the Developing World: Policies for Their Efficient and Equitable Growth*, New York: Oxford University Press.

LLOYD, P. (1973). "The Yoruba: An Urban People?" in *Urban Anthropology*, Southall, A., ed., Oxford, UK: Oxford University Press.

MABOGUNJE, A. (1992). *Perspective on Urban Land and Urban Management Policies in Sub-Saharan Africa*, Washington, DC: World Bank.

MABOGUNJE, A. (1962). *Yoruba Towns*, Ibadan, Nigeria: Ibadan University Press.

MEHRETU, A. (1993). "Cities of Sub-Saharan Africa," in *Cities of the World*, Brunn, S. & Williams, J., eds., New York: Harper Collins.

MCMASTER, J. (1991). *Urban Financial Management: A Training Manual*, Washington, DC: World Bank.

MHLOYI, M. (1987). "The Proximate Determinants and Their Socio-Cultural Determinants: The Case of Two Rural Settings in Zimbabwe," in *The Cultural Roots of African Fertility Regimes. Proceedings of the Ife Conference*, Ife, Nigeria: Obafemi Awolowo University.

NGOM, T. (1989). "Appropriate Standards for Infrastructure in Dakar," in *African Cities in Crisis*, Stren, R.E. & White, R.R., eds., Boulder, CO: Westview Press.

O'CONNOR, A. (1983). *The African City*, New York: Africana Publishing Company.

ONIBOKUN, A. (1989). "Urban Growth and Urban Management in Nigeria," in *African Cities in Crisis*, Stren, R.E. & White, R.R., eds., Boulder, CO: Westview Press.

PANKHURST, R. (1961). "Menelik and the Foundation of Addis Ababa," *Journal of African History*, 2:103–117.

PELLOW, D. (1991). "The Power of Space in the Evolution of an Accra Zongo," *Ethnology*, 38(4):414–450.

POTTS, D. (1985). "Capital Relocation in Africa: The Case of Lilongwe in Malawi," *The Geographical Journal*, 151(2):182–196.

RICHARDSON, H. (1978). "Growth Centers, Rural Development and National Urban Policy: A Defense," *International Regional Science Review*, 3(2):133–152.

RONDINELLI, D. (1983). *Secondary Cities in Developing Countries*, Beverly Hills, CA: Sage.

SAWERS, L. (1989). "Urban Primacy in Tanzania," *Economic Development and Cultural Change*, 37(4):841–859.

SETHURAMAN, S. (1981). *The Urban Informal Sector in Developing Countries*, Geneva, Switzerland: ILO.

STEEL, W. (1981). "Female and Small-Scale Employment under Modernization in Ghana," *Economic Development and Cultural Change*, 30(1):153–168.

STREN, R.E. (1989a). "Administration of Urban Services," in *African Cities in Crisis*, Stren, R.E. & White, R.R., eds., Boulder, CO: Westview Press.

STREN, R.E. (1989b). "Urban Local Government," in *African Cities in Crisis* Stren, R.E. & White, R.R., eds., Boulder, CO: Westview Press.

STREN, R.E. (1992). "African Urban Research since the Late 1980s: Responses to Poverty and Urban Growth," *Urban Studies*, 29(3/4):533–555.

STREN, R.E. & WHITE, R.R. (1989). *African Cities in Crisis*, Boulder, CO: Westview Press.

TODARO, M. (2000). "Urbanization, Unemployment, and Migration in Africa: Theory and Policy," in *Renewing Social and Economic Progress in Africa*, Ghai D., ed., New York: Martins Press.

UMEH, L. (1993). "The Building of a New Capital City: the Abuja Experience," in *Urban Development in Nigeria*, Taylor, R., ed., UK Aldershot: Avebury.

UNITED NATIONS (1984). *Land for Human Settlements*, Nairobi, Kenya: UNCHS.

UNITED NATIONS CENTER FOR HUMAN SETTLEMENTS (UNCHS) (HABITAT) (1999). *Privatization of Municipal Services in East Africa: A Governance Approach to Human Settlements*, (UNCHS), Nairobi, Kenya: UNCHS.

UNITED NATIONS CENTER FOR HUMAN SETTLEMENTS (UNCHS) (HABITAT) (2001a). *Cities in a Globalizing World*, London: Earth Scan Publications.

UNITED NATIONS CENTER FOR HUMAN SETTLEMENTS (UNCHS) (HABITAT) (2001b). *The State of the World Cities*, Nairobi, Kenya: UNCHS.

UNITED NATIONS DEVELOPMENT PROGRAM (UNDP) (1992). *Human Development Report*, New York: Oxford University Press.

UNITED NATIONS DEVELOPMENT PROGRAM (UNDP) (2001). *Human Development Report*, New York: Oxford University Press.

WEEKS, J. (1975). "Policies for Expanding Employment in the Informal Urban Sector of Developing Countries," *International Labor Review*, 111(1):1–14.

WORLD BANK (2000). *World Development Report*, New York: Oxford University Press.

WORLD BANK (2001). *World Development Report*, New York: Oxford University Press.

WORLD RESOURCE INSTITUTE (1996). *World Resources: 1996–1997*, New York: Oxford University Press.

ZACHARIAH, K. & CONDE, J. (1981). *Migration in West Africa: Demographic Aspects*, New York: Oxford University Press.

11

Geography, Gender, and Development in Sub-Saharan Africa

Ibipo Johnston-Anumonwo

UNDERSTANDING THE GEOGRAPHY OF SUB-SAHARAN AFRICA FROM A GENDERED PERSPECTIVE

What similarities or differences are there in the status of women and men from one part of Sub-Saharan Africa to another? In what way does the regional landscape reflect the presence of women? Where do African men and women live and work and why? Do gender roles vary geographically within the continent? How does a gendered perspective improve our understanding of the human geography of Sub-Saharan Africa? These are a few questions that geographers seek answers to in an attempt to obtain a more complete understanding of the geography of Sub-Saharan Africa. Geography typically adopts a broad approach in describing and explaining differences from place to place, yet the discipline has only belatedly recognized patterns and processes of gender differentiation.

Gender can be interpreted to mean socially created differences between men and women. These differences also mean inequalities, with women being decidedly more disadvantaged. In more recent academic discourse on gender, there is considerable effort to correct this bias by highlighting information about women and documenting gender disparities. Since gender differences are created socially, the status of women relative to men varies spatially. Given the vastness of the continent and the diversity of Sub-Saharan societies, one has to guard against simplistic overgeneralizations about the geography of gender. This chapter is, therefore, written only as an introduction that summarizes some patterns and processes of gender differences and outlines other issues worthy of geographic inquiry in the future.

From the geographer's perspective, it cannot be overemphasized that there are important variations across Sub-Saharan Africa. In a continent this huge, differences in government policies, economic conditions, cultural values, and attitudes about women make it almost an impossible task to generalize about gender issues. Further, the status of African women cannot be summarized by any single indicator, in light of varying colonial and postcolonial histories and of the multiple and changing roles of women throughout the continent. To enable us to discuss the geography of gender in the region in a manageable way, two themes have been selected as useful organizing concepts for the chapter. They include (1) similarity within diversity and (2) change and crisis.

As a brief illustration, the question "Where are women and men located in Africa?" could be examined by studying their distribution patterns across the continent. One possible indicator is the sex composition (females as a percentage of the total population), which in Sub-Saharan Africa, is approximately 50.6. Geographers eagerly seek deviations from generalized patterns; for example, although men do not dominate in absolute population numbers in most Sub-Saharan African countries, they typically outnumber women in cities across the region. A common consequence, therefore, is that a higher percentage of women live in rural areas.

However, tempering this common generalization is the fact that women also predominate in some African urban settings, and the rate of female migration from rural to urban areas is now faster than ever before. These two aspects of the sex composition—areal variation and societal change—are not just central principles in geography as a discipline, but critical considerations in any analysis of the geography of gender in Sub-Saharan Africa. Rural–urban differences in sex ratios stem largely from the region's colonial history, when mostly men went to cities seeking waged employment while a woman's mobility was more restricted. Contemporary rapid female migration into cities stems largely from postcolonial, urban-oriented development strategies adopted by many governments at the expense of rural economies and from the resultant agricultural crises, accompanied by rural depopulation.

The complexity of variations in Africans' responses to crises is also central in describing and explaining the continent's spatial dynamics. Understanding the geography of gender, therefore, requires an understanding of the differential effects that capitalist development policies have had on African men and women. Indeed, an analysis of gender and geography is very much like an analysis of development and gender, because the issues that are critical for development in Sub-Saharan Africa—resource use, population, settlement patterns, agriculture, employment, political trends, social welfare, household dynamics—are also of immense practical significance for the well-being of all African women and men. The contributions of geographers to the study of gender confirm this overlap between geography and development in the Sub-Saharan context.

Following a brief summary of the general literature of geography and gender in Sub-Saharan Africa, the remainder of the chapter is divided into six sections, starting with the issue that constitutes the most pressing challenge for the region: development. First, a historical account of the development process in Sub-Saharan Africa focuses on the sources of gender-based inequalities. These gender inequalities are especially pronounced in rural areas where the majority of Africans obtain their livelihoods in the agricultural sector. The second section documents the experiences of rural women and their important roles in sustaining Africans in an increasingly fragile environment. Postcolonial policies have generated rapid social changes, the geographical manifestations of which are detailed in the third section, which reviews demographic, migration, settlement, and urban patterns and their bearing on women's experiences. Gender inequalities are certainly not limited to rural areas, and the nature of regional differences is discussed in the fourth section. African women did not remain, and have never remained, passive in the face either of patriarchal domination or of imperial domination, yet their dynamism, resistance, struggles, and achievements are often underrepresented in written and visual accounts of the continent that depict African women stereotypically as helpless mothers of dying children. The last two sections of the chapter thus aim to (1) recognize political, social, and economic accomplishments of African women in spite of overwhelming odds and (2) identify ways for conceptualizing and studying the human geography of Sub-Saharan Africa in such a way that gender differences and human welfare disparities are reduced.

Geographic Scholarship of Gender in Africa

Information available about Africans, especially women, is very fragmentary, in spite of recent efforts to improve database systems. Nevertheless, the scanty evidence available suggests the existence of gender inequalities, largely in a descriptive manner that shows geographical variations across the continent. These aggregate data do not reveal the varied responses of African women and their strategies for challenging the inequities. Rather, African women's initiatives are better captured in case studies that emphasize local contexts and responses. Several empirical accounts are now contained in geography books and journals, and a small but increasing number of authors are African.

Edited volumes by Momsen and Townsen (1987), Momsen and Kinnaird (1993), and Brydon and Chant (1989) offer case studies on gender issues in Africa. Much of the published research on gender consists of localized case studies that describe changes in the social and economic conditions and the burdens that women face in their daily lives. Typically, microlevel data are utilized; only occasionally are macroanalyses presented. Partly because of the infancy of the subdiscipline, the research on geography and gender in Sub-Saharan Africa overwhelmingly is descriptive rather than theoretical, and it reflects the theme of similarity within diversity. Many existing studies are place specific; therefore, a picture of the spatial variations in gender issues across the continent usually emerges. But the focus on the basic living conditions and the role of women exposes a strong common thread of female subordination.

Mapping the World of African Women: What Counts?

A pioneering atlas by Seager (1997) portrays variations in the living conditions of women in Africa and other world regions. The atlas highlights the difficulties of obtaining and presenting data about women and the implications thereof. The available data typically consist of conventional measures like fertility, life expectancy, literacy, and employment; much of the information about women's activities remains either unaccounted for or discounted. For example, there might be censuses of a man's cattle, or of the extent of water pipes in a community, or of the gallons of pipe-borne water in a community, but there are no censuses of the hours a woman spends tending her husband's cattle, or of the miles she walks to the wells, or of the number of buckets she carries on her head. Women's extensive activities in the informal economy are also notoriously underreported. These tendencies leave women invisible in national statistics. In some instances, even when gender-specific data are collected in censuses and surveys, the published documents do not always provide breakdowns by sex, which makes it difficult to examine detailed information. From the modest information available, much of which is development specific, the situation of African women compares unfavorably to that of women in other parts of the world (Neft and Levine, 1997). Sub-Saharan Africa's slice of the "development pie" is small, and African women's share of that piece is smaller still. It is appropriate, therefore, to start the study of geography and gender in Sub-Saharan Africa by focusing on women and development.

WOMEN AND DEVELOPMENT IN SUB-SAHARAN AFRICA

After colonialism ended in Africa, leaders of the newly independent states attempted to promote development in their countries through growth-oriented policies. (See Chapter 8 on development.) During this "first development decade," the 1960s, there was expansion of production (measured in terms of growth in Gross National Product, or GNP), but other socioeconomic indicators of development were not improving. At the end of the decade, it was very clear that the basic needs of the majority of Africans were not being met; food production levels, in particular, began to decline. Even though some members of African society clearly benefited from economic growth and modernization, several other groups became victims of development. By the 1970s, neocolonial deprivation had simply replaced colonial deprivation. This was also around the time that the first set of academic works documenting the relative disadvantages of women in African countries (and other developing countries) were produced. The seminal work in this respect was Boserup's *Women's Role in Economic Development*, published in 1970. Boserup highlighted the fact that official statistics and definitions do not fully recognize the work of women in subsistence activities or consider their unpaid domestic, subsistence, and informal work as gainful employment. For African economies, Boserup argued that development had actually harmed women over all and that women were worse off than before. She specifically cited women's limited geographic mobility as an important factor in their curtailed access to modernization and urbanization opportunities. Other early studies confirmed her general thesis that the benefits of growth, modernization, and development had accrued mainly to the male half of the continent.

Adverse Effects of Capitalist Development on African Women's Access to and Use of Productive Resources

A spate of studies showed how, during the colonial and postcolonial periods, women became disadvantaged more than men because, among other things, they are disproportionately engaged in traditional occupations that are replaced by modern establishments.

The introduction of cash crops opened greater income-generating opportunities for men, while women's production of subsistence food crops remained less profitable. Also, as men left rural areas to seek incomes in the urban-industrial sector, women's workloads and responsibilities increased tremendously.

In precolonial Africa, there was no universal equality in access to resources; gender stratification was characteristic. Although precolonial inequality in the control of resources existed on many levels, African women nonetheless had greater access to resources than they did under colonialism, largely because parallel female and male authority structures often helped to protect women's interests. There were few colonial policies that benefited women—or African men, for that matter. Colonialism in Africa generated some beneficial changes, but Rodney (1981), among others, showed that the "supposed benefits of colonialism" are largely overshadowed by the adverse consequences. African societies experienced the imposition of taxes, forced labor, and, often, the forced development of cash crops and other products to meet the ruling country's needs for its own factories. Most colonial policies adversely affected men and women, but they were more detrimental to women in the sense that they discriminated systematically against their access to many critical new resources, such as Western education and waged labor. Avenues or opportunities that women had previously used to ensure access to land or agricultural and domestic labor were narrowed considerably. African women lost a lot of the political, social, and economic power that they had had in the precolonial era. Racist apartheid policies worsened the lot of Black women in South Africa. The resultant decline in the position of African women has contributed directly to their lack of access to critical productive resources to date, be they land, labor, livestock, education, skills, equipment, jobs, wages, or credit. Probably in no other sector of the economy is the impact of capitalist expansion in Africa more damaging to women than in the agricultural sector.

AGRICULTURE AND RURAL ENVIRONMENTAL DEGRADATION

Unlike in most parts of the world, women, not men, are the primary food producers in Sub-Saharan Africa. According to official statistics, women in Africa do three-quarters of all agricultural work, in addition to cooking, cleaning, child care, and other domestic work. In the mid-1990s, 75 percent or more of economically active women in over 30 African countries worked in the agricultural sector. (See Figures 11.1 and 11.2.) This generalization that agriculture is feminized does not apply everywhere. For example, in parts of west Africa, men play prime roles in agricultural production. Yet, across the continent, women remain key actors in agricultural work. For each 100 men working in agriculture, 71 women do. In at least a dozen Sub-Saharan countries, there are more women than men in the agricultural labor force, among them Burundi, Congo, the Democratic Republic of Congo, Liberia, Malawi, Mozambique, Senegal, Sierra Leone, Zambia, and Zimbabwe (Seager, 1997).

The very survival and overall well-being of Africans, therefore, depend critically on women's work. A high percentage of African women, however, have little control of agricultural land, and policy makers and agricultural-development planners frequently ignore the work of rural women. Today, insufficient food production ranks high among

Women and Development in Sub-Saharan Africa

Figure 11.1 Women working on a farm in Angola (Courtesy of Nicholas Alipui).

Figure 11.2 Woman and girl operate a Volanta hand pump.

Sub-Saharan Africa's most pressing development issues. The agricultural crisis is one of many dimensions; it goes beyond the commonly mentioned environmental factors and includes socially created and maintained inequalities, especially gender-based ones.

Human-Induced and Environmental Influences on Agricultural Production

Since the 1960s, African-led governments have continued to concentrate on the production of cash crops. The need to pay for imports and loan repayments (since the 1980s, especially) forced governments to concentrate on cash-crop production, often on the best farming land. Food crops, such as maize, also known in east Africa as women's crops, are sometimes ploughed over to make room for cash crops such as sugar cane, cocoa, cotton, and coffee. Well-intentioned government efforts toward agricultural improvements did not necessarily benefit women. In Tanzania, for example, villagization created extra physical hardship for women, in the form of longer trips between homes and fields (McCall, 1987). Colonial and postcolonial inequitable policies, combined with deeply rooted African male traditions of domination, compound the deprivation of women. For example, income from the sale of cash crops or any surplus food crops is controlled, in many instances, by men and is often not used for purposes that would be in the best interest of the family (family maintenance, clothes, fees, health care). Ardayfio-Schandorf (1993) and Mwaka (1993) reach similar conclusions about the reduced status of African women, the negative consequences for the environment, and the overall deteriorating well-being of Africans in general.

Droughts and other disasters worsen the problem of declining food supply. Government repression, military dictatorships, and civil wars (often associated with external intervention) contribute to a decline in local food production and to an increase in the numbers of displaced persons in Sudan, Ethiopia, Angola, and Mozambique, to mention a few examples. (See Chapter 7 on population.) The combined adverse effects on agricultural production due to colonial land alienation, prolonged wars of liberation, and the loss of men, and thus, their labor are far reaching. "When armies march there is no harvest." Hence, as the number of displaced Africans rose in the 1980s and 1990s (out of which two-thirds of the adult refugee population on the continent consisted of women), food production declined precipitously. In 1985, food production in Africa was only about 90 percent of the food production in 1975; in each of those years Asia and Latin America were producing substantially more. In fact, Sub-Saharan Africa stands out as the only region where per capita food production has declined. (See Chapter 13 on agricultural development.) If agriculture is to expand significantly, development planners and policy makers must take into account the value of rural African women.

Unfortunately it remains true, 30 years after the seminal work of Boserup, that the contributions of African women, like those of women worldwide, are overlooked, undervalued, and underreported. Alternative methodologies, such as time-budget studies, help to correct widespread misconceptions about women's duties. A geographer in a west African rural setting detailed women's daily activity patterns and found that, contrary to popular perception, they spend more time farming in the fields than with their trading activities (Spiro, 1987). In the 1990s, women carried out well over half of such agricultural tasks as planting, weeding, transporting, and sorting. One Kenyan farmer laments "there are many women like me, we are wives, but we feel like widows"—a testimony to the fact that an

Women and Development in Sub-Saharan Africa

increasing fraction of African households are maintained by women (maybe up to one-third in some countries). Despite increases in female-headed households in Sub-Saharan Africa, the stereotypical assumption of a male-headed household is still widespread, and development resources continue to be channeled disproportionately to men.

Unequal Allocation of Resources within the Household

To fully understand both the relative alienation of African women from productive resources and the multifaceted dimensions of the African agricultural crisis, one must keep in mind that the household structure also determines a woman's access to resources. Throughout Africa, there are examples of how colonialism exacerbated the differences between men and women. As land was becoming a privatized commodity, titles for land were given mostly to men, rarely to women. Men decide ownership, control, and use of land in many cases. In some parts of Sub-Saharan Africa, women either own farmland or have usufructary or user rights; however, they lack control over their labor and time. In many west African countries, such as Senegal or Burkina Faso, women often point out that the family has all the land it needs, but that there is never enough time to cultivate all the fields. Women must work on their husbands' fields first, before moving to the other fields that they typically have to cultivate without any assistance from men (Figure 11.1). Apart from work on the fields, a woman fetches water (Figure 11.2) and firewood, cooks, and takes care of the children, leaving little time from the day.

Men have more discretionary time resources and greater access to labor-saving devices in the form of bicycles and tractors. Studies have shown that, following local and foreign investment, production methods become modernized and exports increase, but the conditions of workers, especially women, do not improve. Samatar (1993) reports that the backbreaking job of weeding on banana plantations in Somalia is dominated by young girls, because managers claim that "no real man is willing to work for slavelike wages" for tedious tasks like weeding. Mwaka's (1993) account of Uganda amply illustrates the drudgery of the work day for most rural African women. Even when labor-saving devices do save energy and time for women, women's full participation in society is constrained by inadequate attention to women's agriculture (Browne and Barrett, 1993).

Most African women perform daily tasks with their hands, not with machines. Men have great control over women's and children's labor; women complain of overwork and constantly aching backs, wishing that the men would help more, but husbands typically hold on to patriarchal views that it is a woman's traditional duty to do the housework. Given this unequal gender division of household work, and the resulting double day, it is easy to understand why rural African women welcome the arrival of newborn baby girls who will eventually assist with the household work. It is less of a wonder, then, why many low-income African women have several children! Increasingly, therefore, geographers and other social scientists rightly adopt the household as a unit of analysis. Intrahousehold gender relations are studied as part of the processes shaping African conditions. Carney and Watts (1991) urge researchers to "open the black box of the household and examine the circuits of economic, symbolic, and cultural capital that lie inside." In so doing, researchers are better able to probe into the processes by which male domination is maintained and resisted by women.

Apart from analytical considerations, there are policy implications associated with the neglect of women. If official reports fail to recognize the realities of African women's responsibilities and opportunities, then women's access to development resources is restricted. This partly explains why, despite the crucial role women play in food production, they continue to be excluded from agricultural modernization and technological efforts. The exclusion of African women from modernization and development contributes to the familiar demographic and spatial dynamics of Africa's population and results in the "statistics of shame," as detailed in the next two sections of the chapter.

DEMOGRAPHIC TRENDS AND SETTLEMENT PATTERNS: GEOGRAPHICAL CONTEXTS

Population, Fertility, and Mortality Trends

If women have greater access to and control of productive resources, they will not need large numbers of children to ensure their socioeconomic status. Among the best documented aspects of African women's lives is childbearing. Fertility data are thus generally available for Sub-Saharan Africa, and, as Chapter 7 shows, women of childbearing age have an average of 5.6 children.

Fertility rates remain high in the region partly because of high infant mortality rates. Infant mortality for many African countries is well above 90 deaths per 1000 live births. Although there is no evidence of gender bias leading to female infanticide in Sub-Saharan Africa, as is the case in some Asian countries, other forms of mortality differentiation exist among African women. Rural African mothers are still more likely to lose their children than urban mothers, mostly because of inadequacy of health services. Studies show strong positive correlations between women's education and infant mortality. It follows, therefore, that infant mortality rates in urban areas (typically with higher percentages of educated women) are lower than in rural areas. Indeed, place of residence is correlated not only with fertility and infant mortality rates, but also with the prevalence of contraceptive use (Kirk, 1998). In the intraurban setting, case studies show that educated African women who live close to health-care services are more likely to use contraceptives and are less likely to have their children die in infancy or die prematurely.

Similarly to trends worldwide, female life expectancy in Sub-Saharan Africa is higher than male life expectancy (50.3 years compared to 47.6 years), but the high rates of maternal mortality in Sub-Saharan Africa are not seen in other regions of the world. Maternal mortality is influenced by demographic variables, such as population size, urban percent of population, death rates, and birth rates, as well as by social welfare indicators, such as calorie intake, access to safe water, and the practice of female circumcision. According to Paul (1993), African women of reproductive age have the highest death risk from maternal causes in the world. In Benin, Cameroon, Nigeria, Malawi, Mali, and Mozambique, one out of five 15-year-old women dies before reaching 45 years of age for reasons related to pregnancy and childbirth (Topouzis, 1990). While the magnitude of maternal mortality is common throughout the continent, there are regional variations. Higher rates of maternal mortality prevail in most west African countries and in the east African country of Eritrea; countries of southern Africa, with the exception of Mozambique and Angola, are characterized by relatively low maternal death rates (Figure 11.3). Angola, Central African Republic, Sierra Leone, Burundi, Rwanda, Ethiopia, Mozambique, and Guinea-Bissau have some of the highest maternal death rates (more than 1300 per 100,000 live births) in Sub-Saharan Africa; relatively low maternal mortality rates are 330 in Botswana, 230 in South Africa, and 210

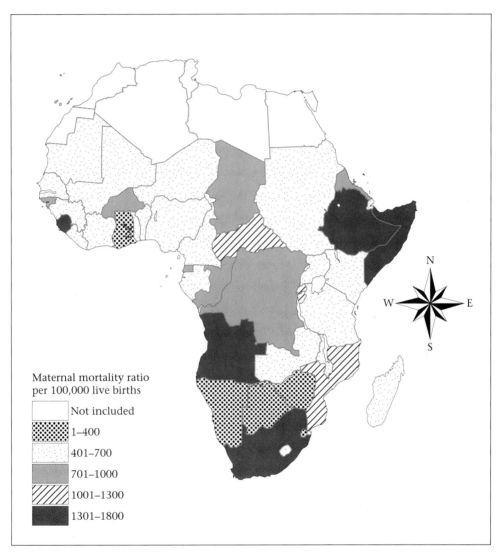

Figure 11.3 Maternal mortality ratio per 100,000 live births, 1990–1999. Compiled from UNDP (2001). *Human Development Report*, Oxford, UK: Oxford University Press and from *African Development Indicators*, World Bank.

in Ghana. Mauritius, with a maternal mortality rate of 50, has sustained one of the best records for Africa.

Aside from deaths related to childbirth, the much-documented spread of the AIDS pandemic is clearly relevant for the continent's demographic future, but death from AIDS has been shown to be socioeconomically based in the sense that the complications from AIDS that lead to death (e.g., infections) are poverty oriented. In Africa, HIV/AIDS is transmitted primarily through heterosexual intercourse, so women are at least equally as afflicted as men. However, in the context of polygamy, and given their worse poverty status, more women than men are subject to the human toll associated with AIDS. Indeed, current statistics reveal that women, including young women, are more affected than men. For every 10 men, twelve African women have HIV/AIDS. African women account for 85 percent of all global female infections. In southern Africa, 1 in 4 women aged 15 to 49 lives with HIV/AIDS. The majority of new HIV infections in Sub-Saharan Africa are occurring among young women aged 15–24. This is the largest growing group with AIDS in the region, and it accounts for nearly 30 percent of all female AIDS cases. In some countries, between 10 and 20 percent of teenage girls are already infected. This means that girls and young women who should be in school or university are sick or are dying in unprecedented numbers, and many more are forced to drop out of school to take care of sick relatives. The additional impact of sexism on the incidence of HIV/AIDS among African women is revealed in the preliminary findings of studies that correlate higher levels of the disease with places where female circumcision is more prevalent. African women's control over their sexuality and fertility is still generally low, but some scholars suggest that variations in female autonomy may contribute to geographical differences in women's ability to avoid being infected with sexually transmitted diseases, especially in an era of HIV/AIDS.

Although the earlier incidence of the disease was highly concentrated in a few countries in east and central Africa, it is not limited to these geographical regions, and the epidemic is now more widespread. There are high levels of HIV/AIDS in west and southern Africa also. But, although HIV/AIDS was reaching epidemic proportions across Sub-Saharan Africa at the end of the 20th century, it is important to note that, even at the beginning of the 21st century, many more Africans, male and female, die from socioeconomic and political factors other than from HIV/AIDS. Thousands of Africans who do not die prematurely vote with their feet to escape civil strife. Rwanda, for example, epitomizes both forms of demographic shifts—unprecedented numbers of premature deaths from AIDS-related illnesses and a massive exodus of men and women from their homes due to civil strife.

Displacement and Migration

The geographical movement of African women, like that of men, arises from both forced and voluntary relocations. In the 1980s and 1990s, African women and children have constituted high proportions of war- and drought-displacement victims. The majority of adult refugees are women, yet this fact is often disregarded in the design of postemergency refugee assistance programs; therefore, the life of African women in refugee camps is also plagued with sexist biases. For example, less vocational training is provided for refugee women (Hall, 1990). Despite the large numbers of people involved, forced migration in Sub-Saharan Africa remains far less prevalent than does voluntary migration for employment-related reasons.

Geographers used to focus their analyses solely on male patterns of labor migration. More recent work highlights the increase in female rural-to-urban migration and cites reasons why women outnumber men in some specific urban settings. Watts (1983) identified female marriage in-migration as a factor that compounds male out-migration and explains the high proportion of women in indigenous African towns. But most female rural migrants, like male counterparts who go to urban destinations, migrate not for marriage per se, but for a broad range of socioeconomic reasons. Wilkinson's 1987 study of Lesotho examined the effects of male migration on women and other family members left behind; other researchers show nowadays that women no longer stay on the farm and that there are regional differences in patterns of rural-to-urban migration (Gugler, 1989). Some observers have noted a decline in the overall rate of rural-to-urban migration in Sub-Saharan Africa (Potts, 1995), but the prevalent trend among men and women is still the pattern of migrating from rural to urban areas.

Common characteristics of female migrants are the tendency to be young, unmarried, and educated, but a sizeable percentage of female migrants are uneducated, and they fill low-skilled urban jobs. The ties that women migrants maintain with their rural origins remain understudied. Among the most significant aspects of urban life for recent arrivals from the rural areas is the precariousness of the lives of the majority of women who depend on trading in the urban informal sector for their livelihood. In what ways are the lives of urban African women different from those of rural women or of urban men?

Differential Access in Urban Settings

With the notable exception of the activities of women traders, the literature (e.g., House-Midamba and Ekechi, 1995; Yeboah, 1998) on urban African women is sketchier than that for rural women. The available research points to the added benefit that women derive in urban areas, be it in economic status, education, health, or access to other kinds of services. Within cities, the educational advantage is important to negotiate bureaucratic hurdles that are not so prevalent in rural settings. Where one lives can be less important than what one knows (or possibly whom one knows) in ensuring a woman's access to such urban opportunities as health care. Cities offer women opportunities to improve their economic and occupational status and to participate in governmental affairs.

Studies that have examined gendered spaces indicate that there is a strong female presence in the west African marketplace. For example, in examining the question of where urban women traders conduct their day-to-day activities, common assumptions about the geographic separation of the home and workplace break down. Accounts of women's activities in more developed countries, usually based on models of separate public and private spheres, are increasingly being challenged as unrealistic.

In the African context (both urban and rural), the inappropriateness of such models in understanding women's daily activity spaces is especially stark. The majority of market women perform their domestic tasks of child care or cooking at the same location where their trading stalls are located; otherwise they "buy" (in cash or kind) these domestic services in the same place. Home-based female traders, such as Muslim women in seclusion, do participate in income-generating trade activities through the labor of their children and wards. They also produce commodities for the local economy in the same spaces where social reproduction takes place. *Social reproduction* refers to reproducing the conditions of life, which, broadly speaking, includes having babies, socializing and educating them, seeking health care, and performing other tasks of social welfare and community development. In addition, when one considers the very high percentage of African women who work for wages as domestic servants, especially in the cities of South Africa, it is clear that the place where many African women conduct their domestic tasks is not the characteristic residence-based kitchen. It is also clear that the social relations at work are not limited to gender disparities, but include distinctions of class and race. Geographic inquiry about the existence and nature of gendered spaces within Sub-Saharan African cities lends itself, therefore, to further exploration of the intricate sociospatial connection, rather than separation, between the spheres of production and reproduction.

Recent statistical sources show many aspects of socioeconomic progress for African women, and the enormous wealth of some African market women has been the subject of several case studies in the past. Among urban women, there is a well-educated and often wealthy elite. Growing numbers of African women hold managerial and political positions at all government levels. For urban African women who are educated and professionally skilled, the evidence, however, shows that there is more gender disparity than parity with urban male counterparts. Such evidence exists because women are more likely than men to be employed in "pink-collar" jobs that are either dead-end or low-paying jobs. Many of these jobs are service occupations in the government sector, such as clerical work, teaching, and nursing.

Under conditions of occupational sex segregation, sex biases in hiring and promotions, economic downsizing, or cuts in public expenditures, women suffer disproportionately; and, as the last hired, they are also more likely to be the first fired. Women who are employed in the formal sector constitute only a minority of the female urban population; the great majority of women are employed in the economically marginal urban informal sector in a wide range of activities that include dressmaking, hairdressing, beer brewing, housekeeping, child care, and, very often, street trading.

Clearly, African women are not members of an undifferentiated social category because, as has been noted, not only are there important differences between rural and urban residents, there are intraurban differences as well. Furthermore, class, ethnic, religious, age, and national differences mean that there are varying categories of women in different parts of the continent. In spite of the diversity of regions and contexts, the common truism is that the nature of development as advocated, practiced, experienced, and continued in many Sub-Saharan countries has played a critical role in keeping African women underprivileged. This is probably why African female scholars are more inclined to adopt a perspective of similarity within diversity when analyzing African gender relations: "regardless of caste, class and ethnic [and geographic] location, African women share common gendered disabilities of subordination and the need to overcome them" (Awe & Mba, 1991, p. 648). Analytical categories and conceptualizations aside, the material conditions of African women suggest that a theme of similarity within diversity is more apt than a theme that emphasizes differences among women; the inequalities between African men and women, though narrowing in some aspects of life, are still very pronounced.

GEOGRAPHY AND GENDER INEQUALITY

Change and Crisis

A prime concern in the geography of gender is to examine changes in women's positions over time and across space. Momsen and Kinnaird (1993) point out that "there has been very limited progress in reducing gender inequality in most social, economic, and political contexts. Moreover, the general economic decline that became characteristic of the 1980s would suggest a continuation of pressure on living standards and resource use for both women and men." The few statistics bear out this trend of limited progress; few African women enroll in science and engineering disciplines at the postsecondary education level or occupy managerial and parliamentary positions. These data and the written sources cited in the chapter so far highlight some changes, but they also show that progress is patchy. The diversity as well as commonalties in the experiences of women relative to men across the continent, as well as the partial progress toward attaining gender equity, can be illustrated through a focus on literacy.

The proportion of literate females in Sub-Saharan Africa relative to the male average has increased from 64% in 1990 to 77% in 1999, indicating a narrowing of the gap between males and females (UNDP, 2001). This progress in adult female literacy rates is a tendency that prevails across the continent, but there are notable regional variations. For example, countries like Guinea-Bissau (31%), Niger (34%), and Benin (43%) still have large gender disparities in adult female literacy. On the other hand, Chad, Sudan, and Mauritania have considerably narrowed the literacy gap between males and females since 1990. In some southern African countries, female literacy rates approach parity with male literacy rates (e.g., Swaziland, Namibia, and South Africa) or even exceed male literacy rates (e.g., Lesotho and Botswana). In spite of this progress, male–female gaps in educational attainment are still high in Guinea-Bissau, Niger, Burkina Faso, and Benin Republic (Figure 11.4).

As was discussed in Chapter 8, the Human Development Index (HDI) is an aggregate measure of average conditions, so it masks differences in human development—for example, between women and men. The United Nations now computes an additional index that captures gender inequalities: the Gender-Related Development Index (GDI). Like the HDI, the GDI measures life expectancy, educational attainment, and income, but it adjusts the results to incorporate gender inequality. In 1998, the GDI was 0.706 for the world and 0.916 for high-income countries, but only 0.459 for Sub-Saharan Africa. Within Sub-Saharan Africa, the GDI for 1999 ranged from relatively high values—0.695 for South Africa and 0.594 for Namibia—to low values of 0.26 for Niger and 0.3 for Burundi. (See Figure 11.5.) In 1997, South Africa created the Commission on Gender Equality to promote gender justice and equal access to economic, social, and political opportunities. Most

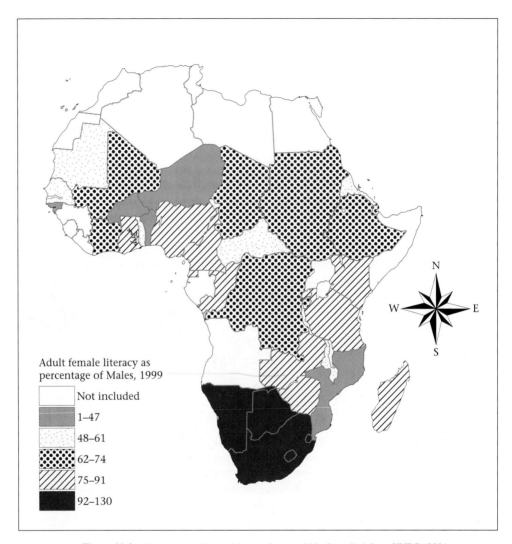

Figure 11.4 Gender Inequality in Literacy Rates, 1999. Compiled from UNDP (2001). *Human Development Report*, Oxford UK: Oxford University Press.

department offices in South Africa now have units called "Gender Focal Points" that are responsible for establishing internal gender policies to enhance gender equality. Other countries that have improved on the GDI include Botswana (0.571), Zimbabwe (0.548), Ghana (0.538), and Kenya (0.512). All four countries have made great strides in supporting the establishment of women's organizations. Kenya has embarked on a number of capacity-building programs to develop a pool of leaders who can coordinate economic literacy and advocacy programs. Ghana is now home to the Gender and Economic Reforms in Africa (GERA) secretariat, a consortium of African women's organizations and researchers that

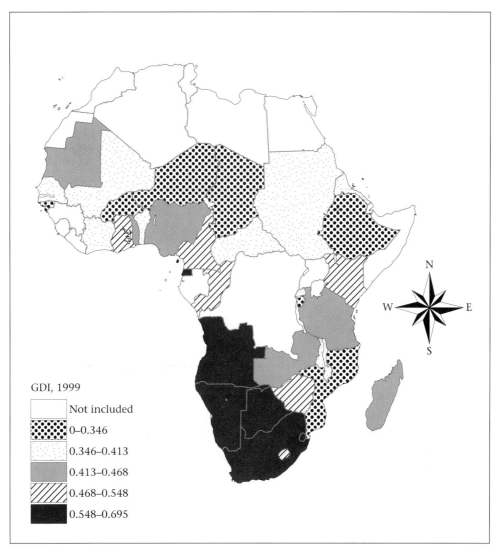

Figure 11.5 Gender-Related Development Index (GDI), 1999. *Source*: Compiled from United Nations Development Program (UNDP) 2001. *Human Development Report*, New York: Oxford University Press.

is geared towards promoting gender equality and justice and expanding the network of women's organizations (Delahanty and Sud, 2001).

Over all, Sub-Saharan African countries need to improve on the GDI; more than half of them have values less than 0.500, showing that women in these countries suffer a double deprivation: low overall achievement in human development and lower achievement compared with men. In a country like Niger, opportunities for women remain very constrained

(UNDP, 2001). In fact nine Sub-Saharan African countries rank at the very bottom of the GDI ranking worldwide, meaning that they have the widest gender gap in human development.

Another indicator —the Gender Empowerment Measure (GEM)—is used to measure the extent to which women participate in economic and political decision making. Unfortunately, current GEM data for Africa are scanty and therefore do not offer much of a basis for analysis and judgment.

The general conclusion is that African women's progress can at best be characterized as patchy and partial. There are also some contradictions. For example, one of the few banks in the world designed exclusively for women, the Kenya Women Finance Trust Bank in Nairobi, can ensure that women receive loans for financing credit-worthy projects, but it can be more difficult for illiterate counterparts to attain a desired ability to read. There are still many African countries where more than half the women are illiterate. How is it that female literacy in Africa still constitutes such a challenge, even as school enrollments for girls have increased faster than for boys? Part of the answer is the prejudice that persists that educating girls is a waste, but a larger part of the answer lies in the conditions of the 1980s and 1990s.

Structural-Adjustment Programs and African Women

For most of Sub-Saharan Africa, and for its women in particular, the 1980s will be remembered as the beginning of structural-adjustment programs (SAPs), the "sapping" years that represented growing hunger and falling standards of living. The effects dragged on into the turn of the century. Partly because of corrupt and inept leadership, ill-fated policies, declining export revenues, and the neglect of vital sectors in their economies (especially agriculture), Sub-Saharan African countries experienced acute economic crises. In an attempt to tackle the burden of foreign debt, African governments were asked by such international lending agencies as the World Bank and the International Monetary Fund to reduce public spending and to pursue structural-adjustment policies (as described in Chapter 8). Such austerity polices would supposedly make the national economies more efficient through such reforms as privatization, devaluing of the local currency, withdrawing of controls on external trade, decreasing of government controls on prices, and reducing of budget deficits.

Several countries implemented structural-adjustment programs, which have had serious negative effects on the poor and have had possibly their harshest effects on women. The International Monetary Fund (IMF) requirements promote maximizing of exports, but this emphasis has resulted in the lowest commodity prices in decades, because expanded production glutted the export market for much of Africa's raw materials, including agricultural products. Consider the irony that, just as countries were devaluing their currencies to provide farmers with expanded markets for their export crops, the price of imported inputs and equipment needed to increase production also rose, making them unaffordable. The price of a tractor is far beyond the reach of a majority of small-scale farmers, who are predominantly women. In spite of the diversity of debt circumstance or government response, the impacts of SAPs on African women show little variation from country to country. Rather, countries are more alike in the widespread retrogressive impacts of economic adjustment policies on women. Case studies in the Democratic Republic of Congo, Malawi, Cameroon, Nigeria, Ghana, and Zambia are unanimous in documenting the plight of African women under economic restructuring as poverty has become feminized.

The feminization of African poverty is evident in the fact that nearly two-thirds of Africa's fast-growing population of the poverty stricken consists of women. Women suffer the most from food shortages, high and rising commodity prices, heavier work loads, and the falling incomes of male family members (Thomas-Emeagwali, 1995). Malnutrition and chronic hunger increased; pregnant women die of anemia; and kwashiorkor, which used to be thought of as a wartime sickness, now threatens homes in times of peace (Eboh, 1990). Another IMF and World Bank recommendation that has had an impact on women is the removal of subsidies on public services such as health and education, which now run on a cost-recovery basis. With the removal of subsidies, the increased income from the sale of farm produce for peasant women farmers dissipates rapidly, because the money can scarcely pay for transportation, health, and educational expenses. Relatively well-off urban women have benefited from opportunities associated with cross-border trading (Peberdy, 2000), but the weight of evidence suggests that the burden of globalization and debt crises falls doubly hard on poor women working in the informal sector (Doane and Johnston-Anumonwo, 2001). Particularly alarming is the fact that there are instances where maternal mortality rates are starting to rise, and school enrollment rates for girls are beginning to fall. In Malawi, Niger, Gabon, Benin, Togo, and Guinea, there was a smaller proportion of women university students in the early 1990s than in the early 1980s (Seager, 1997). One devastating impact of the AIDS epidemic is that infected girls are more likely to drop out of school, a reversing of decades of slow, but steady progress in female education. Such reversals amount to nothing less than the underdevelopment of the African human-resource base. Under these conditions, have women just been passive survivors? How do African women respond to change and crisis? The answer is rooted in African women's social and political participation in past struggles against precolonial and colonial domination and in their continuing resistance to other forms of postcolonial oppression.

AFRICAN WOMEN AS AGENTS OF CHANGE

African societies are characterized by a long history of female solidarity organizations, which remain extensive and influential. All across the continent, there are different forms of formal and informal support networks among women that build on and contribute to women's consciousness about their conditions. These range from community-development organizations to farm-labor groups, resource-conservation movements, market women's associations, professional organizations, religious or social clubs, secret societies, and rotating credit clubs. Equally as diverse as the goals of these associations are the gains. For example, there are a host of female-led initiatives centered on geopolitical concerns of autonomy, representation, and territorial integrity.

Geopolitics and Women's Initiatives

In addition to well-known female political leaders in the history of Ghanaian, Egyptian, and Nigerian societies, to name a few examples, there were several other female-led resistance movements during colonialism that rebelled against different forms of exploitative rule, such as unfair taxation and land alienation. A well-known example is the Igbo women's riots. Other revolutionary movements in Guinea-Bissau, Mozambique, Angola, Zimbabwe, and Eritrea had strong female involvement. The successes of many of these political acts of resistance are due to the determination and sacrifices of women at several levels. Especially with the historic collapse of apartheid, the organization efforts of South African women in the multiple struggles against racism and sexism stand out as one hallmark of the unwavering strength of African women's solidarity. Clearly, most of these resistance struggles by organized self-help groups attempting to gain control over resources from governments or from men are contested not at high-profile levels, but at the grassroots and household levels.

In contemporary times, rural women's complaints of men's mismanagement of land and money produce varying levels of redress from male-dominated community leadership. In Kenya, for example, corporate power and the capitalist imperative was so strong that, despite the reluctantly granted concession for use of family plots to cultivate maize, the tractors were back the following year, ploughing over women's maize fields. The increased participation of women in leadership training programs has often led to more successful results. In Zimbabwe, for instance rural women have gained more rights, including the right to have their own bank accounts. There are also signs of change at the government level; most African governments are now committed (on paper) to full equality for women. By 1996, all African countries with the exception of Djibouti, Mauritania, Mozambique, Somalia, Sudan, and Swaziland had both signed and ratified the UN Convention on the elimination of all forces of discrimination against women—CEDAW. Many women's organizations now receive state support and sponsorship, and several African countries have established women's bureaus. Uganda and Nigeria established such programs as Rural Women's Credit Program, Better Life for Rural Women, and Peoples' Bank, which offer loans to women's cooperatives. Significant hallmarks in African women's leadership in national economic and political arenas are their representation in the activities of the South African Women's Budget Initiative and the Uganda Women's Caucus (UNDP, 2000). Furthermore, with or without state support, African women continue to act as agents of change in restoring the African environment.

Women's Role in Restoring the African Environment

Several studies by geographers emphasize that African women are not passive in the face of changes that negatively affect them and that they are themselves dynamic agents of change throughout the continent (Carney, 1993; Schroeder, 1993). Kenya's well-known self-organized women's groups (now numbering several thousands) have become agents of comprehensive rural development through their participation in farm and tree-planting activities

African Women as Agents of Change

(Hyma & Nyamwange, 1993). Wangari Maathai has emerged as an internationally recognized female advocate championing the Green Belt Movement in Kenya. It is not surprising that African women are increasingly active in environmental conservation programs (e.g., Rocheleau, *et al.*, 1996), because they are much more likely than men to be the ones whose responsibility is to obtain firewood for domestic fuel. Therefore, women have a greater interest in the long-term productivity of food and fuel resources (Figure 11.6).

In sum, although African women have been marginalized in various ways by economic development, they have not been marginal in generating development and progressive change in their countries. They continue to respond to challenges of change and crisis in their lives, from control over sustenance to control over sexuality. Some women-led and women-centered groups in Sub-Saharan Africa often ignore equity issues regarding redistribution of wealth and issues of specific interest to women, but there are important exceptions. For example, a newly organized women's group, Women in Nigeria (WIN), has been very outspoken in standing up for women's rights. Other professional groups with continentwide membership are involved in tasks that document and inform the activities of such women advocacy groups as the African Women Organization for Research and Development (AWORD). Given increasing female political representation in government and in educational institutions, there is hope that women's concerns will receive due attention in policy and scholarship.

Figure 11.6 Women's groups work on soil conservation and land reclamation in Kenya.

ENGENDERING THE GEOGRAPHIC STUDY OF SUB-SAHARAN AFRICA

This chapter has attempted to depict African women's lives as ones in which, in addition to becoming more visible in modern professions and political positions, they are the primary food producers and homemakers. Furthermore, they participate extensively in the marketing and distribution of household produce and are often responsible for the early training of their children—but they do not benefit equitably. Given this reality, the contribution of African women to the economic and social development of their countries cannot be overemphasized, and the attainment of equity for women in Sub-Saharan Africa will continue to be an important development issue.

Gender and Development: Rethinking the Connections

There is some basis for optimism that gender-sensitive programs and research are increasing in Sub-Saharan Africa. Since the 1970s, which witnessed the UN Decade of Women in 1975 and a rethinking of growth-first development models, development experts who advocate "growth with equity" have emphasized the need to develop a country's human-resource base, including the female half of the population. The UN Decade of Women enabled women, the neglected, but vital resource, to become more visible, and attention was drawn to the work and status of African and other women in less developed countries. There were incipient attempts to recognize and correct the tendency for development projects to exclude or harm women. Much of the effort focused on two issues, education and employment, or, in other words, on economic production. The call was to integrate women into development, yet the redirecting of resources to women was disappointingly very slow. For example, of all the money spent on agriculture by the United Nations in 1982, projects for women farmers received less than 1%.

The early 1980s represented the period of reconceptualizing women in development. The previous emphasis on integrating women into African development had missed a number of points. First, calls for increasing productivity ignored African women's existing high levels of productive participation in their societies' agricultural and nonagricultural sectors; second, the emphasis on production neglected social reproduction. Failure to consider social reproduction meant ignoring the lopsided gender-based division of labor within the home. Yet, the division of labor in the home affects women's work outside the home. Furthermore, the tendency to favor the expanding of women's waged income through commercial agriculture or education did not sufficiently question the inferior economic returns, the irrelevant forms of training that African women and girls were receiving, or the nature of sex segregation in the work force. For example, it is not uncommon for agricultural extension workers to be male and to speak to a gathering that is almost exclusively male. Indeed, if women's participation in the waged-labor force increases, but their subordinate status in the home is unchanged, the nature of women's economic independence is questionable.

The relevant question, then, is, into what kind of development should African women be integrated? Assumptions that the kind of capitalist development that African countries are experiencing is the kind of development in which women should expand their participation overlooks the potential drawbacks of that type of development on women. One should question the type of development that is being advocated. Integrating Sub-Saharan women into dependent development is not in their interest.

One must be skeptical of a capitalistic tendency to view African women as an "untapped resource" and an ethnocentric attitude that sees African women just as subjects of development. The image that African women are victims of development is also one sided, and they should not be thought of only as exhausted mothers of malnourished children who are in dire need of development aid. Women in Sub-Saharan Africa should be recognized in their true roles as active contributors to and sustainers of development and as agents of change.

African Development: Recognizing and Empowering Women

Nowhere is the expression "Third-World Second Sex" truer today than for African women. Yet the incredibly subordinate status of African women and the crisis state of African agriculture can be reversed only by tackling the twin problems of neocolonialism and patriarchy simultaneously, not one before the other. The hope for African development in general, and for expanded African agriculture in particular, involves an equalizing of the gender division of labor and an abandoning of dependent-development policies. In 1985, women from across the world joined African women in Nairobi, Kenya at the conference commemorating the end of the Decade for Women and emphasized that economic development policies and patriarchy work hand in hand in causing both the African woman's disadvantaged status and the inability of African economies to meet their food requirements domestically.

The scholarly evidence, including studies on the geography of gender in Sub-Saharan Africa (e.g., Sowden, 1990; Barrett, 1995), indicate that multiple forms of deprivation and inequity stifle expanded agricultural productivity and human welfare. There is now widespread agreement that women must be at the center of development policies and planning. The belief of one African (woman) farmer that "independence for a farmer means having good land and the means to work the land" is a good analogy to what independent development for Sub-Saharan Africa should entail. This can be interpreted to mean having access to and control of a productive resource and the autonomy to use the returns for an improved quality of life. Such an opportunity should be open to all Africans, regardless of gender, to ensure a more positive future.

With their long-tested productive energies, their numerical strength in food production and food handling, and their tenacity and commitment even in times of crisis, women offer Sub-Saharan Africa a major opportunity to reorganize its agricultural policies and related programs to ensure long-term self-reliant development. Accordingly, fundamental rethinking and action must be undertaken to eliminate the marginalization of women and the inequalities they endure, especially in the neglected rural domain. With commitment and vision the development opportunity offered by its women can become Africa's reality (Aidoo, 1988). This observation remains true today.

A Gendered Agenda for Future Geographic Research

The extent to which future geographic research in Sub-Saharan Africa will inform the question of gender and contribute to the attainment of equality for women will depend on a number of conceptual and methodological initiatives. Geographers and other scholars using geographic metaphors have emphasized that employing conventional tools of analysis that are nongendered to examine human geography leads us to map a landscape peopled almost exclusively by men, thus excluding half the humans in human geography and ignoring the world of women. This chapter has shown that African women's conditions and status as a "social periphery" should be understood as socially created; the situation is neither natural nor traditional. Therefore, a nongendered perspective for studying the patterns and processes of African conditions not only provides an incomplete understanding of Africa's human geography, it constitutes a disservice to efforts toward the generating of an informed basis for African women's emancipation.

Centers for women's studies have been established in at least four locations in Africa: Tanzania, Nigeria, Uganda, and Botswana (Awe & Mba, 1991); others exist in South Africa and Ghana. It is very important to continue the collecting and presenting of basic descriptive data, because these are often absent for Africa, but, if the results of research are to yield substantive changes in women's lives, it is more important that the data collected include policy-relevant variables. The data and methodology for undertaking geographic research on gender issues in Sub-Saharan Africa should be varied. For instance, government records need to be supplemented with surveys, fieldwork, interviews, narratives, and oral sources. Some geographers have demonstrated the value of alternative techniques.

Geographic work on gender issues in Sub-Saharan Africa needs to be expanded, but it is especially urgent that the purpose of future geographical inquiries in the region focus on making such studies relevant for the African woman's attainment of equity. As African women scholars continue to emphasize, research about Africa must be on issues that women themselves identify as relevant and critical to their well-being. Problems of nutrition, infant mortality, illiteracy, health-care delivery, and skill training are of central importance in women's lives, and African feminists clearly recognize that economic exploitation remains a primary force in their oppression. In this sense, the improvement of social and spatial accessibility for rural and urban women's lives should be an important research agenda for geographers.

In addition, power relations in the region, from the household to the international level, heavily undermine the productive and reproductive experiences of women. By recognizing that, in the experiences of African women, production versus reproduction and public versus private are not opposite categories, geographers can avoid simplistic terminology and concepts that currently exist in the discipline in their study of gender issues in Sub-Saharan Africa. Finally, despite the common oppression that African women face, geographers must increasingly include variables like class, ethnicity, and age in their analyses, in addition to their usual recognition of place differences, since these socioeconomic and demographic attributes shape African women's gendered experiences differently.

Much of the geographic research on Sub-Saharan Africa is still at the stage of describing gender roles, and little work exists on explaining power differentials between men and women (i.e., gender relations). Feminist geographers point out that, if geographic research is to be instrumental in alleviating gender inequalities, the focus of analysis has to be on understanding power relations in different arenas, from the household level right up to the international level. As the geographic study of gender in Sub-Saharan Africa expands, feminist geographers need to devote more attention to articulating the evidence from several empirical case studies to formulate theories of African women's changing circumstances. The best kinds of conceptual scholarship would of necessity be holistic in perspective, in the sense that international economic systemic changes are identified as paramount in the ways they interact with African structural situations to marginalize women and affect their coping capacities. The challenge for African women is to make sure that they continue to strive for, gain, and retain power, authority, and control of productive resources in their homes and societies. The right kind of education and the right kind of income-generating activities will lead to this. Examples abound showing that African women are capable of improving their opportunities and social position when they take control of their own lives. The challenge for Africans, women and men, is to unite as one against any leadership that espouses dependent development. Independent and sustainable development in Sub-Saharan Africa cannot occur without women's independence. As Africans unite to ensure a more productive and equitable existence, perhaps the next and greatest challenge of all is for women and men in the industrialized world to support all Africans in their struggle for autonomous and sustainable development.

KEY TERMS

Gender relations
Gender inequality
Feminization of poverty
Gendered spaces

Women's bureau
Social reproduction
Gender-Related Development Index
Gender Empowerment Measure

DISCUSSION QUESTIONS

1. In what ways has development adversely affected African women?
2. What factors account for geographical variations in Sub-Saharan Africa's sex ratio?
3. How does a gendered perspective improve our understanding of the factors contributing to food shortages in Sub-Saharan Africa?
4. To what extent is the food crisis in Sub-Saharan Africa human induced?
5. Illustrate, with examples, how patriarchy and capitalism hinder the advancement and mobility of African woman.
6. Is the image of African women as passive helpless victims one sided?
7. Discuss the notion that development in Sub-Saharan Africa cannot occur without the independence of African women.
8. How have African women been active in combating environmental degradation?

9. Produce two maps for Africa, one on the geography of women and one on the geography of gender. Describe the patterns, noting similarities and deviations between places where women's conditions are best and places where women's conditions relative to men's are worst.
10. What kinds of policy-sensitive research might geographers undertake to improve the welfare of African women?

REFERENCES

AIDOO, A. (1988). "Women and Food Security: The Opportunity for Africa," *Development*, 2, 3:56–62.

ARDAYFIO-SCHANDORF, E. (1993). "Household Energy Supply and Women's Work in Ghana," in *Different Places, Different Voices: Gender and Development in Africa, Asia and Latin America*, Momsen, J. & Kinnaird, V., eds., New York: Routledge.

AWE, B. & MBA, N. (1991). "Women, family, state, and economy in Africa," *Signs: Special Issue Editors*, 16(4):848–9.

BARRETT, H. (1995). "Women in Africa: The Neglected Dimension," *Geography*, 80:215–224.

BOSERUP, E. (1970). *Women's Role in Economic Development*, London and Baltimore: Allen and Unwin.

BROWNE, A. & BARRETT, H. (1993). "The Impact of Labour Saving Devices on the Lives of Rural African Women: Grain Mills in the Gambia," in *Different Places, Different Voices: Gender and Development in Africa, Asia and Latin America*, Momsen, J. & Kinnaird, V., eds., New York: Routledge.

BRYDON, L. & CHANT, S. (1989). *Women in the Third World: Gender Issues in Rural and Urban Areas*, New Brunswick, NJ: Rutgers University Press.

CARNEY, J. (1993). "Converting the Wetlands, Engendering the Environment: The Intersection of Gender with Agrarian Change in The Gambia," *Economic Geography*, 69(4):329–348.

CARNEY, J. & WATTS, M. (1991). "Disciplining Women? Rice, Mechanization, and the Evolution of Mandinka Gender Relations in Senegambia," *Signs*, 16:651–81.

DELAHANTY, J. & SUD, G. (2001). *Toward Participatory Economic Reforms in Africa: A Quest for Women's Economic Empowerment*, Ottawa, Canada: The North-South Institute.

DOANE, D.L. & JOHNSTON-ANUMONWO, I. (2001) "Contending Analyses of African Women in the Informal Sector and Responses to Debt Crises," in *Issues in Africa and the African Diaspora in the 21st Century*, by Asumah, S.N. & Johnston-Anumonwo, eds., Binghamton: Global Publications at Binghamton University.

EBOH, M. (1990). *The Impact of the Debt Crisis on Nigerian Women*, International Geographical Union, Study Group on Gender and Geography, working paper, 15.

GUGLER, J. (1989). "Women Stay on the Farm No More: Changing Patterns of Rural-Urban Migration in Sub-Saharan Africa," *The Journal of Modern African Studies*, 27:347–52.

HALL, E. (1990). "Vocational Training for Women Refugees in Africa," *International Labor Review*, 1:91–107.

HOUSE-MIDAMBA, B. & EKECHI, F.K., eds. (1995). *African Market Women and Economic Power: The Role of Women in African Economic Development*, Westport, CT: Greenwood Press.

HYMA, B. & NYAMWANGE, P. (1993). "Women's Role and Participation in Farm and Community Tree-Growing Activities in Kiambu District, Kenya," in *Different Places, Different Voices: Gender and Development in Africa, Asia and Latin America*, Momsen, J. & Kinnaird, V., eds., New York: Routledge.

KIRK, D. (1998). "Fertility Levels, Trends, and Differentials in Sub-Saharan Africa in the 1980s and 1990s," *Studies in Family Planning*, 29:1–22.

MCCALL, M. (1987). "Carrying Heavier Burdens but Carrying Less Weight: Some Implications of Villagization for Women in Tanzania," in *Geography of Gender in the Third World*, Momsen, J. & Townsend, J., eds., London: Hutchinson.

MOMSEN, J. & KINNAIRD, V., eds. (1993). *Different Places, Different Voices: Gender and Development in Africa, Asia and Latin America*, New York: Routledge.

MOMSEN, J. & TOWNSEND, J., eds. (1987). *Geography of Gender in the Third World*, London: Hutchinson.

MWAKA, M. (1993). "Agricultural Production and Women's Time Budgets in Uganda," in *Geography and Gender in the Third World*, Momsen J. & Townsend, J., eds., London: Hutchinson.

NEFT, N. & LEVINE, A.D. (1997). *Where Women Stand: An International Report on the Status of Women in 140 countries, 1997–1998*, New York: Random House.

PAUL, B. (1993). "Maternal Mortality in Africa, 1980–87," *Social Science and Medicine*, 37(6):745–52.

PEBERDY, S. (2000). "Border Crossing: Small Entrepreneurs and Cross-Border Trade between South Africa and Mozambique," *Tijdschrift voor Economische en Sociale Geografie. Journal of Economic and Social Geography*, 91:361–378.

POTTS, D. (1995). "Shall We Go Home? Increasing Urban Poverty and Migration Processes," *The Geographical Journal*, 161:245–264.

ROCHELEAU, D., THOMAS-SLAYTER, B., & WANGARI, E., eds. (1996). *Feminist Political Ecology: Global Issues and Local Experiences*, New York: Routledge.

RODNEY, W. (1981). *How Europe Underdeveloped Africa*, Washington, DC: Howard University Press.

SAMATAR, A. (1993). "Structural Adjustment as Development Strategy? Bananas, Boom, and Poverty in Somalia," *Economic Geography*, 69(1):25–43.

SCHROEDER, R. (1993). "Shady Practice: Gender and the Political Economy of Resource Stabilization in Gambian Gardens/Orchards," *Economic Geography*, 69(4):349–365.

SEAGER, J. (1997). *The State of Women in the World Atlas*, London: Penguin Books.

SOWDEN, C. (1990). "The Statistics of Shame," *The Geographical Magazine*, 62:4–7.

SPIRO, H.M. (1987). "Women Farmers and Traders in Oyo State, Nigeria: A Case Study of Their Changing Roles," in *Geography of Gender in the Third World*, Momsen, J. & Townsend, J., eds., London: Hutchinson.

THOMAS-EMEAGWALI, G. (1995). *Women Pay the Price: Structural Adjustment in Africa and the Caribbean*, Trenton, NJ: Africa World Press, Inc.

TOPOUZIS, D. (1990). "The Feminization of Poverty," *Africa Report*, July/August.

UNITED NATIONS DEVELOPMENT PROGRAM (UNDP) (2001). *Human Development Report*, New York: Oxford University Press.

UNITED NATIONS DEVELOPMENT PROGRAM (UNDP) (2000). *Women's Political Participation and Good Governance: 21st Century Challenges,* New York: UNDP.

WATTS, S. (1983). "Marriage Migration, a Neglected Form of Long-Term Mobility: A Case Study from Ilorin, Nigeria," *International Migration Review*, 17:682–98.

WILKINSON, C. (1987). "Women, Migration and Work in Lesotho," in *Geography of Gender in the Third World*, Momsen, J. & Townsend, J., eds., London: Hutchinson.

YEBOAH, I. (1998). "Geography of Gender Economic Status in Urban Sub-Saharan Africa: Ghana, 1960–1984," *Canadian Geographer*, 42:158–173.

12

Medical Geography of Sub-Saharan Africa

Joseph R. Oppong

INTRODUCTION

Medical geography analyzes the spatial variations of such health-related issues as disease ecology and health-care services, and it seeks to explain *who* gets *what, where,* and *why*. An integrative subdiscipline, medical geography applies geographic concepts and techniques to health-related issues, drawing on ideas and techniques from other disciplines—including epidemiology, sociology, economics, anthropology, psychology, zoology, entomology, botany, parasitology, and health administration. *Disease ecology* studies how human behavior, in its cultural and socioeconomic context, interacts with environmental conditions to cause morbidity (the condition of illness) and mortality (Meade and Earickson, 2000). Tracing the linkages between *disease agents* (the pathogen, germ, or parasite—a disease's causative organism), *disease vectors* (transmitters of diseases, such as mosquitoes, ticks, and fleas), and their *hosts* (the organism infected by a disease agent) is accomplished through a variety of methods, including remote sensing and geographic information systems (GIS) (Gatrell and Löytönen, 1998).

Human ecology and vulnerability are two conceptual frameworks commonly used in medical geography research. The human ecology model views disease as resulting from the interaction of three major variables: genetics, environment, and cultural practices. For example, the group of diseases known collectively as cancer are caused by inherited genetic factors, environmental exposure (chemicals, radiation, and viruses), and human behavior

ENDEMIC DISEASES OF SUB-SAHARAN AFRICA

AIDS—The African Catastrophe

Sub-Saharan Africa has 83% of the world's total AIDS deaths and 87% of its HIV-infected children; nowhere else has the impact of HIV–AIDS been more severe. The 21 countries with the highest HIV prevalence in December 2000 were in Africa. Botswana, with an adult HIV rate of 35.8% was the country worst affected; it was followed by Zimbabwe, with 25.06% (Figure 12.3). In Namibia, Zambia, and South Africa, one in four adults were HIV positive.

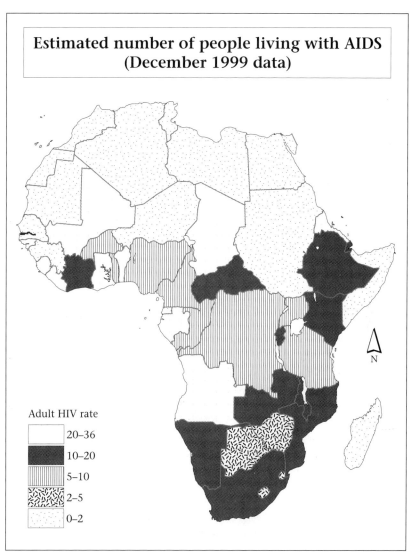

Figure 12.3 Traditional healers are a major source of health care for rural residents. Residents travel several miles and sometimes wait long hours to meet with traditional healers.

In more than 10 African countries, the adult prevalence rate exceeded 10% (UNAIDS and ECA, 2000). Nevertheless, the devastation of HIV–AIDS is not uniform throughout the region. Scattered over the continent, even in countries of relatively low HIV prevalence, are districts, villages, or ethnic groups coping with high levels of infection that are several times the national average. Compared with eastern and southern Africa, west Africa has suffered less, but the situation is changing rapidly. Côte d'Ivoire (10.6%) is among the 15 countries in the world worst affected; in Nigeria, by far the most populous country in Sub-Saharan Africa, about 5% of adults aged between 15 and 49 were living with HIV in December 2000. By killing adults in the prime of their working and parenting life, HIV–AIDS slaughters the workforce, impoverishes families, and orphans millions. Clearly, HIV–AIDS poses a major development crisis.

Heterosexual transmission predominates, but Africa's HIV pandemic comprises miniepidemics with different characteristics in prevalence trends, in those affected, and in the opportunistic diseases observed. Urban areas seem to have been hit the hardest—although, in some places, this conclusion might be an artifact of the data-collection process, since testing facilities are seldom available in rural areas. Rural HIV often remains silent and invisible, because of poor health infrastructure, restricted access to health facilities, and inadequate testing and surveillance. It is fueled by poverty, food insecurity, gender inequality, migration, war, and civil conflict and is accelerated by migration, trade, refugee movements, and strengthened rural–urban linkages (Kalipeni and Oppong, 1998).

The geography of the disease is changing. Initially, it was especially severe in east and central African cities; today, southern Africa is the heart of the epidemic. Why west Africa has such low HIV–AIDS rates compared with eastern and southern Africa is not fully understood. An often-cited reason is the dominance of a different strain of the virus, HIV-2, which does not transmit as easily as HIV-1, the virus responsible for AIDS in the rest of Africa. Moreover, infection with HIV-2 is less likely to cause AIDS. In addition, the HIV genetic subtypes that predominate in Africa, A and C, are more aggressive and easier to transmit than the B subtype, which predominates in Europe and North America. West Africa, as a whole, has an unusually high number of reported HIV-2 cases compared with HIV-1, and Senegal and Guinea-Bissau have consistently higher HIV-2 prevalence rates, yet Ghana, Nigeria, and Côte d'Ivoire have a higher occurrence of HIV-1 cases.

Commercial sex, sexually transmitted diseases, condom neglect, and the lack of circumcision are critical factors that fuel the HIV epidemic. A cultural context where men see their virility as compromised by using condoms and where women are reluctant to insist on condom use (on account of their position in society) promotes HIV spread. Unfortunately, simply telling people to use a condom in this context has little effect, because it ignores the context of masculine and feminine sexual identities. Sexually transmitted diseases are common. In some countries, such as Côte d'Ivoire, commercial sex work is widespread, and most female commercial sex workers are HIV positive. In others, economic need compels many women to provide sexual services in exchange for material and monetary favors. Unnecessary and excessive blood transfusion of poorly screened blood compounds the problem further. Truck drivers are major HIV carriers; they engage in paid sex with local women, making every truck stop a potential regional epicenter. Nowhere is this more visible than in east Africa, where the truck route from Malawi to Durban, South Africa became known as the "Highway of Death." During the late 1990s, an estimated 92% of truck drivers visiting Durban along that route were HIV positive. The military and police are also major sources of HIV spread. In South Africa, Cameroon, and Zimbabwe the HIV infection rate among the military is 3–4 times higher than among civilians. Likewise, HIV among peacekeepers returning from Liberia and Sierra Leone was several times higher than the rate among the general population in these countries.

Already overwhelmed by the numerous tropical diseases, the health system cannot cope with this new and massive onslaught of AIDS cases. In 1997, AIDS patients occupied 41% of all hospital beds in Abidjan, and AIDS-related costs absorbed 11% of the total public health budget (UNAIDS and ECA, 2000). Treatment costs for HIV–AIDS are expected to consume more than 60% of government health expenditure by 2005 in Zimbabwe. In the widespread absence of adequate blood-screening equipment, unhygienic medical practices, including lack of sterilization of equipment, simply put recipients of health services at increasing risk. With these problems, the medical system itself may be an important source of infection (Figure 12.4). In the developed world, antiretrovirals are delaying the development of full-blown AIDS, extending the lives of people living with AIDS, and preventing vertical transmission from mother to newborn children, but, in many African countries, these medications remain, sadly, only a dream—expensive beyond hope. In South Africa, President Thabo Mbeki has questioned the causal link between HIV and AIDS and the efficacy of antiretroviral drugs such as AZT.

Figure 12.4 Problems with primary health care. Short of needles, a primary health-care worker boils her needles to sterilize them for reuse. Multiple usage of injection needles without sterilization may be an important source of the spread of hepatitis and HIV–AIDS.

Efforts to produce generic versions of AIDS drugs are underway in some countries. Also, a few countries, such as Uganda and Zambia, have been fairly successful in reversing the AIDS epidemic. (See Discussion Box 12.1.) However, by and large, the AIDS drum beats on, unhindered, despite the occurrence of localized efforts—for example, the Nyumbani orphanage in Nairobi cares for about 75 HIV-positive children orphaned by the pandemic.

Structural adjustment and HIV–AIDS Structural-adjustment programs might be contributing to the spread of HIV infection in African countries. According to UNAIDS (2000), effective prevention measures include access to condoms, prophylaxis and treatment of opportunistic infections, including sexually transmitted diseases (STDs), sex education at school and beyond, access to voluntary counseling and testing, counseling and support for pregnant women, efforts to prevent mother-to-child transmission, and access to safe drug-injecting equipment. Unfortunately, economic restructuring makes it extremely difficult to accomplish these goals. First, government expenditure reductions mean reduced funding for diagnosis and treatment of STDs and for blood screening, thus producing poor hygienic practices in clinics (e.g., inadequate sterilization of equipment), which can also spread HIV. Second, structural-adjustment programs intensify economic

---

DISCUSSION BOX 12.1:
AIDS SUCCESS STORIES

- Uganda remains the only African country to have reversed its epidemic. Its extraordinary effort of national mobilization pushed the adult HIV prevalence rate down from around 14% in the early 1990s to 8% in 2000. Even rural areas, which are frequently among the last to evidence signs of either the advent or the reversal of an HIV–AIDS epidemic, have shown a reduction in HIV rates. In some areas of rural Uganda, for example, HIV infection rates among teenage girls dropped to 1.4% in 1996–97, from 4.4% in 1989–90. This was matched by a fall in teen pregnancies.
- In 1993, HIV rates among young women in Lusaka, Zambia exceeded 25%, but they have been almost halved in just six years by effective HIV prevention. Premarital sex is losing popularity: Only 35% of young women in Lusaka reported premarital sex in 1996, compared with 52% in 1990. Male sexual abstinence has also risen. In 1998, more than half of young unmarried men reported no sex in the past year compared with just over a third two years earlier. Additionally, the frequency of casual sex is decreasing, as shown by a fall in the proportion of men reporting two or more casual partners in the past year.

Source: http://www.unaids.org/fact_sheets/files/successes_eng.doc.

recessions, increasing inequality and poverty rates. In Ghana, for example, the proportion of the population classified as poor increased from 43% in 1981 to 54% in 1986 and to 55% in 1997 (World Bank, 1997). Poor people engage in more risky behavior and lack access to health-care resources, such as STD control, that could lower their risk of contracting HIV. Finally, erosion of women's real incomes and increasing poverty, typical (and arguably temporary) results of economic reform, intensify gender inequality, weaken women's ability to negotiate sexual relations and practices, and thus increase their vulnerability. Increased rural–urban spatial interaction under economic restructuring might enhance the spread of HIV.

Can Nigeria avoid a major AIDS explosion?   Years of neglect and of a lack of committed political leadership in Nigeria have resulted in a major HIV–AIDS explosion. Behind the official 2001 prevalence estimate (5.4% of the adult population, or 2.6 million people) are some sobering statistics: An estimated 4.9 million Nigerian adults will be carrying the AIDS virus by the year 2003 (Akinsete, 2000; Oyo, 1999). Women in their twenties have the highest rate of HIV infection. Prevalence in pregnant women ranged from 0.5% in the northeastern state of Yobe to 21% in Otukpo, a town in the north central state of Benue, with little difference between urban and rural areas. Benue, where HIV prevalence rose from 2.3% in 1995 to 16.8% in 1999, is the state worst affected (Akinsete, 2000). HIV is prevalent in all six geopolitical zones; within every zone, hotspots exist where rates are increasing. HIV prevalence in the 20–24 age group ranges from 4.2% in the southwest zone to 9.7% in the north central zone, which includes Abuja, the national capital. Among young adults aged 15–19, HIV prevalence ranges from 2.8% in the northeast zone to 8.4% in the north central zone. Hotspots include Akwa Ibom, Benue, Ebony, Kaduna, Lagos, and Tariba. High mobility among Nigeria's population is expected to spread the high prevalence now localized within hotspots to other areas rapidly (Falobi, 1999).

The threat of a major HIV explosion in Nigeria is very real for several reasons. There is very little knowledge about the disease. Female adolescents are more afraid of unwanted pregnancy than of getting AIDS (Ogara, *et al.*, 1998). Increasing poverty and high rates of inflation (in 1996, the Naira was at only 6% of its 1985 value) has caused the number of commercial sex workers and of people who exchange sex for material benefits to rise, particularly at oil-exploration locations (Faleyimu, *et al.*, 1999). Moreover, a high level of premarital and extramarital sexual activity exists, usually occasioned by the need for material or economic assistance, particularly among the younger wives in polygamous marriages. In addition, sex education is still opposed in parts of the country, and stigmatization of, and discrimination against, people with AIDS is widespread (Akinsete, 2000). Untreated sexually transmitted diseases, a major factor in the spread of HIV, is common (Decosas, 1996). A community survey of genital-tract infection found that, among 158 girls aged 17–19 years in a rural area, 44% had a current genital-tract infection, yet less than 3% had ever sought treatment (Brabin, *et al.*, 1994). Besides, young girls who have little education, vocational training, or capital often have few survival choices but to have sex with older men (Decosas, 1996).

Until Obasanjo became president, Nigeria lacked the political leadership necessary to confront the AIDS crisis. The first National Conference on AIDS in Nigeria was held in

Abuja, between December 15 and 18, 1998, even though AIDS was first identified in the country in 1985. Previously, government activities had been restricted to the severely underfunded National AIDS/STDs Control Program established in 1987. The three-day conference was attended by many participants from government agencies and institutions, but, unfortunately, the head of state, General Abdulsalami Abubakar, was absent and also neglected to send a personal representative. Besides the gap in leadership, the conference revealed that a significant percentage of blood and blood products were transfused without screening in almost all hospitals nationwide (Akinsete, 2000; Chikwem, *et al.*, 1997; Odujinrin and Adegoke, 1995).

Effective political leadership in Nigeria is critical to stem further spread and devastation. President Obasanjo's leadership is promising so far. Supporting this commitment with resources to facilitate research and programs could enable Nigeria to postpone or prevent a major disaster. The supplementary budget allocation for AIDS control for 1999, which exceeded the total budget for AIDS from 1996–1998 (UNAIDS, 2000), is a good beginning, but more is required. Without major radical programs, Nigeria faces a major epidemic with widespread disastrous consequences.

Other Diseases

Malaria Malaria is the most serious disease in Africa transmitted by a vector—in this case, the mosquito. It kills more than 900,000 people each year in tropical Africa, is the major cause of infant and juvenile mortality, and is also responsible for many miscarriages and for many babies born underweight. According to the 1986 publication of the World Health Organization (WHO), the total "cost" of malaria—for health-care treatment, lost production, and so on—was about $800[1] million for tropical Africa, and the figure was expected to exceed $1.8 billion in 1995. The *Anopheles gambiae* mosquito is the vector of the parasite, and the disease manifests itself as a recurrent fever and anemia with chills. Victims not only are deprived of energy, but also become increasingly vulnerable to other diseases.

Malaria has been on the rise in recent years and growing resistance to various drugs used in controlling it, such as chloroquine, is becoming widespread. Sub-Saharan Africa has the highest levels of endemism in the world. Transmission is intense and perennial in forest or savannah regions at altitudes up to 1000 m (3,280 ft), having an average rainfall over 200 cm (80 inches) per year (Figure 12.5). At altitudes over 1500 m (4,921 ft) with rainfall below 100 cm (40 inches) per year, endemism decreases, leaving room for periodic epidemic outbreaks. Ecological and meteorological factors, including the cycle of heavy rains, have led to serious outbreaks in Botswana, Burundi, Ethiopia, Kenya, Madagascar, Rwanda, Sudan, Swaziland, Democratic Republic of Congo, and Zambia.

Eradication efforts worldwide have not been very successful. Massive campaigns in Sri Lanka and India seemed to eliminate the mosquito, at least for a while, but, lately, malaria cases have been reported again in these countries. In 1978 alone, India reported 60 million cases of malaria. Chloroquine-resistant malaria strains are becoming widespread in

---

[1] Monetary unit is in US dollars.

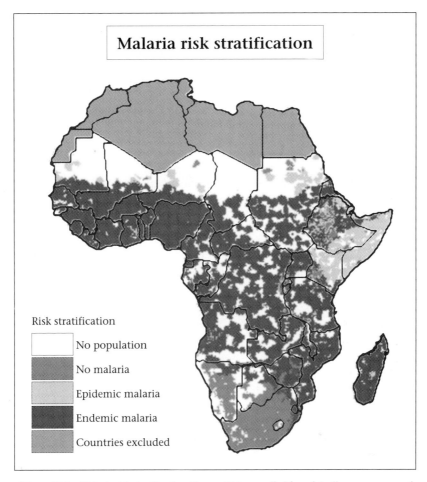

Figure 12.5 Malaria risk stratification. *Source:* Data compiled from *http://www.mara.org.za/*

Africa, and mosquitoes that can resist most of the widely used pesticides have been reported. South Africa is leading the way in developing new research and technologies to combat malaria. (See Discussion Box 12.2.)

Unfortunately, political and economic turmoil exacerbate the devastation of malaria. WHO estimates that up to 30% of the 960,000 people who die of malaria every year in Africa are from countries affected by serious conflict, war, or natural disaster. Indeed, malaria often kills more people in the aftermath of conflict, war, or natural disaster than the number killed during the actual emergency, according to the global Roll Back Malaria (RBM) movement (WHO Press Release, 2000.)

Trypanosomiasis The tsetse fly transmits *sleeping sickness*, or *trypanosomiasis*, during its sucking of blood from an infected animal or person; while doing so, it injects the single-celled agents, or trypanosomes. In the fly's body, these trypanosomes reproduce, and,

> **DISCUSSION BOX 12.2:**
> **SOUTH AFRICA BATTLES AGAINST MALARIA**
>
> South Africa is leading Africa's offensive against malaria, using the latest research and technology at the Malaria Research Center (MRC). First started in 1980, the initial malaria-vector research focus has been extended to insecticide evaluations, research on the malaria parasite, and drug resistance in malaria.
>
> Before large-scale intervention in 1938, the disease pattern was characterized by dramatic interannual variation, with severe epidemics as far south as Durban and Pretoria. By dint of sustained, focused research, South Africa has mitigated malaria's devastating effect on economic development, particularly ecotourism and agriculture. By December 2000, the high-risk areas were mostly border areas, emphasizing the importance of regional, rather than country-specific approaches to malaria.
>
> MRC aims at providing cost-effective control, based on sound scientific principles and using cutting-edge technology. At MRC's health GIS center (HGC), GIS and Global Positioning System are being used not only for malaria surveillance, but also to track drug resistance and delineate risk areas. This pioneering work against malaria demonstrates the extreme usefulness of GIS for health research and is already being applied to other diseases, including tuberculosis.

when the fly next bites a person or animal, it spreads the infection. In humans, sleeping sickness begins with a fever, followed by a swelling of the lymph nodes. Next, the inflammation spreads to the brain and spinal cord, producing the lethargy and listlessness that gives the disease its name. When livestock get the disease, the consequences are similarly severe, with the sick animals withering and dying within a year of infection. For some animals, particularly those raised in tsetse-infested territory, trypanosomiasis does not kill, but instead causes low weight gains, anemia, and high spontaneous abortion rates.

This is a disease of rural areas, hence, many infected people go undiagnosed and untreated and eventually die in the villages. Such deaths are usually not reported. According to WHO (1986), only one-tenth of the people at risk are under surveillance, and 25,000 new cases are identified each year. The only drug for treatment, melarsoprol, exposes 10% of the people who receive it to serious side effects, from which about 1000 people die each year. The most promising line of attack appears to be killing the vector (the flies) in massive eradication campaigns.

Yellow fever Yellow fever, transmitted by mosquitoes, has a long history in west Africa, where it killed Europeans for centuries in what became known as "the white man's grave." Fever, headaches, backaches, and vomiting are some characteristics of yellow fever.

Sometimes, unchecked vomiting leads to death. In less acute cases, the virus attacks the liver, jaundice occurs, and the deposition of bile pigment colors the eyes and skin quite yellow. Once the disease is contracted, there is no treatment, and it has to run its course. Eradicating it is a difficult proposition, because yellow fever affects monkeys and several species of small forest animals. Immunization of humans may be the only feasible solution.

Schistosomiasis *Schistosomiasis* (bilharziasis) is a parasitic disease that leads to chronic ill health. During human contact with water—in the course of everyday activities, such as swimming, washing, and fishing—the parasites enter a person via body openings. Infected individuals pass the eggs in their excrement. Once in water, the eggs open to release the parasite. The parasite ultimately finds its snail host and divides into thousands of new parasites, which are excreted by the snail into the surrounding water. Many development projects in Sub-Saharan Africa involving irrigation schemes have inadvertently introduced schistosomiasis into populations previously not exposed to the parasite. The reservoirs of large dams, such as the Akosombo in Ghana, the Kainji in Nigeria, and the Kariba in Zimbabwe, are major transmission sources and thus the core of endemic areas. Internal bleeding (evidenced by blood in urine), loss of energy, and pain are characteristics of the disease, although it is not fatal. It is endemic today for more than 200 million people worldwide, most of them residing in Sub-Saharan Africa (WHO, 2000). Deep wells with hand pumps to provide water for laundry and bathing usually reduce incidence in women and children.

River blindness River blindness, or onchocerciasis, is one of the major health problems of the savanna belt in west, central, and east Africa. River blindness is caused by a parasitic worm, *Onchocerca volvulus*, and transmitted by a small black fly that carries the disease from one person to the next. (See Discussion Box 12.3.) The black fly lays the parasite's eggs in the skin of humans. After hatching, the larvae mature into threadlike adult worms that live in nodules under the skin. The infant worms escape through the walls of the

---

DISCUSSION BOX 12.3: BLINDED BY A FLY

Isa Tanko used to earn a living as a farmer in northern Nigeria. Now blind and led by his son, he begs motorists for alms during traffic jams in sprawling Lagos.

"I became blind 10 years ago and was forced to leave my village and come to the city to beg for alms (in order) to feed myself and my family," Tanko explains.

Like millions of other able-bodied persons who live and farm along riverbanks in Nigeria, Tanko lost his eyesight to river blindness. The Volta River basin, covering parts of Benin, Burkina-Faso, Côte d'Ivoire, Ghana, Mali, Niger, and Togo, provides the black fly an ideal environment in which to lay its eggs and wreak its havoc.

nodules and migrate to all parts of the human body. They have even been found in tears, sputum, urine, and vaginal secretions.

Infant worms cause havoc in the human body. Their presence causes rashes and itching, and the skin becomes swollen, often with depigmentation that leaves white patches (known graphically as "crocodile skin," "lizard skin," and "leopard skin"). Sometimes, infected persons suffer from genital swelling, loss of weight, and debilitation. Eventually the infant worms get into the eyes; when they die, the victim becomes blind. In 1974, WHO (1990) estimated that 1 in every 10 people living in the 700,000 km^2 (270,270 mi^2) of Ghana's Volta Basin was infected. In northern Ghana, this disease blinds a large percentage of the adult villagers.

River blindness is not randomly distributed. It affects primarily those populations living near swiftly moving streams, the preferred breeding sites of the insect vector. Villages near such watercourses sometimes have adult blindness rates as high as 30%. Gradually, such high incidences of infection and blindness lead to the abandonment of river valleys. The population retreats for several miles beyond the usual flight range of the fly. In Nigeria, more than 100,000 km^2 (38,610 mi^2) of agriculturally rich and economically viable land have been deserted for fear of infection. Also, 3.3 million cases of the disease were reported in Nigeria, where more than 100,000 people have gone blind in the past few years. According to WHO (1990), 60% of all cases in west Africa and 30% to 40% of the world's cases are in Nigeria. Onchocerciasis is a chronic, not a fatal, disease, though many of its victims become totally disabled and have to be supported by their communities for years.

Guinea worm (dracunculiasis) Guinea worm is a painful and debilitating disease that is still a major health problem. It is widespread in Nigeria and other west African countries and is contracted from drinking water, usually from a stagnant pond containing the minute cyclops infected with parasitic larvae of *Dracunculus medinensis* (Hunter 1996; 1997a). The cyclops are then dissolved by the gastric juices in the human gut, and the larvae escape into the tissues. The female worm matures; about 10 to 12 months after ingestion, it emerges, causing a painful abscess. When the abscess is submerged in water, the guinea-worm embryos are released into the pond to continue the transmission cycle. Many afflicted people have multiple worms and are afflicted year after year. Sufferers may be incapacitated for as long as three months, and the disease often has a serious impact on agriculture and other activities. Among the communicable diseases afflicting Africans, guinea worm is perhaps the only one that can be eliminated completely by the provision of a constant, reliable supply of safe drinking water.

Unfortunately, the "walk-away syndrome" characterizes many guinea-worm donor–host projects (Hunter, 1997b). After infrastructure (e.g., wells) is established, program advisors walk away with technical satisfaction, while ignoring postimplementation issues, because they assume wrongly that regional authorities possess adequate technical and financial capacity and are sufficiently prepared to maintain the program.

Buruli ulcer Buruli ulcer has become an important cause of human suffering in Africa since 1980. Caused by *Mycobacterium ulcerans (M. ulcerans)*, an organism from the family of bacteria that causes tuberculosis and leprosy, Buruli ulcer destroys the skin and

Endemic Diseases of Sub-Saharan Africa

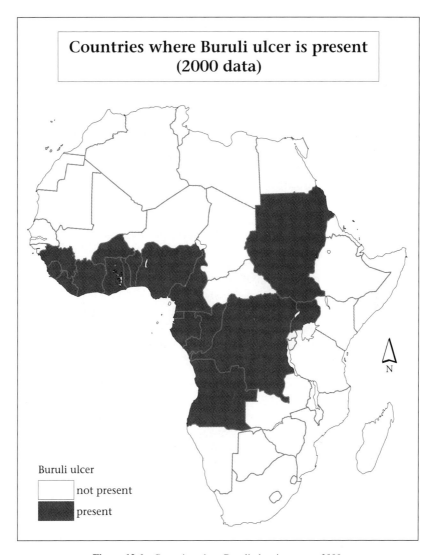

Figure 12.6 Countries where Buruli ulcer is present, 2000.

underlying tissues and causes deformities. Most patients are women and children who live in rural areas near rivers or wetlands, but little is known about the mode of transmission. In Africa, it was first detected in Uganda, but is now present in all countries along the Gulf of Guinea (Figure 12.6). In Côte d'Ivoire, up to one-sixth of the population in some villages is affected. Comparably high rates have been reported in Benin, Ghana, Cameroon, and the Democratic Republic of Congo, but underreporting may be a big problem.

Buruli ulcer often starts as a painless swelling in the skin. A nodule develops beneath the skin's surface, one teeming with mycobacteria. Unlike other mycobacteria, *M. ulcerans*

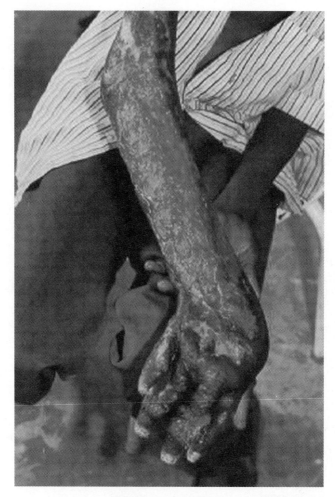

Figure 12.7 Buruli ulcer.

produces a toxin, which destroys tissue and suppresses the immune system. Massive areas of skin and sometimes even bone are destroyed, causing gross deformities, which, on healing, leave extensive scarring that can cause restricted movement of limbs and other permanent disabilities (Figure 12.7). Because Buruli ulcer is minimally painful, those affected do not seek prompt treatment. The only viable treatment in 2001 is surgery to remove the lesion, followed by a skin graft if necessary. Clearly, this is a costly procedure, one not widely available to those affected by the disease, many of whom tend to live in areas with limited access to health services. Patients often seek treatment too late and suffer frequent and severe complications and prolonged hospitalization. (For additional information, go to *http://www.who.int/gtb-buruli/*).

Ebola The Ebola virus is a member of a family of viruses known as Filoviridae. The usual hosts for most of these viruses are rodents or arthropods (such as ticks and mosquitoes).

In the case of the Ebola virus, the natural host for the virus is unknown. The Ebola virus was discovered in 1976; it was named after the Ebola River in Zaire, where it was first detected. Three main varieties are known so far: Ebola Sudan, Ebola Zaire, and Ebola Reston. A fourth strain, Ebotai, also affecting both humans and monkeys, has been identified in Côte d'Ivoire. Ebola Sudan and Ebola Zaire affect humans and monkeys; Ebola Reston harms monkeys, but not humans.

After the exposure to the virus, onset of symptoms (usually fever and muscle pains) begins in 4 to 21 days. Depending on the particular virus, the disease can progress until the patient becomes very ill with respiratory problems, severe bleeding, kidney problems, and shock. Ebola Zaire is fatal in about 90% of the cases; Ebola Sudan is fatal in about 60%. Currently, no cure or vaccine for these viruses exists. Ebola kills its victims by slowly "dissolving" their organs, blood cells, and connective tissue, causing massive and usually fatal internal hemorrhaging.

Ebola is spread through close personal contact with a person who is very ill with the disease. In previous outbreaks, person-to-person contact occurred frequently among hospital care workers or family members who were caring for an ill person infected with the Ebola virus. Transmission of the virus has also occurred from hypodermic needles reused in the treatment of patients. Reusing needles is a common practice in developing countries, such as the Democratic Republic of Congo and Sudan, where the health-care system is underfinanced. Ebola viruses can also be spread through sexual contact.

Previous outbreaks of Ebola appear to have continued only as long as a steady supply of victims came in contact with body fluids from the infected. The epidemics were resolved by teaching the local population how to avoid spreading the disease and by improving conditions at hospitals in affected areas (unsterilized needles and syringes were a major factor in the 1976 outbreak in the Democratic Republic of Congo). Ebola's lethality may also serve to limit its spread: Victims die so quickly that they have little chance to spread the infection.

This is by no means an exhaustive list of all the diseases and illnesses afflicting Africans. Others, such as cerebrospinal meningitis, cholera, tuberculosis, and, lately of increasing importance, such degenerative diseases as hypertension and cancer, also deserve the attention of medical geographers. However, at this juncture, we need to turn our attention to the most notorious source of all infections: sanitation and hygiene.

SANITATION AND HEALTH IN SUB-SAHARAN AFRICA

Most Sub-Saharan African countries have serious sanitation and health problems in both rural and urban areas. Major sanitation problems include management of domestic and human waste and access to safe water (Figure 12.8). For example, in Ghana in 1988, 44.3% of the population did not have an adequate safe water supply and 69.6% lacked appropriate sanitation (WHO, 1990). In urban areas, 93% of the population had access to safe water, compared with only 39% in the rural areas. Similarly, 64% of the urban population had access to appropriate sanitation, compared with only 15% for rural areas.

Management of refuse and human waste is a major problem. In some urban areas, household waste is usually carried from households and dumped at a sanitary site, from

Figure 12.8 Sanitation and health problems. Open drainage lines are frequently dumping grounds for garbage and human waste.

which it is collected and dispatched for final disposal. The weakest link in the system is transportation from these sanitary sites. Frequent breakdowns of refuse vehicles make it extremely difficult to handle the waste generated at these sites. Commenting on Accra, the capital of Ghana, Garbrah (1986) notes that several unauthorized dumping sites have been created and many unfenced or open sites have been turned into uncontrollable refuse-dumping sites. The frequent outbreaks of cholera in Nima and Sukura in Accra demonstrate the magnitude of the health risk. Sanitation conditions in other countries are equally deplorable, particularly in the urban areas. Lagos in Nigeria is particularly infamous.

The management of human excrement is equally problematic. Transportation of it from houses and public lavatories is woefully inadequate. Consequently, it is sometimes dumped indiscriminately into drains and open spaces. When public toilets are not properly serviced, open-space and roadside defecation becomes rampant. Drainage lines meant to carry liquid wastes often become dumping grounds for refuse and human waste. During heavy rains, runoff can contaminate sources of drinking water. It is, therefore, not surprising that water-related diseases, such as diarrhea and gastroenteritis, are quite rampant.

In many rural areas, toilet facilities are extremely limited. Open-air defecation and pit latrines are the norm. Some pit latrines are roofed with aluminum sheets, leaves, or thatch (and usually surrounded with a dwarf wall to assure privacy), others are not. During the rainy season, the latrines overflow, creating various kinds of biological contamination and attendant health problems. Many towns, and even some large cities, have exposed sewage systems.

The pattern of diseases seen in health facilities reflects these realities. HIV–AIDS and malaria accounted for 30% of the total disease burden in Africa in 1999, compared with only about 10% worldwide (Table 12.1). Measles, maternal conditions, and congenital abnormalities, although very important causes of morbidity in Africa, do not rank in the top 10 causes of disease globally. Similarly, the 2 leading causes of death in Africa, HIV–AIDS and malaria, together account for 40% of all deaths there, but for less than 10% worldwide. Infectious diseases continue to be the major cause of death, but the pattern is changing—heart disease, the leading cause of death worldwide, is now the 9th-leading cause of death in Africa. Similarly, cerebrovascular disease, the 2nd-leading cause of death worldwide, is now the 8th-leading cause of death in Africa. Simple sanitary and preventive measures, such as immunization and health education, are important in addressing the major causes of morbidity and mortality. However, without a focused attack on the raging poverty, their effectiveness will be minimal. Poverty undermines individual

TABLE 12.1 Leading Causes of Death, Africa and Global, 1999

	Africa			Globally	
Rank		% of Total	Rank		% of Total
1	HIV/AIDS	20.6	1	Heart disease	12.7
2	Malaria	10.3	2	Cerebrovascular disease	9.9
3	Acute lower respiratory infection	9.1	3	Acute lower respiratory infection	7.1
4	Diarrheal diseases	7.3	4	HIV/AIDS	4.8
5	Perinatal conditions	5.9	5	Chronic obstructive pulmonary disease	4.8
6	Measles	4.9	6	Perinatal conditions	4.2
7	Tuberculosis	3.4	7	Diarrheal diseases	4.0
8	Cerebrovascular disease	3.2	8	Tuberculosis	3.0
9	Ischaemic heart disease	3.0	11	Malaria	1.9
10	Maternal conditions	2.4			

Leading Causes of Disease Burden, Africa and Global, 1999

	Africa			Globally	
Rank		% of Total	Rank		% of Total
1	HIV/AIDS	19.9	1	Acute lower respiratory infection	6.7
2	Malaria	9.9	2	HIV/AIDS	6.2
3	Acute lower respiratory infection	8.5	3	Perinatal conditions	6.2
4	Diarrheal diseases	6.5	4	Diarrheal diseases	5.0
5	Perinatal conditions	6.5	5	Ischemic heart disease	4.1
6	Measles	4.7	6	Unipolar major depression	4.1
7	Maternal conditions	3.4	7	Cerebrovascular disease	3.5
8	Tuberculosis	2.3	8	Malaria	3.1
9	Congenital abnormalities	1.7	12	Tuberculosis	2.3
10	Road traffic accidents	1.7			

Source: World Health Report, 2000. WHO.

health, the health system, and individual experiences with health care. Let us turn our attention to these questions now: What are the medical resources available to deal with the health problems? How are they distributed? Where are they located and who has access to them?

THE HEALTH-CARE SYSTEM

The organization of health services in most Sub-Saharan African countries follows a common pattern. Almost without exception, the government, through its ministry of health, pays for and operates a network of public facilities. The system follows a hierarchical structure, with a few major urban hospitals at the top (teaching hospitals), a second tier of regional hospitals, a third tier of rural health centers and maternal and child health clinics, and a final layer of rural subcenters served by resident or itinerant paraprofessional health workers. Specialty hospitals for leprosy, tuberculosis, and other illnesses are also maintained, plus public facilities for mental health services and, occasionally, for dental care. In many countries, the military, police, and big companies (both public and private) have their own specialized hospitals that cater to employees and their relatives.

There is a significant presence of charitable institutions and religious missions that provide medical care. Usually, these groups do not provide systematic geographical coverage, but where they exist, they provide services that would otherwise not be available. In the Democratic Republic of Congo, it is estimated that mission groups provide up to 75% of the country's rural modern health services and employ 30% of the country's health personnel (Lashman, 1975). In Botswana, local missions and charitable groups were estimated to account for 4.2% of all health expenditures in 1976 (local and national governments provided 78.8%). More than 97% of charitable expenditures were in rural areas, compared with 70% for government expenditures. Mission hospitals in Ghana account for 30% of Ghana's hospital beds and are partially integrated into the ministry of health (Morrow, 1983).

Nearly every country also sustains a private sector that provides fee-for-service medicine. In fact, many countries allow public physicians, nurses, and midwives to supplement their incomes by maintaining private practices. Invariably, shortages of medication, supplies, personnel, and equipment characterize the health-care system of most Sub-Saharan countries. Also, most modern medical facilities are located in urban areas, very few in rural areas. Frequently, traditional medicine is the only form of medical care available to rural residents. (This is discussed in detail later in this chapter.)

Private practice takes the form of maternity clinics, a clinic with a qualified medical practitioner, or a pharmacy. Private practitioners usually concentrate in the urban centers, where their patrons are. Their choice of location exacerbates rural–urban disparities in access to health facilities (Figure 12.9). It is quite normal for pharmacies, besides their normal role of selling, to prescribe and administer medication. In a recent issue of *West Africa* (February 12–18, 1990), the Pharmaceutical Society of Ghana warned that the ministry of health would prosecute some drug houses for selling restricted drugs (antibiotics, paracetamol, and chloroquine) without prescription. We will return to the problem of fake drugs later in this chapter.

The Health-Care System

(a)

(b)

Figure 12.9 Urban–rural contrasts in pharmaceutical services. Pharmacy of a private clinic in the urban area of Accra (*top*). Contrast with pharmacy of rural area (*bottom*). Notice empty shelves.

Figure 12.10 Lack of road transportation. In many rural towns and villages, transportation is rare. In some villages, no vehicle passes for days; in others, there is one vehicle once a day.

Traditional healers are mostly in the rural areas, where they provide treatment for all kinds of ailments. They usually specialize in healing specific ailments and establish a reputation for that. One healer might specialize in childhood diseases, another in treating snake bites, another in barrenness and other obstetrical problems. Traditional healers have no huge advertising bills. They usually put up a signboard detailing the practice and specialization areas, prefixing their names with the title "Doctor." They operate mainly by word-of-mouth communication, usually from satisfied customers. Traditional healers never lack patrons. Their location in rural areas means that sometimes they are the only source of care for rural residents.

A fourth, less recognized group of health practitioners is the itinerant drug vendor. They carry both traditional and Western medical products for sale, in the open market or in bags from door to door in villages, towns, and offices, proclaiming a cure for all diseases. A particularly successful one in Ghana announced himself thus: "Dr. Danquah has brought his cure for all diseases, buy it to cure your rheumatic pains, for impotency and for women to bring forth plenty of children" (Twumasi, 1988). Drug sale on public transportation, especially for long-distance trips, is also quite common.

Occasionally, itinerant drug vendors are known to sell prescription drugs as much as other dubious substances. Most often, they administer injections to treat common ailments such as malaria. Despite the great risk involved in receiving treatment from them (such as overdose, wrong prescriptions, and the fact that often they use one needle several times, without sterilization, for different people), these medicine vendors are very popular in the rural areas, especially where transportation is lacking (Figures 12.10 and 12.11). This is

(a)

(a)

Figure 12.11 Poor roads in rural areas. Rural transportation is difficult. Many rural areas are connected by footpaths alone or only by poor roads such as this one (bottom) in Suhum district, Ghana. Motor transport is rare.

because they bring "health care" right to the doorsteps of people. Sometimes, they are not available when needed.

Spatial Imbalance in Health Services

Throughout Sub-Saharan Africa, health facilities are distributed inequitably. Rural areas with the highest population and greatest need have the fewest health facilities and services. Most health facilities and expenditures serve urban residents. In Côte d'Ivoire, expenditures for the University Hospital in Abidjan amounted to about 40% of the total health budget (Lasker, 1981). In 1981, there was one government physician for 66,000 people in the western region, compared to one for 4000 in Abidjan. The story is similar in Zambia, where expenditure patterns favor urban "line of rail" areas and three large hospitals (Freund, 1986). The provinces share 41% of the recurrent allocations, but the university teaching hospital alone uses 40% of the Zambian health budget.

In Nigeria, Okafor (1982) observed that, in Bendel state, urban population was a far more important factor than total population in explaining the distribution of hospitals, hospital beds, and doctors. The Oredo local government area, containing the state capital, but only 6.6% of the population, had 26.4% of all hospital beds and 59.3% of the state's doctors. Similarly, Stock (1985) reported that, in Kano state, there were only 204 doctors for a population of 9.5 million—one doctor per 46,500 people. Only 30 doctors, all of whom were state employed, served the 8.7 million people living outside Kano city, a doctor–patient ratio of 1:289,000. Meanwhile, a single hospital in metropolitan Kano had 73 doctors. Over all, Kano city, with only 10% of the population, had 85% of the medical officers (Stock, 1986).

Other examples include Senegal, where the entire health-post network reaches only 20% of the rural population (McEvers, 1980). In the Democratic Republic of Congo, fewer than 25% of all facilities are in the rural areas, where about 90% of the population lives (Schoepf, 1986).

There is overwhelming evidence to show that a spatial imbalance in the distribution of health facilities persists throughout most of Africa and that the few facilities in rural areas are in poor shape. They are improperly maintained, are inadequately staffed, and lack enough medical supplies. Sometimes, basic sterilization of equipment is absent, disposable syringes are reused, and injection needles are not changed for each new patient (IDS study group, 1981). The potential for the spread of AIDS in such circumstances should be obvious.

The Rainy-Season Accessibility Problem

The rainy season compounds the rural accessibility problem in some places, making seasonal inaccessibility a standing problem (Figure 12.12). Moreover, during the dry season, when the roads are generally motorable with little difficulty, availability of transportation is scarce. Lasker's (1981) description of the problem in Côte d'Ivoire fits many countries perfectly: In some villages, no car passes for days; in many other villages, there is a taxi that arrives once a day.

The Health-Care System

Figure 12.12 Problems with rural accessibility during rainy season. Rainy season travel is a nightmare. Treacherous roads often halt motor transportation. Imagine emergency medical services along such roads! Incidentally, the rainy season is the peak of the malaria season.

During the rainy season, the unavailability problem is compounded by bad roads. Road conditions deteriorate so much that, in many parts, access is frequently limited to walking along muddy paths and sometimes wading through water. Motor travel becomes extremely difficult. Long waiting periods, frequent vehicle breakdowns (caused in part by bad roads full of pot holes), prohibitive passenger fares, or the complete absence of transportation are typical characteristics. These difficulties substantially diminish the incentive to travel. Incidentally, in many places, this is the peak time for malaria incidence.

Health-Facility Locational Strategies and Location–Allocation Models

Current health-facility locational strategies throughout Sub-Saharan Africa exacerbate, rather than amend, existing spatial imbalances in access to health-care facilities. Spatial analytical methods, such as location–allocation modeling, indispensable to efficient service-delivery planning, are not being used (Rushton, 1988). Health-facility locations frequently are selected via political maneuvers rather than on the basis of demonstrated need or service-efficiency considerations. A limited, mainly evaluative application of location–allocation modeling indicates that, properly used, these models could provide invaluable information for improving geographic access.

Bypassing and Utilization of Health Facilities

Primary health-care (PHC) facilities in Sub-Saharan Africa, at least as advocated by WHO, are hierarchically organized to facilitate an efficient use of health resources. A health-care hierarchy is founded basically on central-place notions, in which certain threshold populations within a given travel distance are provided with a specified level of care (Phillips, 1990). Sometimes, the hierarchy is difficult to identify precisely, given the coexistence of public, private, and charitable providers in formal health care with the many types of traditional practitioners. A three-level hierarchical system comprising a district hospital, or Level C; several health centers, or Level B; and community clinics, or Level A is the norm. These systems are successively inclusive: higher level facilities offer all the services provided at lower level facilities, plus services that are not provided at lower levels. The primary goal of such hierarchical systems is to ensure efficient organization and use of personnel and other health resources. Using minimally skilled health workers to treat minor health problems frees high-level resources and skilled personnel to handle complicated problems at higher level health facilities.

In practice, however, high-level facilities are usually overwhelmed with minor ailments that could have been treated at lower level facilities. District medical officers are swamped with many cases of simple illnesses resulting in unduly long waiting periods, congestion at the higher levels of the hierarchy, and low usage at the lower levels. This phenomenon, ignoring proximate facilities in favor of more distant facilities, is called *bypassing*.

The bypassing phenomenon makes the health-delivery system inefficient for several reasons. Low-level or nonexistent facilities reinforce the polarization of resources and are unable to offer adequate services. For example, the argument might be made that it is useless to employ and pay a full-time health worker in a village if very few patients are seen each time. With only a part-time or nonsalaried health worker, not only do opening times become erratic and frequently unpredictable but the use of the facility drops further. Because such facilities lack drugs and other supplies, their attractiveness is further diminished. The PHC literature is replete with cases of declining utilization due to inadequate resources and poor services (Freund, 1986; Van der Geest, 1982; Annis, 1981; Lasker, 1981). There is little reason to consult a health-care provider who cannot offer treatment for most of people's health problems (Sauerborn, Nougtara, and Diesfeld, 1989).

Two main forms of bypassing can be distinguished: spatially rational[2] and spatially irrational. When a high-order service is required, bypassing a low-level facility that does not offer the particular service is completely spatially rational. Spatially irrational bypassing might occur because a person misperceives the order of service required. For example, a person requiring a low-order service might mistakenly believe that a higher order service is required and thus go to the corresponding facility. Another instance might occur when users perceive higher level facilities to provide better lower order services (i.e., a surgeon might bandage a lacerated finger better). Backward referral, if practiced, would discourage irrational bypassing.

---

[2]The term "rational" is used in the classical economics sense. A spatially rational person knows which facility offering the appropriate level of service is closest and uses it.

Health care depends on many necessary imported inputs, such as vaccines, drugs, and pharmaceuticals, and on medical supplies and equipment. Fiscal difficulties, particularly unfavorable trade balances, mean a limited ability to import these essentials (Vogel, 1993). As interest accrues on the national debt, governments devote more funds toward paying off their debts. Usually, this means decreasing funds for "nonpressing" activities, including primary health. In fact, some people argue that the hidden agenda of SAPs is to disengage ministries of health and thereby reallocate money to debt servicing.

When governments are compelled to cut back on health expenditures, poor rural residents are those who suffer most; urban-based hospitals and programs, usually with significant political support, emerge unscathed. Programs that benefit rural residents, such as primary health care and health education, typically suffer the biggest cuts. Other cuts also affect rural residents adversely. For example, reduction in the transportation and fuel budget, in a period of escalating fuel and spare-parts prices, means limited rural travel by health workers and, consequently, less access to better trained health workers for rural residents. The rich can afford to go to urban centers for care, but the poor have nowhere to go and usually patronize itinerant drug vendors. Privatization, by increasing costs, deters many ill people from seeking care, causing them to die quietly at home, and thus delays awareness and response to new epidemics. Moreover, the resulting inflation and subsequent depreciated incomes demoralize health workers. Many resort to other activities to supplement their meager income; others leave the health services altogether. "Borrowing" official supplies for personal use becomes widespread practice (Oppong and Toledo, 1995). Difficult financial times are a major cause of the massive brain drain of experienced health workers to greener pastures elsewhere. Thus, cutbacks in health-sector spending produce large staff layoffs and significant salary reductions, mere consulting clinics without drugs and sometimes without medical staff, and out-migration of doctors. Similarly, user fees and cost-recovery programs have caused people to delay treatment or do without it, resulting in higher disease and death rates.

Per capita health expenditure and national debt data reveal a lot about the poor shape of the health system in many African countries; they fail to show the widespread implications of the worsening economic crisis. For example, the data do not reveal that, because of declining real incomes, the physician stock in Ghana decreased from 1700 in 1981 to 800 in 1984 (Vogel, 1988). Moreover, the measure of health expenditure per capita has all the limitations of any per capita indicator. That the per capita health expenditure in Zimbabwe ($18.61) in 1985 was about three times that in Togo ($6.89) does not necessarily mean that Zimbabwe nationals are three times better off. The critical determinant is how the expenditure is distributed. If the expenditures in Zimbabwe are spent primarily on a few sophisticated facilities reaching only few people, those in Zimbabwe unserved by this system could be worse off than their Togo counterparts if expenditures there are spread evenly (Vogel, 1993).

The Bamako Initiative and Health in Sub-Saharan Africa

Faced with the challenge of providing quality health services out of very limited resources, African ministers of health announced the Bamako Initiative in September, 1987. The main goal is to provide health services with essential drugs through revolving funds that were to be managed with full participation of the communities. A key component of the Bamako

Initiative is cost recovery. Users are charged a small fee that can be used toward purchasing drugs or improving health-care services. This can be seen as either revenue generation for the ministry of health (Ellis, 1987) or a means to reduce unnecessary usage of health-care services (Mwabu, 1990) and, thus, improve efficiency. Others see it as an unfair and very regressive tax strategy that forces poor rural residents to pay for untrained and poorly equipped rural health workers, while subsidizing services by trained health workers for urban residents.

User fees make health care unaffordable for many people and usually reduce access to health services (Waddington and Enyimayew, 1989). When Swaziland imposed user fees, attendance at health facilities declined by 32% in government facilities and 10% in mission clinics. Visits for childhood vaccination and treatment of sexually transmitted diseases fell by 17% and 21%, respectively. Other studies show that user fees cause people to avoid or delay treatment and could lead to higher morbidity and mortality rates. Given the health, social, and economic costs of excluding so many people, especially the rural poor, from access to basic services, is it worth it? Difficult and unpleasant as these alternatives might be, most African countries have no choice but to accept the dictates of their lenders and hope for economic miracles. Currently, many countries are exploring different strategies for health-care financing, including nationwide health insurance policies. This is going to be the wave of the future.

CONCLUSION

The geographical location of Sub-Saharan Africa causes many health problems. Frequent local and regional famines, persistent entrenched poverty, and malnutrition (particularly of children) are common factors that contribute to the widespread nature of diseases. Sadly, it is the rural poor—without modern health facilities, good sanitation, and water sources—who suffer most from these problems. The current AIDS pandemic sweeping the continent makes the future uncertain, particularly in Uganda, Malawi, and Rwanda. A radical reorientation of AIDS policy in Sub-Saharan Africa that ties prevention to economic development is clearly necessary.

Spatially imbalanced health-facility systems, usually the legacy of colonialism, are the norm throughout the continent. Radical measures are necessary to improve rural care and to reduce bypassing to urban facilities. Location–allocation models will be very useful in providing ways to improve geographic access and use limited health resources effectively. Unfortunately, it is doubtful whether African health policy makers have the political will to adopt optimization measures in health-resource allocation decisions. Entrenched political interests make this unlikely.

Indigenous medicine, the only form of medicine available and acceptable to many rural residents, must be encouraged. Truly indigenous privatization of health care could begin here. In fact, privatization of health care appears inevitable in the future. Privatization is assumed to ensure efficiency, but unfortunately, when the profit motive is high, ethical considerations are easily set aside. Will this lead to a proliferation of dangerous fake drugs, multiple use of injection needles without sterilization, and other unhealthy practices? Who will speak for the rural poor?

Finally, like any other African problem, health issues cannot be separated from wider issues: widespread poverty, power imbalances, and economic development. Without an economic miracle, the disease and health-care future of Sub-Saharan Africa will continue to be largely in the hands of those who control African economies. The person who pays the piper calls the tune, even if it is unpleasant.

KEY TERMS

Disease ecology
Trypanosomiasis
Disease agent
Schistosomiasis
Epidemic
Ebola
Host
Endemic
Traditional medicine
Malaria
Morbidity
Guinea worm
Etiology

HIV–AIDS
Vulnerability
Location–allocation model
Bypassing
Structural adjustment
Mortality
Yellow fever
Disease vector
Onchocerciasis
Pandemic
Buruli ulcer
Endemic region

DISCUSSION QUESTIONS

1. To what extent is geographic location responsible for Africa's health problems?
2. "Poverty, not human behavior, is the major factor for HIV spread in Africa." Explain why you agree or disagree.
3. River blindness and Guinea worm are both related to river water. What are the major differences between them?
4. AIDS in Africa is an excellent example of a disease that spreads by hierarchical diffusion. Explain what this means. In what way is the medical system itself an important factor in the spreading of HIV–AIDS?
5. "African health problems are simply sanitation and water problems." Explain why you agree or disagree.
6. Bypassing is the inevitable result of an imbalanced health-care delivery system. By identifying and explaining the different forms of bypassing, suggest how the problem might be solved.
7. Is African traditional medicine obsolete?

8. "Itinerant drug vendors are a mixed blessing in rural Africa." Explain why you agree or disagree. Suggest ways of improving their efficiency in light of privatization efforts.

9. "Colonialism is to blame for Africa's health problems." Discuss this statement.

REFERENCES

ADEMUWAGUN, Z. et al. (1979). *African Therapeutic Systems*, Waltham, MA: Crossroads Press.

AKINSETE, I. (2000). "Nigeria Country Report," Paper Presented at the Foundation for Democracy Regional Conference on Strategies to Combat the Spread of HIV–AIDS in West Africa, Abuja, Nigeria: June 4–9, 2000.

ALUBO, S. (1994). "Death for Sale: A Study of Drug Poisoning and Deaths in Nigeria," *Social Science and Medicine*, 38(1): 97–103.

ANNIS, S. (1981). "Physical Access and Utilization of Health Services in Rural Guatemala," *Social Science and Medicine*, 15D: 515–23.

BRABIN, L., KEMP, J., OBUNGE, O.K., et al. (1994). "Reproductive Tract Infections and Abortion Among Adolescent Girls in Rural Nigeria," *The Lancet*, 346:530–536.

CHIKWEM J.O., MOHAMMED, I., OKARA, G.C., UKWANDU, N.C. & OLA T.O. (1997). "Prevalence of Transmissible Blood Infections among Blood Donors at the University of Maiduguri Teaching Hospital, Maiduguri, Nigeria," *East African Medical Journal*, 74(4):213–6.

DE BLIJ, H. & MULLER, P.O. (1994). *Geography: Realms, Regions and Concepts*, 7th ed., New York: John Wiley.

DECOSAS, J. (1996). "HIV and development," *AIDS 10* (Suppl 3), S69–S74.

DOWNING, T.E. (1991). "Vulnerability to hunger in Africa: A climate change perspective," *Global Environmental Change: Human and Policy Dimensions*, 1(5):365–380.

DUNN, F. (1976). "Traditional Asian Medicine and Cosmopolitan Medicine as Adaptive Systems," in *Asian Medical Systems*, Leslie, C., ed., Berkeley, CA: University of California Press.

ELLIS, R. (1987). "The Revenue Generating Potential of User Fees in Kenyan Government Health Facilities," *Social Science and Medicine*, 25:995–1002.

FALEYIMU, B.L., OGUNNIYI, S.O., URBANE, L.A., FALEYIMU, A.I. (1999). "Sexual Networking and AIDS Education in the Workplace and the Community: The Case of Oil Locations in Nigeria," National HIV Prevention Conference (United States), August 29–September 1, (abstract no. 158).

FALOBI, O. (1999). "Pre-ICASA 4: Services Under Strain—Nigeria," International Conference on AIDS and STDs in Africa.

FREUND, P. (1986). "Health Care in a Declining Economy: The Case of Zambia," *Social Science and Medicine*, 23(9):875–88.

GATRELL, A.C. & LÖYTÖNEN, M. (1998). *GIS and Health*, Philadelphia: Taylor and Francis.

GARBRAH, B. (1986). "Population and Environment in Ghana with Reference to Accra," *Proceedings of the Ghana National Conference on Population and National Reconstruction, Population Impact Project*, Legon, Ghana: University of Ghana.

GOOD, C. (1987). *Ethnomedical Systems in Africa*, New York: Guilford Press.

GREEN, E. & MAKHUBU, L. (1984). "Traditional healers in Swaziland: Toward Improved Cooperation between the Traditional and Modern Health Sectors," *Social Science and Medicine*, 18:1071–1079.

INTERNATIONAL DEVELOPMENT STUDIES STUDY GROUP (1981). Health needs and health services in rural Ghana, *Social Science and Medicine,* 14G:41–57.

HUNTER, J.M. (1996). "An Introduction to Guinea Worm on the Eve of Its Departure: Dracunculiasis Transmission, Health Effects, Ecology and Control," *Social Science & Medicine*, 43(9):1399–1425.

HUNTER, J.M. (1997a). "Geographical Patterns of Guinea Worm Infestation in Ghana: An Historical Contribution," *Social Science & Medicine*, 44(1):103–22.

HUNTER, J.M. (1997b). "Bore Holes and the Vanishing of Guinea Worm Disease in Ghana's Upper Region," *Social Science & Medicine*, 45(1):71–89.

JANZEN, J. (1978). *The Quest for Therapy in Lower Zaire*, Berkeley: University of California Press.

KALIPENI, E. & OPPONG, J.R. (1998). "The Refugee Crisis in Africa and Implications for Health and Disease: A Political Ecology Approach," *Social Science and Medicine*, 46(12):1637–1653.

LASKER, J. (1981). "Choosing Among Therapies: Illness Behavior in the Ivory Coast," *Social Science and Medicine*, 15A:157–68.

LASHMAN, K.E. (1975). *Zaire*, Washington, D.C.: U.S. Dept. of Health, Education, and Welfare, Office of International Health, Division of Planning and Evaluation.

MACLEAN, U. (1971). *Magical Medicine, A Nigerian Case Study*, Baltimore: Penguin Books.

MATTHEWS, S.A. (1990). "Epidemiology Using a GIS: The need for caution," *Computer, Environment and Planning*, 17:213–221.

MCEVERS, N. (1980). Health care and the assault on poverty in low income countries, *Social Science and Medicine*, 14G:41–57.

MEADE, M. & EARICKSON, R. (2000). *Medical Geography*, New York: Guilford Press.

MINISTRY of HEALTH, GHANA (1978). "A Primary Health Care Strategy for Ghana," Accra, Ghana: Ministry of Health.

MORROW, R. (1983). "A Primary Health Care Strategy for Ghana," *Practicing Health for All*, in Morley, D., Rohde, J., & Williams, G., eds., Glen Toronto: Oxford University Press.

MWABU, G. (1990). "Financing Health Services in Africa: An Assessment of Alternative Approaches," World Bank Working Paper Series, 467, Washington, DC: World Bank.

ODUJINRIN, O.M., & ADEGOKE, O.A. (1995). "AIDS: Awareness and Blood Handling Practices of Health Care Workers in Lagos, Nigeria," *European Journal of Epidemiology*, 11(4):425–30.

OGARA, A.I. NNAMANI, J.I. & AMEH E.L. (1998). "Knowledge, Attitude and Practice related to HIV/AIDS, and STDs in Rural Schools in Nigeria," International Conference on AIDS, 12:428 (abstract no. 23433).

OKAFOR, S. (1982). "Policy and Practice: The Case of Medical Facilities in Nigeria," *Social Science and Medicine*, 16:1971–77.

OPPONG, J. & TOLEDO, J. (1995). "The Human Factor and the Quality of Health Care," *Review of Human Factor Studies*, 1(1):80–90.

OYO, R. (1999). "Health-Nigeria: HIV Spreads at the Rate of One Person per Minute," *Inter Press Service*, December 6, http://www.aegis.com/news/ips/1999/IP991205.html.

PARKER, R.G. (1996). Empowerment, community mobilization and social change in the face of HIV/AIDS. *AIDS*, 10(3):27–31.

PHILLIPS, D. (1990). *Health and Health Care in the Third World*, Essex, England: Longman Scientific and Technical.

POLGAR, S. (1962). "Health and Human Behavior: Areas of Interest Common in the Medical and Social Sciences," *Current Anthropology*, 3:159–205.

RUSHTON, G. (1988). "Location Theory, Location–Allocation Models and Service Development Planning in the Third World," *Economic Geography*, 64:97–120.

SAUERBORN, R., NOUGTARA, A., & DIESFELD, H.J. (1989). "Low Utilization of Community Health Workers: Results from a Household Interview Survey in Burkina Faso," *Social Science and Medicine*, 29:1163–74.

SCHOEPF, B. (1986). "Primary Health Care in Zaire," *Review of African Political Economy*, 36:54–58.

SEN, A. (1981). *Poverty and Famines,* Oxford: Clarendon Press.

STOCK, R. (1986). "Disease and Development" or the underdevelopment of health: a critical review of geographical perspectives on African health problems, *Social Science Medicine*, 73(7):689–700.

STOCK, R. (1982). "Distance and Utilization of Health Facilities in Rural Nigeria," *Social Science and Medicine*, 17:563–50.

STOCK, R. (1985). "Health Care for Some: a Nigerian Study of who gets what where and why." *International Journal of Health Services*, 15(3):469–484.

TWUMASI, P. (1988). *Social Foundations of the Interplay between Traditional and Modern Medical Systems*, Legon, Ghana: Ghana Universities Press.

UNITED NATIONS. (1988). Global Review of the Economic Situation and Its Repercussions on Health Status: Health Care Services and Policies, New York: United Nations.

UNAIDS (2000). "International Partnership against AIDS in Africa," *Weekly Bulletin*, 35 (November 27), http://www.unaids.org/africapartnership/bulletin/apb271100.html.

UNAIDS and Economic Commission for Africa (ECA). (2000). *AIDS in Africa: Country by Country*, Africa Development Forum 2000, Geneva, Switzerland: UNAIDS, http://www.unaids.org/wac/2000/wad00/files/AIDS_in_Africa.htm.

UNICEF (1990). *Economic Crisis, Adjustment, and the Bamako Initiative: Health Care Financing in the Economic Context of Sub-Saharan Africa*, New York:UNICEF.

UNICEF (1996). *The State of the World's Children*: New York: Oxford University Press for UNICEF.

VAN DER GEEST, S. (1982). "The Efficiency of Inefficiency: Medicine Distribution in Southern Cameroon," *Social Science and Medicine*, 16:2145–53.

VOGEL, R.J. (1988). *Cost Recovery in the Health Sector: Selected Country Studies in West Africa,* Technical Paper No. 82, Washington, DC: World Bank.

VOGEL, R.J. (1993). *Financing Health Care in Sub-Saharan Africa*, Westport: Greenwood Press.

WADDINGTON, C.J. & ENYIMAYEW, N. (1989). "A Price to Pay: The Impact of User Charges in Ashanti-Akim District, Ghana," *International Journal of Health Planning & Management*, 4:17–47.

WEST AFRICA (1990). February Issue 12–18: 239.

WORLD BANK (1995). *Country Briefs*, Washington, DC: World Bank.

WORLD BANK (1997). *World Development Report*, New York: Oxford University Press.

WORLD HEALTH ORGANIZATION (WHO) (1978). *Primary Health Care*, Geneva: WHO.

WORLD HEALTH ORGANIZATION (WHO) (1986). *Evaluation of the Strategy For Health for all by the Year 2000: 7th Report on the World Health Situation,* Washington, DC: Pan American Health Organization.

WORLD HEALTH ORGANIZATION (WHO) (1995). *The World Health Report,* Geneva, WHO.

WORLD HEALTH ORGANIZATION (2000). *The World Health Report 2000,* Geneva: WHO.

WHO (2000). Third of African malaria deaths due to conflict or natural disaster, WHO Press Release WHO/46, 30 June 2000, http://www.who.int/inf-pr-2000/en/pr2000-46.html.

WORLD HEALTH ORGANIZATION (1990). International Water and Sanitation Decade, Geneva: WHO.

WYLLIE, R. (1983). "Ghanaian Spiritual and Traditional Healers' Explanation of Illness: A Preliminary Survey," *Journal of Religion in Africa*, 14:46–51.

13

Agricultural Development in Sub-Saharan Africa

Godson C. Obia

INTRODUCTION

Agriculture plays an important role in the economies of African countries and in the livelihoods of most Africans. It supplies food for subsistence and provides income for rural populations. Agriculture is the main engine for social and economic development for the majority of Africans and is important for the conservation of land and natural resources. "Agriculture accounts for a large share of the Gross Domestic Product (GDP) (32% of Africa's GDP), employs a large proportion of the labor force with 50% to 75% of the population engaged in agriculture" (Figure 13.1), "and represents a major source of foreign exchange (20% of exports)" (Africa-Europe Faith and Justice Network, 2000). The development of agriculture is, therefore, essential to economic growth and forms a crucial foundation for improving people's livelihoods, providing employment, and ensuring food security and food availability, as well as helping countries to maintain a favorable balance of payments. Against a backdrop of globalization and biotechnology, changes in African agriculture are being precipitated by global economic and technological realities; often, they do not favor the rural farmers, who constitute a majority in the population.

The thrust of this chapter is to examine African agriculture from the perspective of agricultural geography. Agricultural geographers study the variations in agricultural production, the physical environmental conditions that set limits on agricultural

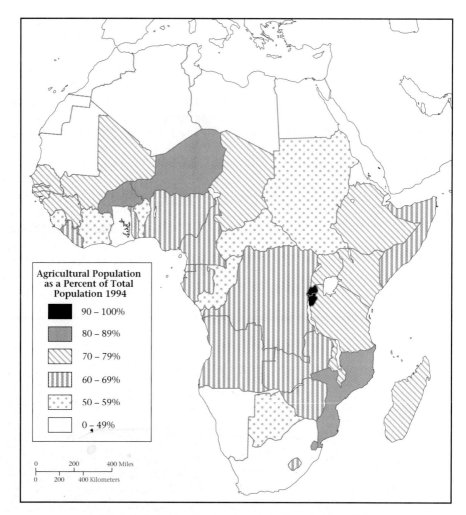

Figure 13.1 Percentage of total population in agriculture, 1994.

practices, and the cultural adaptations to and modifications of the physical environment to promote human agricultural production systems. The degree of modification often varies according to customary beliefs, preferences, technological capabilities, educational attainment and training, economic opportunity, and other political and policy factors (Rubenstein, 1989), including the forces of globalization and biotechnology. The variations show the variety of farming system types and crops, the preferred methods and techniques, the intensity of agricultural production, the relative profitability of production, and the extent of resource conservation, of environmental degradation, and of mitigation factors.

We will begin by examining the major farming systems of Africa: the traditional subsistence system and the commercial and cash-crop systems of the agricultural economy. The characteristic features of these farming systems are presented, particularly the types of farming associated with each agricultural system and the dual nature of the agricultural economy. The chapter proceeds to an evaluation of production trends in Sub-Saharan Africa, to shed light on regional deficiencies, on the relative value of exports and imports, and on the implications for policy. The issues associated with declining production trends—rural infrastructure, land degradation, research and development, inappropriate technologies, the Green Revolution, pricing and marketing policies, institutional arrangements, and the participation of women in agricultural production—are then discussed. In addition, the effects of structural-adjustment programs and of World Trade Organization (WTO) regulations on agriculture are examined. The chapter then concludes by evaluating such policy options as upgrading rural infrastructure, developing appropriate technologies and agrobased industries, increasing community-based cooperatives and participation, and designing integrated rural-development projects.

FARMING SYSTEMS

The term *farming system* has a variety of definitions. It is used here to denote levels of agricultural production that are wedded in a social, political, economic, and environmental context (Turner II and Brush, 1987). In this context, an area adopts a farming system that reflects the stage of technological development of the people and their adjustments to physical environmental conditions (Udo, 1983). There are two broad classes of farming systems in Africa: the traditional and the cash crop or commercial. Both traditional and commercial agricultural systems are dynamic systems that respond to changes in the social, political, economic, and environmental conditions of an area. For any given area, therefore, these changes can result from the increasing pressure of population, from the adoption of new techniques of production, from the vagaries of climate and environmental conditions, and from a host of institutional factors, such as land-tenure and land-ownership systems.

Traditional Farming Systems

Traditional agriculture comprises cultivation systems, pastoralism, and fishing activities (Table 13.1). Most traditional farming activities are undertaken solely for subsistence. A subsistence-farming family should be able to produce almost all the goods it needs, usually without any significant surplus for sale. Such a traditional system of production usually involves extensive use of land and long fallow periods. In an ideal situation, land is abundant and population is generally small and sparsely distributed, giving traditional farmers an opportunity to manipulate physical environmental resources for the production of crops and livestock. The concept of an ideal subsistence farmer in contemporary Sub-Saharan Africa is, however, a myth. Population growth has had the effect of reducing the per capita land area available to subsistence farmers. In areas of high population density, former subsistence-farming methods, which depended

TABLE 13.1 Peasant Agricultural Systems

Allan[1]	Boserup[2]	Benneh[3]	Morgan[4]	Ruthenberg[5]
Pastoral.	—	—	Pastoral.	Grazing.
Shifting cultivation.	Forest fallow.	Shifting cultivation.	Shifting cultivation.	Shifting cultivation.
Recurrent cultivation.	Bush fallow.	Bush fallow.	Rotational bush fallow.	
Semipermanent cultivation.	Short fallow.	Planted fallow.	Semipermanent cultivation.	Semipermanent cultivation.
Permanent cultivation.	Annual cropping	Permanent small-scale cultivation.[6]	Permanent cultivation.	Ley farming; permanent rain-fed cultivation.
—	Multicropping	—	—	Perennial cultivation.
—	—	Floodland cultivation	Floodland cultivation	Irrigation farming

[1] Allan, W. (1965). *The African Husbandman*, Edinburgh, Scotland: Oliver and Boyd.
[2] Boserup, E. (1965). *The conditions of agricultural growth: the economics of agrarian change under population pressure*, New York: Aldine Publishing Company.
[3] Benneh, G. (1972). "Systems of Agriculture in Tropical Africa," *Economic Geography*, 48, 244–257.
[4] Morgan, W. B. (1969). "Peasant Agriculture in Tropical Africa." In: Thomas, M. F. and Whittington (eds.). 1969. *Environment and Land Use in Africa*. London: Methuen & Co Ltd., pages 241–272.
[5] Ruthenberg, H. (1971). *Farming systems in the tropics*, Oxford: Clarendon Press.
[6] Includes compound farming, mixed farming, specialized horticulture, and tree cropping.

Source: Knight, G. & Newman, J. (1976). *Contemporary Africa: Geography and Change*, Englewood Cliffs, N. J.: Prentice-Hall.

on the extensive use of land, are no longer practical. Modifications in extensive farming systems vary from region to region. For example, in the more densely settled areas, rotational bush fallow is modified to include rotations around a fixed settlement or the development of permanent cultivation that uses household manure to improve the quality of the soil (Udo, 1983). Permanent, intensive production has the advantage of increasing the productive capacity of farmers, but mismanagement can be destructive to the land. According to the World Resources Institute (1989):

> Permanent cropping and cattle grazing deplete soil fertility or cause soil erosion if they are too intensive. As soil fertility declines, crop yields fall and more forest must be cleared to maintain the level of food production. When yields drop too low the land is abandoned.

Traditional farmers have low technical skills and employ low levels of technological inputs in their production. Farm tools are simple in their design and use (Figure 13.2). In most tropical farming areas, hoes, machetes, axes, and several kinds of digging tools are used to clear the vegetation, prepare the land for cultivation, and plant the crops.

Farming Systems

Figure 13.2 Rural farmers employ simple tools, such as machetes, to clear brush and undergrowth.

TABLE 13.2 The Importance of Agriculture to Sub-Saharan Economies

	Agriculture Population as percent of Total Population, 1990	Agriculture Exports as percent of Total Exports, 1990	Agriculture Imports as percent of Total Imports, 1990	Share of Total Imports Financed by Agriculture Exports (%), 1990
Angola	70		31	
Benin	61	42	17	17
Burkina Faso	84	35	19	20
Burundi	91	93	10	30
Cameroon	61	42	20	38
Central African Republic	63	19	19	20
Chad	75	63	6	19
Côte D'Ivoire	56	66	25	106
Equatorial Guinea	55	19	18	13
Ethopia	75	76	22	21
Gambia	81	43	35	7
Ghana	50	41	15	31
Guinea-Bissau	79	48	28	13
Kenya	77	68	7	32
Liberia	70	18	24	27
Malawi	75	91	8	66
Mali	81	85	21	60
Mozambique	82	28	21	4
Niger	87	14	18	10
Rwanda	91	86	16	34
Senegal	78	24	28	16
Sierra Leone	62	12	63	11
Sudan	60	98	19	43
Tanzania	79	62	5	20
Togo	70	32	23	16
Uganda	81	86	3	25
Zaire	66	11	26	17
Zambia	69	2	3	2
Zimbabwe	68	48	3	61

From FAO (1992). *The State of Food and Agriculture 1992*, Rome, Italy: Food and Agriculture Organization of the United Nations.

The human effort is therefore quite substantial—in fact, paramount. As a result, African traditional agriculture remains labor intensive—that is, it employs people more than machines during production. Some 70% or more of the population of most Sub-Saharan African countries is engaged in agriculture (Figure 13.1); the vast majority of these people are found in the traditional sector (Binns, 1992).

As a result of the number of people engaged in agriculture, of the value of agricultural exports as a percentage of total exports, and of the share of total imports financed by agricultural exports (Table 13.2), agriculture can rightly be called the backbone of the economy of most countries of Sub-Saharan Africa.

Farming Systems

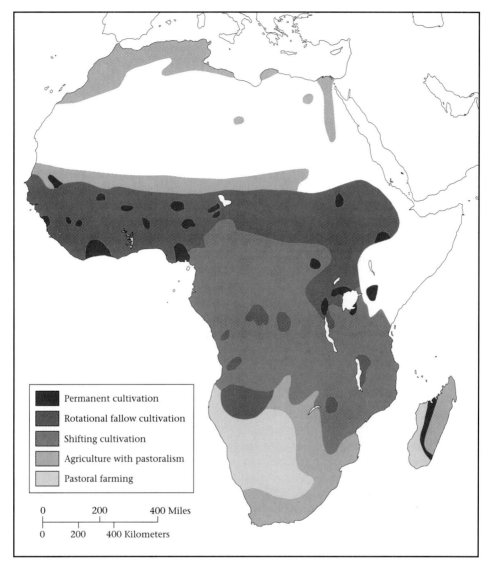

Figure 13.3 Traditional systems of agriculture. [Adapted from Griffiths, I. L. (1994). *The Atlas of African Affairs*, 2d ed, London: Witwatersrand Press, p. 125.]

Shifting cultivation Most countries of Sub-Saharan Africa engage in shifting cultivation, or one of its variations (Figure 13.3). *Shifting cultivation* is the rotation of land rather than crops. Farmers cultivate a piece of land for a few years and leave it fallow for a long period, while they move to another area of good soils. When the available cultivable fields become too distant from the village, the settlement also is moved

(Udo, 1983). Fallowing the land is a natural process of allowing the soil to regenerate the nutrients and structure lost during cultivation. In tropical environments, forest and savanna soils are generally poor. Rotating the land used in cultivation is, therefore, a prudent means for the replenishing of soil quality without the introduction of external sources of manure or fertilizer. Population growth has, however, become a major threat to this way of life in the forest and savanna zones. As population increases, cultivable land must be used for expanding villages, as land is allotted to new households for housing and compound cultivation. Such expansions usually take place on the outskirts of the village or on the village-ward margins of shifting cultivation land. In extreme cases of population pressure, some settlements are established on village land used for shifting cultivation, though at the risk of community conflict.

Slash and burn agriculture For shifting cultivators, the controlled use of fire is an important component of crop production; this practice is commonly called *slash-and-burn* agriculture (Figure 13.4). Farmers can incorporate fire into the process of crop production in two primary ways. First, the vegetation is cleared and then burned. Second, in some areas, farmers burn the standing vegetation in the area of cultivation

Figure 13.4 Slash-and-burn agriculture in the west African rain forest.

before clearing the land, which is a method commonly used in areas of forests, where clearing is cumbersome because of the thickness of the vegetation. Both methods create a deep layer of ash that enhances the fertility of the soil. Immediately following the clearing, farmers plant a variety of crops usually native to the regions of cultivation (de Blij and Muller 1994).

Intercropping *Intercropping*, or the practice of intermixing several types of crops in one field, is an important characteristic of shifting agriculture (Jackson and Hudman, 1990). By intermixing crops, the farmer provides a cover for the soil, one closely resembling the original vegetation that existed before burning and clearing. Because of this practice, the intensity of soil erosion and leaching is reduced. Most shifting cultivators add variety to their production by raising domestic animals—chickens, goats, sheep, pigs, and cattle (Udo, 1983).

Rotational bush fallow *Rotational bush fallow* is a modification of shifting cultivation, one that recognizes the influence of population increase in areas of shifting cultivation. As population increases, the farmed area is reduced and long rotations become infeasible. In rotational bush fallow, the cultivated area rotates around a fixed settlement and farmland (Binns, 1992). Fallow periods are shorter; they often vary from four to over seven years. The shorter fallow period does not give the land enough time to revert naturally to forest or woodland cover and to rebuild the fertility lost through years of cultivation (Morgan, 1969). Maintaining soil fertility and good crop yields has become the major challenge of farming in areas of rotational bush fallow. In areas of high population density, such as the Igbo settlements of Imo and Abia states in Nigeria, Benneh's (1972) *rotational planted fallow* has become an inevitable adaptation. In it, farmers plant "special fallow plants" to improve soil fertility and to provide poles for staking yam vines (Morgan, 1969; Udo, 1983).

Permanent farming *Permanent farming* requires no fallowing of the land. It is found mainly in areas where there is available land within settlement compounds and in areas where extreme population pressure makes cultivation without fallow mandatory. Without fallow, the land lacks a natural means of building up fertility and must depend on other sources of fertilization to maintain reasonable yields. Benneh (1972) developed the following comprehensive classification of permanent farming: compound farming, mixed farming, specialized horticulture, tree cropping, and floodland cultivation. Two types—compound and mixed farming—are discussed next as part of traditional farming systems.

Compound farming *Compound farming* involves the intensive growing of fruits and vegetables within or close to the walls of compound dwellings. Cultivation near the settlement makes crops easily accessible to farmers, affords farmers the opportunity to protect crops from theft and pest destruction, and takes advantage of fertile compound land (Morgan, 1969). Household and compound wastes are used to fertilize the soil, making it possible for permanent cultivation to take place once or twice a year. Especially in the rain-forest areas of west and central Africa, compound farming is the primary source of a variety of vegetables, fruits, herbs, and some highly valued staples, including special varieties of yams and cocoyams. In the high-population-density areas in the Abia and Imo states of Nigeria, for example, the increased intensity of cultivation on compound farmland assures higher yields. The Ngwa people of Abia State, Nigeria, have added yet another adaptive cultivation system within the compound land—the use of part of the compound land for growing garden eggs, which are highly demanded in the urban centers. This "new farming," as some farmers like to refer to it, is the prevailing practice among younger, educated people residing in the rural areas. Many of these farmers have other occupations—teaching, civil service, and private sector employment—and practice this type of farming mainly to supplement their incomes.

Mixed farming *Mixed farming* combines the raising of livestock with the crops described under compound farming. Mixed farming exploits the symbiotic relationship between animal and crop production: animal wastes are used to fertilize cropped areas, and animals feed on grains and crop wastes. To avoid crop damage, farmers might construct special fences made of sticks to keep out livestock or might house the livestock in stables. The Igbo of the eastern states of Nigeria employ fence enclosures or huts (or both) to separate the livestock from crops; the Wakara of the Ukara island in the southeastern part of Lake Victoria in Tanzania keep their cattle in huts (Binns, 1992).

Pastoralism *Pastoralism* is the dominant land use in the grasslands of Sub-Saharan Africa. (See Figure 13.2.) It is found primarily in the semiarid regions of west, east, and north central Africa, where conditions are too dry for cropping and where water and natural forage are available for only a few months in the year (Udo, 1983). Migration of pastoralists and their livestock is a common feature of pastoralism, as pastoralists are in constant search of water and good grazing land for their animals (Figure 13.5). This migratory process is of two types—nomadism and transhumance.

Nomadism *Nomadism* involves the irregular, yet cyclic, movement of pastoralists and their herds (de Blji and Muller, 1994). Nomadic migrations, contrary to widely held misconceptions, are not random, but take place along defined routes and follow frequently repeated travel patterns (Udo, 1983). Nomads often establish settlements at selected locations in their travel routes, giving them the opportunity to cultivate some grains that are essential to their diets. Because of the combination of cultivation and herding, Udo (1983) argues that the purely nomadic herder does not exist in Africa.

Transhumance *Transhumance*, involving regular movement between dry-season and wet-season grazing lands, is practiced by most of the pastoralists. In the dry season, pastoralists temporarily settle close to sources of water. Where rivers are not available, pastoralists dig wells to gain access to water. In extreme cases, they dig water holes right in the river bed, for easy access to the water table. This practice is common in the steppe or savanna areas, where seasonal rivers completely dry out in the dry season.

Farming Systems

Figure 13.5 Masai cattle herders measure their wealth in terms of the number of cattle they possess.

Several pastoral groups exist in Africa, each with its unique and distinguishing characteristics. However, a general classification of pastoralists can be based on mobility and permanence of settlement and on the extent of food cultivation to augment needed products. Binns's (1992) examples of pastoral groups include the following:

1. The *Masai*, found mainly in eastern Africa (particularly Kenya), are cattle herders who adapt to the meager vegetation of the semiarid steppes and savannas by moving frequently to locate new sources of pasture. The Masai in Kenya commonly exchange milk and meat for grains with Kikuyu cultivators. Although the Masai often own a large number of cattle, beef is hardly ever a part of their diet; they depend, instead, on milk and blood, which are supplemented with grains received in exchange for meat.

2. The nomadic herding *Tswanas* of Botswana have to adapt to the predominantly semiarid and arid environment of Botswana. The herding of cattle is done in distant cattle posts by young men, while other members of the family practice sedentary farming to provide food crops needed by the family (Udo, 1983).

3. The *Fulani*, found mostly in the northern states of Nigeria, have a distinctive pastoral lifestyle. Their unique variations depend on the type and degree of nomadism involved and to the extent to which pastoralism is mixed with crop production. Based on these distinctions, four main groups of Fulani pastoralists can be identified (Binns, 1992; van Raay, 1975):

   a. The true Fulani nomads, or Bororo, practice transhumance—that is, they move south in the dry season, and north in the wet season, in search of water, pasture, and markets and to avoid cattle-disease infestations. Because of their frequent movements, these Fulani do not have permanent homes, but must depend on "temporary tents or wooden structures with grass thatch" (Udo, 1983).

   b. Seminomadic Fulani have fixed settlements near cultivated areas. The elders stay throughout the year and are joined by the rest of the group during the start of the rainy season, when crops are planted. This group is characterized by herds smaller than those of the Bororo.

   c. Semisettled Fulani regard cropping and livestock as equally important. Their herds are smaller than those of the Bororo and those of the seminomadic group, and their pastoral activities are done near their permanent homesteads for most of the year, except in the dry season, when they move south to avail themselves of more pasture and water.

   d. Settled stock owners are a part of the Fulani pastoral culture. The main activities of this group revolve around daily trips from the permanent homesteads by men and boys, who take small herds of livestock for grazing. This type of pastoralism is no longer the exclusive occupation of the Fulani. Cattle can be bought by wealthy businesspeople who then pay the Fulani in money, or kind, to take care of such livestock.

Pastoralists in Sub-Saharan Africa face a wide variety of environmental, economic, and political problems. Primary among these problems are (1) the environmental constraints of drought, desertification, land degradation, and the loss of grazing land; (2) the economic issues associated with the loss of herds and wealth, competition from other land uses for limited grazing land, and the pressure to change to a less extensive type of pastoralism because of limited space; and (3) the political realities of uncontrolled nomadic pastoralism, as some governments pursue policies to limit grazing land and settle the nomads. The pastoral lifestyle is seriously under attack and is, therefore, undergoing revision and change, but the problems are far from being solved. According to the World Resources Institute (1993), overgrazing is responsible for 49% of the soil degradation in Africa. Consequently, as grazing land declines in quality and quantity, a "Tragedy of the Commons" sets in, whereby uncontrolled grazing causes more deterioration of the resource base and the death of humans and animals, as well as massive out-migrations. Some governments see this dilemma as their cue to intervene and create a system of integrated management of grazing lands.

Farming Systems

What are governments and international aid agencies doing, and how well are they succeeding? In Kenya, Zimbabwe, and Botswana, to name a few countries, governments are designating some rangeland as part of the national park system, thus restricting pastoralists to a limited area. Attempts to provide water to reduce uncontrolled migrations and transhumance have met with limited success. In extreme cases, overgrazing leads to rapid deterioration of the land, and the land is abandoned. Under such circumstances, pastoralists see no choice but to drive their herd into national park lands. Aid agencies have developed projects "to raise the economic productivity of range use systems by promoting the use of homogenous herds of cattle" (World Resources Institute, 1985). Many of these assistance programs have failed, because of "the export of inappropriate technology" (World Resources Institute, 1985). It seems obvious that, together, governments, international aid agencies, and pastoralists have to reach a compromise that recognizes the conditions of African rangelands and understands the cultural attributes of pastoralists.

Cash-Crop and Commercial Farming

Dividing types of farming systems into traditional and commercial may suggest that the two systems are mutually exclusive. This is hardly the case for most farmers in Sub-Saharan Africa, and it is done only to enhance pedagogy. In reality, although it seems easy to separate farming into these categories, it is more problematic to group farmers similarly. The same farmer can practice traditional farming—in all its forms—and still raise livestock and grow cash or other commercial crops to provide food for the family and raise the level of family income for needed purchases. In areas where acute population pressure reduces the per capita land area available to sustain most forms of traditional farming, farmers adopt more intensive land uses (often involving cash-crop and commercial agricultural production) in association with a limited practice of subsistence or traditional farming. This type of "semicommercial" farming is different from the true practice of commercial agriculture commonly found in the industrialized countries of the world.

Cash-crop production   Cash cropping can involve "the utilization of indigenous crops already part of the agricultural system, such as the oil-palm, ... or the introduction of exotics, the most notable example being that of cocoa" (Gleave and White, 1969). They distinguish three types of cash cropping: tree cropping, annual cropping for export, and the raising of food crops for cash sale.

Tree cropping   Tree cropping is common in the tropical rain forests of west and equatorial Africa (Figure 13.6). In southern Ghana, Côte d'Ivoire, Cameroon, Togo, and the states of southeastern Nigeria, crop-bearing trees are planted and nurtured within family settlements to enhance family incomes and to increase the productivity of a given area of land. The type of trees generally found in the forest zone of Africa include oil palm, wine

Figure 13.6 Farmers spray pesticides on cash crops, such as cocoa, with little protection from fumes.

palm, coconut, orange, African pear, oil bean, mango, cocoa, kola nut, banana, plantain, and rubber. These trees produce highly valued income-yielding products, such as fruits, nuts, oils, beverages, and construction material. In some African cultures, rights of ownership of tree crops are not transferable in cases where the land is pledged or leased. (See Discussion Box 13.1.) It is, therefore, common practice for the person to whom land is pledged to have rights that exclude the use of crop-bearing trees and to agree to make no permanent alterations to the land, including the cutting down of existing trees or the planting of new ones.

Expansions and improvements in tree-crop production are being undertaken by African farmers, especially by small-scale peasant-type farmers (Udo, 1983). These expansions reflect the lucrativeness of tree-crop production. Improvements include the rehabilitation of the existing stock of oil palm, cocoa, and rubber trees, some of which are in small-holder plantations. Although improvement techniques vary significantly, common features include the pruning of trees to remove diseased and dead components, the clearing of unwanted vegetation and weeds around trees, and the replanting of dead or diseased trees. Expansions can require the establishing of small plantations of trees,

> **DISCUSSION BOX 13.1:**
> **CONDITIONS OF LAND PLEDGE**
>
> "Among the Igbos of Nigeria, a pledge is the most common type of land tenure. It is a transaction in which a lender of money obtains a temporary usufructuary interest in a piece of land as security for his loan. The land remains perpetually redeemable by the borrower or pledgor on repayment of the loan.... The pledgee may not plant permanent trees or build on the land. Economic trees like oil palm and iroko may or may not be part of the pledge. There is no compensation for improvement effected by the pledgee in the land." (Uchendu, 1965)

such as oil palm and cocoa plantations. These new plantations are taking steps toward the rehabilitation of overused shifting cultivation land, which yields amounts of food crops too meager to justify continued crop production. Since trees take several years to mature, the farmer trades off present use for future benefits when the trees mature for harvesting. To raise needed short-term income, it is customary practice to plant such shade-tolerant crops as cocoyams within the plantation. This is common in the cocoa and oil-palm plantations of west Africa, particularly in Nigeria, Cameroon, Ghana, and Côte d'Ivoire.

An important part of cash-crop production is the cultivation of food crops for cash sale. This is done right along with other subsistence-farming activities. Let me illustrate how this works by drawing on my personal experience as a young man growing up in a farming family. (See Discussion Box 13.2.) Not many farmers in the region practice subsistence farming exclusively. Crop production provides subsistence needs for families and generates a cash income as well. As population pressure reduces the amount of land available for cultivation, farmers adopt creative methods to produce more per unit area of land. In situations of acute shortage of cultivable land, when limited crop production is used solely for subsistence, the farmers try to augment their incomes through diversification into nonfarm activities. Special attention is given to crops that mature annually, to meet the steady needs of the export sector and of African urban markets. Such crops include the tree crops already discussed. Because the primary objective of this type of farming is to sell the crops off the farm, it can be considered a commercial agricultural enterprise. One notable difference is that it is difficult to separate these farmers from those who constitute the subsistence-farming population. They are usually of small scale in operation. In Sub-Saharan Africa, groundnuts, oranges, palm oil and kernel, cocoa, and bananas fit into this mode of production.

Although peasant farmers are usually responsible for the actual cultivation, harvesting, and, to a limited extent, processing of the crops, most of the sale and distribution is

> **DISCUSSION BOX 13.2:**
> **PERSONAL EXPERIENCE**
>
> My father practiced a combination of shifting cultivation, rotational bush fallow, and cash-crop agriculture, revolving around the production of yams—the king of the crops. But some yams were more "king" than others. In the land most distant from the village, hardier types of yams were grown in combination with maize, pumpkins, cassava, and melons. The land was usually fertile after a long fallow period. In some instances, the land was left fallow for such a long period that the regenerated trees were tall enough to be used for staking the yam vines. In this polycultural environment, the yams were tenderly cared for, through constant tending of the vines and intensive weeding, usually done by women. The yields were always high. In the intermediate distance, the land was always less fertile because of more frequent cultivation and the consequent reduction in fallow time. Crops planted in these fields were inferior types of yams, requiring less fertile soils and, very often, fetching less money when sold. This was a necessary adjustment to the overcultivation that sometimes is the outcome of population pressure. In the farmland closest to the village and that in the compound, the most prestigious of the yams were planted, in association with special types of cassava, cocoyams, assorted vegetables, and tree crops. Livestock also were raised to form a veritable mixed crop and livestock operation.
>
> Although the crops mentioned served to maintain the subsistence needs of the family, the surplus production was treated in several ways: (1) some of the crops were stored in barns for consumption in the off-season; (2) part was consumed immediately; and (3) a large quantity was stored for sale during the high-demand, off-season period and for replanting during the next planting season. This guaranteed adequate food supply all year round, while providing reasonable income for other family needs, especially for education.

handled by middlemen or agents. The middlemen purchase the products from individual farmers, transport them, and market them to their established customers in the cities. This arrangement is predominant in villages that lie either in the urban fringe or within reasonable travel distance of the city. In some locations close to the large metropolitan areas of Sub-Saharan Africa, the middlemen enter into negotiations that guarantee them the rights to future harvests of crops. For example, in areas within a 20-mile radius of Aba (the largest city in Abia State, Nigeria), some farmers have sold the rights to the harvests from their plantain, banana, orange, and pineapple orchards. The economics of such transactions favor the middlemen in the long run, but farmers derive short-term gain by using the payments to ensure their family well-being.

Commercial farming Commercial agricultural production is dominated by export-commodity production and by the production of raw materials for the urban-industrial sector. There are two types of commercial agriculture in Sub-Saharan Africa: quasi- and large-scale commercial agriculture. The dominant one is *quasi-commercial production*. Subsistence farmers perform a dual function in agricultural production: They produce most of the food they consume, while raising crops, livestock, and fish for sale to augment family income. Trading patterns in rural agricultural settings suggest that production for sale off the farm entails more than the disposal of excess family food commodities. Some subsistence farmers devote some land to the production of such specialty items as cabbage, carrots, lettuce, and flowers (especially roses) specifically for sale in the urban markets. These products are not traditional to African agriculture, but are grown in response to an increasing demand from the educated elite, who have acquired a taste for these exotic crops. In farming villages surrounding major cities, farmers have established a niche in the urban market and are continuing to expand as demand increases.

Quasi-commercial production At another level of quasi-commercial production, subsistence farmers are engaged in small-scale production of raw materials for local industries and export crops. In west and east Africa, for example, subsistence farmers might own small-scale oil-palm, cocoa, coffee, and tea plantations. Intercropping on such land provides both the food necessary for subsistence and raw materials for sale to factories and to overseas markets. Oil mills, for instance, purchase palm fruits from small-scale producers, then process them for palm oil and kernels, which are sold in large quantities to soap-making factories and vegetable-oil manufacturers in domestic and international markets.

Large-scale commercial agriculture *Large-scale commercial agriculture* consists of plantations, grain production, livestock production for meat and milk, commercial harvesting of forest resources, and intensive farming in the urban fringe of cities. These types of farming benefit from imported technology and skills and are often capital intensive; in fact, forestry and plantation agriculture are both capital and land intensive. During the colonial period and the early years of independence, tree-crop plantations for cocoa, oil palm, rubber, coffee, and tea, to name a few, were established and maintained to ensure a reliable supply of industrial raw materials for European industries. The economies of west and equatorial Africa depended on the export of these crops to generate revenue and to earn foreign exchange for needed imports. Since then, plantations have continued to be an integral part of commercial agriculture. Today, plantations are owned by governments and private entrepreneurs. Many countries still export a large portion of their plantation crops, but some engage in processing or preprocessing, in order to reduce the cost of shipping raw materials and to increase the export value of commodities.

In the grassland areas of Africa, livestock, cotton, groundnuts, millet, and sorghum are produced on a large scale, often with the help of irrigation. This type of production is predominant in the wet and dry savanna zone of Sub-Saharan Africa—from the groundnut farms of Sene–Gambia, through the groundnut and cotton farms of northern Nigeria, to the irrigated cotton and grain farms of the Sudan and the Horn of Africa. The production of these crops is essentially a monocultural operation, a type of practice partly responsible for rapid depletion of soil nutrients, so good land management strategies must be adopted.

Lack of water, among other factors, is a major limitation on agricultural production in the arid environments. Irrigation projects are helping to increase production by expanding cultivable land, sometimes into marginal lands. Farms are owned predominantly by wealthy commercial farmers, some small-scale producers, and the government. Individuals and governments also own and operate livestock ranches and dairy farms. These are not to be mistaken for the livestock of pastoralists. The primary goal of these commercial farms is to produce meat, hides and skins, and dairy products, especially for urban markets and for the shoe and leather industries. These operations are run by skilled professionals trained in agricultural science and veterinary medicine. Yet another type of commercial agriculture is the large-scale, intensive production of vegetables, fruits, flowers, and dairy products, often associated with the periurban fringes of cities.

PRODUCTION TRENDS IN SUB-SAHARAN AFRICA

Sub-Saharan Africa faces the challenge of sustaining its agricultural production. It must deal with the difficult task of managing its agricultural resources effectively in order to meet the needs of its burgeoning population. Agricultural production in Sub-Saharan African is the economic mainstay for approximately 70% of the population. (See Table 13.2.) For some countries, the percentage of the population engaged in agriculture can be as high as 90%, as is the case in Burundi and Rwanda. Although the percentage of the labor force engaged in agriculture is declining, the proportion is still higher than in most developing regions of the world. Food crops represent the main type of agricultural production, as the primary motive of most farmers is to provide food for themselves and their families. The inevitable need to raise foreign exchange makes the production of export crops an important component of the agricultural activities in most Sub-Sahara African countries.

There is a persistent trend of decline in per capita agricultural production in Sub-Saharan Africa. According to the US Department of Agriculture, total agricultural production in Africa grew at a compound rate of 2.20% from 1950 to 1985, but only at the rate of 1.72% from 1976 to 1985. Agricultural production per capita for the two periods was 0.31% from 1950 to 1985, −1.04% from 1976 to 1985. Tables 13.3 and 13.4 illustrate the vagaries of per capita agricultural production. According to Lele (1981), three-fourths of the population of tropical Africa might have experienced a decline in

Production Trends in Sub-Saharan Africa

TABLE 13.3 Africa's per Capita Production of Major Food Products (Kg per capita)

Years	Wheat	Maize	Millet & Sorghum	Rice Paddy	Cassava	Pulses
1961–65	21	54	64	19	102	13
1970	22	54	57	11	104	14
1973	22	45	45	18	108	13
1974	21	67	50	18	111	13
1975	21	62	51	19	109	13
1976	26	57	49	19	102	12
1977	20	64	45	19	105	12
1978	22	64	49	18	102	11
1979	20	52	42	19	99	11
1980	18	58	44	18	100	11
1992	20	37	37	21	109	10
1993	19	52	36	23	108	10
1994	23	53	38	22	103	10

Source: Europa Publication Limited, *Africa South of the Sahara 1995 Volume*, London: Europa Publications Limited; FAO Production Yearbooks, Rome.

TABLE 13.4 Africa's per Capita Production of Major Cash-Crop Products (Kg per capita)

Year	Coffee	Tea	Cotton	Cocoa	Tobacco	Sugar Cane
1950	1.3	0.10	3.1	2.3	0.6	70
1955	1.6	0.10	2.9	2.1	0.6	60
1960	2.9	0.20	3.3	3.1	0.7	80
1965	3.7	0.23	3.4	2.8	0.8	80
1970	3.7	0.33	3.7	3.0	0.6	120
1973	3.6	0.40	3.2	2.5	0.6	130
1974	3.3	0.4	3.1	2.6	0.6	130
1975	2.8	0.37	2.8	2.5	0.6	130
1976	2.9	0.40	2.5	2.1	0.6	140
1977	3.0	0.45	2.9	2.1	0.7	140
1978	2.4	0.45	2.7	2.0	0.6	140
1979	2.6	0.40	2.5	2.2	0.7	130
1980	2.5	0.42	2.6	2.1	0.7	120
1981	2.7	0.41	2.4	2.2	0.5	140
1982	2.5	0.44	2.3	0.7	0.6	140
1992	1.2	0.45	2.0	1.9	0.7	97
1993	1.4	0.50	2.0	1.9	0.7	90
1994	1.5	0.50	1.9	1.9	0.6	99

Source: Food and Agricultural Organization of the United Nations, FAO Production Yearbooks, Rome. FAO. (1995).

food production since the 1970s. In spite of efforts to improve agricultural technology and increase investments in large-scale commercial agriculture, several factors act as deterrents to per capita production increase.

Population is, undoubtedly, a key factor. Among world regions, Africa is experiencing the greatest proportional increase in population. (See Chapter 7, on population.) As population increases, the per capita land resources required to sustain the extensive subsistence production of the labor force in agriculture are declining. During the 1960s, agricultural production grew at a rate of 2.7% a year, about the same as population. Thereafter, agricultural growth slowed considerably, averaging 1.4% from 1970 to 1985, about half the rate of population growth (World Bank, 1989).

The trend in food exports and imports mirrors the tragedy of agricultural production—decreasing export production and increasing food imports to supplement domestic production declines. In the early 1960s, when many African countries became independent, Africa was a net exporter of an average of 1.2 million tons of staple food (Eicher, 1992), with west Africa accounting for all of Africa's net exports of staple foods (mainly noncereals, such as groundnuts and palm oil) at independence (Paulino, 1987). The 1970s saw a disappearance of this trend of net exports in staples. Eicher (1992) summarizes the trend as follows:

1. In the mid-1970s, Africa was importing 2.5 million tons of staple food a year, with west Africa contributing half of total imports.
2. The negative trends from 1960 to the mid-1970s reflected west Africa's change from exporter to net food importer.
3. West Africa's decline in food exports resulted from the low growth rate of millet and sorghum production and the decline of groundnut production—the three dominant crops in the region.

The decline in food exports would not be so critical if agricultural production met domestic demands for food. Accelerated levels of food imports, especially for crops in which Africa has previously enjoyed a comparative advantage, underscore the failure of economic-development policies in some countries. Caught in a web of competing demands for scarce resources, decision makers seem to have taken the road of rapid, large-scale industrialization at the expense of agricultural production. Nigeria is an archetype of countries in this category. Nigeria, among other countries, imports large amounts of palm oil from Malaysia to meet its growing domestic demand. According to Eicher, the experience is "humbling to Nigeria because at independence in 1960, it was the world's leading producer and exporter of palm oil" (Eicher, 1992). Faced with food shortages, many countries experience serious problems of rising food imports, rising prices, and decreased productivity (Binns, 1992). But government policies are only partially responsible for the worsening conditions in agriculture. Recurrent droughts in Africa, especially in the Sahelian zone, imperil lives and agricultural production, despite the tenacious effort of countries in this zone to combat the problem.

Export-crop production is important to the economies of most African countries. Although the quantity and quality of production vary by country, cocoa, coffee, rubber, sugar, seed cotton, tea, groundnuts, tobacco, and palm oil, among other items, contribute significantly to each country's export earnings. For example, in 1988, coffee accounted for 97% of Uganda's total export earnings and 82% of Rwanda's. For Ghana, cocoa provided 59% of total export earnings in 1987; groundnuts made up 81% of Gambia's earnings in 1986. For many other export-dependent countries, the percentages are just as high. Such a dependency makes African countries vulnerable to price slumps in the world market. The devastating results are felt in many countries.

Livestock production trends in Africa are constrained by a lack of accurate headcounts for producing countries. Most of the livestock is raised by nomadic people, and the nomadic lifestyle precludes long-term settlements that make accurate censuses of livestock possible. Moreover, owners of livestock are unwilling to supply the necessary information, for fear of taxation (Udo, 1983). Production trends, therefore, are based on available data that combine actual headcounts and estimates.

In 1994, there were, by estimate, 192 million cattle, 385 million sheep and goats, 21 million pigs, and 976 million chickens (Table 13.5). Between the years 1979 to 1981 and 1994, there was an 11% increase in cattle, a 19% increase in sheep and goats, a 108% increase in pigs, and a 73% increase in chickens. By 1999, Sub-Saharan African farmers raised 223 million cattle, 446 million sheep and goats, 27 million pigs, and about 1 billion chickens. The increase in livestock numbers could reflect several factors. First, the increase in cattle could be the result of recovery from the droughts in the 1970s and 1980s, which dealt a severe blow to the cattle-herding economy of Sub-Saharan Africa. It could also indicate an adjustment mechanism on the part of nomadic herders, who might keep more cattle so that, when droughts reoccur, they will have a higher chance of survival. (Ironically, excessive numbers of cattle can exacerbate the degradation of already fragile resources of land and water in dry areas.)

Chickens, pigs, sheep, and goats are, however, a different story. Their increase perhaps represents a diversification in agricultural production, especially in areas where population pressure is so high as to be prohibitive of extensive land-use practices. Raising these livestock, in combination with limited crop production, constitutes a prudent

TABLE 13.5 Livestock Populations in Africa, 1979–99 (thousands of head)

Years	Cattle	Horses	Goats	Pigs	Sheep	Chickens
1979–81	172,011	3671	138,228	10,144	180,465	564,000
1992	189,658	4688	168,452	18,923	206,843	930,000
1993	189,244	4763	171,468	20,484	205,785	953,000
1994	192,180	4758	176,089	21,080	208,845	976,000
1999	223,343	4863	205,639	27,017	240,342	1,142,000

Source: FAO. (1994) and (1999). *Yearbook of Production*, Rome, Italy: FAO.

use of scarce land resources. In densely populated rural areas suffering from extreme land scarcity, the intensification of livestock production, especially chicken, pigs, and goats, is an adjustment that is, perhaps, long overdue. Chicken production is experiencing rapid growth, because it does not demand large amounts of land, capital, and labor. Very few rural producers use commercially produced chicken feed; rather, they use discarded grains and other crop residues to feed the chickens. The meteoric rise in the number of pigs, between 1981 and 1999, is another reflection of the attempt to diversify agricultural production. Like chickens, pigs are raised almost entirely on family-produced garbage. Although initial cost of purchasing the pigs can be daunting, the small operating cost and the high expected returns continue to lure some farmers into this line of production.

Fish production in Sub-Saharan Africa is concentrated in riverine, coastal, and lacustrine areas. Western and eastern Africa rank highest in Africa in terms of total fish catch; compared with other world regions, however, they rank among the lowest (Figure 13.7). With a total catch of about 3.2 million metric tons in 1993, Africa produced 3.2% of world total catches. Some comes from northwest Africa, but most of this catch comes from Sub-Saharan Africa. In 1993, Africa as a whole was second to Asia in catches from inland waters, with over 10% of the world total (Figure 13.8). In terms of nominal catches, however, inland and marine catches were about the same, with 1.8 (1.98 tons) and 1.5 metric tons (1.65 tons) for inland and marine-water catches, respectively. Despite the apparent parity in marine- and inland-water catches, inland-water fishing employs more people than marine fishing and, thus, contributes more to the economies of affected areas. According to O'Connor (1991),

> There is some large-scale fishing off both the Atlantic and Indian Ocean coasts, but this takes place from the major ports and so does not form part of the rural economy. Some employment is provided by this activity, but far more people are occupied in small-scale fishing from canoes in inshore waters, in coastal lagoons, in lakes and in rivers.

Marine fishing is still underdeveloped; it involves a small number of relatively large fishing vessels. In the past, the fishing operations and vessels were owned, almost exclusively, by foreign companies. Today, African businesspeople, especially in Nigeria and Ghana, are joining foreign companies to invest in trawler fishing, and some governments are providing the needed infrastructure for the expansion of marine fishing (Udo, 1983).

Inland fishing resources are the backbone of the economies of many communities living near coasts, rivers, lakes, and lagoons. Important inland fishing areas include the Niger Delta, the middle Niger Valley, and Lake Chad in west Africa; some coastal areas of eastern Africa, for example, the island of Zanzibar; and the rivers and coast of equatorial, southern, and Southeastern Africa. Fishing is a specialized activity for some groups or communities in Sub-Saharan Africa, who practice it full time or part time.

Generally, the operations are small in scale, typically requiring no sophisticated fishing techniques and equipment. From small canoes, fishermen catch fish with small nets.

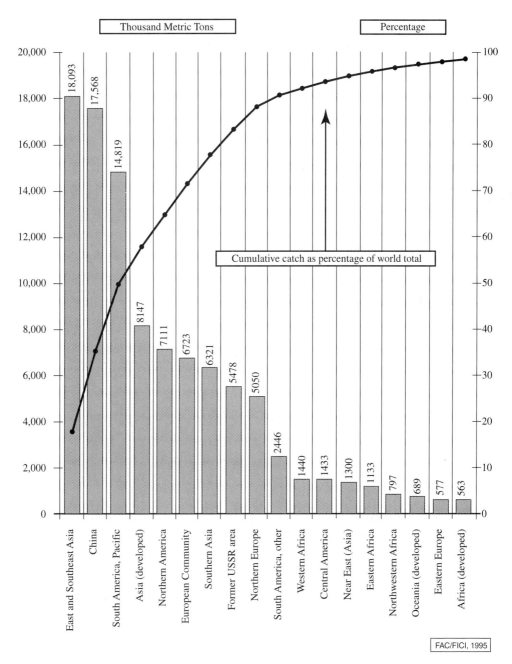

Figure 13.7 Total catch by principal regions in 1993 and cumulative catch as percentage of the world total. (From FAO (1995). *FAO Yearbook of Fishery Statistics 1993*, Volume 77, Rome, Italy: Food and Agriculture Organization of the United Nations.)

by Continents in 1993

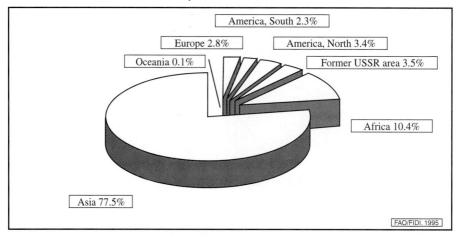

by Continents, 1970–1993

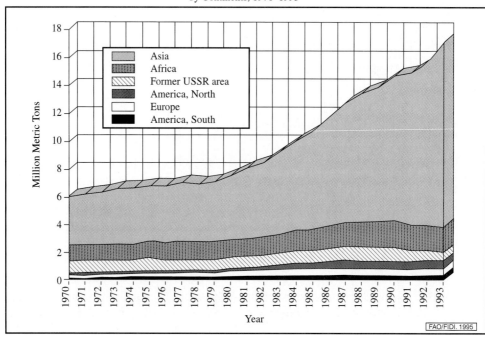

Figure 13.8 Catches in inland waters, by continents. (From FAO (1995). *FAO Yearbook of Fishery Statistics 1993*, Volume 77, Rome, Italy: Food and Agriculture Organization of the United Nations.)

TABLE 13.6 Africa, Nominal Fish Catches in Inland Waters by Species Groups, 1987–93

Group		1987	1988	1989	1990	1991	1992	1993
Freshwater	11	87,472	85,247	109,828	110,180	104,719	82,956	82,121
	12	222,617	254,039	235,369	287,764	298,228	278,991	248,381
	13	977,951	1,022,193	998,190	1,026,474	990,250	981,563	1,020,668
Diadromous	22	169	107	110	89	52	72	123
	23	1068	1238	1175	1500	1806	1602	1686
	24	93,032	68,211	77,182	90702	85,999	86,525	90,345
	25	280,944	335,476	351,224	375,514	289,064	311,115	311,444
Marine	33	1324	1198	1183	3368	5368	2462	3846
	34	7964	10,417	18,728	21,067	21,719	13,470	19,604
	35	4059	2217	2465	2500	2440	2400	2290
Crustaceans	41	6472	5048	6475	6474	6122	6,074	4886
	42	59	66	55	101	216	110	318
	45	3531	3075	4723	4723	6375	4689	5732
Molluscs	51	16	6	18	13	20	20	20
	56	–	26	10	6	8	7	10

Source: FAO. (1993). *Yearbook of Fishery Statistics*, Volumes 76 and 77, Rome, Italy: FAO.

Notes:

11 Carps, barbels, and other cyprinids

12 Tilapias and other cichlids

13 Miscellaneous freshwater fishes

22 River eels

23 Salmons, trouts, smelts, etc.

24 Shads, etc.

25 Miscellaneous diadromous fishes

33 Redfishes, basses, congers, etc.

34 Jack, mullets, sauries, etc.

35 Herrings, sardines, anchovies, etc.

41 Freshwater crustaceans

42 Sea spiders, crabs, etc.

45 Shrimps, prawns, etc.

51 Freshwater molluscs

56 Clams, cockles, arkshells, etc.

Most of Sub-Saharan Africa's nominal fish catches in inland waters are predominantly of the freshwater and diadromous species groups (Table 13.6). From 1987 to 1993, tilapia and carp were the dominant types of freshwater fish. Diadromous species, which migrate from salt to fresh water environments, have gained some prominence in the fishing industry of Sub-Saharan Africa. Salmon, trout, smelt, shad, and other diadromous fishes remain important to the fishing industry, in terms of annual nominal catches.

PROBLEMS OF AGRICULTURAL PRODUCTION

Rural Infrastructure

The inadequate infrastructure in Africa's rural areas compounds the problems of agricultural production, storage, transportation, and distribution and of the expansion of cultivable land through irrigation. Many African countries are plagued by the lack of effective links to the cities—the primary market for most agricultural commodities or the bulk-gathering point for export crops. Moreover, existing transportation networks are poorly maintained, causing untold problems related to the evacuation of farm produce, especially during the rainy seasons. (See Chapter 9.) Without access to both urban and rural markets, exchange and trade are limited. Perishables, such as vegetables and fruits, are especially affected. Delay can cause these products to be lost or, if sold, to fetch only a small fraction of their optimum monetary worth.

Rural agricultural producers lack adequate storage for their crops. This is one reason that seasonal food shortages persist. Robust harvests often provide more than enough food either for family consumption or sale. Because of the relative "oversupply" during the harvest season, food is abundant and prices are low. In contrast, without efficient storage systems, food is scarce and costly in the off-season. Even where such traditional storage systems as mud granaries, barns, clay pots, and metal drums are made available, they often expose the crops to destruction by pests and theft. The losses attributable to these sources reduce the quantity and quality of stored grains, root crops, and vegetables needed for sale or replanting. To guarantee the availability of seeds for the next farming seasons, farmers often conserve whatever is left for planting, thus reducing the family food supply and its income from farm sales.

Water and electricity are also scarce in rural Africa. In the savanna grasslands of Sub-Saharan Africa, grain production is confined to areas receiving enough rainfall and to areas with access to irrigation water from rivers, dams, and lakes. Grazing is also a function of water availability, as pastoral nomads move constantly in a search of suitable pasture and water. Electricity might not appear to be a key factor in rural peasant production, but it is. Rural areas rarely are connected to the national electric power grid. Services are concentrated in the cities, and extension to rural communities is often not a high priority for the government. Progressive communities use self-help efforts to bring electricity to their population. Without power, basic mechanization processes to increase agricultural intensity and yields are not deployed. Processing and preprocessing of agricultural commodities is usually reserved for a small minority of wealthy entrepreneurs, who buy raw produce at low cost for higher profits after processing.

Institutional Factors

Institutional arrangements, such as land-tenure and land ownership, can affect agricultural development. Traditional land tenure systems are characterized by communal ownership, in which the land belongs to the community. Individuals who are members of the community have rights to use the land. Community members who hold title to the land do so on a funding limits research productivity. Since the mid-1960s, there has been a 35% decline in total research output, and African countries are spending only an average of 2.7% of their operating budgets on research (Morna, 1991). The second problem is that research centers tend to be islands unto themselves, often prescribing solutions that are scarcely relevant to the majority of rural people who surround them. It is not uncommon to find that new techniques and methods of production developed at these institutions are applauded for their efficacy in improving agricultural production, while most subsistence farmers, even near the centers, still practice primitive agriculture. The lack of diffusion and adoption can result from low levels of education and from lack of technical skills among farmers or from the paucity of extension agents trained to diffuse the innovations to needed areas. The third problem relates to how agricultural advances and new techniques are distributed. Technological advances follow a hierarchical diffusion process, in which people and locations with skills and know-how will receive available knowledge and techniques. Areas and people lacking in these characteristics, even if they are located close to the centers of innovation, are bypassed.

In addition to the research production within Sub-Saharan Africa, the continent receives technical training and personnel assistance from foreign countries and international agencies, such as the World Bank. Such assistance is largely targeted to the needs of individual countries or to selected projects within countries or regions of Sub-Saharan Africa. These assistance programs often come with strings attached. In his critique of World Bank programs, Carl Eicher argues that the initiative by the World Bank to build a consensus among African policy makers and donors for an Africa-wide capacity building, with emphasis on economic policy and management, has met with limited success, because

> this orientation is excessively narrow and is tied to the Bank's current fixation on structural adjustment programs, rather than on Africa's long-term needs. Again, the Bank is providing financial but not intellectual leadership in a key problem area (Eicher, 1992).

Another reason for the failure of research and technical solutions is the African attitude toward technology and innovation. Some experts have argued that, for technology to be employed in creating acceptable levels of rural development, it must be seen as "an attitude of mind, not an assemblage of artifacts" (Achebe, 1982, cited by Stamp, 1989). After attaining independence, most African countries began a march toward increased prosperity, based on the euphoria of technological emancipation. Technical solutions to agricultural and industrial problems were projected to lead Africa to increased productivity and consumption. The design of economic development plans was, therefore, predicated upon the urgent need to achieve rapid industrialization and mechanization. In the tradition of the center–periphery models employed successfully by Europe and North America, the concentration of technology in the cities was expected to erode rural backwardness by infusing new techniques of production into rural agricultural production. In the race for rapid economic development, many African countries see the amassing of "artifacts" as an end, rather than as a means to the *crucial* end: meeting the basic production needs of the population. As a result, millions of dollars[1] worth of sophisticated machines and equipment,

---

[1] Monetary unit is in US Dollars.

modern production methods, and large-scale production systems are concentrated in a few hands and so are not accessible to the majority of the population responsible for most of the production in agriculture. Peasant farmers, for lack of better alternatives, persist in their low-yielding, traditional production methods.

Governments, International Agencies and Agricultural Markets

From the colonial period to the present, African governments have influenced agricultural production and consumption through policy and other types of direct intervention. Colonial governments encouraged the expansion of export-oriented agriculture to the detriment of food production. This type of expansion led to attempts "to confine nomadic pastoralists to limited grazing land" in an effort to increase the amount of arable land available for export crops—an attempt that probably exacerbated overgrazing (West Africa, 1989). The demand for export crops invariably shifted emphasis from food production to the production of plantation crops, such as cocoa, oil palm, rubber, coffee, tea, cotton, and peanuts. In postcolonial Africa, global price fluctuations, rather than government intervention, have been mostly responsible for the deemphasis on plantation agriculture. For example, the economies of countries from Senegal to Cameroon, Sudan to Tanzania, and Democratic Republic of Congo to Angola have been greatly affected by declining revenues from plantation crops.

The majority of Sub-Saharan African governments pursue policies to increase or diversify agricultural production, especially food production. Under these policies, programs are designed either to increase total food production in a country or to provide incentives for primary school, secondary school, and college graduates to contribute to agricultural production. In the 1960s, "farm-settlement" or "land-settlement" schemes were established in some countries (e.g., in Nigeria and Tanzania) to increase food and commercial-crop production. In the 1970s, 1980s, and 1990s, some governments experimented with programs designed to mobilize college graduates for agricultural production and to ensure self-sufficiency in food production. The "Operation Feed the Nation" programs in Nigeria and the "Operation Feed Yourself" program in Ghana are examples of such programs. A shortcoming of these programs is that, despite good intentions, they take resources away from rural, traditional producers. Moreover, many people engaged in the programs see them as stepping stones to other, nonagricultural jobs.

Governments also intervene in agricultural production by influencing market prices for agricultural commodities. Marketing boards have played a major role in controlling agricultural prices, especially the price of export crops. They serve as middlemen between export-crop producers and export crop-purchasers, who are usually from foreign countries. As government agencies, the boards set the prices of agricultural commodities that small-scale commercial and peasant producers are to receive—such prices are usually substantially lower than market price—while they sell to overseas customers at world-market prices.

In a number of African countries, marketing boards provided a direct mechanism for suppressing the prices producers receive relative to world-market prices, via the granting to marketing boards of monopoly power over the pricing and marketing of cash crops (Lovelace, 1998). In addition to suppressing the prices that producers receive, marketing boards have been criticized for being inefficient in distributing agricultural inputs, with some producers receiving fertilizer after harvest (Shepherd and Farolfi, 1999; Krueger, 1992). Furthermore, some critics have argued that marketing boards limit competition among producers and deny them the opportunity to gain marketing experience (Bauer, 1963).

Some experts, however, see the advantages that marketing boards provide, arguing that marketing boards "enabled forward sales to occur between overseas buyers and exporters, exporters and the marketing board, and ultimately through pre-announce prices to domestic producers" (Lovelace, 1998). Marketing boards have also been seen as having some effect in stabilizing incomes, if not prices, from year to year (Hodder and Gleave, 1992). In an effort to reform their economies, many African countries have accepted conditions set forth by the World Bank and International Monetary Fund (IMF) in their structural-adjustment programs (SAPs). In compliance with such reform programs, some governments have withdrawn their control of marketing boards and have privatized their management.

Governments and international agencies also intervene in agricultural production by infusing needed capital and technical resources into production. With the help of agencies like the IMF, the World Bank, the Food and Agriculture Organization (FAO) of the United Nations, the European Community, and other such organizations, governments push for agricultural modernization and intensification. In some projects, attempts are made to conserve the natural environment while increasing the quantity and quality of agricultural production. More and more, governments are forced to accept unfavorable structural adjustments to their economies in order to qualify for international loans. Critics of structural adjustment cite high inflation rates in some countries as a problem, especially since this reduces the purchasing power of the loans or grants received for agricultural improvements. A difficulty with government attempts to improve agricultural production is that, oftentimes, the majority of poor farmers have limited education and skills to benefit from technical programs; thus, the advantages go to a new breed of educated farmers.

The World Trade Organization and Agriculture in Sub-Saharan Africa

Agricultural production in Sub-Saharan Africa is not immune to the vagaries of global economic conditions and the associated biotechnological changes. As such, countries in the region are affected by policies imposed by the developed market economies, which seek trade liberalization and economic globalization. Measuring the impact of these changes is daunting, but the case of WTO, with its associated Agreement on Agriculture (AOA) and the Trade-Related Aspects of Intellectual Property Rights (TRIPS), provides ample evidence of some of the global challenges facing agriculture in Sub-Saharan Africa.

WTO was created in 1995 to replace the General Agreement on Tariffs and Trade (GATT), which had been in existence since 1947, as the organization overseeing the multilateral trading system. WTO is a rules-based, member-driven organization in which the member governments make all decisions and the rules are the outcome of negotiations among members. The governments that had signed GATT were officially known as "GATT contracting parties." Upon signing the new WTO agreements (which include the updated GATT, known as GATT 1994), they officially became known as WTO members. WTO is the only global international organization dealing with the rules of trade between nations.

The agreements are negotiated and signed by the bulk of the world's trading nations and ratified in their parliaments. The goal is to help producers of goods and services, exporters, and importers conduct their business.

Two WTO agreements have profound effects on agriculture in Sub-Saharan Africa. The first agreement—AOA—seeks to promote free trade by obliging country members to reduce their support for agriculture by providing market access, reducing domestic support for agriculture, and removing export subsidies. Some critics, such as the Africa-Europe Faith and Justice Network (AEF), argue that, rather than bringing African member countries into lockstep with their more developed counterparts, AOA is widening the competitive gap between African members and members from industrialized countries. Thus, as African countries have opened their borders to food imports, they face difficulty in exporting their agricultural products to the industrial world, because the European Union (EU) and the United States (USA) have maintained their subsidies. The undesirable effects on Africa are summarized as follows:

> The AOA has favored the transnational companies (TNCs) but has become a great threat to small farmers and rural ecologies in Africa, which [have] is now suffering its impact. African markets are often flooded with European and American cheap subsidized agricultural products (mainly meat and cereals) that [cause the decline of] local prizes, deteriorating the livelihood of farmers and threatening local food security (AEFJN, 2001).

The critics also contend that AOA is an unfair trade agreement, which allows low-priced, sometimes poorer quality food imports to replace local and traditional foods. Moreover, WTO has been accused of using AOA to promote a shift toward export or cash-crop production, or both, and away from production for meeting local needs. The drive for expanding production land for exports is closely associated with the loss of land for subsistence production, especially among women farmers. Perhaps on the basis of these anomalies, some African countries, among them Kenya, Nigeria, Uganda, and Zimbabwe, have joined other developing countries to call for a *special and differential* treatment for developing countries, because of the importance of agriculture in these countries.

WTO is also affecting agriculture in Sub-Saharan Africa through the stipulations contained in TRIPS. By depriving farmers of the right to produce, select, exchange, and sell their own seeds, TRIPS threatens centuries of indigenous knowledge in Sub-Saharan Africa about the cultivation, nurturing, protection, and selection of seeds for genetic improvement. TRIPS grants patents on "inventions," including patents for genetically modified (GM) plant varieties, organisms, genes, and proteins. WTO members have a system of intellectual property rights for plant varieties. What potential problems does TRIPS pose for farmers in Sub-Saharan Africa?

- Will small farmers be able to compete with TNCs? Rich TNCs have a capacity that small farmers do not have for privatizing seeds and genetic resources by securing patents and by increasing the number of genetically modified organisms (GMO) for sale and for cultivation. What will be the cost to farmers who are forced to buy seeds from a company, if they are not allowed to reuse, exchange, and sell seeds from the seeds companies or must pay the companies if they violate the agreement?

- What might be the loss to biodiversity? Since GM seeds favor the practice of monoculture, what will be the impact on areas of polyculture, most of which are in the developing countries, especially in Sub-Saharan Africa? One speculation is that the use of GM seeds will cause the loss of biodiversity by discontinuing the practice of growing a variety of crops as insurance against the vagaries of climate.
- How will farmers cope with increased costs? As in the case of hybrid crops, if GM plants are programmed to produce only with the help of certain chemicals, will the small farmers be able to afford the added costs, and will they be dependent upon the companies with the seed and chemical patents?
- Will local communities with centuries of traditional knowledge be compensated, or will TRIPS promote *biopiracy*? The south contains the bulk of the earth's genetic plant resources (90%), yet a small number of northern corporations control over 86% of plant patents worldwide and are driving the rush to patent plant genetic resources from the south (AEFJN, 2000). The countries of the south have, for centuries, cared for those plants and discovered and applied their efficacy as medicines and food resources. If countries from the north, which have undue technological and financial advantage, take the genetic material from the plant for private gain, does TRIPS have a provision for compensating the owners of the native plants, or will it amount to stripping those local communities of their centuries of traditional knowledge?

POLICY OPTIONS

Upgrading Rural Infrastructure

Infrastructural development policy should have three foci. The needs are (1) to improve the linkages between rural agricultural producing areas and the urban centers of food and raw-material consumption, (2) to expand linkages between agricultural producing areas and their local markets, and (3) to facilitate the environment of production, storage, and distribution of agricultural commodities.

The majority of Africans, who are living and working in rural areas, particularly in agriculture, are still waiting for governments to deliver on the grand promises made to improve rural well-being by connecting rural and urban regions in a veritable network of mutual relationships. Polarization between urban and rural areas means that rural infrastructure has deteriorated through wanton neglect and poor maintenance. Although most African countries have more miles of roads today than at independence, many of these roads are in serious disrepair, thus inhibiting interrural and rural–urban linkages. These linkages are vital to the successful operation of agriculture and deserve policy changes to enhance them.

The loss of already harvested crops, the lack of irrigation infrastructure, and the inadequate facilities for housing and transporting livestock pose problems for agricultural areas. Cooperative grain storage bins and the extension of irrigation canals in dry areas are examples of areas needing public policy support. To reduce cost constraints, policies should emphasize cooperation rather than individual efforts.

It is important to mention that, after decades of waiting for government help (with little success), many rural villages and agricultural producing areas are embarking on massive self-help projects to improve rural infrastructure. In Nigeria, Kenya, Tanzania, and other African countries, rural people are undertaking projects to construct roads, to connect water and electricity, and to build schools, churches, and other public buildings. Although inter-village rivalries are leading to the proliferation of infrastructure in rural areas, the small-scale, village-specific nature of these projects is creating unnecessary duplication and excessive costs. Government policies should be aimed at increasing cooperation and reducing cost. For example, it is not necessary for each village to have its own electric transformer, a water borehole, and a water storage tank, as often happens in some countries. Government can provide basic infrastructure for a larger region or group of villages; even if a tax system is imposed, the cost will be less than the cost presently incurred per capita for these projects.

Education

One of Africa's problems is that, despite proclamations to the contrary, educational systems are still elitist, technocratic, and urban biased. Whether it is by fault or design, the attainment of educational qualifications is perceived as a preparation to escape the tedium of rural life. Policy for rural development has always suffered from the "identification of agriculture with lack of education and the latter with nonagricultural employment" (Ega and Famoriyo, 1988). The irony is that "while urban-based technological innovations are expected to diffuse to the rural area for improved production and well-being, those who stand to benefit from the new changes are making a speedy and mass exodus from the rural areas" (Obia, 1994).

African nations should educate rural people on the techniques of sustainable agriculture. Because old methods of production are firmly ingrained in the rural culture, governments must provide incentive programs to help with the experimentation of new productive systems. Experience suggests that regimented, return-to-land policies are not the answers, as evidenced by the failure of Farm Settlements in Nigeria and the Ujamaa program in Tanzania. Incentives should include providing adequate infrastructure to rural areas, rewarding the initiative to experiment with new techniques by giving farmers access to special funds and capital, using agricultural extension programs to teach combined methods of yield improvement and land conservation, and helping farmers find new markets for experimental crops that are not part of the domestic staple.

Developing Appropriate Technologies

A major problem with technological transfer in Africa is the tendency to adopt conventional development paradigms, without putting them within the context of Africa's socioeconomic, political, cultural, and ecological conditions. According to Smith (1983, also cited in Bell, 1986),

> Throughout the world, economic growth and social change have taken place within very varied political and cultural contexts, and hence such apparently uniform developments as technical advance, rising material standards and output, rapid urbanization, the commercialization of agriculture, the redistribution of wealth and the provision of a wider range of social services

for the population, take on a different significance and have consequences according to this political and cultural context.

There is inadequate understanding of this context, and, without it, technological innovations have been applied to environments that possess neither the readiness nor the knowledge to foster desired technological change. For technology to be employed effectively in creating acceptable levels of rural agricultural development, it must be seen as "the raw material created out of historical experience, which, in turn, recreates society" (Stamp, 1989). Sub-Saharan African countries have failed in the task of incorporating such historical experience in recreating their society.

What policy options exist? Technological diffusion to rural areas must recognize the role of women in production and consumption, and any constraints on full-scale participation should be corrected. Making technology available for use in rural areas guarantees neither incorporation into rural production nor effective use. One primary reason often cited for the ineffective application of technology to rural production and consumption is the lack of support for women; women are the primary force behind rural production. As Agarwal (1986) argues in the case of wood-stove adoption in India:

> The status of women within the household could be a significant factor in wood-stove adoption, especially where adoption requires cash expenditure, by virtue of the fact that although women are the potential users of the innovation, and therefore in the best position to assess its advantages and disadvantages, it is men who usually handle the household cash and make decisions on how it is spent...

Credit policies should give women access to capital, for the purchase of new technologies, or as an income supplement while they are experimenting to incorporate the new technology into their production system. Uganda's Rural Farmers' Scheme, Zimbabwe's Agricultural Finance Cooperation, Malawi's Mudzi Bank, and Kenya's Agricultural Finance Corporation are examples of institutions that are helping to extend credit to farmers, especially women.

Technology for rural agricultural production should be relevant to the status of the users (especially educational status and training, as well as skills and awareness). Where needed, education, training, and awareness programs should accompany any attempts to introduce technology to rural farmers. Policies should encourage the use of technologies that increase productivity and protect the environment. Incentives or cross-compliance should be written into such policies, to reward those who practice systems of production with built-in mechanisms for protecting the environment.

Developing Alternative Energy Sources

Most rural people in Africa depend on fuel wood for their domestic energy needs. It was once common for the farmland near settlements to provide more than an adequate supply of fuel wood for a family; only on rare occasions did fuel-wood gathering extend into unfarmed wooded areas. Today, women and children travel many miles in search of wood—an exhaustive and frustrating effort that is not always rewarded. Without choice, desperate rural women gather material that should be left to cover or fertilize the soil, thus exacerbating the problems of soil erosion, leaching, and soil nutrient depletion. This condition creates a vicious cycle in which poverty begets land degradation and low productivity, which, in turn, create an environment for poverty to fester.

Although some development efforts are extending electricity and water to rural areas, especially through community participation, the heart of the problem remains unresolved. Poor rural people can least afford the expense of conventional energy, leaving them forced to continue with their usual practices, which in turn create environmental degradation and poverty. Future development must identify clear alternatives to fuel wood to curb the impact of wood gathering on productivity and the environment. Such development should be targeted to the women, who are the major primary workforce in rural areas.

Agro-Based Industries

Processing or preprocessing of agricultural products is a necessary condition for increasing rural family incomes from the sale of agricultural products and for diversifying the rural economy. This is one of the success stories in African peasant agriculture. In the Abia state of Nigeria, for example, the cooperation between colleges of agriculture and technology and the farming community is yielding benefits. The extraction of oil from the palm fruit is now carried out with small-scale oil-palm presses, which saves time and labor cost over manual processes. The same press is also used for the processing of cassava, the main staple in the state. Not only are costs of processing reduced, but also the women who traditionally perform this tedious chore are left with more time to tend to agricultural production or to trading activities, in order to augment family income. Moreover, the shortage of labor arising from the migration of able-bodied young men to the city is being remedied by the use of these machines. In addition to large-scale oil mills, wood-processing plants, and grain mills, agro-based industries are providing added value to agricultural products, by helping farmers to sell or export processed or preprocessed products rather than raw materials.

Community-Based Cooperatives

The idea of community-based cooperatives is perhaps as old as Africa itself. Mutual assistance for various kinds of activities is solidly entrenched in most African cultures. In agricultural villages, the peasantry relies on cooperative communal labor from members of the immediate and extended families and from friends for land clearing, planting, weeding, harvesting of crops, and tending to livestock. At the broader community level, road construction and maintenance, public construction of school and church houses, and harvesting of community-owned plantation crops were done through community-based cooperative action.

Today, the acquisition of higher educational standards and training and the need for appropriate employment in the urban centers have reduced the pool of able-bodied young men in the rural areas. This condition has created a scarcity of agricultural labor, thus making labor competitive and, sometimes, a limitation on farm productivity. Now that the male labor force is reduced, the bulk of farmers are women. From east to west Africa, from equatorial to southern Africa, women's groups are forming community-based cooperatives to share their labor for agricultural activities, to raise capital to diversify the rural economy, and to acquire land for large-scale farm enterprises. Policy initiatives for such cooperatives used to begin at federal, state, or local governments, but are increasingly coming from grassroots organizations initiated by village women. Cooperatives are important in stimulating rural agricultural production and in encouraging participatory decision making for rural development.

Integrated Rural Development

Resource conflicts are common in Africa, particularly between villages, between communities, or between ethnic groups. Land conflicts are especially sensitive, because land confers upon a community certain inalienable rights. To take such rights away is tantamount to a declaration of war. Therefore, bringing villages or communities together for common projects involving the redistribution of land, as with most rural agricultural projects, requires a firm and clear policy that is closely monitored by a neutral agency of government. This will focus the attention of the communities on the important issue of building a cooperative strategy to promote development. Integrated development programs are not new in Africa, but they need to be expanded into areas where physical or environmental resources cross community or state boundaries. A good example is the development of river basins or large forest reserves. There is no doubt that Sub-Saharan Africa needs to harness the power of rivers and lakes for hydroelectricity, irrigation, fishing, and recreation. Yet many rivers cross community, regional, state, or national boundaries. A dam constructed in one state, for instance, could leave a lake that floods land in another state or region. An integrated project will provide the mechanism for mediating conflicts, so that a greater good can be realized through comprehensive, regional development strategies.

African countries are already reaping the benefits of intranational or international integrated projects. Nigeria's River Basin Development Authorities provide a good example of integrated rural development projects. The entire river basin, rather than individual states or communities, is designated as the focus of multifaceted economic development, such as irrigation, fishing, crop and livestock production, and the provision of drinking water. Because of government intervention, policies define the system of land acquisition and compensation, as well as the location of facilities within the project. When conflicting international interests are involved, conflict resolution can become a tough challenge. The purpose of this discussion is not to list all available integrated river projects in the study area, but to point them out as examples of regional and international cooperation to stimulate economic development, especially in rural, nonurban areas that possess these resources.

Integrated rural development at the level of villages and communities is overdue in many parts of Africa. The proliferation of village-centered infrastructural development projects increases the urgency of the need for integrative programs that use the resources—human and natural—for more effective large-scale projects. Policies should address the cost-effectiveness of larger projects, which can provide such amenities as water, electricity, schools, and public transportation to several villages or communities.

CONCLUSION

This chapter has articulated the necessary dichotomy between traditional and commercial agricultural development in Sub-Saharan Africa. It has shown that traditional and commercial agricultural systems respond to changes in social, political, economic, and environmental conditions of nations. The analysis of agricultural production supports the perception of a trend of long-term decline. Factors responsible for this decline include inadequate rural infrastructure, several institutional arrangements such as land tenure and land ownership, land degradation, rapid population growth, the use of inappropriate technology, limitations in research and development directed at improving production, and various constraints imposed by governments, international agencies, and agricultural markets. The chapter suggests various options for improving the quality and quantity of agricultural production: improving rural infrastructure, using education to teach proper land ethic, developing appropriate technologies and alternative energy to reduce land degradation, and increasing agricultural production by encouraging community cooperatives. The importance of focusing agricultural policy on alternative production methods and techniques are emphasized, in order to expand levels of agricultural production, reduce population pressure on agricultural land (especially on marginal land), and ameliorate environmental degradation.

KEY TERMS

Traditional farming systems	Commercial farming
Permanent cropping	Marine fishing
Shifting cultivation	Rural infrastructure
Mixed farming	Land degradation
Intercropping	Green revolution
Rotational bush fallow	Rural infrastructure
Compound farming	Appropriate technologies
Pastoralism	Agro-based industries
Cash-crop farming	Community-based cooperatives
Semicommercial farming	Integrated rural development

DISCUSSION QUESTIONS

1. Describe the main farming systems of Sub-Saharan Africa.
2. What types of farming make up the main farming systems discussed in question (1)?
3. Discuss the similarities and differences between commercial farming in Sub-Saharan Africa and the United States.
4. Identify the major nomadic groups in Sub-Saharan Africa and describe the characteristics of each.

5. What are the key constraints to pastoral nomadism and how do nomads adjust to these constraints? How are these constraints and adjustments different from, or similar to, those in developed countries?
6. Give an account of the agricultural production trends in Sub-Saharan Africa for crops, livestock, and fish.
7. What are the main reasons for the decline or stagnation of agricultural production in Sub-Saharan Africa, and what steps can the countries of Sub-Saharan Africa take to improve the quality and quantity of their agricultural production?
8. Give a critical evaluation of the impact of WTO in promoting or hindering agricultural production.
9. In what ways can African universities and research institutions help to improve the state of food production and distribution?
10. How does population affect agricultural production in Africa?

REFERENCES

ACHEBE, C. as quoted in STAMP, P. (1989). *Technology, Gender, and Power in Africa*, Ottawa, Canada: IDRC.

AFRICA-EUROPE FAITH AND JUSTICE NETWORK (AEFJN) (2001). "Impact of TRIPS and the WTO on the Agriculture in Africa," AEFJN, *http://www.aefjn.org/english/publications/agri_e0102.htm*.

AGARWAL, B. (1986). *Cold Hearths and Barren Slopes: The Woodfuel Crisis in the Third World*, Riverdale, MD: The Riverdale Co., Inc.

ALLAN W. (1965). *The African Husbandman*, Edinburgh, Scotland: Oliver and Boyd.

BARBIER, E.B. (1989). "Sustaining Agriculture on Marginal Land: A Policy Framework," *Environment*, 31(9):12–17, 36–40.

BAUER, P.T. (1963). *West African Trade*, London: Routledge and Kegan Paul.

BELL, MORAG. (1986). *Contemporary Africa*, London: Longman.

BENNEH, G. (1972). "Systems of Agriculture in Tropical Africa," *Economic Geography*, 48:244–257.

BINNS, J.A. (1992). "Traditional Agriculture, Pastoralism and Fishing," in (1992), *Tropical African Development*, Gleave, M.B., ed., Essex, UK: Longman Scientific and Technical.

BOSERUP, E. (1965). *The conditions of agricultural growth: the economics of agrarian change under population pressure*, New York: Aldine Publishing Company.

DE BLIJ, H.J. & MULLER, P.O. (1994). *Geography: Realms, Regions, and Concepts*, New York: John Wiley & Sons.

EGA, L.A. & FAMORIYO, S. (1988). "Basic Contradiction between Rural Development Strategies and Rural Welfare in Nigeria," *International Journal of Environmental Studies*, 31:157–165.

EICHER, C.K. (1992). "African Agricultural Development Strategies," in *Alternative Development Strategies in Sub-Saharan Africa*, Stewart, F., Lall, S., & Wangwe, S., eds., New York: St. Martins's Press.

EUROPA PUBLICATIONS LIMITED (1995). *Africa South of the Sahara*, London: Europa Publications Limited.

FAO (1995). *FAO Yearbook of Fishery Statistics 1993*, Volume 76, Rome, Italy: Food and Agriculture Organization of the United Nations.

FAO (1995). *FAO Yearbook of Fishery Statistics 1993*, Volume 77, Rome, Italy: Food and Agriculture Organization of the United Nations.

FAO (1995). *FAO Yearbook of Production 1994*, Volume 48, Rome, Italy: Food and Agriculture Organization of the United Nations.

FAO (2000). *FAO Yearbook of Production 1999*, Volume 48, Rome, Italy: Food and Agriculture Organization of the United Nations.

FAO (1992). *The State of Food and Agriculture 1992*, Rome, Italy: Food and Agriculture Organization of the United Nations.

GLEAVE, M.B. & WHITE, H.P. (1969). "Population Density and Agricultural Systems in West Africa," in *Environment and Land Use in Africa*, Thomas, M.F. & Whittington, G.W., eds, London: Methuen.

GRIFFITHS, I.L. (1994). *The Atlas of African Affairs*, London and New York: Witwatersrand University Press.

HARDIN, G. (1968). "The Tragedy of the Commons," *Science*, 162:1243–1248.

HODDER, G.W. & Gleave, M.B. (1992). "Transport, Trade, and Development in Tropical Africa," in *Tropical African Development*, Gleave, M.B., ed., Essex, UK: Longman Scientific and Technical.

HUNGER PROJECT (1988). "Farmers' Unions in Zimbabwe," *The African Farmer*, 1(July):22.

HUNGER PROJECT (1988). "Africa's Regional Research Capacity," *The African Farmer*, Number 1 (July), page 33.

JACKSON, R.H. & HUDMAN, L.E. (1990). *Cultural Geography: People, Places, and Environment*, St. Paul, MN: West Publishing Company.

KNIGHT, G. & NEWMAN, J. (1976). *Contemporary Africa: geography and change*, Englewood Cliffs, N. J.

KRUEGER, A. (1992). *Economic Policy Reform in Developing Countries*, Cambridge MA: The MIT Press.

LELE, U. (1981). "Rural Africa: Modernization, Equity and Long-term Development," *Science*, 211(Feb):547–553.

LOVELACE, J.A. (1998). "Export Sector Liberalization and Forward Markets: Managing Uncertainty During Policy Transitions," *Africa Economic Analysis*, http://www.afbis.com/analysis/financial_markets.htm.

MELLOR, J.W. (1998). "The Intertwining of Evironmental Problems and Poverty," *Environment*, 30(9):8–13, 28–30.

MORGAN, W.B. (1969). "Peasant Agriculture in Tropical Africa," in *Environment and Land Use in Africa*, Thomas, M.F. & Whittington G.W., eds., London: Methuen & Co Ltd.

MORNA, C.L. (1991). "Scholarly Research Drying Up at African Universities as Financial and Political Problems Force Cutbacks," *Chronicle of Higher Education*, August 7:A28.

OBIA, G.C. (1993). "Brain Drain and African Development: A Descriptive Framework for Deriving Indirect Benefits," *Journal of Third World Studies*, X(2):74–97.

OBIA, G.C. (1994). "Technological Change and Rural Development in Africa: Constraints and Prospects," unpublished paper presented at the annual AAG Meeting, San Francisco, California.

O'CONNOR, A. (1991). *Poverty in Africa: A Geographical Approach*, London: Belhaven Press.

ORAM, P.A. (1988). "Land Capacity and Environmental Fragility," *Environment*, 30(9):10.

PAULINO L.A. (1987). As used in EICHER, C.K. (1992). "African Agricultural Development Strategies," in Stewart, F., Lall, S., & Wangwe, S., eds., *Alternative Development Strategies in Sub-Saharan Africa*, New York: St. Martins's Press.

RUBENSTEIN, J.M. (1989). *The Cultural Landscape: An Introduction to Human Geography*, New York: Macmillan Publishing company.

RUTHENBERG, H. (1971). *Farming systems in the tropics*, Oxford: Clarendon Press.

SHEPHERD, A. & FAROLFI, S. (1999). *"Export Crop Liberalization in Africa—A Review,"* Rome, Italy: FAO Agricultural Services Bulletin 135.

SMITH, A.D. (1983). *State and Nation in the Third World*, Brighton, UK: Wheatsheaf; as used in BELL, M. (1986). *Contemporary Africa*, London: Longman.

STAMP, P. (1989). *Technology, Gender, and Power in Africa*, Ottawa, Canada: IDRC.

THOMAS, M.F. & WHITTINGTON, G.W., eds. (1969). *Environment and Land Use in Africa*, London: Methuen & Co., Ltd.

TURNER II, B.L. & BRUSH, S.B. (1987). *Comparative Farming Systems*, New York: The Guilford Press.

UCHENDU, V.C. (1965). *The Igbo of Southeast Nigeria*, New York: Holt, Rinehart and Winston.

UDO, R.K. (1983). *The Human Geography of Tropical Africa*, London: Heinemann Educational Books.

VAN RAAY, J.G.T. (1975). *Rural Planning in a Savanna Region, the Case of Fulani Pastoralists in the North Central State of Nigeria*, Rotterdam: University of Rotterdam, as used in BINNS, J.A. (1992). "Traditional Agriculture, pastoralism and fishing," in (1992). *Tropical African Development*, Gleave, M.B. ed., Essex, UK: Longman Scientific and Technical.

WEST AFRICA (1989). "The Crisis in African Agriculture," *West Africa*, January 30–February 5:36–38.

WORLD BANK (1989). Sub-Saharan Africa: from crisis to subtainable growth: a long-term perspective study, Washington, DC: World Bank.

WORLD RESOURCES INSTITUTE (1985). *World Resources 1986: A Report by The World Resources Institute, in Collaboration with The United Nations Environment Programme and The United Nations Development Programme*, New York and Oxford: Oxford University Press.

WORLD RESOURCES INSTITUTE (1992). *World Resources 1992–93: A Report by The World Resources Institute, in Collaboration with The United Nations Environment Programme and The United Nations Development Programme*, New York and Oxford: Oxford University Press.

WORLD RESOURCES INSTITUTE (1993). *World Resources 1993–94: A Report by The World Resources Institute, in Collaboration with The United Nations Environment Programme and The United Nations Development Programme*, New York and Oxford: Oxford University Press.

14

Industry, Business Enterprises, and Entrepreneurship in the Development Process

Barbara Elizabeth McDade

INTRODUCTION

Economic geographers study the spatial organization and distribution of economic phenomena, including the spatial dynamics of industrial activity, business enterprises, and entrepreneurial development. This chapter focuses on the problems and prospects of industrialization in Sub-Saharan Africa, with specific reference to the potential for African entrepreneurship, privatization, and small-business development.

A great majority of Sub-Saharan Africans continually engage in an array of business activities. Such activities are carried out within the organization of business enterprises created by entrepreneurs for the purpose of buying and selling goods in an exchange economy. More attention is now being given to the importance of these private business enterprises and their owners and entrepreneurs. Their ability to supply the demands of people to achieve the goals of development is being increasingly understood, and policies are shifting toward allowing the private sector to play a more dominant role in nation building. However, the function of the state, in providing the appropriate framework of a regulatory business climate and an appropriate infrastructure, is still recognized as central to development. Public and private sectors are seen as partners rather than antagonists, and the latter is beginning to emerge as the basis of economic development.

PROBLEMS WITH INDUSTRIAL AND MANUFACTURING ENTERPRISES

Sub-Saharan African industrial enterprises are plagued by a host of problems. The major ones outlined here include the spatial imbalance in the distribution of large-scale industries; weak inter- and intrasectoral linkages; highly protectionist policies; lack of diversification; weak institutional capacity; high degree of public ownership; and a failure to develop human capabilities to support, nurture, and promote African entrepreneurship in small- and medium-sized enterprises.

Spatial Imbalance in Large-Scale Industries

There is an urban bias in the distribution of large-scale, capital-intensive industries in Sub-Saharan Africa. This occurs because major cities offer urbanization, communication, and agglomeration economies. Industries save on production, transportation, and transaction costs as they benefit from urban amenities, better telecommunication networks, face-to-face transactions, a more skilled labor force, and closer proximity to complementary industries and service centers. Sub-Saharan African cities offer these attributes. Industries have failed to be enticed by government incentives designed to encourage them to move to smaller towns and rural communities, mainly because of the lack of a social, physical, and technological infrastructure in these areas, where, at times, even communication systems so basic as telephones remain inoperable over extended periods.

In Zimbabwe, for example, 75% of all manufacturing occurs in Harare, the capital, and Bulawayo, the second largest city. In Ethiopia, 70% of fixed assets in total manufacturing are concentrated in the largest towns. Kenya, Nairobi (with only 5.3% of the total population) accounts for 54% of its country's formal-sector manufacturing earnings (Riddell, 1992). This uneven pattern of industrial development denies rural areas opportunities for employment, training, and apprenticeship. It also perpetuates the already uneven terms of trade that exist between the agricultural and industrial sector.

Weak Inter- and Intrasectoral Linkages

Sub-Saharan African industries are often criticized as lacking dynamism essentially because they are "delinked" from other sectors of the economy, especially the agricultural sector. Most of the produce from the agricultural sector either is exported or is locally consumed, leaving very little for industrial processing. Strengthening intersectoral linkages generates the necessary multiplier effects and positive feedback mechanisms to ensure a more self-sufficient economy. Most countries are naturally inclined to strengthen linkages by promoting more agroprocessing industries, but, as Steel and Sarr (1983) point out, a country that has a comparative advantage in the production of food might not necessarily have a similar advantage in the processing of food, unless it has the right infrastructure and production mechanisms in place.

Bagachwa and Stewart (1992) have concluded that the magnitude of consumption linkages in rural areas, measured as the expenditure of farm incomes on locally produced consumer goods and services, is relatively weaker in Africa than in Asia. In a specific example, a 10% increase in rural household incomes led to an increase of 13% in consumption of nonagricultural goods and services in Sierra Leone or northern Nigeria, compared with 20% in Muda, Malaysia.

Furthermore, intrasectoral linkages among Sub-Saharan industries in the form of *backward* (demanding inputs from other sectors or industries) and *forward* (supplying goods or materials to other industries) linkages are weak. This is because most of the industries are oriented toward basic goods that have very few backward linkages, on account of the low level of industrialization and the import dependency on inputs. In the Côte d'Ivoire, for example, 54% to 58% of material inputs purchased for manufacturing industries were imported (Riddell, 1992). The opportunity to generate forward linkages is also limited; most basic consumer products are exported or locally consumed where a domestic market exists.

Highly Protectionist Policies

The majority of Sub-Saharan African countries adopted import-substitution policies immediately after independence. The intentions were (1) to replace manufactured goods formerly imported with domestic goods, (2) to protect infant industries from foreign competition, (3) to enhance the utilization of local resources, (4) to raise employment opportunities, and (5) to stimulate local demand for local products, thus expanding local industry and enhancing economies of scale in production.

In the process, several tariffs and quotas were erected to protect local industries, and governments ended up spending huge amounts of capital to subsidize the import of intermediate and capital goods required by domestic and foreign companies (who were the principal beneficiaries). The net effect of import substitution, aside from being urban biased, was that it encouraged greater inefficiency, endorsed the production of lower quality goods, and created higher production costs. Moreover, small-sized countries (Togo, Gambia, Rwanda, Burundi, Lesotho, and so on), plagued by the legacies of colonialism, had very limited domestic markets to support the expansion of local industries.

Ironically, import dependency increased in some countries that had adopted import-substitution policies. Gulhati and Sekhar (1982) report that Kenya experienced a 10% increase in its coefficient of imported inputs in 1970, as there were large upward shifts in its import dependence for food, textile raw materials, wood, rubber, and chemicals. Another study in Ghana showed that import substitution increased rather than decreased the cost of foreign exchange: One-quarter of manufacturing output suffered a net loss of foreign exchange in 1967 to 1968 (Steel, 1972).

Certainly, the concept of import substitution sounds logical for countries that are beginning to take off in their quest for industrial success. After all, the objective is to nurture and protect industries until they become self-sufficient and competitive. Moreover, Sub-Saharan African countries, at the time of independence, were eager to catch up with the international community, and industrialization was seen as the key, even if it meant relegating agriculture to a lower priority status. Industry, compared with agriculture, was seen as having stronger linkages, higher productivity levels, higher wages, and greater marketing prospects. Unfortunately, the overemphasis on protecting industry meant that very little investment was channeled into the agricultural sector. A declining agricultural sector, in turn, meant lower farm incomes to support industrial goods and fewer raw materials to complement agroprocessing industries. Import substitution could have fared better if it had been more selective and better managed. Selecting industries that have the potential to succeed and compete is a much better policy than one based on blanket protection.

In view of the problems associated with import substitution, some countries, with the aid of World Bank and International Monetary Fund (IMF) adjustment programs, have developed more export- and outward-oriented industrialization programs to stimulate foreign investment and technology and to expose industries to a more competitive environment. Mauritius created an Export Processing Zone that grants to export-oriented enterprises tax incentives on earnings, tax exemptions on imported machinery and equipment, free repatriation of profits, liberal work permits, and export financing at preferential rates. As a result, export-oriented industries have expanded and grown, providing over 80% of all manufacturing jobs and 50% of export earnings (Meier and Steel, 1989; World Bank, 1989). Other African countries, like Ghana, Senegal, Mozambique, Namibia, Malawi, Côte d'Ivoire, and Kenya, have Export Processing Zone packages similar to Mauritius's. In Zimbabwe, an expanded export revolving fund, designed to guarantee access to foreign exchange, provide tax-reducing export incentives and search for new markets, led both to a 68% increase in nontraditional manufactured exports in 1984 and continued expansion into 1987 and 1988 (Riddell, 1992).

Some might question the feasibility of an export-oriented approach, given the inability of some African countries to compete at the international level. The evidence does show, however, that some African countries are gaining ground. Garment manufacturers in Botswana sell fashion designs in London and New York, and custom-made parts for old automobiles from Zimbabwe and textiles from the Côte d'Ivoire are finding their way to European markets (World Bank, 1989).

Lack of Diversification and Expansion

Another problem facing Sub-Saharan Africa's industrial sector is its concentration on basic consumer and food-processing goods. Such goods have been vulnerable to price declines and, in some instances, have low income elasticities (the extent to which demand for a product changes with respect to a change in income). Successful diversification, beyond this stage of infancy, requires research, development, and testing; increased efforts toward product innovation; and the ability of Sub-Saharan governments and institutions to create the necessary support structure and enabling environment. It also requires the ability of governments to search for new markets locally, regionally, and internationally, through trade missions and bilateral and multilateral negotiations.

Weak Institutional Capacity

Most Sub-Saharan African countries lack the appropriate industry-related institutional structure and enabling environment to support and sustain industrial development. What is required, for instance, are institutions for industrial standards, testing, export support, quality assurance, training, technology information, and research and technical extension, as well as assistance to investors, suppliers, and subcontractors (Lall, 1992). In Korea, for example, industry associations provide technical, counseling, and information services, and they assist in the implementation of government policy and export targeting. Sub-Saharan governments can remedy these deficiencies by upgrading the physical and technological infrastructure, improving information network systems, improving access to credit, and providing more science and technical training programs to enhance human capabilities.

The problem and policy analysis so far suggests a need to reduce disparities in the distribution of industries between urban and rural areas, strengthen inter- and intrasectoral linkages, increase the competitive advantage of industries by means of export-oriented approaches, diversify and expand on the industrial base, and upgrade the capacity of industry-related institutions to facilitate development in industrial and manufacturing enterprises. Two other issues that are central to the debate about industrialization in Sub-Saharan Africa—expanding the private sector and promoting African entrepreneurship and small business—are dealt with in the remainder of the chapter.

PUBLIC VERSUS PRIVATE ENTERPRISES IN SUB-SAHARAN AFRICA

The Background to State-Owned Enterprises

Upon achieving independence from colonialism, Sub-Saharan African governments looked for political and economic ideologies to guide development strategy and industrial policy. In the early years of independence, various African leaders embraced socialism, on the ground that it was consistent with traditional African communalism, values, and institutions. Ghana, Guinea, Senegal, and Tanzania, among others, set about to implement a strategy of development and industrialization based the concept of African Socialism. This translated into state or government ownership and operation of major industrial enterprises, such as steel mills, oil refineries, and fish- and food-processing plants, and also control of the purchase and sale of the primary export crop.

Ghanaian philosopher Kwame Gyekye (1988) feels it is incorrect to equate traditional African communalism with Marxian-based socialism. He says this equation ignores the fact that ownership of private property and individual prerogative are "traditional" values that are consistent with many African cultural systems. African communalism is essentially a socioethical doctrine, dealing with human, cultural, and social relationships. Socialism is essentially an economic arrangement that involves public control of the dynamics of the economy.

On the other hand, capitalism, with its emphasis on individualism and corporate profit, seemed to embrace economic and social values that were discordant with traditional African culture. Also, capitalism was associated with colonialism and the economic and political domination of Sub-Saharan Africa by foreign powers. While espousing capitalism, colonial governments in Sub-Saharan Africa sanctioned discriminatory practices and enacted laws that either prevented or severely restricted Africans from participating in most formal-sector business and economic activity within their own boundaries, reserving these rights for Asians and Europeans. In South Africa, capitalism was associated with the oppressive apartheid system. Proponents argue that to identify capitalism with the malevolent periods of African colonial and postcolonial history is misleading and gives only a part of the total picture. Economist Walter Williams (1994) contends that the restrictive system of apartheid controls actually was hostile to true capitalism. Williams says that apartheid fostered not capitalism but, instead, a *welfare state* for whites that undermined free enterprise, a basic tenet of capitalism.

Such practices undermined the development of a formal private sector in Sub-Saharan Africa and deprived indigenous Africans of experience in establishing and operating modern business enterprises. Therefore, at independence, there were few African businesspeople with the large amounts of accumulated capital required to invest in major private enterprises. On the other hand, some countries, while not avowedly "capitalist," pursued economic policies that produced a modified version of the liberal market economy with substantial state regulation and direction characteristic of the capitalist economies of the United States and Western Europe (Young, 1982). Among these countries were Nigeria, Côte d'Ivoire, Democratic Republic of Congo, Cameroon, Morocco, and Malawi.

Because there were insufficient domestic private investment resources, some leaders, such as Kwame Nkrumah, made the national government the major financial investor in economic growth. He believed that development could be achieved faster by using the substantial resources and power of the state, rather than by waiting for private capital to be accumulated. The reasoning behind this was that government, being the representative of the people, would act in the best interests of society as a whole, rather than in the narrow interests of private corporations and individual capitalists. State-owned companies and parastatals were presented to the public as common property, owned collectively by all of the citizens of the country. The people as a whole could then share in the profits and benefits of economic growth. This idea continues to have supporters. Recently, one of the national workers' unions in Ghana protested the Ghanaian government's divestiture of public companies and the sale of major state-owned enterprises (SOEs). They chanted, "Private—it's *theirs*. Public—it's *ours*."

Postindependence private-sector development lagged for other reasons. Private foreign investment was marginal, because foreign companies were skittish about the high financial risks they felt were inherent in doing business in new and evolving African economies. But, while risk aversion was hindering private investment in the region, the Cold War waged by the former colonial powers in the decades after World War II fostered public investment in the African Continent. European, American, and Soviet governments competed for the allegiance of various countries via offers of assistance with development financing. The industrialized nations also sought to protect their existing foreign investments and to preserve or solicit proprietary concessions to Sub-Saharan Africa's natural resources. Funding for development projects was meted out for the purpose of retaining political or economic influence in the region, rather than with the goal of helping Africa achieve sustainable economic development. Since a lot of the foreign "aid" to African governments was then (and is even now) in the form of loans, it has caused severe economic repercussions. In some cases, the net outflow is greater from "poor" countries in the form of debt servicing than the aid they receive from the "rich" countries.

Public enterprises, in spite of employing the majority of Africans in the formal sector, have not fared well. A sample study of 12 west African countries revealed that 62% experienced net losses and that 36% had negative net worth (Meier and Steel, 1989). The inefficiency associated with public- and state-owned enterprises has also contributed to the underutilization of African industry. For instance, a state-run shoe factory in Tanzania once operated at about 25% of capacity (World Bank, 1989). Generally, such enterprises are plagued by high debts, overstaffing, highly centralized bureaucracies, and inept management. The poor performance of these enterprises has prompted some Sub-Saharan African governments to look toward expanding the private sector.

Expanding the Private Sector in Sub-Saharan Africa

Privatization is on the rise. More than 8500 state-owned enterprises (SOEs) around the world have been privatized since the early 1980s. Between 1993 and 1997, there were 2804 transactions involving the privatization of public enterprises in Sub-Saharan Africa, totaling about $5 billion (World Bank, 1998). Of these transactions, 25% were in the manufacturing sector, 20% in the service sector, and 18% in the agricultural production and processing sector. Privatization means investing in private-sector businesses, large and small; expanding their role in the national economy; and reducing government ownership of enterprises and parastatals, which could mean selling SOEs such as an electric utility, an airline company, a fish cannery, a railroad system, or a textile factory to a private corporation or individual. Advocates of this approach believe that privatization is necessary to revitalize Sub-Saharan African economies, promote economic growth, and produce sustainable development that will improve the quality of life for the majority of Africans.

The international finance community is leading efforts to expand private-sector activities in Sub-Saharan African countries. Shortly after the first free elections in South Africa in early 1994, the British pledged 1.25 billion pounds in trade support and assistance to develop businesses in that country. In 1987, the International Finance Corporation (IFC) launched a major new initiative to help develop private enterprise in Africa through a program called the Africa Project Development Facility (APDF), sponsored jointly with the UN Development Program (UNDP) and the African Development Bank (ADB). One of the objectives of this program is to help indigenous African businesses compete against the large foreign multinational companies for development projects in Africa. In addition, the IFC assists African entrepreneurs to develop sound investment projects and to find financing for them. Small- and medium-sized businesses that ordinarily would not qualify for direct IFC investment receive benefit through subcontracting arrangements with larger companies.

The motivation behind privatization is that the incentive for efficiency and productivity is higher when it is driven by profits that go to individual owners rather than to the state. In addition, it reduces the pressure on public capacity and finances by allocating the provision of a larger proportion of products and services to private companies. The public sector also derives revenues from taxes on sales—sometimes, the equivalent of thousands or billions of dollars, depending on the enterprise.

The World Bank acknowledges that most privatization success stories come from high- and middle-income countries. The poorer the country, the less likely it is that privatization will produce the hoped-for benefits: efficiency, increased revenues, and profits. This is attributed to the fact that privatization is more likely to produce positive results when the company operates in a competitive market, with a favorable business environment, a strong financial system, and adequate infrastructure. This economic environment is rarely found in most Sub-Saharan African countries.

One of the reasons African governments have been reluctant to sell large state-owned enterprises is that privatization often means laying off employees for the sake of operational efficiency. Usually, this is done because many public enterprises sometimes have larger staffs than required to carry on their business effectively due to political considerations in hiring or the state's interests in providing employment for its citizens. Therefore, governments should pay special attention to the needs of workers that might become unemployed. This attention could be in the form of employee ownership schemes, unemployment benefits, or the redeployment or retraining programs to reduce the social costs of privatization.

To be successful, privatization programs should be accompanied by a total package of economic reforms designed to open markets to private companies to supply products and services that were previously provided by state-owned monopoly companies (such as airlines, railroads, and telephone service), to remove price- and exchange-rate distortions produced by government fiats for reasons other than consumer demand, and to encourage the development and expansion of the rest of the private sector through free entry into the market. But, even when companies are privatized, regulation of their activities should be put into practice in the public interest. This is particularly true for monopolies or in industries that are difficult to enter for such reasons as high start-up costs—for example, the steel and petroleum-refining industries.

SOEs are not always easy to sell. Private investors might well be reluctant to take on failing companies with bloated staffs that must be reduced or compensated, ones suffering from high debts or accounts payable to suppliers or ones facing uncertain competitive-market conditions. On the other hand, the state could easily be the "loser" when it must sell a property in which it has invested considerable public funds at prices below its real market value when it is being pressured to divest its SOEs. Under pressure from the World Bank, Ghana had to liquidate 12 of 18 SOEs listed for sale, because it could not find private investors to buy them in the time period mandated.

Africa accounted for 35% of the sale of SOEs worldwide in 1988. This divestiture activity in Africa has not been spread evenly throughout the continent: Over all, west African states and the French-speaking countries have accounted for one-half to three-fourths of the divestitures either completed, underway, or at the planning stage. Out of these, 30% of the divestitures have been of state-owned manufacturing enterprises, such as steel firms, canneries, or textile firms. Often, these types of manufacturing firms already have established markets for their products. Only 10% of divestitures have been of commercial and service enterprises, such as airlines, communications, and other transport services. Apparently, there are certain types of businesses that are not yet attractive to the private sector. In addition, there is also a continuing debate among African countries about whether the public or the private sector can more equitably deliver certain goods and services. Raging debates about whether to choose private over public enterprise are not as relevant as devising appropriate strategies to enhance both sectors. African governments can still maintain a select number of public enterprises that are viable and beneficial by improving their management performance and investment portfolios. Also, several countries have had a long tradition of joint ventures between foreign companies and local parastatals that they can build on. Private enterprises can be encouraged by improving the business climate and, more importantly, by nurturing African entrepreneurship and small-scale enterprises.

AFRICAN ENTREPRENEURSHIP AND SMALL-SCALE INDUSTRY

Schumpeter (1934 and 1961) introduced the concept of entrepreneurship into modern economic development thought, and he stated that entrepreneurial activities might include the introduction of a new product, improvements in and adapting existing products to new markets, and the gaining of access to new sources of raw materials and inputs. Entrepreneurship also requires investment initiative, risk taking, and innovation in product design and business organization. In developing countries, the entrepreneur is often seen primarily as a mobilizer and coordinator. The entrepreneur in this sense is able to organize people and their activities into more efficient utilization of resources (Kennedy, 1988). The entrepreneurial process is, therefore, concerned with expansion and creation of new wealth.

Entrepreneurs perform many and varied activities, from creating a new high-technology industry in Silicon Valley, California to opening an artisanal shop in Burkina Faso. The interest in entrepreneurship and the attention it is now being accorded come from the belief that entrepreneurial activities help to achieve such specific objectives of development as the creation of jobs, the increasing of household income, and the enlargement of the revenue base in a national economy, which has led to a closer examination of the entrepreneurial function in Sub-Saharan Africa.

Characteristics of African Entrepreneurs

Research on entrepreneurship in the United States and Europe views innovation as a technological event. However, in Africa, in addition to primarily inventive functions, the process of creative innovation can include obtaining adequate financing, adapting Western technology to local needs, making input substitutions, maximizing the factors of production, and minimizing costs.

One study of African businesses concluded that entrepreneurs in the United States and other developed countries can learn from the experiences of entrepreneurs in west Africa (Diomande, 1990). Most African entrepreneurs plunge into business with a bare minimum of capital or resources, and some become quite successful in spite of their austere start-ups. The relative abundance of resources in the advanced economies could cause an entrepreneur to become indifferent to the need to conserve and use resources more efficiently; rarely is this the scenario among African entrepreneurs. African entrepreneurs in the informal sector rarely choose their workers from a formally educated and technologically skilled labor force, so they secure personnel through apprenticeships, in tutelage arrangements, and from connections with family members and acquaintances. They

African Entrepreneurship and Small-Scale Industry

raise start-up and operating capital from community resources such as *tontine*, a rotating borrowing system involving social groups. Operating several enterprises simultaneously is criticized by Western observers as spreading oneself too thin, but African entrepreneurs, choose to diversify their investments rapidly, in the belief that operating several types of enterprises guards against risks common in the economic climate of many African countries.

A problem facing entrepreneurial enterprises is that very few graduate into larger entities. More research needs to be done into why so few promising African enterprises ever evolve into large-scale organizations. In Nigeria and in most of west Africa, where trading is almost a universal occupation, there are comparatively few large African trading firms, and, though there are numerous artisans engaged in innovation and production, there are few large African manufacturing firms.

Governments in some countries, such as Nigeria and Kenya, have been consistent in pursuing policies that promote indigenous African enterprise. People in traditional cultures do make changes in their lifestyles and behavior when they see the changes as beneficial. Research on both modern and traditional business owners in Botswana, Kenya, Ghana, Malawi, Sierra Leone, Tanzania, and Zambia, among others, provides ample evidence of the presence of entrepreneurship in Sub-Saharan Africa. (See Discussion Boxes 14.1 and 14.2.)

Some scholars point out that, although there is no shortage of "entrepreneurial drive" in Africa and many entrepreneurs take advantage of opportunities, there is still a critical lack of management and technical expertise to support the entrepreneurial capability required to establish, organize, and expand modern industries (Lall, 1992).

Small-Scale Enterprises

Small and *micro* enterprises (see Table 14.2 for size classification) make up the vast majority of businesses in Sub-Saharan Africa, as they do in most countries. However, because of the lack of large-scale industries with linkages into the domestic economies, the functions that these small enterprises perform are even more critical to the domestic economy. Small and micro enterprises provide goods, services, employment, and income for the majority of Africans. Studies have shown that they account for two-thirds or more of industrial employment, contribute from 26% to 64% of total manufacturing value added, and account for 3% to 8% of the GDP in Africa (Liedholm, 1992). A vast number of small industries are also located in rural areas and provide opportunities for nonagricultural employment. Examples of small-scale activities include clothing production, wood production, furniture making, metalworking, and food production.

Small industries, especially in the so-called informal sector, continue to emerge and sustain themselves in spite of the ups and downs of economic conditions in Sub-Saharan Africa. In addition to contributing to economic stability in a country, healthy small and medium-sized enterprises contribute to political stability. By providing income and employment and by supplying people's wants and needs, they help to promote some sense of economic security in the population. They help form a more democratic economic base, because they decentralize and distribute economic resources and wealth. This is not to discount the importance of large businesses in generating economic growth and

> **DISCUSSION BOX 14.1: BUSINESS PROFILE OF MOPIPI LIMITED, GABERONE, BOTSWANA**
>
> Mrs. Norah Mmonkudu Glickman is owner and manager of Mopipi Limited, which manufactures knitwear, dresses, and school and military uniforms. Mrs. Glickman started the company in 1973 with two employees; it now employs 65 people with annual turnover of $300,000. Her father was a smallhold farmer; she left secondary school at the form-three level (approximately ninth grade, in US terms) to get married.
>
> The company is quite an integrated operation. Shirting material comes from Malawi, cloth for dresses from the Far East. The Botswana government supplies the material for the military uniforms. She bought her equipment from South Africa, where she had worked for four years as a knitting-machine operator in a small factory before returning to Botswana to start her own business. Mrs. Glickman has plans for upgrading the quality of her products and for expanding into wider markets. She has been consulting with experts in Sweden and Germany and would like to establish a joint venture if the government eases its restrictions on foreigners operating in the clothing business.
>
> All of the company's workers are trained on the job. Most of the employees are women and mothers. They are paid well above the minimum wage (currently $0.52[1] per hour). The owner describes her employees as reliable and hard working. To start her business, she was able to obtain a guaranteed loan of about $1700 from the National Development Bank. Her first factory was located at an industrial site of the Botswana Enterprise Development Unit. Finance has always been a constraint on expansion. Mrs. Glickman says banks in Botswana are very conservative in avoiding risk. She says there is no venture capital facility that she could tap into.
>
> Mrs. Glickman has contacted import agents to tap into the US and European markets, and the prospects for firm orders look good. European and American importers are looking for alternative sources of supply, because their traditional Far Eastern suppliers are reaching their import quota ceilings.
>
> [1] Monetary unit is in US dollars.
>
> *Source:* Marsden, K. (1990). *African Entrepreneurs,* Washington, DC: The World Bank.

development—a dynamic national economic system consists of a network of small, medium-sized, and large enterprises that have some degree of linkage and interdependence.

For the past three decades, small-scale enterprises have been promoted as the necessary components for economic development in African countries. Both cross-sectional and time-series data suggest that the industrialization process begins with the rapid growth of small enterprises, some of which then expand into medium- and large-scale

> **DISCUSSION BOX 14.2: BUSINESS PROFILE OF AGBEMSKOD ENGINEERING LIMITED, ACCRA, GHANA**
>
> Owner and general manager Godfred Kwaku Agbemenya is a graduate of the University of Science and Technology in Kumasi, having earned a Bachelor of Science degree in mechanical engineering. He is a partner in business with his brother and father. Agbemskod Engineering was started by the father in the 1950s. The company is a medium-scale business and employs 35 people. The company manufactures such agricultural-processing machinery as cassava graters, meal and flour mills, fruit and vegetable crushers, and fufu machines. It also provides consulting services in mechanical engineering and machining.
>
> The elder Mr. Agbemenya was a blacksmith in his village, moved to Accra in the 1950s, and worked as a "shade-tree" auto mechanic, according to his son. The business has been located at its present site in an industrial park in Accra since 1974. It is an organized and efficient operation and has a manufacturing and assembly factory, warehouse, and display area in front for retail sales. Godfred said that, in earlier years, the factory was a training ground for young apprentices, but now that production has become more sophisticated, to meet higher standards of product quality, they must hire personnel already skilled, rather than train novices.
>
> The business has been moderately successful, but access to funds for capital investment in production equipment and expansion has always been a problem. For example, the general manager pointed out, he is currently operating with a 40-year-old lathe and another machine that is almost 60 years old. However, he has been reluctant to apply for loans, for fear of a fluctuating economy and high interest rates. He feels that the best help for medium-scaled businesses like his is for the government to create a positive economic environment for locally owned companies by stabilizing the economy and lowering interest rates and by providing some protection for indigenous industries to get established before they have to compete, even in the local market, with imports from larger, well-established foreign companies.

firms. Successful industrialization must be based on indigenous enterprise. Therefore, expansion of small enterprises and, where appropriate, establishment of linkages with medium- and large-scale indigenous firms would produce more fully integrated national economies. Most statements on national economic policy in Sub-Saharan countries include the promotion of such enterprises. However, while alluded to, they are often neglected when the actual implementation of economic development policy is initiated.

TABLE 14.2 Classification of Enterprises by Size

Category	Full-Time Employees
Micro	0–3
Very small	4–9
Small	0–29
Medium Sized	30–99
Large	100+

Source: Steel, W. & Webster, L.M. (1991). *Small Enterprises Under Adjustment in Ghana,* Washington, DC: The World Bank.

Although *small* has been proclaimed as *beautiful* in economic development planning (Schumacher, 1973), *bigger* in terms of large-scale businesses has received the most attention and investment.

Some of the reasons for this neglect are obvious. African countries have looked to industrialized countries as models on which to style their own development. To many African leaders, large-scale, modern factories, which were the foundation of the developed nations' economic advantage, are easy to identify with progress, and it is easy for political and business leaders to quantify the effect of one or two large industrial enterprises on a local or national economy in terms of production and employment. As a consequence, small and medium-sized enterprises are lost in the shuffle. For politicians the world over, bringing one large enterprise to a community gets more acclaim than the nurturing of several small-scale enterprises.

In many Sub-Saharan countries, large industries have weak or nonexistent functional linkages with small, local enterprises. This lack of connectivity has created a dual economic system, one in which large-scale industries are linked to external foreign inputs and markets and the small indigenous enterprises to the local economy. The external linkages could consist of imported production inputs, expatriate labor skills, and repatriated profits. Local internal linkages, on the other hand, are more likely to include indigenous raw materials and other inputs, local people as employees, and revenues invested within the domestic economy. Large-scale industries are able to stimulate growth in other sectors of an economy only if they are linked with local enterprises and activities. Economic maturity is reached only when a diversified and multifunctional business network has developed in which local businesspersons have the dominant influence.

Small-scale enterprises can be more conducive to economic development, because their requirements are probably more suited to a Sub-Saharan African country's current conditions of capital availability and labor skills. They use local inputs and supply products and services to the domestic market, and they retain a local orientation within the community. Given the complexity, unpredictability, and turbulence of the economic environment in many African countries, small-scale enterprises may have a comparative advantage over large-scale enterprises in terms of flexibility, innovativeness, adaptability, and ability to survive.

Small-scale industries require a number of essential inputs. Although the availability of adequate finance has previously been seen as central to economic growth, capital is actually less critical to small-business success than other inputs (Malecki, 1991). Such inputs as information, business knowledge, management expertise, and technical education appear to be more important to the success of modern small-scale enterprises. These factors also allow traditional, micro, and informal-sector enterprises to break through and expand.

CONCLUSION

For industrialization to "take off" and become an economic generator, a conducive macroeconomic environment is required, plus an adequate supply of capital or direct investment, adequate infrastructure, and a motivated, skilled-labor supply. In the current global labor market, firms are becoming less interested in merely finding "cheap" labor and are now seeking an efficient, skilled-labor force.

Increasingly, it is being understood that one of the major constraints on development and expansion of modern businesses in Sub-Saharan Africa is lack of access to information. This includes not only the absence of information, but the inability to exploit technology and develop human capabilities. This is reflected in the deficiencies in experience, expertise, and training.

Skills training centers or associations in which entrepreneurs can network and share information should be supported. National education systems, including both public and private institutions, should provide what entrepreneurs feel they need to sustain and improve their businesses. In addition, professional and service skills training should encourage business development and entrepreneurial activity. Investment in education is required to produce a skilled, educated labor force, which is an essential input for industrial and enterprise development.

A review of the financial services available to these enterprises is essential in any policy that purports to promote them. Policies should be developed to make it easier for credible businesses to raise capital from such financial institutions as the business and industrial development agencies, commercial banks, and nongovernmental organizations.

Even though private entrepreneurs take advantage of export opportunities, it is governments that actually seek to negotiate and maintain favorable terms for their countries' industries and products in the international market place. Regulations and facilities can promote easier access to export markets for the products of small business. Processing agencies can serve mentors to train entrepreneurs on how to enter international trade. Government can also review or revise rules, investment codes, and regulatory requirements so that they facilitate, rather than retard, the process of business and industrial development. Entrepreneurs are constantly on the lookout for opportunities. For example, as Discussion Box 14.3 shows, a recent change of demand in US retail garment and household products markets has created a window of opportunity for African enterprises.

DISCUSSION BOX 14.3: RESURGENCE OF CULTURAL IDENTITY AMONG AFRICAN AMERICANS CREATES US OPPORTUNITIES FOR AFRICA'S CLOTHING INDUSTRY

Until recently, very few finished products from Africa were seen in retail stores in the United States. Most African clothing and crafts were sold in small Afrocentric stores and boutiques or at festivals and special events. However, the resurgence of cultural identification with Africa among African-Americans has created a major market niche for African manufacturers of garments and home products. African-Americans with a population of 30 million and purchasing power of $223 billion[1] are the largest minority group in the United States.

According to a World Bank report, a window of opportunity has emerged for African manufacturers to supply major US retailers like J.C. Penney, Montgomery Ward, Pier 1 Imports, K-Mart, and Dayton Hudson. This began with the demand by African-Americans for authentic clothing, accessories, and household items from Africa. When "authenticity" is a key component of consumer demand, African firms have a unique advantage over Asian or European manufacturers. This entry into the US retail market is providing African firms with the technical and marketing expertise to expand into the larger American market for clothing and other products. Export prospects for African manufacturers in the United States are increasing sharply. (See Table 14.3.)

[1] Monetary unit is in US dollars.

Source: Biggs, T., et al. (1994). *Africa Can Compete!* Washington, DC: The World Bank.

TABLE 14.3 Current and Potential Size of US Afrocentric Merchandise Market

Sales Outlet	Annual Sales ($ millions)
Afrocentric retail stores	120
Large wholesalers*	96
Festivals	30
Black Expo	19
Major department stores	4
Total	**269**
Estimated Potential Size	**540**

*Such as Cross Colours, Spike's Joint, Sotiba.

Such as J.C. Penney, Montgomery Ward, Dayton Hudson, K-Mart.

From Biggs, T. et al. (1994). *African Can Compete!* Washington, DC: The World Bank.

KEY TERMS

Manufacturing value added
Import substitution policy
Agroprocessing industries
Small-scale industries
Public enterprises
Consumer goods industry
Backward linkages

Intersectoral linkages
Export-oriented policy
African entrepreneurship
State-owned enterprises
Business climate
Forward linkages

DISCUSSION QUESTIONS

1. Describe the major characteristics of industries in Sub-Saharan Africa.
2. What are the major problems confronting Sub-Saharan African industries?
3. Describe the major characteristics of African entrepreneurs.
4. How viable is the private sector's role in industrial development?
5. Should public enterprises be totally replaced by private enterprise in Africa?
6. What, in your opinion, should be the major components of an appropriate industrial policy in Sub-Saharan Africa?

REFERENCES

BAGACHWA, M. & STEWART, F. (1992). "Rural Industries and Rural Linkages in Sub-Saharan Africa: A Survey," in *Alternative Development Strategies in Sub-Saharan Africa,* Stewart, F., Lall, S., & Wangwe, S., eds., New York: St. Martin's Press.

BIGGS, T., et al. (1994). *Africa Can Compete!* Washington, DC: The World Bank.

DIOMANDE, M. (1990). "Business Creation with Minimal Resources: Some Lessons from the African Experience," *Journal of Business Venturing,* 5:191–200.

GULHATI, R. & SEKHAR, U. (1982). "Industrial Development Strategy for Late Starters: The Experience of Kenya, Tanzania, and Zambia," *World Development,* 10(11):949–972.

GYEKYE, K. (1988). *The Unexamined Life: Philosophy and the African Experience,* Accra, Ghana: University of Ghana Press.

KENNEDY, P. (1988). *African Capitalism: The Struggle for Ascendancy,* Cambridge, MA: Cambridge University Press.

LALL, S. (1992). "Structural Problems of African Industry," in *Alternative Development Strategies in Sub-Saharan Africa,* Stewart, F., Lall, S., & Wangwe, S., eds., New York: St., Martin's Press.

LIEDHOLM, C. (1992). "Small-Scale Industries in Africa: Dynamic Issues and the Role of Policy," in *Alternative Development Strategies in Sub-Saharan Africa,* Stewart, F., Lall, S., & Wangwe, S., eds., New York: St. Martin's Press.

MEIER, G. & STEEL, W., eds. (1989). *Industrial Adjustment in Sub-Saharan Africa,* New York: Oxford University Press.

MALECKI, E. (1991). *Technology and Economic Development,* Essex, UK: Longman.

MARSDEN, K. (1990). *African Entrepreneurs,* Washington, DC: The World Bank.

RIDDELL, R. (1992). "Manufacturing Sector Development in Zimbabwe and the Côte d'Ivoire," in *Alternative Development Strategies in Sub-Saharan Africa,* Stewart, F., Lall, S., & Wangwe, S., eds., New York: St. Martin's Press.

SCHUMACHER, E.F. (1973). *Small Is Beautiful: Economics as If People Mattered,* New York: Harper and Row.

SCHUMPETER, J. (1934 & 1961). *The Theory of Economic Development,* New York: Oxford University Press.

STEEL, W. (1972). "Import Substitution and Excess Capacity in Ghana," *Oxford Economic Papers,* 24(2):212–240.

STEEL, W. & SARR, B. (1983). *Agro-Industrial Development in Africa,* Abidjan, Côte d'Ivoire: African Development Bank.

STEEL, W. & WEBSTER, L.M. (1991). *Small Enterprises Under Adjustment in Ghana,* Washington, DC: The World Bank.

STEWART, F., LALL, S., & WANGWE, S., eds. (1992). *Alternative Development Strategies in Sub-Saharan Africa,* New York: St. Martin's Press.

UNITED NATIONS INDUSTRIAL DEVELOPMENT ORGANIZATION (1991). *International Yearbook of Industrial Statistics,* Aldershot, England: Edward Elgar.

WILLIAMS, W. (1994)."South Africa: Nelson Mandela's Greatest Challenge," *Reader's Digest,* September:78–84.

YOUNG, C. (1982). *Ideology and Development in Africa,* New Haven: Yale University Press.

WORLD BANK (1989). *Sub Saharan Africa: From Crisis to Sustainable Growth,* Washington, DC: World Bank

WORLD BANK (1992). *African Development Indicators,* Washington, DC: The World Bank.

WORLD BANK (1998). *African Development Indicators,* Washington, DC: The World Bank.

THE GLOBALIZATION PROCESS: OPPORTUNITY OR THREAT?

It is assumed that globalization provides enhanced opportunities for human advancement. Greater integration into the world economy is held to provide African businesses and industries greater access to new markets and to an extensive network of production units and technological innovations. As a result, African economies have the potential to expand and diversify, thus creating more opportunities for human development. This of course presumes that African governments have the appropriate infrastructure and institutional capacity to support socioeconomic initiatives associated with globalization. UNDP (1999) points out that global markets, technology, and solidarity can expand choices and improve living conditions for people everywhere; however, the benefits of globalization spread unequally, concentrating power and wealth in advanced countries and in a select group of people and transnational corporations, while marginalizing nations in Sub-Saharan Africa.

The concept of globalization is nothing new. Henriot (2000) points out how globalization has led to the penetration of Africa by outside forces and resulted in negative social consequences for its people. He identifies four stages of penetration, beginning in the 16th century: (1) the period of slavery that drained Africa of its human resources; then (2) colonialism; (3) neocolonialism; and, currently, (4) globalization. In the 1970s, a group of developing countries, through a special session of the UN General Assembly, called for a New International Economic Order to eradicate the negative consequences of colonialism and establish international relations based on the principles of mutual cooperation, equality and solidarity. What distinguishes the current era of globalization from the era of the 1970s and 1980s is that the current pace of globalization has accelerated, as a result of technological developments, the Internet, and an easing of regulations to facilitate the flow of goods and services. Certainly, this does present an opportunity for a lagging country to accelerate its own pace of development, but the global community could learn from past mistakes and focus on a more humane approach to a globalization process that achieves accelerated economic growth, poverty reduction, and a sincere engagement with Sub-Saharan Africa. It is essential to develop a globalization process that is devoid of exploitation, marginalization, and paternalism, one that is inclusive and integrative.

It is not surprising that the advanced countries continue to benefit from the opportunities of economic and technological globalization. Also included among the list of beneficiaries are (1) the newly industrializing countries of east Asia and (2) the Big Emerging Markets: Argentina, Brazil, Mexico, China, India, and Thailand. Recent reports published by the World Bank (2000) and UNDP (2001) demonstrate that these economies are strengthening their links in the global marketplace as they experience significant growth in exports, foreign investment activity, and technological innovation. Conspicuously absent from this list are Sub-Saharan Africa countries, which are not faring well in the new global economy. As Discussion Box 15.1 shows, Sub-Saharan African countries account for a very small portion of world exports and continue to attract the lowest amount of foreign direct investment of any region in the world. The countries continue to lag behind in the development of a technological infrastructure and in the creation of a business environment and institutional capacity that is conducive to investment. As long as these problems linger,

DISCUSSION BOX 15.1:
GLOBALIZATION: HOW IS AFRICA FARING?

Financial Flows

Sub-Saharan Africa attracts the lowest amount of foreign direct investment (FDI) of any region in the world. In 1998, Sub-Saharan Africa received only $4394 million, 2.57% of the total amount of FDI that went to low- and middle-income countries. Latin America and the Caribbean (40.1%) and East Asia and the Pacific (37.5%) were the primary beneficiaries. Nearly half of Africa's FDI allocation went to oil- and mineral-producing countries: Nigeria, South Africa, and Angola. In 1998, Sub-Saharan Africa accounted for $250 million, 6.3% of the portfolio investment flows related to bond issues purchased by foreign investors from low- to middle-income countries. In the same year, net private capital flows going to Sub-Saharan Africa were $3,452 million, compared with $126,854 million in Latin America and the Caribbean and $67,249 million in East Asia and Pacific.

Aid

Although Sub-Saharan Africa received the largest proportion of aid in 1998 ($14,186 million, out of the $54,742 million to low- and middle-income countries), aid capita declined from $30 in 1993 to $21 in 1999.

Trade

The share of merchandise exports by Sub-Saharan Africa in world trade declined from 3.79% in 1980 to 1.46% in 2000. Africa continues to experience slow growth in export and import activity when compared with Latin America and Asia. Between 1990 and 1999, the average annual growth rate of exports in Africa was only 1%, compared with 8% in Latin America and 7% in Asia. Imports in Africa grew at an annual rate of 4%, compared with 11% in Latin America and 6% in Asia.

Information, Science, and Technology

The majority of countries in Sub-Saharan Africa are classified as "marginalized" in terms of their technological achievement index (an index developed by UNDP to describe a country's ability to (1) create and diffuse technology and (2) build a critical mass of human skills. (See Chapter 8.) The 29 member states of the Organisation for Economic Co-Operation and Development (OECD), representing postindustrial economies, contain 97% of all Internet hosts; 92% of the market in production and consumption of computer hardware, software, and services; and 86% of all Internet users. By contrast, the whole of Sub-Saharan Africa contains only 2.5 million Internet users, less than 1% of the world's on-line community. (See Chapter 9.)

there will continue to be much skepticism about the benefits of globalization, prompting some criticism about the whole process.

Part of the skepticism is linked to concerns about growing inequalities and increasing poverty worldwide. While today's globalization has resulted in increased trade, expanded markets, and new technologies, Birdsall (1998) points out that the ratio of average income of the richest country in the world to that of the poorest has risen from about 9 to 1 at the end of the 19th century to at least 60 to 1 today. Income inequalities continue to widen in Asia, Eastern Europe, Latin America, and certainly Africa. In Latin America, the income ratio of the top 20% of earners to the bottom is about 16 to 1. The World Bank (2001a) acknowledges that, despite impressive economic growth in many developing countries, absolute poverty worldwide is still increasing, especially in Sub-Saharan Africa. For example, the number of poor people in the world (excluding China and the North America) living on less than $1.00[1] a day increased from 1183 million in 1987 to 1199 million in 1998. In Sub-Saharan Africa, the number of poor people increased from 217 million to a substantial 291.9 million, accounting for nearly half of the continent's population.

Much of the skepticism surrounding the impact of globalization on Africa can also be traced to some of the negative impacts of structural-adjustment programs (SAPs) outlined in Chapters 8, 11, and 12. SAPs were initiated to adjust malfunctioning economies, liberalize trade, encourage privatization, assist in debt restructuring, and promote greater economic efficiency in Sub-Saharan Africa, but there have been some negative social and human impacts. For example, neoliberal policies geared toward trimming government bureaucracies often have not been supplemented with job retraining or redeployment programs. Government cuts in health and education have driven up costs, making it difficult for poor families to secure a good education and decent health care. Also, currency devaluations and strict credit restrictions have had the effect of raising the prices of consumer goods and agricultural inputs, thus harming women and children who rely on food trading and production. Chapter 11, for example, points out that, although relatively well-off urban women have benefited from opportunities associated with cross-border trading, the weight of evidence suggests that the burden of globalization and increasing debts falls doubly hard on poor women working in the informal sector. Chapter 12 further notes that government expenditure reductions have translated into reduced funding for diagnosis and treatment of sexually transmitted disease (STDs) and blood screening, producing poor hygienic practices in clinics (such as the inadequate sterilization of equipment) and thus contributing to the spread of HIV.

SAPs, along with globalization, are institutionalized in policies designed by such powerful institutions as the International Monetary Fund (IMF), the World Bank, the World Trade Organization (WTO), and transnational corporations. Such policies result in unanticipated social and human consequences, because very few Africans get to participate in the decision-making and policy-design processes. Input from African professionals is needed to avoid the pitfalls of SAPs and globalization. We are not suggesting that African countries disengage from SAPs as no viable alternatives are available; rather, we suggest that programs

---

[1] Monetary unit is in US dollars.

of this nature be augmented with policies and programs that are geared towards the preservation of human dignity. The Programme of Action to Mitigate the Social Cost of Adjustmant (PAMSCAD) program in Ghana, for example, is a step in the right direction; so are the Heavily Indebted Poor Countries (HIPC) and poverty-reduction initiatives mentioned in Chapter 8.

Further skepticism about globalization revolves around issues of political sovereignty and cultural assimilation. From a geopolitical standpoint, there is concern that African states will become subservient (if they already are not) to the powerful influences of the multinational companies that in effect constitute "new corporate states." This subjugation, in conjunction with global regulatory systems, could potentially erode the control that African states should have over their own domestic laws and policies. There is also concern about the homogenizing effect that globalization has, via multimedia, Internet, telecommunication networks, and the numerous consumer goods that come out of Europe and the United States, on traditional cultural value systems in Africa. The syncretistic nature of African cultures brought into being by colonialism (as described in Chapters 4 and 6) will only be reinforced as the forces of globalization transcend and make porous the boundaries of African states.

IMPACTS OF WORLD TRADE NEGOTIATIONS ON AFRICA

Trade is an essential aspect of the globalization process. Trade expands the global market for domestic producers, compels them to be competitive, and provides opportunities to adopt and develop new innovations and technologies. Sub-Saharan Africa's successful integration into the global economy is predicated upon its access to world merchandise markets. Unfortunately, the current evidence suggests that, although developing countries as a whole have expanded their trade, that African countries are still marginalized. Certain global pacts and institutions—the General Agreement on Tariffs (GATT) and the WTO—were designed to provide countries, including African countries, with opportunities to access world markets by engaging in a series of negotiations to liberalize trade by reducing tariff barriers and minimizing such discriminatory trade practices as dumping (selling export goods at low cost), import–export quotas, and export subsidies. GATT went into effect in 1947, with the mission of orchestrating the expansion of world trade. Since its inception, a "round" of trade negotiations has been held approximately every five to six years. The Kennedy Round (1964–1967) was followed by the Tokyo Round (1973–1979), the Uruguay Round (1986–1994), and, recently, the Millennium Round in Seattle.

The Uruguay Round (the eighth round of trade negotiations) was comprehensive and ambitious, because it took on a much broader agenda, beyond mere tariff reductions, one that included the elimination of agricultural subsidies, the protection of intellectual property rights, the broader liberalization of services, and the elimination of quotas on some manufactured products—to mention a few. The Uruguay Round also resulted in the creation of WTO, whose functions include facilitating trade reform, providing a forum for trade liberalization negotiations, resolving trade disputes, and reviewing national trade policies to ensure compliance with international rules (World Bank, 2000). WTO remains the only international body mandated to establish rules that govern bilateral and multilateral trade negotiations between countries. As of July, 2001, WTO had 142 members, including 37 from Sub-Saharan Africa (Ethiopia, São Tomé and Principe, and Sudan have observer status). China's recent decision to join will have many implications for world trade, especially for trade in textiles.

As a result of the tariff-reduction activities of GATT–WTO, the value of world merchandise exports expanded from $58 billion in 1948 to $5.47 trillion in 1999 (WTO, 2001a.) Asia and Latin America have both benefited from the expansion of world trade. Asia increased its share of world merchandise exports from 15% in 1973 to 26.7% in 2000; Latin America increased its share from 4.7% to 5.8% during the same period. Africa, on the other hand, continues to lag behind both regions: its share of world merchandise exports dropped from 4.8% in 1973 to 2.3% in 2000. In 2000, the lion's share of Africa's merchandise exports went to Western European markets (57.5%), followed by North America (15.2%) and Asia (7.7%). Intra-African merchandise exports accounted for only 5.9%. The leading traders were South Africa and Nigeria, accounting for nearly 43% of the continent's merchandise export. Other players on a minor level include Côte d'Ivoire, Gabon, Cameroon, Kenya, and the Congo Republic. Table 15.1 shows that major exports consist of fuel and other minerals, while major imports comprise the manufactured goods. Africa's continued reliance on a few primary commodities makes it susceptible to fluctuating international commodity prices and other external shocks. Furthermore, as Chapter 8 demonstrated, problems with African trade are linked to declining terms of trade (importing more and exporting less), a limited manufacturing capacity, weak financial sectors, poor transportation and communications infrastructure, weak governance, and a lack of technical, financial and management capacity to manage the full slate of mechanisms and responsibilities required for successful regional and global integration.

Africa's trade problems are further compounded by the fact that, historically, there has been very little representation of African interests in trade negotiations and dispute resolution. The consensus among African representatives is that there is no level playing field when it comes to an integrated framework here.

TABLE 15.1 Merchandise Imports of the European Union from Africa by Product, 2000

Total Merchandise Imports	Percentage Share in 2000
Mining Products	57.8
Fuels	54.7
Ores	2.0
Nonferrous metals	1.1
Manufactures	25.0
Clothing	11.2
Other semimanufacturers	4.8
Agricultural Products	17.0
Food	13.0
Raw materials	3.9

Source: World Trade Organization (WTO). (2001a). *International Trade Statistics Yearbook,* Geneva: WTO.

The Uruguay Round set the stage for a series of trade negotiations and disputes that will have far-reaching effects on the ability of Sub-Saharan Africa to access global markets. They include reaching agreements on agricultural subsidies, protecting intellectual property rights, liberalizing a range of services, eliminating quotas on some manufactured products, and rationalizing labor standards and environmental standards.

Negotiating Trade Reforms in Agriculture

The controversy surrounding reforms in agriculture centers around market access, domestic policies that protect farmers, and export subsidies that make export prices artificially competitive. As Chapter 13 indicates, agriculture is the backbone of several African economies, accounting for 50% to 70% of the labor force and 32% of the GDP. A major problem confronting Africa's agricultural sector is the reluctance of the European Union (EU) and United States to eliminate export subsidies that favor "dumping" (selling a product at a price lower than its real value) of cheaper food products, which threatens local African farmers. The alternative for EU countries and the United States (revising their protective domestic agricultural policies) would have some internal political repercussions.

The objective of the Uruguay Round was to promote fairer competition by eliminating as many agricultural trade distortions as possible and by requiring member countries to commit to tariff reductions and a number of "binding" agreements. Developed countries, for example, were required, over a 6-year period (1995–2000), to reduce their agricultural tariffs and the value of their export subsidies by an average 36% (WTO, 2001b). Developing countries, on the other hand, were given 10 years (1995–2004) to reduce their tariffs and subsidies by 24%. Least developed countries, including several from Sub-Saharan Africa, did not have to make commitments to reduce tariffs or subsidies. In Zimbabwe, one of a few exceptions, tariff rates were reduced on 8.8% of its agricultural imports (World Bank, 2000). Chapter 13 suggests that the Agreement on Agriculture (AOA) is an unfair trade agreement, which allows low-priced, sometimes poorer quality food imports to replace local and traditional foods. Moreover, WTO has put more of an emphasis on export and cash-crop production without considering opportunities for expanding agricultural production to meet local needs. The drive toward expanding production land for cash-crop exports is closely associated with the loss of land for subsistence production, especially among women farmers.

Member countries also engage in a process referred to as "binding," which is a commitment not to increase tariffs on imported goods above a listed rate (the bound rate). If a country breaks its binding commitment, by increasing its tariff beyond the bound rate, it would have to negotiate a compensation package with its affected trading partners. Developing countries increased the number of bound rates on imported goods from 21% of product lines to 73%. The Uruguay Rounds further required all member countries, including those from Africa, to bind their tariff lines in agriculture. For example, before the Uruguay Rounds, Zimbabwe bound its tariff rates on only 35.9% of its agricultural imports; after the Uruguay Rounds, it agreed to bind its rates on 100% of its agricultural products. Senegal agreed to bind its tariff rates on 58.3% of all its imports (100% on agriculture) after the Uruguay Rounds, compared with 39% before the Uruguay Rounds. However,

Senegal did not make any commitments towards reducing its tariff rates, since it was not obligated to do so.

Trade-Related Aspects of Intellectual Property Rights (TRIPS)

In order to deter illicit practices, such as piracy, bootlegging, and the production of counterfeit goods, the industrialized countries saw a need to create an agreement between themselves and the developing countries that involved the protection of patents, copyrights, and trademarks associated with inventions, research design, and the testing of new products. The agreement was to affect an array of products, including integrated industrial designs, pharmaceuticals, plant varieties, wines and spirits, and agricultural chemicals, to name a few. Thus, during the Uruguay Round, when WTO was formed, the Trade-Related Aspects of Intellectual Property (TRIPS) was implemented. The goal was to develop a more harmonious system of protecting intellectual-property rights by creating a uniform set of international rules. Consequently, developing countries were given 5 years to adjust their laws, standards, and practices to conform with TRIPS agreements. Less developed countries were given 11 years (WTO, 2001b), and more developed countries were given 1 year. African countries have petitioned for an extension to allow for enough time to develop the administrative capacity and the right institutional mechanism to adjust their laws and regulations.

The UNDP (1999) has observed that the TRIPS agreement is unfairly tilted in favor of multinational corporations, because it consolidates their dominant ownership of technology and increases the cost of transferring technology to developing countries, thus potentially restricting access by the latter to generic drugs and appropriate computer software. The UNDP, for example, points out that, prior to the TRIPS agreement, some countries—for example, China, Egypt, and India—allowed patents on pharmaceutical processes, but not on final products; this approach encouraged domestic industries to produce generic drugs that were cheaper than those bearing the original brand names. As a result, prices for some drugs in Pakistan, which administers patents, are more than 10 times as high as drug prices in India. The TRIPS agreement requires 20-year patents on both processes and products. WTO member countries, such as India, are required to modify their patent laws to conform with the TRIPS agreements. This requirement has an adverse impact on those African countries that relied on India for a generic version of the drug AZT that costs less than half the annual price tag, $3000, charged by the multinational company.[2] Also, conforming to the TRIPS agreement will likely suppress current efforts among developing countries to bring to market new generic drugs for AIDS, as Chapter 12 points out.

Another adverse effect of the TRIPS agreement on Africa is that it does not acknowledge the rights of local communities to their traditional, customary, and indigenous knowledge, which opens the door for multinational pharmaceuticals and agribusinesses to

---

[2] The head of the Nyumbani AIDS orphanage in Nairobi was among the prominent figures now working in Africa who put public pressure on the brand-name multinational producers of these antiretrovirals in a successful attempt to get them to reduce their price to $350 per year.

gain access to, and exploit, the indigenous resources of local communities. Another example provided by the UNDP (1999) refers to how a leading pharmaceutical company exploits Madagascar's rosy periwinkle, a plant species containing anticancer properties, with hardly any profits going back to Madagascar. Chapter 13 raises a question about whether local communities with centuries of traditional knowledge will be compensated or whether TRIPS will promote *biopiracy*. The developing countries in the south contain the bulk of the earth's genetic plant resources and have for centuries cared for those plants and discovered and applied their efficacy as medicines and food resources. If countries from the north, which have undue technological and financial advantage, take genetic materials from plants for private gain, does TRIPS have a provision for compensating the owners of native plants, or will it amount to stripping those local communities of centuries of traditional knowledge? Given these adversarial examples, it is incumbent upon African countries to safeguard the rights and knowledge base of indigenous communities.

Chapter 13 also raises some concerns over potential problems that TRIPS poses for farmers in Sub-Saharan Africa. These include (1) the extent to which small farmers will be able to compete with transnational corporations (TNCs) that have the capacity to privatize seeds and genetic resources by securing patents and by increasing the number of genetically modified organisms (GMO) for sale and for cultivation, and (2) the high cost to small farmers who are forced to buy seeds from a company or purchase chemicals that are tailored for certain genetically modified plants.

General Agreement on Trade and Services (GATS)

The General Agreement on Trade and Services (GATS) was a significant part of the Uruguay Rounds; it commits the signatories to conducting periodic negotiations that encourage the liberalization of such sectors as finance and banking, insurance, transport and telecommunications, tourism, health, and building and construction services. Two major service sectors excluded from GATS are government-authorized services and air transportation. In Sub-Saharan Africa, value-added services account for about 50% of the GDP (World Bank, 2001b). Business, professional services, and transportation or telecommunication services (or both) are vital complements to agricultural and industrial activities. GATS requires member countries to develop a list of service sectors that will be committed to interested foreign service providers. According to the WTO (2001b), these services can be traded to foreign service providers via four modes of supply: by supplying a service, such as international calls, from one country to another (cross-border supply); by consuming a service, such as ecotourism, in another country (consumption abroad); by allowing a firm, such as an insurance company, to set up a branch in another country (commercial presence); and by allowing individuals to travel to supply a service, such as computing, to another country (movement of natural persons). Another important principle underlying GATS is the "most-favored-nation" clause—if a member country grants favorable treatment to a foreign service provider, it must extend the same treatment to all other GATS members.

Africa has the potential to enhance and improve its service delivery capacity by exchanging information with other service providers and by benefiting from the transfer of emerging technologies associated with various service sectors. According to information compiled from the European Commission Directorate General for Trade (2001), the majority of Sub-Saharan African countries have committed to liberalizing services in the tourism, business, and financial sectors. In contrast, the environmental and distribution service sectors received the fewest commitments. Sierra Leone scheduled commitments in 10 service industries, and Senegal and Ghana each scheduled 6, but Chad and Mauritania each scheduled only one commitment, to the travel and tourist service industry. The majority of scheduled commitments in travel and tourism granted market access to hotel and restaurant chains, travel agents, and tour operators. Kenya, which is well known for its tourism industry, made broad commitments to grant full market access and national treatment to foreign providers of all travel and tourism services through a commercial presence (USITC, 1999).

Some African countries have also scheduled commitments in the telecommunications, health, business, financial, and education sectors. Telecommunications services include enhancements in voice telephone, telex and facsimile services, electronic mail, and voice and data transmissions over privately leased circuits. For example, Senegal scheduled commitments to address local, long-distance, and international telecommunication services by using cellular and satellite technologies. Senegal's national provider, SONATEL, will continue to monopolize telecommunications until the end of 2003, when the government will open up local and long-distance services to other bids. In regards to health services, some countries, among them South Africa, will grant full market access to foreign medical practitioners and human health-service providers who offer these services through a commercial presence. Business services, such as accounting, auditing, and bookkeeping, are also encouraged in some countries—one being Zambia, which grants access through a commercial presence, consumption abroad, and cross-border supply. In the financial sector, Kenya, Malawi, and Mozambique, among others, have considered liberalizing their banking sectors completely. In education, Ghana is one of the countries encouraging the delivery of secondary and other education services via a commercial presence, cross-border supply, and consumption abroad.

Trade Negotiations on Manufactured and Industrial Goods

African interests regarding trade agreements on manufactured and industrial goods revolve around (1) avoiding further commitments to tariff reductions on industrial goods, (2) preserving preferential tariff treatment for less-developed countries to access markets in developed countries, and (3) resolving disputes surrounding the issue of textiles and clothing. There is concern among African countries that further tariff reductions on imported industrial products will cause local industries to lose their market share and employment base and, therefore, shut down. Kenya, in an effort to shift from a policy of import substitution to a policy of outward orientation and trade liberalization, reduced its average tariff rate from about 42% to 34%; however, Buffie (2001) points out that its textile, cement, glass, tobacco, and beverage industries have all struggled to compete with imported products. Likewise, Senegal lost about one-third of its manufacturing after reducing the average rate of protection from 165% in 1985 to 90% in 1988. Under such circumstances, it is important for African countries to study the short- and long-term consequences of tariff reductions before making any commitments.

African countries would rather be in a position to gain market shares for their industrial products. The Generalized System of Preferences (GSP) provides them with an opportunity to do just that. Through bilateral and multilateral negotiations, developed countries grant import preferences to a select number of goods exported from African countries (and other developing countries). For example, the European Community, for an unlimited period of time, provides duty- and quota-free treatment for all products originating from least developed countries (LDCs), except for arms and ammunition. Special provisions apply to rice, sugar, and fresh bananas, where customs duties will be phased out by 2009 (UNCTAD, 2001). Japan also grants duty- and quota-free treatment to LDCs, including a 40% tariff reduction on refined copper from the Democratic Republic of Congo. The United States grants duty-free treatment to 1783 products from LDCs—excluding textiles, watches, footwear, handbags, and luggage, to mention a few. In addition, the Africa Growth and Opportunity Act (AGOA) of 2000, mentioned in Chapter 8, provides duty-free treatment to a wider range of African products for a longer period of time, compared with the GSP scheme. The concern among African countries is that developed countries tend to ignore the GSP schemes by increasing tariffs on some manufactured goods. The GSP schemes are seen by Africans as mechanisms to correct imbalances in the global trading system. Therefore, African countries would like to preserve these rights, rather than see them erode under new WTO rules.

The issues surrounding the textile and clothing industries relate to a phasing out of import quotas that have been imposed by textile producing countries since the early 1960s. Beginning in 1974, under a Multifibre Arrangement (MFA), countries could protect their local textile and clothing industries by imposing quantitative restrictions or quotas on similar imports if there was a perceived threat. The WTO Agreement on textiles and clothing, which phases out quotas by 2005, replaced the MFA. This allows enough time for such major exporters and importers as China (a new member), the United States, Korea, Germany, and the United Kingdom to make the necessary adjustments. Although Africa is not considered a major player, South Africa, Zimbabwe, Lesotho, Madagascar, Mali, Kenya, and Mauritius, among others, have active textile and apparel industries that could potentially access US and European markets as the phaseout of quotas begins.

Other Trade-Related Issues

Other WTO trade-related matters that will affect African countries include negotiations on sanitary standards for products, labor standards, environmental standards, and information technology. One issue that requires further clarification by WTO members involves the defining of uniform sanitary and phytosanitary standards. This requires countries to adhere to safety and sanitary standards by labeling their export products and by divulging pertinent information about any contaminants that their products might contain. Meeting internationally prescribed standards will undoubtedly be a costly venture for countries in Sub-Saharan Africa, because it involves increased investments in improving production, storage, and transportation methods as well as compliance procedures (Wilson, 2000).

The issues surrounding environmental management and preservation are also complex, because they involve the international harmonization of national regulations on such matters as reduction of greenhouse gases, deforestation, water pollution, etc. It looks even more complex when one considers the different sociocultural values and administrative capacities of African countries in relation to environmental-asset management. WTO member countries are further encouraged to embrace core labor standards that prohibit forced labor, slavery, the exploitation of child labor, and discriminatory employment practices and ones that protect a worker's rights to unionize and to bargain for better working conditions (Maskus, 1997). Developed countries would like to see trade sanctions imposed on countries that violate these core standards.

WTO has further developed agreements to eliminate the duties on, and to ensure market access for, information-technology products such as computers, telecommunications, semiconductors, semiconductor manufacturing equipment, software, and scientific instruments. As of February 2001, there were 56 participants in the information-technology agreements (ITAs), accounting for over 93% of world trade in information-technology products. None of the participants was from Africa.

The range of trade-related issues addressed during the Uruguay Round serve as a framework for future trade negotiations, among them being the Millennium Round initiated at Seattle, Washington. However, as Sinclair (2000) points out, the Seattle meetings set the tone for what will continue to be a concerted effort on the part of developing countries, including those from Sub-Saharan Africa, to demand greater market access for their products, stronger enforcement of antidumping rules, an extension of TRIPS implementation deadlines, and further clarification on core labor and environmental standards. For Sub-Saharan Africa, the challenge remains building the technical and administrative capacity to participate effectively in negotiations that are beneficial to the interests of Africans.

AFRICA'S PROSPECTS FOR GREATER INTEGRATION INTO THE GLOBAL ECONOMY

The evidence presented in Discussion Box 15.1, if taken at face value, might make Africa's prospects for global integration appear slim. Its share of world trade has declined in the last 20 years; it continues to attract the lowest amount of foreign direct investment (FDI) of any region in the world; and it continues to lag behind the rest of the world in technological innovation and advancement. Although this presents a gloomy picture, it is not an excuse to languish in misery and abandon all efforts toward progress. There are a number of success stories that demonstrate Sub-Saharan Africa's ability to marshal its human and physical resources to transcend social and economic impediments. As Chapter 8 points out, both Uganda and Mozambique have demonstrated that countries perceived to be "at-risk" or "beyond help" can overcome substantial development problems. Each country has overcome significant obstacles and has embarked on comprehensive socioeconomic development initiatives to steer its economy in the right direction. More importantly, each country has developed a set of poverty-reduction strategies designed to empower the impoverished via health and education programs and community-based development initiatives. There are similar success stories in other countries: In Mali, Tanzania, and Benin and in the rural communities of Kenya, Cameroon, Senegal, and Zimbabwe, successful community initiatives are receiving support from nongovernmental organizations. Moreover, the series of democratic reforms taking place in African countries in response to donor conditions and stipulations (as described in Chapter 5) is encouraging.

For the global community, the real challenge becomes developing a more humane globalization strategy wherein Sub-Saharan African countries can simultaneously take advantage of the benefits of globalization, preserve their cultural heritage, and develop the capacity to promote and sustain democratic reforms to consolidate their political and economic sovereignty. A more humane globalization strategy focuses on creating opportunities for the poor, who are primarily women, small-scale farmers, and children who live in poverty. Given this perspective, how can African countries best position themselves to take advantage of the benefits of globalization? What are the prerequisites for a successful integration into the global economy?

A major prerequisite for a successful transition into the global economy remains the need to upgrade the physical, social, and technological infrastructure. Chapter 9 provides evidence of the deteriorating conditions of road, rail, and water transportation. It is encouraging, however, that some progress has been made along the lines of Regional and Pan-African highways. The eventual completion of the coastal West African, Trans-Sahelian, and Trans-East African highway systems should help facilitate much-needed trade transactions among African countries. The evidence from Chapter 10 suggests a need to upgrade the urban infrastructure (solid-waste and wastewater disposal systems, pipelines, electricity, urban roadways) if African cities are to become the engines of economic growth and structural transformation.

Upgrading the social infrastructure includes a commitment to investing in the poor by providing them with access to affordable education, health care, and economic opportunity. This is consistent with developing a more humane globalization strategy. About 36 Sub-Saharan African countries are eligible for the World Bank/IMF poverty-reduction and growth-facility program (PRGF), which includes an education and health component. Several African countries have proposed education reforms that stress improving primary universal education, reviewing the curriculum to ensure quality and relevance, and streamlining the management and utilization of educational resources. Other educational priorities include expanding science and technology and business and communications skills; introducing cost-sharing mechanisms in secondary and tertiary education; exploring avenues for distance education; and promoting vocational, technical, and entrepreneurial training. In the health sector, the major focus is on dealing with HIV–AIDS, endemic diseases, and infant and child mortality, as was discussed in Chapter 12. Developments in information and communication technologies should also help in managing health care. HealthNet is an Internet-based resource that provides doctors all over the world with access to medical libraries. Doctors in Africa can use this resource to keep up with new developments in malaria, AIDS, and disease ecology research. The African Networks for Health Research & Development (AFRO-NETS) constitutes an advocacy group of medical practitioners and institutions that focus on mobilizing health resources, disseminating and sharing research findings, utilizing information technology to network, and building the capacity to address health problems in Africa.

Given the current state of affairs highlighted in Discussion Box 15.1, developing a technological infrastructure is a major challenge to African countries. As Chapter 9 points out, the challenge is one of building digital bridges to overcome spatial divides. This challenge includes (1) developing an infrastructure that can accommodate the growing number of Internet Service Providers (ISPs) and (2) encouraging public access to computer and communication technologies via community-based phone-shops, telecenters, Internet cafes, and network learning centers. The Economic Commission for Africa (ECA) is sponsoring the African Information Society Initiative (AISA), which calls on African experts to prepare an action plan to assist Africa's entry into the Information Age and to develop appropriate institutional frameworks and human, information, and technological resources for a sustainable build-up of an information society in African countries. UNESCO is sponsoring a Regional Informatics Network for Africa (RINAF) to improve the communication capabilities among African research institutions, to create a group of technicians specialized in the management of African network services, and to develop a number of regional centers that are capable of managing such services for Africa. Furthermore, the WTO's General Agreement on Trade and Services has encouraged some African countries to make a commitment to enhancing and upgrading telecommunication services, as was discussed earlier in this chapter.

Successful integration into the global economy is further predicated on good governance, founded on the principles of decentralization, civic engagement, transparency, and accountability. Chapter 5 points to the turbulent past of African governments. Under persistent pressure from donor agencies and humanitarian organizations, several African countries, since the 1990s, have embraced democratic reforms that advocate pluralism, participation, decentralization, representation, accountability, and the rule of law. However, as events in Sierra Leone and the Sudan demonstrate, the democratization process is by no means universal. Chapter 5 further states that, while most countries have been more successful in opening their political systems to multiparty competition, they have been less successful creating appropriate constitutions that endorse the rule of law; in other words, the separation of power, administrative accountability, the freedom of speech and assembly, and a hegemony of the civilian over the military.

Beside building appropriate democratic institutions, Sub-Saharan African countries need to develop an enabling environment that is conducive to local and foreign investment. This involves developing the appropriate institutional mechanisms and technical capacity to support business and entrepreneurial activity. For example, Chapter 14 stressed the need to support business and industrial development by providing institutions for industrial standards, testing, export support, quality assurance, training, research, and technical extension. Such an enabling environment is especially relevant as African countries shift from a policy of import substitution to a more outward-looking industrial policy. Furthermore, African governments need to provide the necessary mechanisms to facilitate privatization and entrepreneurial activity—two critical ingredients in the globalization process. Governments can do this by providing such support services as information-referral networks, financial management, accounting and legal assistance, and performance-monitoring guidelines. In addition, governments can minimize the constraints that frequently frustrate the privatization process—for example, by avoiding too much bureaucracy and red tape and by avoiding delays in payments to private contractors.

Promoting greater intra-African trade could serve as mechanism for greater integration into the global economy. The prerequisites for regional cooperation—liberalizing trade via tariff reductions, creating incentives for investment, reforming legal and regulatory requirements, streamlining banking and financial institutions, building the administrative capacity and technical capacity, and establishing institutions to coordinate regional initiatives—in many respects parallel the preconditions for global integration. Developing the capacity to enhance regional trade could serve as a learning experience and could very well prepare African countries for the rigidities of the global economy. There are concerns that regional trading arrangements, such as protective tariffs against nonmembers, could undermine global trading negotiations aimed at liberalizing trade. However, greater regional cooperation among African nations can only enhance, rather than hinder, African efforts in seeking greater global integration.

Successful global integration goes hand in hand with economic diversification. In the agricultural sector, this implies a shift from a reliance on one or two cash crops to consider opportunities in nontraditional export activities, such as the horticultural sector. African fruits and vegetables, such as avocados, mangoes, pineapples, and green beans, are finding markets in Europe and Asia. There is also an emerging market for flowers from some countries, such as Zimbabwe and Uganda. Diversification in the industrial sector hinges on a government's ability to develop a strong entrepreneurial base, to expand small- and medium-sized businesses, to support privatization efforts, and to develop an outward-looking, export-oriented approach. As Chapter 14 reveals, successful diversification requires the ability and commitment of governments Sub-Saharan Africa to create the necessary support structure and enabling environment. It further requires the ability of governments to search for new markets locally, regionally, and internationally, through trade missions and through bilateral and multilateral negotiations.

Debt restructuring becomes an even more pressing matter if these countries are to succeed in implementing the aforementioned strategies. Various scenarios have been offered, ranging from outright write-offs to conditional debt restructuring. The debt problem in Africa has reached crisis proportions: governments are paying more for servicing foreign debts than they do for the health or education of their own people. The twin crises of debt and poverty have essentially paralyzed African countries to the point where it is extremely difficult to sustain the kind of effort required to promote social, political, and economic reforms, let alone make an aggressive move towards globalization. The World Bank's HIPC initiative is designed to ease the burden of African countries; however, some have argued that it is not enough. Any form of debt relief must be tied to educational and health reform, especially for the poor. In other words, African countries that qualify for debt relief must demonstrate a sincere commitment to targeting relief funds to programs that expand opportunities for the poor and for populations with special needs.

CONCLUSION

It is tempting to get caught up in a debate about the threats and opportunities that globalization presents. Ultimately, what matters is the need for African governments to develop an enabling environment that provides opportunities for the realization of human potential. There are some encouraging signs that point to an "African Renaissance," as some have suggested. Democratic reforms in the 1990s have embraced the principles of good governance and expanded political freedoms; some countries (among them Uganda, Mozambique, and Mali) once written off have overcome severe impediments to record impressive economic growth rates; rural communities and self-help organizations are marshaling their energies and resources to finance community-based projects; an increasing number of women's organizations are actively promoting gender justice and equal access to economic, social, and political opportunities; and international donor agencies are becoming increasingly sensitive to the plight of the poor. These promising signs are shadowed by some lingering problems: high debt burdens, growing poverty, and the threat of HIV–AIDS. These problems often seem insurmountable, but opportunities exist for Africa to tap into its vast reservoir of human and natural resources and to create policies that are not solely growth inducing, but also pay more attention to issues of poverty, human equity, and social and economic justice. International efforts (bilateral and multilateral) are welcome, but they must be well intended. More importantly, they must embody strategies that provide the poor with opportunities to achieve their full potential.

KEY TERMS

Globalization
World Trade Organization
Generalized System of Preferences (GSP)
General Agreement on Trade and Services (GATS)
Trade-Related Aspects of Intellectual Property Rights (TRIPS)

Agreement on Agriculture (AOA)
General Agreement on Trade and Tariffs (GATT)
Foreign Direct Investment (FDI)
Uruguay Round

REFERENCES

A.T. KEARNEY, INC. (2001). "Measuring Globalization," *Foreign Policy*, January/February: 56–64.

BIRDSALL, N. (1998). "Life Is Unfair: Inequality in the World," *Foreign Policy*, Summer: 76–93.

BUFFIE, E. (2001). *Trade Policy in Developing Countries*, Cambridge, NY Cambridge University Press.

EUROPEAN COMMISSION DIRECTORATE GENERAL FOR TRADE (2001). "Opening World Markets for Services," *The European Commissions "Info-Point" on World Trade in Services*, http://gats-info.eu.int/gats-info/gatscomm.pl.

HENRIOT, P. (2000). "Globalisation: Implications for Africa," *Sedas Bullettin on Net*, http://www.sedos.org/english/global.html.

MASKUS, K.E. (1997). "Should Core Labor Standards be Imposed Through International Trade Policy?" *Policy research working paper: 1817*, World Bank, Washington, DC.

SINCLAIR, S. (2000). "The WTO: What happened in Seattle? What's next in Geneva?" *Canadian Centre for Policy Alternatives Briefing Paper Series: Trade and Investment*, 1:(2), pp. 1–13.

UNITED NATIONS CONFERENCE ON TRADE AND DEVELOPMENT (UNCTAD) (2001). *Generalized System of Preferences: Handbook on Special Provisions for Least Developed Countries,* Geneva: UNCTAD.

UNITED NATIONS DEVELOPMENT PROGRAM (UNDP) (1999). *Human Development Report: Globalization with a Human Face,* Oxford, NY: Oxford University Press.

UNITED NATIONS DEVELOPMENT PROGRAM (UNDP) (2001). *Human Development Report: Making New Technologies Work for Human Development,* Oxford, NY: Oxford University Press.

UNITED STATES INTERNATIONAL TRADE COMMISSION (USINTC) (1999). *General Agreement on Trade in Services: Examination of the Schedules of Commitments by African Trading Partners,* Publication 3243, Washington, DC: USITC.

WILSON, J. (2000). *The Development Challenge in Trade: Sanitary and Phytosanitary Standards,* Washington, DC: World Bank

WORLD BANK (2000). *The World Development Report 1999—2000: Entering the 21st Century,* Washington, DC: World Bank.

WORLD BANK (2001). *Can Africa Claim the 21st Century?* Washington, DC: World Bank.

WORLD BANK (2001b). *The Uruguay Round: Data Files, http://www.worldbank.org/research/trade/how2use.html.*

WORLD TRADE ORGANIZATION (WTO) (2001a). *International Trade Statistics Yearbook,* Geneva: WTO.

WORLD TRADE ORGANIZATION (WTO) (2001b). *Trading into the Future: the Introduction of the WTO,* Geneva: WTO, *http://www.wto.org/english/thewto_e/whatis_e/tif_e/tif_e.htm.*

Recommended Web Sites on Africa

Please note that the content and format of web sites may change over time. In case of any changes, go to the original URL address and retrace your steps.

COMPREHENSIVE WEBSITES

Africa: South of the Sahara: http://www-sul.stanford.edu/depts/ssrg/africa/guide.html
Provides comprehensive coverage on African countries. Includes a search engine that links students to the African internet guide and a photo library.

Africa Action: http://www.africaaction.org/index.shtml
Incorporates the Africa Fund, Africa Policy Information Center (APIC), and the American Committee on Africa. Provides a wealth of information on health, AIDS, debt, action and advocacy networks, and policy documents. Advocates political, economic and social justice in Africa. Also features country-specific data links.

African Studies Center (UPENN): http://www.sas.upenn.edu/African_Studies/AS.html
Features country-specific information, multimedia archives including city maps and satellite imagery, and useful web links.

Columbia University African Studies: http://www.columbia.edu/cu/lweb/indiv/africa/cuvl/
Features a Virtual Library on Africa with links to electronic journals, databases, literature, films and videos, and a range of topics.

http://newafrica.com/
Wealth of information on culture, health, ecotourism, country statistics, and the environment. Excellent information on mining and business development.

Africa Focus: Sights and Sounds of a Continent: http://africafocus.library.wisc.edu/
Features thousands of digitized visual images and sounds from all over Africa including buildings and structures, cities and towns, landscapes, religion, and more.

AllAfrica Global Media: http://allafrica.com/
Features special reports and news on a range of social, political, and economic issues.

National Geographic on Africa: http://www.nationalgeographic.com/africa
Features maps, lesson plans, virtual tours, and video clips.

The Index on Africa: http://www.afrika.no/links/
Links to a range of topics including history, politics, development, conflict, business, environment, and tourism.

The Library of Congress Country Studies: http://lcweb2.loc.gov/frd/cs/cshome.html#toc/
Topics include the historical setting, society and its environment, the economy, government and politics, and national security.

BY TOPIC

Physical Environment of Africa

African Environment: http://newafrica.com/environment/
Biodiversity, climate change, endangered species, legislation, impact assessments, pollution, and conservation. Also includes several articles and reports on environmental issues.

African Conservation: http://www.africanwebsites.net/
Conservation databases, interactive knowledge base & discussion forums.

African Data Dissemination Service: http://edcintl.cr.usgs.gov/adds/adds.html
Reference maps and digital data on crop use, hydrology, digital elevation models, rainfall charts, vegetation, and agro-climatic zones in Africa. Also linked to the **Famine Early Warning System network: http://www.fews.net/.** Developed by USAID to monitor drought and famine conditions.

Database on African Lakes: http://www.ilec.or.jp/database/map/regional/africa.html
Information on surface area, volume, water usage, land use, pollution levels, and fish catch.

African History

**African Timelines:
http://www.cocc.edu/cagatucci/classes/hum211/timelines/htimelinetoc.htm**
Features accounts on Ancient Africa, African Kingdoms, Slave Trade, Colonialism, and Post-Independent Africa.

The Story of Africa—Living History:
http://www.bbc.co.uk/worldservice/africa/features/storyofafrica/1chapter1.shtml
Chronological account of ancient civilizations, traditional religions, African kingdoms, and African cultures.

Wonders of the African World: http://www.pbs.org/wonders/
Description of Black kingdoms of the Nile, major civilizations, holy lands, and the Swahili coast.

African Culture

African Cultures: http://www.teko.kherson.ua/africancultures/
Music and dance, cuisine, religion, languages, literature, arts and museums.

Languages of Africa (Ethnologue country index):
http://www.ethnologue.com/country_index.asp?place=Africa
A compendium of languages in each African country.

African arts and crafts: http://www.africancrafts.com/artist/main.htm

African music and dance: http://www.cnmat.berkeley.edu/~ladzekpo/

African Political Systems

African Governments: http://www.gksoft.com/govt/en/africa.html
Detailed description of the types of African governments.

Population Studies in Africa

Population Council in Africa: http://www.popcouncil.org/africa/africa.html
Population and family planning activities.

Population Information Network of Africa:
http://www.un.org/Depts/eca/divis/fssd/popin/startframe.htm

African demography (country specific) http://www.pop.upenn.edu/world/africa/

Urban Geography of Africa

Africa Cities and Towns: http://www.newafrica.com/cityguides/

Cities of Africa: http://library.thinkquest.org/16645/the_people/cities_west.shtml

U.N. Habitat (United Nations Human Settlements Programme)
http://www.unhabitat.org/default.asp
Best practices of urban management, planning, and governance.

Development in Africa

World Bank Group-Sub-Saharan Africa: http://lnweb18.worldbank.org/afr/afr.nsf
Wealth of information and publications on economic and human development, poverty, capacity building, transportation policy, environmental strategy, and indigenous knowledge for development.

Africa Internet Connectivity: http://www3.sn.apc.org/africa/
Information & Communication Technologies (ICTs) Telecommunications, Internet and Computer Infrastructure in Africa.

United Nations Development Program:
http://www.undp.org/dpa/publications/regions.html#Africa
Publications on poverty, inequality, globalization, and human development. Includes human development reports **(http://www.undp.org/hdro/).**

Economic Commission for Africa: Regional Integration:
http://www.un.org/Depts/eca/ reg.htm
Provides examples of regional economic organizations in Africa.

USAID in Africa: http://www.usaid.gov/regions/afr
Features various U.S. development assistance programs in Africa.

Gender and Women Studies in Africa

Women in Development Network (WIDNET): http://www.focusintl.com/r1a.htm.

The African Women's Development and Communication Network (FEMNET): http://www.africaonline.co.ke/femnet/default_e.htm.

Medical Geography and Health in Africa

United Nations Program on HIV/AIDS: www.unaids.org/africapartnership.

AIDS in Africa http://www.washingtonpost.com/wp-dyn/world/issues/aidsinafrica/index.html (Washington Post special on AIDS in Africa).

African Networks for Health Research & Development: http://www.afronets.org.

World Health Organization: http://www.who.int/
Information on tropical diseases and epidemiological fact sheets for each country.

Agricultural Development in Africa

African agriculture: http://www.newafrica.com/agriculture/
Information on cash crops, food markets, and agricultural policy.

AFROL Agriculture in Africa:
http://www.afrol.com/Categories/Economy_Develop/agriculture.htm
Articles on food production, food security, cereal prices, desert encroachment and food staples.

Business and Industrial Development in Africa

Yale Africa Guide InterActive: Business and Industry in Africa:
http://swahili.africa .yale.edu/links/Business_and_Industry_in_Africa/
Links to sites on export activity, and business and finance.

Recommended Web Sites on Africa

African Business Network: http://www.ifc.org/abn/index.htm
Business, economic, and investment profiles of African countries; articles; and emerging markets.

Mbendi Information for Africa; http://www.mbendi.co.za/index.htm
Information on African oil and industry.

African Business Information Services: http://www.afbis.com/
Information and analysis on business and economics in Africa.

Electronic Journals on Africa

African Affairs: http://www3.oup.co.uk/afrafj/contents/
African News: http://www.peacelink.it/afrinews.html
Africa Recovery: http://www.un.org/ecosocdev/geninfo/afrec
Africa Studies Quarterly: http://web.africa.ufl.edu/asq
EcoNews Africa: http://www.econewsafrica.org/ (sustainable development)
Internet Journal of African Studies: http://www.brad.ac.uk/research/ijas/
Jenda: a journal of culture and African women studies: http://www.jendajournal.com/jenda/
Journal of African Economies: http://jae.oupjournals.org/current.shtml
Journal of Sustainable Development in Africa: http://publicpolicy.subr.edu/africa.htm
Science in Africa: http://www.scienceinafrica.co.za/
South African Geographical Journal: http://www.egs.uct.ac.za/sagj/

MULTIMEDIA

Africa Online: http://www.africaonline.com/site/africa/index.jsp
CNN: http://www.cnn.com/WORLD/#africa
New York Times: http://www.nytimes.com/pages/world/africa/index.html
Christian Science Monitor: http://www.csmonitor.com/world/africa.html
The Economist: http://www.economist.com/world/africa/index.cfm
C-Span Online: http://www.c-span.org/international/

Photo Credits

2.3 © Sharna Balfour; Gallo Images, CORBIS;
2.4 © Gregory G. Dimijian, Photo Researchers, Inc.;
3.3 © Dr. William Osei
4.2 © MIT Collection, CORBIS;
4.6 © Frank Fournier, Woodfin Camp & Associates;
6.7 © Joseph R. Oppong;
6.10 © Joseph R. Oppong;
9.1 to 9.5 © Joseph R. Oppong;
9.7–9.8 © Joseph R. Oppong;
9.10a–b © Dr. Abdul Alkalimat & Brian Zelip;
10.4 © Philippe Maille, Photo Researchers, Inc.;
10.5a © Daniel Laine, CORBIS;
10.5c © Peter Johnson, CORBIS;
10.5d © Hubertus Kanus, Photo Researchers, Inc.;
11.6 © Betty Press, Woodfin Camp & Associates;
12.7 © Joseph R. Oppong;
12.13 © Joseph R. Oppong;
13.2 © Joseph R. Oppong;
13.4 © Joseph R. Oppong;
13.5 © Georg Gerster, Photo Researchers, Inc.;
13.6 © Joseph R. Oppong;
All other photos are courtesy of Samuel Aryeetey Attoh, Editor.

Index

A

AAF-SAP, 226
Abidjan, 4, 20, 88, 240, 243, 251, 259, 296, 331, 348
Aborigines, 84
Abstinence
    postpartum
        fertility rates, 172–173
Abuja, 259, 291–292, 297, 333–334
Acacia albida, 73
Accra
    internal structure, 265–269
    residential land use, 267
Acute disease
    definition, 327
Addis Ababa, 258
    rainfall, 30
    temperature, 30
Adinkra cloth
    symbolism, 150–153
Adornment, 150–158
Adulis, 255
AEC, 221
AFDL, 119
AFORD, 125
Africa
    migration from Gondwana, 13
African Alternative to the Structural Adjustment Programs for Socio-Economic Recovery and Transformation (AAF-SAP), 226
African capitalism, 120, 213–214
African cities
    internal structure, 258–269
African Common Market, 221
African culture, 138–147
    vs. European culture, 159
African Growth and Opportunity Act (AGOA), 220
African Information Society Initiative (ASIA), 439
African National Congress (ANC), 127
African Networks for Health Research & Development (AFRO-NETS), 438
African socialism, 120
African Women Organization for Research and Development (AWORD), 317
Africa Project Development Facility (APDF), 414
Afrikaans, 7, 146
Afro–Asiatic family, 145–146
Afromontane vegetation, 41
AFRO–NETS, 438
Afrotropical biogeographical realm, 34
Aged
    modernization, 160–161
AGOA, 220
Agreement on Agriculture (AOA), 395
Agricultural density, 168
Agricultural development, 363–401
Agricultural drought, 69
Agricultural markets
    government, 393–395

449

Agricultural production
  alternative energy, 399
  community-based cooperatives, 400
  education, 397–398
  environmental influences, 304–305
  green revolution, 390–391
  human-induced, 304–305
  institutional factors, 388
  integrated rural development, 400–401
  land degradation, 388–389
  problems, 386–391
  research and development, 391–396
  rural infrastructure, 386–388, 397
  technology, 398–399
    inappropriate, 389–390
Agricultural trade reforms, 432–433
Agriculture. *See also* Farming systems
  causes, 54–57
  large-scale commercial, 378–379
  slash and burn, 369–370
  women, 302–304
  WTO, 395–396
Agro-based industries, 399–400
AIDS/HIV, 329–334
  cause, 330
  Côte d'Ivoire, 330
  military, 331
  Nigeria, 333–334
  police, 331
  structural adjustment, 332–333
  truck drivers, 331
  women, 308
Air transport, 241–244
Albedo effect, 69
Alfisols, 44
Alhajj, 142
Allah, 86
Alliance for Democracy (AFORD), 125
Alliance of Democratic Forces for the Liberation of Congo (AFDL), 119
Alternative energy
  agricultural production, 399
Altitude, 29
ANC, 127
Ancient civilization, 79–82
Angola, 5
  political instability, 116
Anopheles gambiae, 334

AOA, 395
Apartheid system, 116
APDF, 414
Arable land, 54–55
Arithmetic density, 166, 169
Ascription, 212
ASEAN, 221
Ashanti, 110, 130
ASIA, 439
Assimilation, 100
Association of South East Asian Nations (ASEAN), 221
Autochthones, 84
Avunculocality, 144
AWORD, 317
Axum, 255

B

Baganda, 149, 171, 175
Bamako Initiative, 357–358
Bamba, Ahmadou, 88
Banda, Kamuzu, 109, 124
Banks
  loans, 280
Bantu, 4, 84–85, 111, 137, 145–146, 160
  languages, 4, 84, 138, 144–147, 158
  migrations, 84–85
Bantustans, 127
Basic needs, 100, 209
BDP, 129
Bedie, Henri, 121
Belgian rule, 100
Benin, 2, 36, 53, 55, 83–84, 108–109, 115, 123, 131, 167–169, 173, 181, 195–196, 199, 203, 222–223, 239–240, 251, 256, 306, 311, 315, 337, 339, 368, 437
Berlin Conference, 6, 100, 103
Biafran war, 100, 109, 114
Bilateralism, 220
Bilateral kinship system, 143
Bilharziasis, 337
Biome, 34
Biopiracy, 434
Boer Trek, 96–97
Bororo, 373
Botswana, 5
  pastoral groups, 373

political stability, 128–131
population distribution, 170
Botswana Democratic Party (BDP), 129
Brain drain, 181–182
Bribery, 162
Bride wealth, 144
Britain
  slavery, 90–96
British rule, 100
Burkina Faso, 2, 4, 55, 57–58, 107–108, 122, 145, 149, 160, 167, 169, 174, 180–181, 203, 234, 239–240, 305, 311, 356, 368, 416
Buruli ulcer, 338–340
Burundi
  mountain gorillas, 61
Bush-fallow cultivation, 54
Bus systems
  government-operated, 235
Bypassing, 350–351, 358

C

Cameroon, 3–4, 7, 17, 23, 30, 36, 41, 44, 52–53, 55, 57, 62, 64–65, 84, 97, 106, 108–109, 141, 147, 149, 161, 167–169, 171, 173, 176, 180, 182, 239–240, 243, 251, 257, 270, 281–282, 285, 306, 315, 331, 339, 356, 368, 375, 377, 394, 413
Cape Agulhas, 2
Cape and Karoo shrubland, 35, 41
Cape of Good Hope, 145
Cape Province, 16, 32, 41, 61
Cape Ranges, 16
Cape Verde, 2, 68, 108, 123, 131, 169, 179
Capitalism, 412–413
  African, 120, 213–214
  women, 301–302
Carrying capacity, 166
Cash-crop farming, 374–379
  per capita production, 380
Cattle
  production, 383
CDI, 283
CDR, 111
Central Africa, 4
Central African Republic, 4, 23, 31, 65, 113, 123, 131, 145, 160,

Index

166, 169, 171, 176, 239, 245, 307, 368
Central business district, 260
Chad, 3–4, 6, 22–23, 31, 54–55, 57, 108–109, 114, 145, 167–170, 175, 239, 245, 259, 275–276, 368, 384, 435
Chad Basin, 23
Change agents
  women, 315–317
Charismatic individual leaders, 121
Charitable institutions
  health care, 344
Chickens
  production, 383
Chief Haruna, 94–95
Chihana, Chakufwa, 124
Children
  economic assets, 174
  fosterage, 174
  malnutrition, 328
  mortality, 176
  spacing, 175
Christianity, 139–141
Chronic disease
  definition, 327
Cities
  African
    internal structure, 258–269
  precolonial
    historical evolution, 254–258
City Development Index, 284
City Development Index (CDI), 283
Civilization
  ancient, 79–82
Civil war. *See* Political instability
Clans, 59, 130
Climate, 23–34
Climatic classification, 30–34
Cloth
  symbolism, 150–153
Coalition for Defense of Freedom (CDR), 111
Coastlines, 19–21
Cobalt, 18, 118
Cocoa, 4, 44, 55, 64, 97, 180, 213, 304, 375–377, 379, 381, 383, 394
Coffee, 4, 44, 55, 64, 97, 197, 205–206, 213, 304, 379, 381, 383, 394
Colonialism
  core–peripheral disparities, 207

non-African culture, 158–160
  traditional African medicine, 353–354
Colonial period, 97
  economic and cultural impact, 97–99
  political and social impacts, 100–101
COMESA, 224
Commercial farming, 374–379, 378
Commercial sex, 330–331
Common market, 221
Common Market for Eastern and Southern Africa (COMESA), 224
Communal land, 148–149
Community-based cooperatives
  agricultural production, 400
Compound farming, 371
Congenital disease
  definition, 327
Congo Basin
  railways, 240–241
Congolian belt, 36
  deforestation, 52
Congo River, 22
  navigation, 244–245
Conservation
  budgets, 65
Contagious diffusion, 137
Continental drift, 13
Continentality
  vs. maritime effect, 28
Continental shelf, 21
Cool Benguela Current, 29
Cooperatives
  community-based
    agricultural production, 400
Copper, 17–18, 97, 118, 168, 204–205, 241, 257, 292, 436
Copper belt, 17, 168, 205
Coptic Christians, 4, 163
Core house, 289
Core-peripheral disparities, 207–214
  colonialism, 207
  economic theories, 207–211
Core regions, 127, 193, 204–205, 210–212, 215
Corruption, 162
Cost
  food, 211
  Internet, 248–249

Côte d'Ivoire, 109
  AIDS, 330
  local government, 287
Counterfeit drugs, 351
Cradle of humanity, 79
Credit, 211
Crude density, 166
Cultivation
  bush-fallow, 54
Cultural conflict
  modernization, 160–162
Cultural diffusion, 136
  obstacles, 137–138
Cultural ecology, 5
Cultural geography, 135–138
Cultural integration, 135
Cultural landscapes, 1, 78
Culture
  African, 138–147
    vs. European culture, 159
  change, 136–137
  hearths, 136–137
  region, 135
Culture area, 135
Customs union, 221
Cybercafes, 249

D

Dahomey Gap, 36
Dakar, Senegal, 20, 168, 185, 238, 243, 265, 285
Dar es Salaam, 21, 111, 204, 239–241, 259, 276, 280, 285–286, 289, 292
Dar Fur, 23
Death
  leading causes, 343
Debt, 197
  restructuring, 440
Deforestation, 49–66
  causes, 54–61
  defined, 50
  effects, 61–64
  response strategies, 64–66
  trends, 51–53
Democratic Republic of Congo, 109, 117–120
Democratization, 108, 123
  Malawi, 123–125
Demographic transition, 176–177
Demographic trends, 306–311

Dependency Burdens, 177
Desert, 40–41
Desert climate, 31
Desertification, 66–73
  causes, 68–69
  combating, 70–71
  defined, 67–68
  United Nations conference, 71–72
Development, 193–228
  basic-needs strategy, 215–216
  definition, 194–201
  economic dimensions, 194–198
  growth with equity, 214–215
  human and social dimensions, 198–200
  interdependent, 218–220
  measuring, 194–201
  poverty-reduction strategy, 215–216
  redistribution with growth, 215
  self-reliance strategy, 216–218
  sociopsychological theories, 212–214
  strategies, 214–220
  structural–institutional theories, 211–212
  technological dimensions, 200–201, 216
Diamonds, 4, 17, 97, 118, 182, 204, 407
Diet, 62
Diffusion barriers, 137
Digital divide, 247–251
Direct bypassing
  health care facilities, 351
Direct rule, 100
Disease
  ecology, 326–328
  leading causes, 343
  sexually transmitted, 330–331
  terminology, 326–328
Displacement
  women, 308–309
Distance
  learning, 251
Distance learning, 251
Distributional effects, 197
Divination, 143
Djouf Basin, 23
Djibouti, 4, 182, 223, 241–242, 251, 316
Doctors in Africa, 438

Dodoma, 204, 259, 292
Dracunculiasis, 338
Dracunculus medinensis, 338
Drainage basins, 22–23
Drakensberg mountains, 16
Dress forms, 150–158
Drought, 68–69, 69, 217, 304–305
Drugs, 351
  counterfeit, 351
  prescription, 346
Drug vendor
  itinerant, 346
Dry lands
  spatial coverage, 68
Dualism
  geographic patterns, 201–206

E

East Africa, 4
  railways, 241
East African highlands, 23
East African Plateau, 16
Eastern Arc forests, 36–37
Ebola, 340–341
Economic Commission of Africa (ECA), 237, 439
Economic Community of West African States (ECOWAS), 221–224
Economic crisis
  health, 355–358
Economic diversification, 440
Economic refugees, 183
Economic union, 221
Economy
  national
    dual, 203–205
    new global, 425–441
ECOWAS, 221–224
Education
  agricultural production, 397–398
  distance, 251
Efficiency-oriented growth, 210
Egypt, 79
Elections
  media, 126
Electricity, 388
Elmina Castle
  Ghana, 91
Elongated lakes, 19

Endemic
  definition, 327
Endemic diseases, 329–341
Endemic regions
  definition, 327
Energy
  alternative
    agricultural production, 399
Energy policies
  forest abuse, 61
Enslavement, 90–96
Entebbe
  climograph, 30
  rainfall, 31
  temperature, 31
Entisols, 45
Entrepreneurship, 416–421
  characteristics, 416–417
Environment
  restoration, 316–317
Environmental degradation
  women, 302–304
Environmental determinism, 136
Epidemic
  definition, 326
Equatorial Guinea, 3–4, 36, 53, 108, 160, 168–169, 180, 368
Equity-oriented growth, 210
Ethiopia, 3–4, 7, 18, 29–31, 40, 44, 51, 55, 57, 65, 79, 93, 97, 101, 108–109, 111–112, 116, 120–123, 131, 139, 141, 149, 161, 166–169, 171, 175, 181–182, 203, 209, 239, 241, 243, 248–249, 251, 304, 307, 392, 431
Ethiopian highlands, 4, 34, 168
Ethnic rivalry
  fertility rates, 175
Ethnic solidarity, 162
Ethnic tensions, 110
EU, 395
Eurocentricity, 90
Europe. *See also* Western influence
  raw materials, 96–97
European culture
  vs. African culture, 159
Europeans
  racist attitudes, 90
European Union (EU), 395
Expansion diffusion, 137
Expectations, 161–162

Index

F

Fadamas, 38
Fair elections
   media, 126
Fake drugs, 351
Family, 143–144
   Afro-Asiatic, 145–146
Family land, 148
Family planning
   at-risk populations, 188
   community-based strategies, 187
   government policy, 185–188
   integrative policies, 186–187
   market-based strategies, 187
   NGO
      private sector partnership, 188
   refugee, 188
Family systems
   extended family, 81, 129, 157, 159–160, 162, 174
Farming systems, 365–379
   traditional, 365–367
Federal system, 120
Female. *See* Women
Female empowerment, 7, 187
Female life expectancy, 306
Feminization of poverty, 321
Fertility rates, 306
   cultural determinants, 173–174
   high
      causes, 172–173
   male-dominated culture, 173–174
   replacement, 178
   total, 171
   trends, 171–175
Fish
   production, 383–386
Fishing
   inland, 386
   marine, 386
Flora and fauna, 16, 61, 71
Flying, 241–244
Food
   per capita production, 380
   production trends, 379–386
Food crops, 376
Food prices, 211
Food supply, 304–305
Forests
   abuse
      population, 60–61

   defined, 50
   degradation
      defined, 50
   Eastern Arc, 36–37
   ecosystem management, 63
   plantations, 65
   rain, 35–37
   soils, 42
   tropical, 50, 52
Fortune-telling, 143
Fosterage
   children, 174
Freehold rights, 149–150
Free trade area, 221
French colonies, 99–100
Fuel-wood consumption
   deforestation, 57–59
Fulani, 373

G

Gabon, 3–4, 18, 21, 30–31, 36, 52–53, 57, 64–65, 108, 139, 160, 167–169, 171, 180, 237, 239, 270, 281, 285–286, 315, 407, 431
Gambia, 2, 21, 55, 73, 99, 108, 131, 169, 245, 368, 383, 410
Gao, 81, 255–258, 256
GATT, 395, 430, 434–435
GDP, 194, 407
Gedi, 257
GEM, 314
Gender. *See also* Women
   African development, 318–319
   geographic scholarship, 300
   geography, 298–301
Gender and Economic Reforms in Africa (GERA), 312–313
Gender Empowerment Measure (GEM), 314
Gender inequality
   geography, 311–315
Gender migration, 273
General Agreement on Tariffs and Trade (GATT), 395, 430, 434–435
Generalized System of Preferences (GSP), 436
Genocide
   Rwanda, 111–113

Geography
   cultural, 135–138
   engendering, 318–321
   gender, 298–301
   gender inequality, 311–315
   historical, 78–102
   medical, 324–359
   population, 165–189
   urban, 254–293
Geopolitics
   women's initiatives, 316
GERA, 312–313
Ghana
   capital, 265–269
   civilizations, 82
   democratization, 125–126
   Elmina Castle, 91
   fuel-wood consumption, 59
   independence, 107
   precolonial cities, 257
Global economy
   capitalist, 99
   integration, 437–440
Globalization, 425–426
   definition, 426
   opportunity *vs.* threat, 427–430
Global urban indicators
   database
      UNCHS, 283
GNP, 194
Goats
   production, 383
Gods
   nature, 142
Gold, 17, 81–82, 84–85, 90, 94, 96–97, 158, 160, 168, 195, 204, 256–257
Gold coast, 96, 160
Gondwana
   migration from, 13
Gondwanaland, 14
Gorillas
   mountain, 61
Government
   agricultural markets, 393–395
   types, 115
Government-operated bus systems, 235
Grassy shrublands
   Karoo, 40

Great Escarpment, 15
Great Zimbabwe, 82, 258
Green Belt Movement, 317
Greenhouse gases, 70–71
Green revolution, 205–206
   agricultural production, 390–391
Gross domestic product (GDP), 194, 407
Gross national product (GNP), 194
Groundnuts, 44, 96, 99, 205, 377, 380, 382–383
Group Areas Act, 127
Growth
   efficiency-oriented, 210
   equity-oriented, 210
Growth poles, 291
GSP, 436
Guenon, 62
Guide, 121
Guinea, 2–4, 17, 36–37, 41, 44, 51, 53, 94, 101, 108, 120, 145, 160, 167–169, 180, 182, 240, 276, 315, 338–339, 368, 412
Guinea-Bissau, 2, 53, 307, 311, 316, 330, 368
Guinea worm, 338
Gyekye, Kwame, 412

H

Hajji, 142
Harambee, 217
Harare, 116, 204, 241, 259, 265, 283, 285, 409
Harbors, 20–21, 45, 233, 244–245
Harmattan winds, 25, 27, 46
Harrod-Domar model, 208
Hausa, 146, 267–269
HDI, 311
Healers
   traditional, 346
Health care
   economic crisis, 355–358
   per capita expenditures, 356
   spatial imbalance, 348
   technology, 351
Health care facilities
   locational strategies, 349
   utilization, 350–351
Health care system, 344–355
   rainy season accessibility, 348–349

Heavily Indebted Poor Countries (HIPC), 430
Hierarchical diffusion, 137, 206
Highland Africa, 15
Highland regions, 33
HIPC, 430
Historical geography, 78–102
HIV. See AIDS/HIV
Hottentots, 145, 176
Houphouet-Boingny, 109, 121
Household
   unequal resource allocation, 305
Housing, 278–281
HPI, 198
Human Development Index (HDI), 311
Human excrement
   management, 342
Human Genome Project, 325
Humanity
   cradle, 79
Human Poverty Index (HPI), 198
Human rights, 186
Humid equatorial climates, 30
Humid temperate climates, 32–33
Hutu, 111–113, 182
Hydrologic drought, 69
Hyperendemic
   definition, 327
Hypoendemic
   definition, 327

I

Ibadan, 100
Ibo, 145
Ifa oracle system, 143
IMF, 225, 314–315
Inceptisols, 45
Incidence
   definition, 327
Income inequalities, 197
Independence
   struggle, 116–117
Indigenous heritage, 78–86
   implications, 86
   Islam, 88
Indigenous medicine, 358
Indirect bypassing
   health care facilities, 351

Indirect rule, 100
Individual private ownership, 149–150
Industrial and manufacturing enterprises
   diversification, 411
   expansion, 411
   institutional capacity, 411–412
   intrasectoral linkages, 409–410
   problems, 409–412
   protectionist policies, 410–411
   trade negotiations, 435–436
Industry
   agro-based, 399–400
   large-scale
      spatial imbalance, 409
   trends, 407–408
Infant mortality, 176
Informal housing settlements, 278
Information-technology agreements (ITAs), 437
Infrastructure
   upgrade, 438
Inland fishing, 386
Institutional care, 161
Integrated rural development
   agricultural production, 400–401
Intellectuals
   brain drain, 181–182
Intercropping, 370
Interdependent development, 218
Interest rates, 280
Internal colonialism, 211
Internal refugees, 183
International dualism, 202–203
International migration, 180–182
   trends, 180–181
International Monetary Fund (IMF), 225, 314–315
International refugee, 180–182
Internet, 247–251
   cost, 248–249
Internet service providers (ISPs), 247–251, 439
Intertropical Convergence, 24
Intertropical Convergence Zone (ITCZ), 25–26
Investment, 211, 439
Iron, 82–84
Iron ore, 4, 204, 240, 407
Irrigation, 40, 44–45, 69, 71, 82, 337, 366, 380, 388, 397, 401

Index

Islam, 82–84, 110, 141–142
  indigenous heritage, 88
  influence, 86–89
  spread, 87–88
  Western heritage, 89
ISPs, 247–251, 439
ITAs, 437
ITCZ, 25–26
Itinerant drug vendor, 346

J

Jenne, 256
Johannesburg
  rainfall, 27
  temperature, 27

K

Kabila, Laurent Desire, 119
Kalahari Basin, 23
Kalahari wooded grasslands, 40
Kampala
  rainfall, 31
  temperature, 31
Kano, 259
Karanga, 82
Karoo grassy shrublands, 40
Karoo shrubland, 33, 35, 41
Katangan rock, 17
Kaunda, Kenneth, 109
Kente cloth
  symbolism, 153–158
Kenya 2–4, 7, 15–16, 18–19, 21, 29, 31, 34, 36–37, 44, 53–55, 65, 72, 97, 102, 108, 115–116, 120–121, 123, 129, 131, 139, 144–145, 147, 149, 161, 166–169, 171, 173–175, 177, 179–182, 185–187, 206, 235, 239–241, 243, 247, 249, 257, 263–265, 273–274, 276, 278, 280–283, 287, 290, 293, 295–297, 312, 314, 316–317, 319, 334, 354, 356, 409–411, 417, 431, 435–437
  altitude, 29
Kenyatta, Jomo, 121
Khartoum, 241, 245, 259, 261–262, 265, 286
Khoisan family, 145

Kikuyu, 117, 144, 147, 173, 175, 211
Kilimanjaro, Mount, 16, 29, 34, 36, 168
Kilwa, 257
Kingdoms
  modern, 84–86
Kinship, 143–144
  bilateral kinship, 143, 162
Kinship relations, 7, 143
Kongo, 83–84, 94, 146, 257
Koran, 142
Kumasi, 257
Kumbi Saleh, 256
Kush, 79–81, 137, 255
Kyoto Protocol, 70–71

L

Labor
  mobility, 180–181
Lagos
  rainfall, 27
  temperature, 27
Lake effect, 30
Lake Malawi, 18–19, 168
Lakes
  elongated, 19
Lake Tanganyika, 18–19, 240
Lake Victoria, 4, 18–19, 22, 30, 168, 171, 240–241, 244, 372
Land
  arable, 54–55
  communal, 148–149
  dry
    spatial coverage, 68
  exploration, 96–97
  family, 148
  state, 149
  stool, 149
  tenure, 148–150
Land degradation
  agricultural production, 388–389
Landforms, 15
Land management, 281–282
Land pledge
  conditions, 376
Land tenure systems, 148–150, 282
  communal land, 62, 135, 148, 282, 389
  family land, 148–149

state land, 149
stool land, 148–149, 282
Language, 144–147
Language branches, 145
Language family, 145
Large-scale commercial agriculture, 378–379
Large-scale industries
  spatial imbalance, 409
Lateritic soils, 42
Latitudinal effect, 27–28
Leaching, 42, 44, 371, 389–390, 400
LDCs, 436
Leaders
  charismatic individual, 121
Least developed countries (LDCs), 436
Legon, 100
Lemur, 61–62
Lesotho, 4, 106, 108, 128, 149, 167–169, 173, 179, 181–182, 195–199, 222–224, 309, 311, 436
Le Vieux, 121
Liberia, 2–4, 30, 44, 51–55, 64, 97, 108–109, 114, 131, 134, 145, 147, 166–169, 173, 176, 180–182, 240, 302, 326, 331, 356, 368, 392
Libreville, 21, 168, 223, 259
Life expectancy
  women, 306–307
Lilongwe, 124, 204–205
Limpopo river, 38
Lingala, 145–146
Lingua franca, 146–147
Literacy, 200
Literate women, 311–312
Livestock production, 381–383
Livingstone, David, 96
Local Government Law, 286
Local voluntary associations
  Tanzania, 287
Logging, 52
  deforestation, 57
Lowland Africa, 15
Low stationary stage, 177
Lumumba, Patrice, 117
Lusaka, 124, 259, 265, 332

M

Maathai, Wangari, 317
Madagascar
  lemur, 61–62
  military coups, 107
Maggia Valley Project, 66
Makerere, 100
Makgadikadi (Makarikari) salt pans, 23
Malaria, 334–335
Malawi, 109
  democratization, 123–125
Malawi, Lake, 18–19, 168
Malay–Polynesian family, 145–146
Mali, 82
Malindi, 256
Malnutrition
  children, 328
Mandela, Nelson, 121
Manganese, 4, 17, 204, 407
Manufacturing
  trends, 407–408
Manufacturing enterprises. *See* Industrial and manufacturing enterprises
Maputo, 21, 204, 285–286
Marine fishing, 386
Maritime effect
  vs. continentality, 28
Market
  agricultural
    government, 393–395
  common, 221
Marriage patterns
  fertility rates, 172
Masai, 38–40, 144–146, 373
Matatus, 235
Maternal mortality, 306–307
Maternity clinics, 344
Matriliny, 144
Matrilocality, 144
Mauritania
  military coups, 107
Mauritius, 2, 53, 161, 167–169, 181, 247, 249, 308, 356, 411
Meat, 62
Media
  fair elections, 126
Medical geography, 324–359
Medical practitioners
  traditional, 352
Medications, 351
  counterfeit, 351
  prescription, 346
Medicine
  traditional African
    colonialism, 353–354
    strengths and weaknesses, 354–355
Medieval civilizations, 82–84
Mediterranean, 41
Meroe, 255
Meteorologic drought, 69
MFA, 436
Mfecane, 85
Migration
  Bantu, 84–85
  gender, 273
  international, 180–182
    trends, 180–181
  refugee, 182–183
  rural-urban, 272
  women, 308–309
Military
  AIDS, 331
Military coups, 107–108
Military revolutions, 122
Mineral resources, 17, 96–97, 99
Minibuses
  privately owned, 235
Mining, 18, 48, 54, 128, 168, 193, 197, 204, 224, 241–242
Miombo woodlands, 38
Missionaries, 86, 97, 99
Mixed farming, 371
Mobutu, Joseph-Desire. *See* Seko, Mobutu Sese
Modernization
  aged, 160–161
  cultural conflict, 160–162
Modernization theory, 207
Modern kingdoms, 84–86
Mogadishu, 20, 83, 116, 168, 204, 256, 259
Mombasa, 21, 83, 90, 239–241, 256–257, 259, 283
Monrovia, 182, 240, 259, 392
Mopane, 38
Mortality
  children, 176
  infant, 176
  leading causes, 343
  levels, 175–176
  maternal, 306–307
Mortgage loans, 280
Mosaic zone, 37–38
Mossi, 4, 81, 83, 94–95, 145–146, 149
Mountain gorillas, 61
Mountains, 15–17
Mountains of the Moon, 16
Mount Kenya
  deforestation, 65
Mount Kilimanjaro, 16, 29, 34, 36, 168
Mozambique, 5
MPLA, 115, 182
MRND, 111
Muhammed, 86
Multifibre Agreement (MFA), 436
Multilateralism, 220–227
  NGO, 227
  structural-adjustment programs, 225–227
Mutesa II, 110
Mwalimu, 121
Mycobacterium ulcerans, 338–340
Mzee, 121

N

NADEOSA, 251
NAFTA, 221
Nairobi, Kenya, 77, 240, 243, 319
Namib Desert, 29, 31, 41, 168
Namibia, 3–5, 5, 29, 31, 102, 114–115, 119, 123, 145, 160, 167–169, 171–172, 176, 179, 184, 203, 241, 247, 311, 329, 411
National Association of Distance Education Organizations of South Africa (NADEOSA), 251
National economy
  dual, 203–205
National Revolutionary Movement for Democracy (NRMD), 111
Nations
  sovereign rights, 186
Nature gods, 142
Navigation, 244–245
N'Djamena, Chad, 31, 259
Neocolonialism, 99
New global economy, 425–441
New towns, 290–291
NGO
  multilateralism, 227
Niamey, 259

Index

Niger
　ecosystem rehabilitation, 66
Nigeria
　AIDS, 333–334
　independence, 107
　pastoral groups, 373
Niger–Kordofanian family, 145
Niger River, 21
　navigation, 245
Nile river, 79
Nilo–Saharan language branch, 145
Nkrumah, Kwame, 121
Nok civilization, 80
Nomadism, 372
Nongovernmental organizations (NGO)
　multilateralism, 227
North American Free Trade Agreement (NAFTA), 221
Northeast Trade Winds, 25
NRMD, 111
Nubia, 79–81, 137
Nursing homes, 161
Nyerere, Julius, 109, 121, 209
Nyika Plateau, 16

O

OAU, 183
Obote, Milton, 110
Ocean currents, 28–29
OECD, 249
Ogaden Desert, 31
Okavango Swamp, 23
Olduvai Gorge, 79
Olodumare, 142
Onchocerca volvulus, 337–338
One-party states, 101, 106–109, 113
Orange river, 4
Organization for Economic Co-Operation and Development (OECD), 249
Organization of African Unity's (OAU), 183
Organization of Rural Associations for Progress, 218
Orisha, 142
Orphans, 330
Osagyefo, 121
Overurbanization, 270
Oyo, 84, 256–258, 333

P

Palm Oil, 96, 377, 379, 382–383
PAMSCAD, 430
Pan-African Economic Community (AEC), 221
Pan-African Highway, 237–239
Pan-Africanism, 220–225
Pandemic
　definition, 326
Pangaea, 13
Pantheism, 138
Park, Mungo, 96
Parliamentary system
　Westminster, 120
Particularism, 212
Pastoralism, 371–372
Paternalistic, 100
Patrilineal system, 143
Patriliny, 143
Patrilocality, 144
Peasant agricultural systems, 366
Permanent farming, 371
Pharmacy, 345
PHC, 350
Physical landscapes, 12–23
Physiological density, 166, 169
Pidgin languages, 147
Pigs
　production, 383
Plantation agriculture, 55
Plateau continent, 14–17
Plate tectonics, 12
Poda-podas, 235
Police
　AIDS, 331
Political geography, 86, 101, 105
Political instability, 106–120
　causes, 110–117
　Rwanda, 111–113
Political landscape, 105–131
Political stability
　Botswana, 128–131
　hope, 122–123
　scenarios, 120–122
　traditional government systems, 129–130
Polygyny
　fertility rates, 172
Population
　age composition, 177–182
　change, 171–185
　data, 167
　density, 166–167
　distribution, 166–167
　geography, 165–189
　trends, 166–170
　urban
　　world regions, 270
Population policies, 186
Portugal, 90
Possibilism, 136
Postpartum abstinence
　fertility rates, 172–173
Poverty, 429
　feminization, 315
　forest abuse, 61
　origin, 99
　political instability, 114
Poverty-reduction and growth-facility program (PRGF), 438
Precambrian rock, 17
Precolonial cities
　historical evolution, 254–258
Preferential Trade Area (PTA), 221–223
Prescription drugs, 346
Pressure systems, 24–27
Prevalence
　definition, 327
PRGF, 438
Primary health-care (PHC), 350
Primate city, 275–276, 291–292, 294
Private enterprises
　expanding, 414–416
　vs. public enterprises, 412–416
Private ownership
　individual, 149–150
Private practice, 344
Programme of Action to Mitigate the Social Cost of Adjustment (PAMSCAD), 430
Prophet Muhammed, 86
Protected areas, 65
PTA, 221–223
Public enterprises
　vs. private enterprises, 412–416
Public housing estates, 279
Public transportation, 285, 289
Pygmies, 63, 137, 176

Q

Quasi-commerical production, 378

R

Railways, 239–241
Rain forests, 35–37
Rank-size regularity, 274
Ras Dashen, 17
Rate of natural increase, 189
Rawlings, Jerry John, 125–126
Raw materials
  Europe, 96–97
Reforestation, 72
Refugees
  economic, 183
  family planning, 188
  internal, 183
  migration, 182–183
  sources, 183
Regional Informatics Network for Africa (RINAF), 439
Regional integration, 220–225
  examples, 222–223
Regional ISP Africa Online, 249
Religion, 139–144
  fertility rates, 173
  traditional, 142–143
Religious conflict, 143
Religious missions
  health care, 344
Relocation diffusion, 137
Remittances, 181
Replacement fertility rate, 178
Réunion, 2–3, 53
Revegetation, 72
Rift Valley, 18–19, 79
RINAF, 439
River blindness, 337–338
Rivers, 21–22
Road transportation, 233–235
Rock
  Katangan, 17
  Precambrian, 17
Rostow's theory of economic growth, 207
Rotational bush fallow, 370–371
RTPCs, 292
Rubber, 38, 44, 55, 64, 97, 205, 376, 379, 383, 394, 410
Rural development
  integrated
    agricultural production, 400–401

Rural dualism, 205–206
Rural infrastructure
  agricultural production, 386–388, 397
Rural Trade and Production Centers (RTPCs), 292
Rural transportation, 233–235, 347
Rural–urban migration, 272
Ruwenzori mountains, 16
Rwanda
  mountain gorillas, 61
  political chaos, 111–113

S

SADC, 223
Sahara Desert, 5, 13, 27, 31, 40, 45, 67–68, 82, 93, 137
Sahel belts, 82
Sahelian zone
  ecosystem rehabilitation, 66
Sahel shrubland, 40
Sanitation, 341–344
São Tomé and Principe, 2
SAPs, 225–227
Savanna grassland, 38–40
Savings, 211
Schistosomiasis, 337
Sclater
  Guenon, 62
Sea-floor spreading, 13
Seasonal inaccessibility
  rural transportation, 233
Seko, Mobutu Sese, 109, 117–120, 121
Selassie, Haile, 121
Self-help, 289
Self-reliance, 216, 219
Semiarid climates, 31
Semidesert, 40–41
Semitic–Hamitic family, 145–146
Senegal, 2, 6, 20–21, 38, 40, 55, 65, 73, 88, 99, 106, 108, 120, 149, 167–169, 173, 181, 185, 203, 238, 240, 243, 245, 247, 251, 265, 275–276, 278, 280–281, 285, 288, 302, 305, 330, 348, 355–356, 368, 394, 411–412, 432–433, 435, 437
  local government, 287

Serengeti Plains, 39
Service centers, 291–292
Settlement patterns, 306–311
Sewage disposal, 282–283, 342
Sex
  commercial, 330–331
Sexually transmitted disease, 330–331
Seychelles, 2–3, 53, 123, 179, 222–223
Shadow towns, 291
Shango, 143
Shariah law, 110
Sharpe, Granville, 96
Sheep
  production, 383
Shifting cultivation, 368–369
Shona, 84, 145
Sierra Leone, 3, 21, 30, 44, 51, 53, 96, 108, 114, 131, 141, 145, 166–167, 169, 176, 182, 235, 302, 307, 331, 351, 368, 410, 417, 435, 439
Site and service programs, 289
Slash and burn agriculture, 369–370
Slavery, 90–96
  oral history, 94–95
Sleeping sickness, 335–336
Slums
  upgrading, 289
Small-scale enterprises, 417–421
Socialism, 209
  African, 120
Social marketing
  family planning, 187
Social reproduction, 310
Soil degradation, 5, 31, 55, 68, 70, 374
Soil management, 44–45
Soils, 41–45, 389
  distribution, 42–45
  lateritic, 42
Soil Taxonomy System, 43
Sokoto, 257, 259
Solid-waste disposal, 284–285
Somalia, 2–4, 20, 31, 37–40, 54–55, 57, 68, 108–109, 113, 116, 134, 141, 145, 149, 167, 169, 181, 239, 241–242, 305, 316, 326, 356
Songhai, 80–82, 82, 145–146
South Africa
  example, 127–128
Southeast trades, 24–25

Index

Southern Africa, 4–5
  railways, 241
Southern Africa Development
    Community (SADC), 223
Southwest trades, 26
Spirit shrines, 163
Squatter
  upgrading, 289
Stanley, Henry, 96
State housing corporations, 279
State land, 149
State-owned enterprises
  background, 412–414
Stereotypes, 86
Sterility
  fertility rates, 173
STHP, 24
Stool land, 149
Strip cropping, 73
Structural-adjustment programs
    (SAPs), 225–227
  multilateralism, 225–227
Students
  brain drain, 181–182
Subfamilies, 145
Sub-Saharan Africa
  developmental context, 8–10
  global context, 10
  location, 2
  physical environment, 5,
    12–45
  regions, 2–5
  size, 2
  sociocultural context, 5–7
Subtropical High-Pressure System
    (STHP), 24
Sudan, 2–4, 6, 15, 22–23, 31, 38, 40,
    44, 55, 65, 67–68, 88–89, 100,
    107–108, 113, 131, 143–145,
    166–167, 169, 171, 173,
    181–183, 239, 241, 259,
    261–265, 290, 326, 334, 341,
    368, 380, 394, 431, 439
Sudan Basin, 23
Sudanic empires, 82
Sustainable development, 21, 64, 66,
    72, 74, 76, 201, 252, 414
Swahili, 146
Symbolism
  adinkra cloth, 150–153
  kente cloth, 153–158
Syncretism, 7

T

TACAS, 243
TAI, 200
Tanganyika, Lake, 18–19, 240
Tanzania, 39
  altitude, 29
  leadership, 109
  local voluntary associations, 287
Tax collecting, 288
Taxis, 235
Tea, 4, 44, 55, 97, 379, 381, 383, 394
Technology
  agricultural production, 389–390,
    398–399
  health care, 351
Technology Achievement Index (TAI),
    200
Telephone, 245–247
Tema, 20, 233, 292
Third African Population Conference,
    185–186
Timber, 4, 37, 52, 61, 97, 205, 233
Timbuktu, 256
  rainfall, 28
  temperature, 28
Time–distance decay effect, 137
TMPs, 352
Togo, 3, 51, 53, 55, 97, 100, 108–109,
    123, 167, 169, 171–173, 175,
    181, 196, 199, 222–223, 227,
    236, 239–240, 249, 274, 278,
    337, 356–357, 368, 375, 410
Total fertility rate, 171
Trade negotiations
  industrial and manufacturing enter-
    prises, 435–436
Trade reforms
  agricultural, 432–433
Trade-Related Aspects of Intellectual
    Property Rights (TRIPS), 395,
    433–434
Traditional African medicine
  colonialism, 353–354
  strengths and weaknesses, 354–355
Traditional farming systems, 365–367
Traditional healers, 346
Traditional medical practitioners
    (TMPs), 352
Traditional religions, 142–143
Traffic Alert and Collision Avoidance
    Systems (TACAS), 243

Transhumance, 372–374
Transition/mosaic zone, 37–38
Transportation, 232–247
  air, 241–244
  political instability, 242
  public, 285, 289
  road, 233–235
  rural, 233–235
  urban, 235–237
Trans-Saharan trade, 81, 84, 87, 90
Tree cropping, 375–378
TRIPS, 395, 433–434
Tropical forests, 50, 52
Tropic of Cancer, 2, 15, 23, 26, 67
Tropic of Capricorn, 2, 23, 27
Tro-tros, 235
Truck drivers
  AIDS, 331
Trypanosomiasis, 335–336
Tsetse fly, 87, 233, 335
Tswanas, 373
Tutsi, 111–113, 182
Twa, 62, 111

U

UDF, 125
Uganda
  crater lake, 20
  denuded hillsides, 56
Ujamaa village program, 217
Ultisols, 44
UNCHS
  global urban indicators database,
    283
UNHCR, 183
Unilineal descent system, 143
United Democratic Front (UDF), 125
United Nations Center for Human
    Settlements (UNCHS)
  global urban indicators database,
    283
United Nations conference
  desertification, 71–72
United Nation's High Commissioner
    for Refugees (UNHCR), 183
University colleges, 100
Uranium, 4, 118, 204, 407
Urban
  informal sector, 275–278
Urban dualism, 205

Urban geography, 254–293
Urban growth
  components, 272–273
  consequences, 274–275
  rates, 280
Urbanization, 19, 71, 148, 161–162, 168, 176, 204, 269, 271, 273, 275, 277, 279, 281, 283, 285, 295, 301, 398, 409
  current trends, 269–285
  problems
    solutions, 285–292
Urban living conditions, 280
Urban management, 8, 164, 275, 283, 287–289, 293–296
  infrastructure provision, 282
  macrolevel strategies, 290–292
  microlevel strategies, 286
  service delivery, 282
Urban models
  comparison, 260
Urban population
  world regions, 269
Urban primary, 274–275
Urban–rural linkages, 273–274
Urban transportation, 235–237
Urban wages, 272
Uruguay Round, 432–433
Usufructuary, 388

V

Value, 161–162
Vegetation, 33–41
  afromontane, 41
  types, 35–37
Vertisols, 44
Victoria, Lake, 4, 18–19, 22, 30, 168, 171, 240–241, 244, 372
Victoriaborg, 266
Virginity, 174
Virunga Mountains, 16
Volta, 21, 94–95, 160, 245, 338

W

Wages
  urban, 272
Walk-away syndrome, 338
Walvis Bay, Namibia, 29, 241, 259
Water, 388
Water Conservation, 5, 72, 74
Water supply, 282–283
Weathering, 42, 44
Web, 247–251
Wegener, Alfred, 13
Welfare state, 413
West Africa, 2–3
  deforestation, 51–52
  railways, 240
Western heritage
  Islam, 89
Western influence, 89–102
  independence, 101–102
  initial contact, 89–90
Westminster parliamentary system, 120
Wilberforce, William, 96
WIN, 317
Wind movement, 24–27
Witchcraft, 143
Witwatersrand, 15, 47, 127, 168, 369
Wolof, 4, 83, 145–146
Women. *See also* Gender
  agriculture, 302–304
  capitalist development, 301–302
  change agents, 315–317
  displacement, 308–309
  empowerment, 187–188
  environment
    degradation, 302–304
    restoration, 316–317
  HIV/AIDS, 308
  initiatives
    geopolitics, 316
  life expectancy, 306–307
  literate, 311–312
  mapping, 300–301
  migrations, 308–309
  structural-adjustment programs, 314–315
  tasks, 305
  urban settings
    differential access, 309–311
Women in Nigeria (WIN), 317
Wooded grasslands
  Kalahari, 40
Woodlands, 37–38
World Bank, 8, 52, 54, 61, 120, 123, 125, 179–180, 209, 214, 237, 243, 251, 270, 273, 282–283, 288–290, 294–297, 333, 355, 382, 393, 395, 405, 407, 411, 414–415, 420, 422, 425, 427, 429–430, 432, 434, 438, 440
World Links for Development, 251
World trade negotiations, 430–431
World Trade Organization (WTO), 436–437
  agriculture, 395–396

Y

Yellow fever, 336
Yoruba cities, 258
Yoruba traditional religion, 142

Z

Zaire. *See* Democratic Republic of Congo
Zambezian, 38
Zambezi River, 21
Zambia
  independent development, 219
  leadership, 109
Zaria, 259
Zimbabwe, 5, 82, 258
  independent development, 218–219
Zomba Plateau, 16
Zulu, 81, 85, 145–146